S0-BAX-353

PERGAMON INTERNATIONAL LIBRARY
of Science, Technology, Engineering and Social Studies
The 1000-volume original paperback library in aid of education,
industrial training and the enjoyment of leisure
Publisher: Robert Maxwell, M.C.

FATIGUE DESIGN

SECOND EDITION

A X21 $28.50

THE PERGAMON TEXTBOOK
INSPECTION COPY SERVICE

An inspection copy of any book published in the Pergamon International Library will
gladly be sent to academic staff without obligation for their consideration for course
adoption or recommendation. Copies may be retained for a period of 60 days from
receipt and returned if not suitable. When a particular title is adopted or recommended
for adoption for class use and the recommendation results in a sale of 12 or more copies,
the inspection copy may be retained with our compliments. The Publishers will be
pleased to receive suggestions for revised editions and new titles to be published in this
important International Library.

INTERNATIONAL SERIES ON THE STRENGTH
AND FRACTURE OF MATERIALS AND STRUCTURES

General Editor: D. M. R. TAPLIN, D.Sc., D.Phil., F.I.M.

Other Titles in the Series

Related Pergamon Journals *(free specimen copies available on request)*

Acta Metallurgica
Canadian Metallurgical Quarterly
Computers and Structures
Corrosion Science
Engineering Fracture Mechanics
Fatigue of Engineering Materials and Structures
International Journal of Solids and Structures
Materials Research Bulletin
Metals Forum
Physics of Metals and Metallography
Scripta Metallurgica

NOTICE TO READERS

Dear Reader

An Invitation to Publish in and Recommend the Placing of a Standing Order to Volumes Published in this Valuable Series.

If your library is not already a standing/continuation order customer to this series, may we recommend that you place a standing/continuation order to receive immediately upon publication all new volumes. Should you find that these volumes no longer serve your needs, your order can be cancelled at any time without notice.

The Editors and the Publisher will be glad to receive suggestions or outlines of suitable titles, reviews or symposia for editorial consideration: if found acceptable, rapid publication is guaranteed.

ROBERT MAXWELL
Publisher at Pergamon Press

FATIGUE DESIGN

by

CARL C. OSGOOD, M.S.

Mechanical and Metallurgical Engineer
Cranbury, New Jersey, USA

SECOND EDITION

PERGAMON PRESS

OXFORD · NEW YORK · TORONTO · SYDNEY · PARIS · FRANKFURT

U.K.	Pergamon Press Ltd., Headington Hill Hall, Oxford OX3 0BW, England
U.S.A.	Pergamon Press Inc., Maxwell House, Fairview Park, Elmsford, New York 10523, U.S.A.
CANADA	Pergamon Press Canada Ltd., Suite 104, 150 Consumers Rd., Willowdale, Ontario M2J 1P9, Canada
AUSTRALIA	Pergamon Press (Aust.) Pty. Ltd., P.O. Box 544, Potts Point, N.S.W. 2011, Australia
FRANCE	Pergamon Press SARL, 24 rue des Ecoles, 75240 Paris, Cedex 05, France
FEDERAL REPUBLIC OF GERMANY	Pergamon Press GmbH, 6242 Kronberg-Taunus, Hammerweg 6, Federal Republic of Germany

First edition 1982

Library of Congress Cataloging in Publication Data

Osgood, Carl C.
Fatigue design.
(International series on the strength and
fracture of materials and structures) (Pergamon
international library of science, technology,
engineering, and social studies)
Bibliography: p.
Includes index.
1. Structural design. 2. Machinery—Design.
3. Materials—Fatigue. I. Title. II. Series.
III. Series: Pergamon international library of
science, technology, engineering, and social
studies.
TA658.2.08   1982   620.1'123   81-17932
AACR2

British Library Cataloguing in Publication Data

Osgood, Carl C.
Fatigue design.—2nd ed.—(International
series on the strength and fracture of materials
and structures).—(Pergamon international
library)
1. Materials—Fatigue 2. Fracture mechanics
I. Title II. Series
620.1'123   TA418.38

ISBN 0-08-026167-1 (Hardcover)
ISBN 0-08-026166-3 (Flexicover)

In order to make this volume available as economically and as rapidly as possible the author's typescript has been reproduced in its original form. This method unfortunately has its typographical limitations but it is hoped that they in no way distract the reader.

Printed in Great Britain by A. Wheaton & Co. Ltd., Exeter

For my wife FRANCES
a true helpmeet in all things

CONTENTS

vii

PREFACE TO THE SECOND EDITION

During the period since the First Edition there have been significant changes and additions in several disciplines - most notably the coming-of-age of Fracture Mechanics as a practical design procedure. Although modern Fracture Mechanics has been under development since the late 1940's, it is only recently that the intense pressure from the engineering community for more useful solutions to the Life Prediction problem in the constructional alloys has forced the extension from linear elastic fracture mechanics (LEFM) into the elasto-plastic regime. The acceptability of this extension is now generally recognized, and valid data is available with which to treat the low-strength, high-toughness, plane stress situation. A Section has been added to illustrate this Procedure, with examples drawn from work in the railroad and heavy structural industries. Several Sections, especially that on Welded Joints with their inevitable flaws, have been rewritten from the Fracture Mechanics viewpoint.

The older, classical Fatigue approach has been expanded to include recent consider-ations such as the statistically-acceptable counting schemes for defining a "cycle"; and the cycle-by-cycle summation of damage. The general coverage of statistical treatments has been made more concise, and that on the Weibull method expanded.

I have found no reason to alter my original statement: "All machine and structural designs are problems in Fatigue because the forces of Nature are always at work and every object must respond in some fashion." In fact, the ever-increasing demand for guaranteed performance and life intensifies the need to regard all design situations as those in which imperfect materials are subject to dynamic loadings.

My hope is that this presentation may help others find solutions to their Fatigue and Fracture Control problems.

Cranbury, New Jersey   Carl C. Osgood
November 1, 1979

PREFACE TO THE FIRST EDITION

All machine and structural designs are problems in fatigue because the forces of Nature are always at work and every object must respond in some fashion. This volume results from a long-standing interest in the question of how to determine the individually characteristic response; it is primarily a discussion of solutions to previous problems as controlled by their particular conditions. A comparison of the methods of life prediction--the basic problem--is included, and I have tried to write for the working engineers, designers, and instructors in these topics. This work is not intended as a text--many excellent treatments already exist for the standard derivations and for the associated disciplines such as stress analysis. My intention is to demonstrate the limitations of some of the accepted methods and to explore the realism and validity of the resulting solutions.

I am greatly indebted to my friends who have given so freely of their time in the discussion of difficult points, and especially to the management of the Astro Electronics Division of the RCA Corporation for their generous support in the production of the manuscript.

The work is presented in the hope that it may help others in their efforts to recognize and solve problems in fatigue.

Carl C. Osgood

Cranbury, New Jersey
January 1970

CHAPTER 1

INTRODUCTION:
SCOPE, LEVEL, LIMITATIONS

It might be said that all stress analyses are basically fatigue analyses, the differences lying in the number of cycles of applied stress. Theoretical dead load may be treated as a single cycle; or, since structures and machines completely free of alternating stress are extremely rare, the dead load may be taken to obtain the mean stress. At the other end of the life spectrum, most of the early work in fatigue and much of that following has been concerned with an endurance limit--a minimum cyclic stress below which life is taken as infinite. A parallel exists between that development of thought which led to the concept of an endurance limit and that which postulated linearity in the stress-strain relationship. Had Hooke used other than iron-base alloys in his experiments, and could the early automotive investigators have been more interested in the light metals than steel, these fundamentals would have been correctly recognized much earlier as approxima- tions and special cases. Such a condition might then have led to earlier recognition of the significance of phenomena discovered only lately, such as low cycle fatigue. But, one must take serious note of the inevitability of hindsight: the writers and investigators in scientific disciplines at any one time can hardly avoid taking advantage of previous work. The present state of knowledge in Mechanics and Strength of Materials is, of course, at the transient peak of the progress curve that started about three centuries ago with Hooke. The integral under this curve represents a vast storehouse of information, much of which is, fortunately, useful in fatigue design.

All of which goes to say that designers have every reason to use the results of previous work, that general knowledge of this work is incumbent upon them, and that they are rarely justified in ignoring it. Not that all the problems are solved and the questions answered, but that much, probably sufficient, information does exist with which to make reasonably comprehensive solutions. The primitive design based on a stress taken from a single S-N curve is usually intolerable.

That portion of the entire design process known as fatigue design can only follow the loads and stress analyses. In general, the presumption is made here that such analyses have been properly done; the relationships and methods are available in numerous excellent works already in print [13-21]. However, the quality of the fatigue design is dependent on many factors, as noted below, not the least of which is the quality and completeness of the loads and stress analyses. There is also an intermingling of the steps, especially the last iteration or two of the stress analysis for a fatigue problem, due chiefly to the need for great attention to detail. Here, the term detail is taken to mean not only details of shape and tolerances on size, but also such items as the difference in fatigue strength in the longitudinal or transverse directions of rolling, the degree of local yielding, second order effects in the calculation of stress concentration factors, and the like. It is rare that a design problem can be satisfactorily solved in separate and distinct steps after the loads analysis is made, especially not by separate groups for the stress analysis and for the fatigue design. For realistic solutions, it is essential that they work together, particularly during those iterations in which the fatigue strength of the material, and the size and shape of the part are being finally fitted to the always somewhat variable stress pattern. In extensive projects, as the design of a new airframe, the sheer numbers of people necessary to get the work done preclude numerous personal interchanges. Thus the lead engineers and designers must carry out this liaison, which requires of them considerable skill in recognition and interpretation.

It is important to distinguish between the results at various stages of the stress analysis and those of application of a damage theory. A stress analysis is required to define the stress history at any critical point in the structure; it is the application of a theory of fatigue damage that defines the status of a crack at the same location as a function of that stress history. Even if a perfect theory of fatigue damage were available, the problem of fatigue life prediction could not be solved until the stress analysis was also completed.

At the start of any effort on life prediction it must be fully recognized that the development of fatigue cracks in metals (and probably in any material) is a statistically random process. The stresses in any structure subjected to natural loads, and the properties and conditions of the materials in these structures, will also vary in a random manner. The integrated influence of all these variables can then yield only wide scatter in the results. At the present level of the state of the art, one should not hope to make a sharply precise prediction--the accuracy, confidence, and conservatism of several of the calculation methods are the subject of the following discussions.

CHAPTER 2

CHARACTERISTICS OF THE
DESIGN APPROACH

<u>2.1   COMPREHENSIVENESS AND CONSERVATISM</u>

Fatigue design, or as some put it, designing against fatigue failure, may have the objectives of infinite life, zero weight, infinite strength, or 100% reliability-- or perhaps all four simultaneously. These objectives must be viewed realistically, for, as Shanley [14] says, "While neither zero weight nor infinite strength is attainable, the former certainly makes the better goal." Treating a design situation as a problem in fatigue means that the lifetime or a fatigue stress level has already been chosen as the controlling factor. To reach such a state of decision, one must have performed sufficient analysis of loads, their modes of application, and the resultant stresses to choose among, at least, the usual failure criteria of ultimate, yield, or fatigue strengths, or percent damage. That is, either the situation does not admit of failure by a single application of stress, thus eliminating both ultimate and yield strengths as criteria, or the probability of failure is so low as to make it an unreasonable assumption. Thus the necessity for considering the degree of comprehensiveness arises, or one might ask: "How sophisticated a solution is really required?"

The present state of design is such that the title of the device under consideration will indicate the general degree to which fatigue considerations are needed, but the real question in attacking the problem is the level with which the known factors and data should be treated. The parameters associated with a design task that is, or may become, a fatigue problem include the following general items:

   Safety: Consequences of a partial or catastrophic failure.
   Realism: In the statement of performance requirements.
   Realism: Of any derived requirements, as minimum weight or maximum
            reliability.
   Cost: Especially the ratio of design cost to part cost, or to return on
            investment.
   Time: Calendar time for the design and test efforts, and for feedback of
            test results to design modifications.
   Level: Of the design effort, that is, computer iterations on graded S-N
            curves from local tests or slide rule treatment of handbook data.

The more specific parameters associated with a fatigue design include:

   Information: Type of service, environments, e.g., specific corrodants.
   Loads: Magnitudes and frequencies.
   Analyses: Extent and accuracy.
   Materials: Availability and applicability of data.
   Materials: Validity of the choice.
   Previous design: Extent of holdover as an element of redesign.
   Manufacturing: Choice of modes and evaluation of damage therefrom.
   Manufacturing: Quality of personnel and equipment.
   Maintenance: Quality of personnel and length of operational periods.

With the possible exceptions of environmental information and maintenance, these latter items should be under the designer's control, but the general ones are often dictated to him.

Assuming the usual situation, that the analysis was sufficiently extensive, accurate, and broadly based to provide a reasonably acceptable picture of the stresses, and that these stresses fall within the general capability of available materials, decisions based on circumstances related to the following are often required. A high-speed pump shaft is to operate in corrosive media such that a nickel or chrome-nickel surface is required, as a first approximation, to give any useful life. A first choice might be a stainless steel shaft, but it turns out that, in the correct compositions, the required fatigue strength results in an inconveniently large diameter. A second choice might be a medium- or high-alloy steel of nominally correct strength and convenient diameter, but to get the corrosion protection chrome plating will be required. Now the degradation in fatigue strength because of plating begins to drive the diameter back up, with a portion of this increase being possibly compensated by a hydrogen bake-out after plating. But how much compensation can be obtained? And is the minimum diameter important enough to specify a material and plating that will probably be more expensive than a single material, plus the time and involvement of a second vendor, or at least a second department? The designer could attempt a compromise by using a degraded fatigue strength for a plated, high alloy steel and gamble on no bake-out, but the conservatism of such an operation might be questioned. He could take a second look at the shaft and bearing configurations, at the shapes and their stress concentration factors, at the possibility of improving the fatigue strength of the single-material shaft by peening, and so on.

Innumerable variations of such cases may be found; each must be solved according to its individual conditions. As mentioned above, while the cost and talent level assigned to a task are beyond the control of the designer, these specific parameters and their treatment are wholly within his control. Indeed, his position is defined as the one who performs the proper treatment for all applicable parameters. Thus the success of the design is strongly dependent on both the level of talent that can be applied within the allowable cost and time, and the extent to which the allotted time and money permit the application of the known information.

Except in its most simple and direct forms, conservatism is a rather subjective matter. The ratio of the tensile stress in a loaded wire to the material's yield strength is readily accepted as a measure of conservatism, whether or not one agrees with the magnitude of any given ratio. But, as examples, the degree of significance to be placed on the value of unity in the Miner summation [30], the manner of interpreting dimensions in the calculation of stress concentration factors, and the allowances made for scatter in the data on fatigue stress are subject to opinion, to the inevitably small sample of the individual's previous experience and to his innate recklessness or caution. Conservatism is a broad topic, not easily generalized, and, as are all philosophies, subject to inconclusive argument. If one cautions the designers to be careful, a frequent tendency is to increase the sections, and thereby the weight, in an effort to reduce nominal stresses but without necessarily improving the design's resistance to fatigue. Attention to detail and the exercise of engineering judgment on realism are probably the most important manifestations of conservatism. The term should not be interpreted in the sense of caution or timidity; it is quite possible that a proper design would indicate working stresses in the plastic range, provided that the accuracy of the loads analysis and consequences of failure had been accepted by all concerned.

Fatigue analysis is often the last step in the design procedure and in many cases it may be necessary to reconsider some of the earlier results, as for example, the degree of optimization of material distribution in the several cross sections (such as the optimum frame area and spacing in stiffened panels). The need and usefulness of such parameters as <u>configuration</u> and material efficiencies are being more widely recognized; when available they provide much insight into the conservatism of the design under scrutiny. Beyond the simple go-no-go test of final success or failure, there is probably no single, direct measure of conservatism acceptable to all. But, as a modification to this seemingly final statement, a comparison is given in Section 3.1.1 of the conservatism inherent in the several methods for the prediction of fatigue life.

2.2   LOADS, STRESS, AND STABILITY ANALYSES

As an obvious prerequisite to fatigue design, the working loads, their modes and frequencies of application, and the resultant stress fields must be known to the required degree of completeness and precision. The fact that fatigue is inherently a dynamic phenomenon makes it imperative that the dynamic loads be known and that any "equivalent static" analyses be handled with the proper degree of conservatism.

Analysis for dynamic loads and stresses can easily become a very involved topic; its treatment is generally left to the several references already recommended. Sufficient discussion is given in the examples below to indicate acceptable degrees of comprehensiveness and conservatism. It is important, however, to note here that not only must the analysis be for dynamic loads, but that a vast scale of lifetimes may be involved. Transportation and processing equipment require long life--2000 rpm is slightly over 10^9 cycles per year--so an endurance limit stress level is indicated. However, many spacecraft undergo little or no significant stressing during their operational life but are severely stressed during the relatively short period of launch, 5-10 min, thereby reducing their fatigue lifetime to about 10^5 cycles of engineering level stress, plus an acoustic exposure of roughly 3×10^5 cycles. Thus appropriately higher fatigue strengths may be taken for these short lifetimes. For spacecraft and missiles, an additional factor for the portion of life consumed during testing is required because the vibration and acoustic test sequences very often represent a major portion of the total lifetime. It is often and truly said that the design of these latter vehicles is directed more toward the requirements of the specification vibration tests than to those of their operational life.

The range of sophistication in the treatment of loads analyses is enormous. In some of the older automotive design efforts (prior to 1940), the maximum load on a critical part, as a steering knuckle, was found by strain-gaging an actual knuckle and driving at various speeds over "rough" roads. The peak load was then translated by means of the part geometry and stress concentration factor to an allowable stress for infinite life, with no reference to the frequency distribution of the load levels. Other, more recent work has indicated that the frequency distribution for such loads is normal or gaussian; extrapolations are then possible on a somewhat more secure base. It is probable that, for automotive parts of that era, such an approach was justifiable both technically and economically. That one could safely apply rigid-body mechanics to this design, that the minimum weight criterion need not be rigorously applied, and the psychologically untenable uproar resulting from the occasionally failed knuckle (however remote the probability) all tend to support this conclusion.

But in more modern applications for land, sea, air, and space vehicles wherein minimum weight, nonrigid-body mechanics, and extreme safety and reliability are sine qua non, the most acceptable approach is now well recognized as the probabilistic one. The nature of load occurrences is statistically random, at least for moving bodies not tied to a ground frame. For stationary, rotating machinery, the variation in the load cycle is known or small or both, but such a limitation may not be safely applied to the design of moving vehicles. For such vehicles, the power-spectral-density (PSD) method of loads analysis is being widely applied, for it is now generally agreed that the source of the loads, such as atmospheric turbulence, road shock, and propellant combustion, is properly represented by a continuous, stationary, random process. The advantages offered by the power-spectral-density method are:

1. It provides for a more realistic representation of the continuous nature of the load than does the assumption of a discrete maximum load for a certain number of cycles. It inherently accounts for variations in the shape and gradient distance of the input function.

2. The power-spectral-density method is readily extended to the construction of curves of frequency of occurence vs. the load level at various locations in the structure for representative operational profiles. It can thus account directly and quantitatively for any differences between new and old designs in regard to the relationship between normal (or old) and "design" (or new) operating conditions.

3. It enables a rational account of structural configurations and response characteristics, that is, if a new design contains, or is suspected of containing, modes of a different frequency or different damping, the changes in response due to these factors may be properly considered.

4. Closely related to the above is the recent development of computer techniques for solving the equations of motion, with which the power-spectral-density method is very compatible. It is now practical to represent the dynamics of complex loading geometrics in sufficient refinement to account for not only the rigid-body motions, but also for those due to elastic compliance, as well as the effects of guidance and control devices.

The high level of talent and the volume of the effort required, as well as computer facilities, tend to make the power-spectral-density method somewhat expensive to apply. Because of this, and of relative unfamiliarity with the process, the varieties of application have been limited. One well-known effort is that presently being implemented by the airframe designers and others under the title of "dynamic or gust loads criteria." The motivation for this work is, of course, to secure a safer and lighter structure. Yet these advances do not in themselves result in direct achievement of this objective--there are still two steps required. The first, and most important, is to modify or redevelop the structural criteria by which a required strength level is established for any given vehicle. The second is to fit the newer methods into the routine by which design loads are obtained and the stress analyses performed, to assure a strength consistency in all the individual elements of the structure. As an example, the conversion of a gust spectrum to a stress spectrum is outlined in Section 3.2.1. In another form, there is greatly increasing activity in power-spectral-density analysis combined with random load testing (by the newer, servo-controlled hydraulic machines) of, for instance, full-scale automotive assemblies.

2.2.1 Stability Analyses

Such analyses and their experimental substantiation constitute a third step closely interrelated with the loads and stress analyses. Stability or interaction equations and diagrams, such as Fig. 2.1, are generally accepted; but the complexity of many of the relationships, especially those for optimized properties like weight or cost, is such as to require developmental testing. As usual the required outputs of

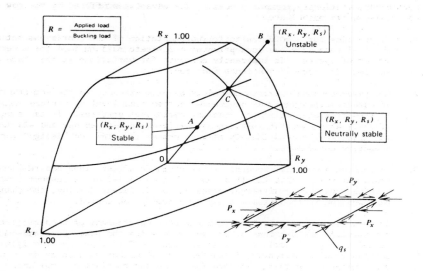

Fig. 2.1   Buckling under Combined Loads.   Figs. 2.1 to 2.7
from W. J. Crichlow [312].

such analyses are the allowable design stresses and margins of safety, and Crichlow [312] has made an admirable presentation of the information flow to produce such results; see Fig. 2.2.   The diagram illustrates particularly well the relationship of the geometrical parameters and fatigue considerations with the basic loads and materials data.   If the structural sizing accommodates the inputs from fatigue, fail-safe, damage tolerance, and crack-growth limits, and shows positive margins, then it may be assumed that all pertinent stress limit requirements have been met.

2.3   DAMAGE AND FAILURE CRITERIA

Fatigue design is the basic method by which dynamic stresses are treated, as well as the types of failure that result from them.   In the information flow diagram for fatigue life assurance, Fig. 2.3, these criteria are contained within the inputs marked Design concept, Design criteria, Performance characteristics, and Mission analysis.   Their expression may take on a variety of forms: as a limiting stress, deflection, acceleration, or exposure time; lowered probability of continued successful operation due to accumulation of damage, and so forth.   The objectives of a design effort are usually stated in terms of a successful test or operation, but since a totally unqualified success is rarely attained, the definition of nonsuccess or the degree of failure is equally important.   It is also advantageous that these definitions and criteria be established early in the program, preferably before the start of the design layout, and surely before the sizing is done.

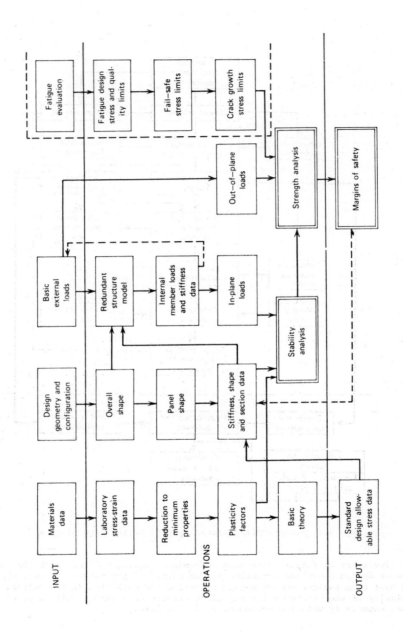

Fig. 2.2 Stability and strength analysis.

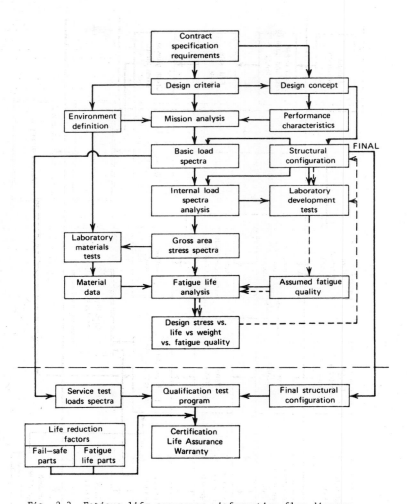

Fig. 2.3 Fatigue life assurance--information flow diagram.

About five modes of failure have been recognized, which, combined with the large number of design operations, results in the matrix of the design process, Fig. 2.4. There must be a failure and/or damage criterion for each item in the matrix. The granularity of the breakdown for any given article or assembly will vary, of course, especially with the seriousness of a failure.

Design Operation / Mode of Failure	Specification	Design Criteria	Design Loads	Design Process	Tests		
					Material and Components	Assembly	Structural Data Delivery
Static strength-stiffness-undamaged	• Requirements • Environments • Performance	• Performance envelope • Factors of safety	• Static • Dynamic • Stiffness requirements	Selection of • Material • Configuration • Sizing	• Materials data • Envir. and process effects • Design allowable • Comp. develop.	• Proof tests • Service loads tests	• Performance strength and operating limits
Deformation undamaged	• Life • Performance	Environment • Thermal • Chemical Factors of safety	• Steady • Cyclic • Temperature	Selection of • Material • Configuration • Sizing	• Materials data • Envir. and process effects • Design allowable • Comp. develop.	• Operating life tests	• Inspection techniques

Residual static stiffness-damaged	• Damage tolerance goals	Fail-safe • Performance envelope • Damage size	• Fail-safe loads • Stiffness requirements	Selection of • Material • Configuration • Sizing	• Materials data • Envir. and process effects • Design allowable • Comp. develop.	• Fail-safe damage tolerance tests	• Inspection techniques • Damage limits • Repair instructions
Fatigue crack Initiation-undamaged	• Life • Routes • Operations	• Route analysis • Fatigue-quality standards	• Life spectrum of operating loads	Selection of • Material • Configuration • Sizing	• Materials data • Envir. and process effects • Design allowable • Comp. develop.	• Full-scale fatigue tests	Inspection • Locations • Techniques Repair instructions
Crack propagation-life-damaged	• Operations • Inspection • Maintenance • Repair	• Inspection techniques • Damage size limits	• Limited spectrum-operating loads	Selection of • Material • Configuration • Sizing	• Materials data • Envir. and process effects • Design allowable • Comp. develop.	• Extended fatigue tests • Arbitrary damage tests	Inspection • Locations • Techniques • Intervals Repair instructions

Fig. 2.4 Matrix of the design process.

With regard to the life and fatigue aspects of design, it should be recalled that the designer has only three basic means of control: (1) selecting the materials and processes, (2) setting the quality of the detail design, and (3) limiting the service stress levels. In the last case he can, of course, only specify the limit and must depend on the operators to observe it; the alternate of designing in fuses or shear pins is frequently intolerable from the operating viewpoint.

Treatments of damaged structure, whether by fatigue accumulation or accident, are required in the consideration of residual strength, remaining life, and size or other description of the damage and its rate of growth with continued operation. General relationships among these parameters are given in Fig. 2.5. The static limit and ultimate design loads for undamaged structure are plotted on the ordinate as are the damage size and the fail-safe loads for a damaged structure. Design

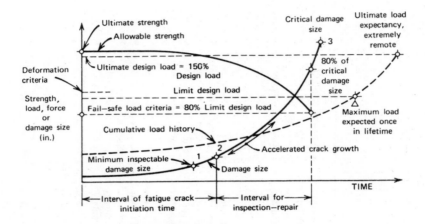

Fig. 2.5 Relationships of applied load history, strength, and damage size with time.

criteria include the condition that zero or positive margins exist at any time during operational life. The allowable strength will decrease as the crack grows with time, the relative rates being roughly as shown. The time to initiate and grow a fatigue crack to inspectable size is given at point 1; growth to the most probable size at discovery requires some additional time, to point 2. The interval from 2 to the critical crack size for unstable growth at 3 is the time available for inspection and repair.

Calculations of residual strength require information on the rate of crack growth as a function of stress level and of instantaneous crack length, particularly required are the conditions for unstable crack propagation, during which growth continues rapidly with no applied stress. Illustrative plots of typical data are shown in Figs. 2.6 and 2.7. There are several expressions now available for handling the data, that for "Stress Intensity" arising from considerations of Fracture Mechanics. The "Stress Concentration" and "Effective Width" equations

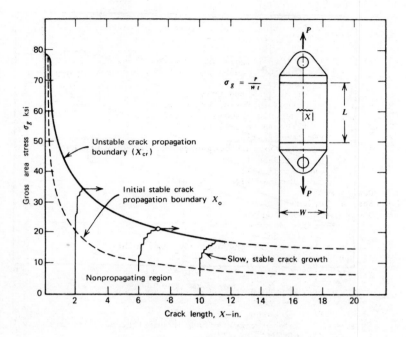

Fig. 2.6 Typical residual static strength data.

also contain empirical constants that may be adjusted to fit the data; thus the design results are relatively insensitive to which expression is used. An illustration of the Effective Width approach in Damage Tolerant design is shown in Section 2.4. This type of design work is considered to be within the emerging discipline of Fracture Mechanics, but there is still debate on the accuracy and efficacy of some of the methods, to the extent that handbook values of, say, the stress intensity factors for 2024-T4, are not yet universally available.

The effect of the fail-safe or damage-tolerant design approach on reliability has been demonstrated in probabilistic terms. If the standard deviation of the logarithm of the number of cycles to failure (first crack) has only slight dependence on the fatigue life, the probability density of fatigue cracking can be represented by a log-normal distribution. The cumulative probability that a fatigue crack needing repair develops in a given time is determined by statistical treatment of large sample data for both series and for parallel configuration structures, and plotted as Fig. 2.8. A series structure is defined as one in which cracking in any one member would result in a failure or less than full operational capability; for a parallel structure the cracking of all members would be required for a failure. The family of curves shown is based on $\sigma = 0.20$, which seems to be a

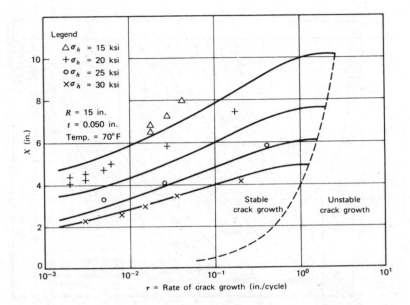

Fig. 2.7   Rate of crack growth in unstiffened cylinders,
titanium 8-1-1 DA.

representative value. As an example, consider a population (fleet or group) of
2000 operating units with a service time of 25% of their service life as demonstrated by fatigue tests, point A. A probability then exists that 2000 x 0.0015
or 3 units would have failed if only one critical member were present. If the
design contained ten such members, point B, the probability of failure enlarges to
30 units; thus it is highly desirable to provide damage tolerant capabilities. If
these structures were large and expensive such as airframes, and not so designed,
the failures or loss of full operational capacity would have been catastrophic
losses--a prohibitive condition.

The curves illustrate two significant phenomena that account for the effectiveness
of damage tolerant design in reducing failure probability to almost negligible
proportions. Convergence of the series configuration curves at low failure probabilities implies that many systems may be placed in series without severely
reducing overall reliability if the reliability of each system is maintained at a
high level. On the other hand, the divergence of the parallel configuration curves
indicates that the period of high reliability can be significantly extended for
each system by introducing only a small number of parallel (redundant) elements or
members as long as the failure probabilities of the added elements remain low.
This condition may be assured by frequent inspection or the replacement of elements.

BEABLE TO USE CHART

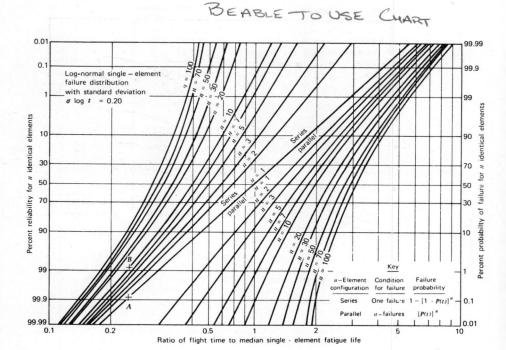

Fig. 2.8   Failure probability for series and parallel
configurations.   From W. T. Shuler [374], Fig. 35.

2.4   SAFE–LIFE AND FAIL–SAFE DESIGNS

In order to guard against failures from unforeseen circumstances, two major
approaches to structural substantiation have been devised.   One is a fatigue
analysis and testing program that attempts to establish a "safe-life" for the
structure under the assumed loading conditions.   This procedure implies that the
life can be predicted and that before the end of this time the structure can be
repaired, replaced, or retired.   If the analysis/synthesis can be accomplished
sufficiently early in the design schedule, deficiencies may be eliminated to
achieve the desired life.   It has also been recognized that, inevitably, some
structural damage and failures would occur and that catastrophic failure is rarely
tolerable.   This recognition led to the concept of a damage–tolerant or "fail–safe"
design in which the damage from those failed component(s) could be temporarily
tolerated by providing alternate load-carrying members and sizing them such that
those remaining after the partial failure would sustain reasonable load levels.

While these concepts are widely discussed and applied, they are being interpreted in many and various ways. Opinions regarding the fail-safe structure vary from the extreme that the structure will be tolerant of any damage that may be inflicted, to the opposite one that such a structure cannot be successfully configured to carry the service load once major failures or damage has occurred. With regard to fatigue, the variations of viewpoint are as great: that the structure will exper- ience no fatigue failures during its useful life, or that safe-life cannot be predicted or a reasonably economic life goal assured. The real goals of the analytical efforts on fatigue and on damage tolerance are, of course, somewhere between these extremes, but they are so interdependent as to defy separate defini- tion and an interpretation of one certainly involves the other. The purpose of a fatigue investigation is to set a design-allowable stress such as to minimize the rate of fatigue damage in service, and to establish suitable inspection and main- tenance procedures where applicable. At present, the state of the art makes improbable the definition of a structure that will be completely free of fatigue failure for a given time under given load conditions, that is, no guaranteed safe-life. A fatigue failure is here defined as a structural failure that would require replacement of a major structural component for safety or which would have a seriously degrading effect on utility. Safe-life is the time period for opera- tion in a known environment with a known probability of exposure to ultimate loads.

The distinction between safe-life and fail-safe is neither formal nor rigorous. The principal considerations in describing the concepts are:

1.   The number of elements in the load path, one or more than one.

2.   Parallel or series arrangement of elements along the load path.

3.   Relative motion between adjacent pieces: strain or displacement.

4.   Consequences of a single-element failure.

5.   Monolithic or built-up forms.

Figure 2.9 illustrates some of the distinctions, chiefly the influence of the number of parallel load-carrying elements. Table 2.1 continues the distinction by listing some common structures, machines, tools, devices, appliances, and so forth in a best-fit category.

As an example of the overlap and non-rigorousness of the classifications, consider the connecting-rod, cap, and crankshaft portion of a reciprocating engine. The rod, cap, and shaft are monolithic single pieces without crack-stoppers of any sort; they fall without question in the safe-life category. But the rod/cap joint, which might be considered as the more troublesome area, can be treated in either category. Many designs incorporate only a single bolt on each side, the failure of which usually stops the engine. Now, suppose that the boss could be modified in shape or lengthened so as to receive three smaller bolts, any two of which could sustain the load upon failure of one. Thus the joint becomes fail-safe, but a mixture of categories results, so that attempts at extension of such con- siderations come to an early halt. One might conceive of slicing the rod into three longitudinal elements, but a fail-safe wrist pin or piston is really diffi- cult to obtain.

It is to be noted that in these considerations the numbers jumped from one to three (really to no limit); thus the question of redundancy or standby arises. Standby usually refers to complete extra entities, as boiler feed pumps or emergency lighting, and redundancy to some larger number of smaller items that may

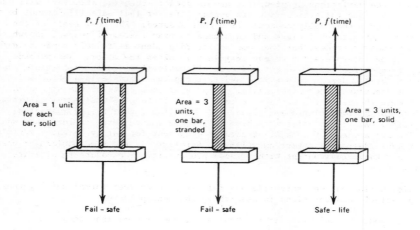

Fig. 2.9   Design categories.

come into or go out of operation (or of "picking up the load") without a switchover. But the use of two items (other than standby) is open to questions of purpose and efficiency.

The mode of the first-element failure is also highly significant in its effect on the design approach decision. In a fail-safe assembly, the result of normal failure is a small deflection and no loose pieces; the structure remains an entity because of the crackstoppers. The sudden release of the strain energy has been shown as relatively unimportant. In the monolithic safe-life designs, failure is usually synonomous with fracture into two or more pieces, with a relatively high probability of their causing stoppage or additional failure by interference. This is not to say that the fail-safe structure cannot be completely fractured by an application of a sufficiently high load or by the working load for a sufficiently great number of cycles, but that the action during and immediately following the first-element failure is quite different. The repair or replacement of safe-life parts is often considered on the basis of size, cost, availability, complexity, down-time, and so forth, with the decision often being made for replacement, while the repair of a fail-safe assembly by replacement of some elements and/or doubling or strapping across the crack is routine.

One major conclusion may be that the fail-safe approach is probably more readily applicable to those objects in which the relative motion of adjacent parts is that of elastic strain only. Its increasing use in airframe design stems not only from its applicability, but chiefly from the necessity to prevent the consequences of a single-element failure. Safe-life design was the earlier of the two approaches and is usually included in many of the elements going into a structural assembly, the overall design of which may be fail-safe. Safe-life design then, is, that activity usually applied to single pieces or sub-units but not to complete assemblies, and which involves fatigue analysis generally and such geometry as affects stress concentration factors.

TABLE 2.1   Safe-Life and Fail-Safe Categories of Design

Safe-Life	Fail-Safe

1. Function and shape.
    Belts
    Shafts
    Gears
    Blading, for turbines and fans
    Impellors
2. Function alone.
    Fluid containers (especially glass)
    Heat exchangers
    Boilers
3. Shape alone.
    Monolithic or inherently one-
      piece parts:
      Forgings
      Castings
      Films, single
      Wires, single
4. Assemblies.
    Motors, electric and hydraulic
      internal/external combustion engines
    Bearings, antifriction
    Weldments [might also be in (3)]
    Tires, single casing
    Domestic appliances
    Processing machinery: food, textile,
      paper, metals, plastics, etc.
	Airframes, but not engines
	Autoframes, but not engines
	Bridges
	Marine hulls (welded or riveted)
	  because of compartmentalization
	Framed buildings
	Tires, multiple casings
	Rope
	Fabrics
	Assemblies: bolted, riveted, or
	  bonded

2.4.1   Safe-Life Design

The term safe-life has often been loosely defined. Recently, and unfortunately, safe-life has been thought to describe the results of fatigue tests and analyses that were performed to prove a structure to be free of fatigue failure. Such is not always the case. The beginning of a fatigue crack, and many will be experienced even in minor tests, is not the end of the useful or safe-life. Proper inspection and maintenance coupled with a damage-tolerant design could lead to an infinite safe-life. To be sure, at the end of an economically useful life many components may have been replaced or repaired, but the overall structure will possess the same integrity as it did initially. Fatigue analyses and testing, then, become tools for the purpose of eliminating design features that cause early failures, as well as a means of identifying potential problem areas for maintenance.

For that special class of structures which, for any reason, do not receive inspection and maintenance during their operating period, the safe-life becomes more directly related to the lifetime derived from fatigue analysis and test. Spacecraft and one-time-use devices are characteristic of this class. Since, for spacecraft generally, the damage accumulated in service is nil compared to that of launch and vibration testing, the required safe-life could be defined as a period equal to the sum of the launch and test times plus such margin as good engineering judgment dictates. Upon the development and use of spacecraft for multiple trips, the safe-life for their structures will become identical in definition to that for present aircraft, involving the disciplines of inspection and maintenance.

If, in a fatigue analysis, one can acquire a reasonably accurate loading spectrum and apply a little intuition and experience in evaluating potentially troublesome areas, the expected service life (until a major failure occurs) is not difficult to determine. The load histories, however, may well depend on a large number of variables, and the resulting life predictions are no better than the assumptions made in determining the load spectra and in using the damage assessment method. In attempts to determine a safe-life for the several major components of an air-frame, some correlation was found between an actual (and crash-free) service life and the fatigue damage ratio (value of n/N) for the components. The results in Table 2.2 indicate the relative severity of the fatigue loading on the several components. For this aircraft the permissible safe damage ratio was set at 0.3; thus some of the components would have lives restricted to less than 30,000 hours. A graphical comparison is given in Fig. 2.10 of the observed life and of the average life estimate for many components in another aircraft [90]. It should be noted that the width of the band necessary to include all observed points results in a service life variation factor of about 5 and an estimated life variation factor of nearly 10. It is probable that the band should become much narrower for future designs as the science of life prediction is further explored and carefully applied. But, as Fracture Mechanics develops, especially in its treatment of the inevitable initial cracks and their propagation, it appears that the Damage-Tolerant approach to design will be more frequently indicated as superior.

TABLE 2.2   Fatigue Damage Ratios for Various Airframe Components

Fatigue Loading Condition	Stress Range	No. of Cycles in 30,000 hr, n	Cycles to Failure, N	Fatigue Damage Ratio, n/N
1.0 g Wing lift	-2000 +14,000	10^4	5×10^4	0.2
Wing, 10 ft/sec gust Cruise at 20,000 ft Cruise at 8,000 ft	±3500 ±3500	2×10^5 6×10^5	6×10^5 6×10^5	0.3 1.0
Fin, 10 ft/sec gust, Cruise at 20,000 ft	±4600	2×10^5	2×10^5	1.0
Cabin pressure loads	0-13,000 0-20,000	10^4 10^4	10^5 2×10^4	0.1 0.5
Landing gear	0-20,000	10^4	2×10^4	0.5
Hydraulic component	0-25,000	10^4	10^4	1.0
Engine mounting tube	±4000	3×10^5		--

SOURCE: H. Giddings [27].

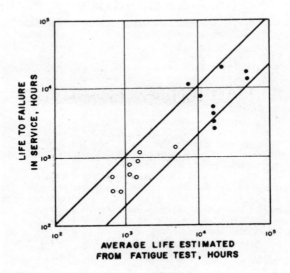

Fig. 2.10 Comparison of service life with estimated life. Skin
cracks, ○; spar cracks, ●. From K. D. Raithby [90],
Fig. 7.

2.4.2   Fail-Safe or Damage-Tolerant Design

Since it is not likely that a completely fatigue-resistant structure would be
economically feasible or that total protection could be obtained against damage
from any source, some service failures must be expected. The possibilities of
failure, or at least overloads, also exist from errors in design, manufacturing,
and maintenance, as well as from corrosion and malfunctions during service. The
recognition of such possibilities requires the provision of sufficient residual
strength and stiffness in the remaining members for continued operation under
reasonably normal loading until the failure can be repaired, or if such is not
possible, for continued operation under conditions less severe than the design
level. Under continued loading, the failure of one element will most certainly
propagate throughout the remaining structure, and, at least for transport vehicles,
the probability exists for encountering loads above normal, and greater than the
strength remaining at that time.

Fail-safe design can thus best serve a useful purpose when supplemented by suitable
inspection and maintenance. For the cases wherein the latter are not possible,
such as spacecraft, fail-safe design provides the best probability of sufficient
survival to continue performance of the mission, but at a degraded level. The
problem that fail-safe design is supposed to prevent is an old one; perhaps the

best of the early descriptions is that by Pippard and Pritchard [23] in 1919:
"The problem of the disabled aeroplane is one that must be faced by the designer.
Any aeroplane should be capable of reaching the ground safely even after certain
parts of it have failed in the air. It is impossible to guard completely against
the effect of breakage of any part...." In 1919 fatigue design was, of course,
even more primitive than it is now; important changes since then include the
following: (1) the direct relating of fatigue and the fail-safe approach, (2)
recognition that a lifetime is only finite, however long, and (3) inspection for
cracks or yielding as preventive maintenance, with the understanding that variation in lifetime can be a function of the frequency and severity of this inspection.

A philosophy for use in fail-safe design might be put somewhat as follows:

1. The structure must have an adequate life, either as a crack-free period, or
    one during which the growth rate of cracks is sufficiently low so as to
    escape detection by adequate inspection procedures.

2. There must be the capability of carrying a predetermined load under a given
    amount of damage, when the latter is known to be present, that is, the design
    must include adequate residual strength.

3. Visual inspection of all critical areas must be possible, in service or after
    tests.

4. Damaged elements are to be repaired by replacement or the use of doublers.
    The large amount of labor and time, plus the high probability of accidental
    damage in replacing a major member in a complex structure often leads to the
    use of doublers or conventional patching.

The fail-safe type of design may be interpreted best in terms of (1) a residual
strength after cracking, and (2) redundant load paths. There is need for considering the degree to which the fail-safe property may be made available; this is
related to the number of load paths possible and to the general size relationship
of the load-bearing elements to each other and to the overall assembly. Each load
path is required to be designed with some extra capacity above its theoretical
share of the ultimate load; upon failure of one path, the total load is assumed to
redivide and to flow around the fracture in the adjacent paths. Thus the concept
of degree of fail-safety exists, is qualitatively proportional to the number of
load paths and the extra load capacity designed into each path, and is dependent
on sufficient ductility being present to permit the realignment required for the
new load paths to become established.

The determination of a proper number of load paths may be very difficult and at
best is usually a nonrigorous solution. Blind application of the fail-safe approach
would lead to the duplication of every load-bearing element at least once, a
condition that may be impossible, undesirable, or unnecessary. Numerical treatment
of load paths depends on the inherent nature of the structure or machine: tire
cords are redundant for obvious reasons, but automobiles are rarely provided with
more than four wheels; tension members in trusses are often redundant but seldom
are the compression members; and aircraft wings usually have more than one spar
but only one skin. Redundant load paths may be obtained by adding members or by
dividing an original entity into several smaller members. There seems to be no
general rule as to which course to take or how far to go along the finally chosen
one; the considerations must include not only the specific technical items such as
loading geometry and material properties, but also the "natural" items of use and
practicality.

Residual strength. The provision of residual strength, or load-carrying capacity in the presence of cracks is a fundamental of fail-safe design. Two criteria are involved: (1) the fail-safe load, and (2) the critical number of part failures. The fail-safe load criterion is related to the risk of exceeding the fail-safe strength remaining after one member has failed. The critical number criterion appears because of the finite probability of collapse of the fail-safe structure due to the fatigue failure (between two inspections) of a critical number--more than one--of the members of such a structure. Since these criteria are dependent on the design, although in a different way, it is not possible to predetermine the portions of the fail-safe failure rate that might be assigned to each criterion. Lundberg [57] suggests an equal division, and that both portions of the failure rate should then be subdivided into smaller parts assigned to each of the subassemblies of the primary structure. A failure rate may be established by setting a safety level goal. As an example, aircraft accident history, together with present design and operational practices, indicate that such a goal might be 300 accidents in 10^9 hours of flight. The division of this number according to causes results in 10 structural failures; of these, one is alloted to fatigue as being the least excusable cause of accident. There is thus implied a failure rate of 10^{-9}.

The calculation of residual strength after cracking is rather involved, and, while present approaches are yielding results of most adequate confidence, a universally accepted method does not seem to be available yet. The works of Hunt [58], Crichlow [59, 382], and Harpur [61] are suggested to illustrate recent examples. The establishment of a proper level of residual strength is, of course, paramount in importance, not only for safety but also from the standpoint of structural efficiency. In the evolution of fail-safe designs, a very high level of residual strength has appeared, approximately 80% of design ultimate, which is no doubt due to both the great talent applied in these design efforts and to the nature of such designs. A number of aircraft wings have achieved, by test, residual strength levels in excess of 80% of design ultimate; the British Air Registry Board [54] has, in effect, placed a floor under this parameter by requiring that in a fail-safe-type design the number of individual elements be chosen so that at least 67% of the design ultimate load can be met with any one element failed, as a design goal.

In regard to the matching of details within the fail-safe design with the nature of the applied stress field, it appears that, for undirectional stress and a load spectrum that includes some high peaks, the optimum structure should be one with multiple load paths separated by span-wise discontinuities or joints as crackstoppers. This condition would probably permit a maximum level of applied stress, since rapid crack propagation could be accepted. The condition that all critical areas are inspectable is required here. For a biaxial stress field, the optimum structure probably has a very large number of individual elements: the wide use of skin and flange doublers; a high percentage of bonded joints, such bolted and riveted joints as are used being designed to develop full compression in the skin; and crackstopping done by reinforcement. A lower allowable stress level than in the above structure would probably be necessary. Finally, for those cases wherein the provision of sufficient multiple load paths creates a weight penalty or other disadvantages in complexity of construction or loss of inspectability, the safe-life structure, with the lowest stress level, turns out to be not only optimum but possibly the last resort.

Essentially the same methods are used to determine that the desired strength tolerance levels are actually provided in designs, and to confirm static strength provisions. Since the analysis is empirical in nature and based on test results, few new tests are needed except where structural designs differ significantly from types already tested. Furthermore, since analysis relates to tests in which damage is simulated while the structure is undergoing design loads, no multiplying factors accounting for dynamic effects of failure under load are necessary. Here, as in static design tests, a margin of safety of zero is permissible.

Prior to initiating analysis, the extent and type of damage and the load levels to be achieved are specified as shown in Table 2.3 These basic criteria are established to ensure that damage may be readily detectable before structural strength is impaired beyond the point of safe operation. Differences in design detail among various aircraft types account for the small variances in criteria applied to them. Once damage tolerance criteria are defined, the structure may be designed to conform.

A rather simple concept serves as the basis of determining residual (ultimate) strength, as of a damaged fuselage skin panel. It is based on a fictitious effective width, W_e, measured ahead of the tips of a crack in the skin, as shown in Fig. 2.11. This approach considers natural crackstoppers such as stringers and

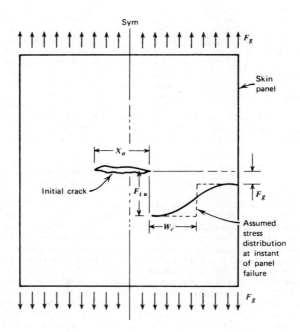

Fig. 2.11 Skin panel damage analysis.

skin splices located perpendicular to the longitudinal axis of the crack. Several typical cracks are shown in Fig. 2.12; this information is then used in conjunction with the type of information given in Figs. 2.13 and 2.14 to predict the residual strength of the damaged panel.

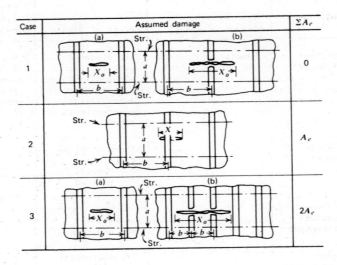

Fig. 2.12   Curved, stiffened panels, with a longitudinal
crack, under internal pressure.

For a fuselage shell, used as a typical example, the residual strength is expressed
as the critical internal pressure, P_{cr}, at which explosive failure of the skin and
frames occurs, and is written as:

$$P_{cr} = \frac{F_g t}{R}$$

where

F_g = the hoop tension

t = the skin thickness

R = the shell radius

The extent of the damage, related to the P_{cr}, is assumed; F_g is determined from
parametric plots of the ratios of effective area and thickness, A_e/t, vs. the
ultimate strength of a frame and that of the skin, F_{tuf}/F_{tus}. The effective area
is an important parameter; generally for channel and zee-section frames this area
is very small because of the stringer cutouts: $A_e = A_f/5$ but increases to $A_e = A_f/3$
if a reinforcing angle can be attached to the frame near the cutout. Predicted
strength is then compared with required strength to determine the existing level
of damage tolerance. Some typical test results are given in Figs. 2.15 and 2.16
for panels with bonded crackstoppers, a major design detail that gives great
promise for a truly fail-safe and efficient structure.

TABLE 2.3   Load Capability of Damaged Pressure Cabins

Aircraft Model and Type	Type of Damage Assumed in Design	Extent of Damage	Ultimate Load Capability of Damaged Structure
C-130 turbo-prop assault transport	Longitudinal crack in pressure cabin	Circumferential ring failed, accompanied by skin crack across both adjacent skin panels.	2.0 g maneuver or 49 fps gust encounter with full cabin operating pressure.
C-140 jet executive transport		Circumferential ring failed, accompanied by 12-in. long skin crack.	
C-141 jet medium logistics transport		Same as C-140.	
C-5A jet heavy logistics transport		Circumferential ring failed, accompanied by skin crack across both adjacent skin panels.	
C-130 turbo-prop assault transport	Circumferential crack in pressure cabin	Longeron failed, accompanied by 36-in. long skin crack	2.0 g maneuver 49 fps gust encounter with full cabin operating pressure differential.
C-140 jet executive transport		Longeron failed, accompanied by 12-in. long skin crack.	
C-141 jet medium logistics transport		Any single longitudinal stringer failed, plus skin crack across both adjacent skin panels.	
C-5A jet heavy logistics transport			

SOURCE: W. T. Shuler [374], Table 1.

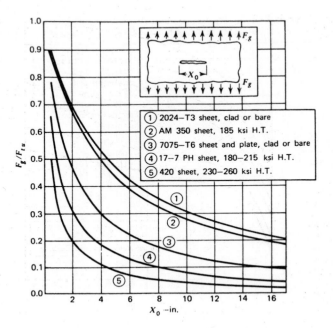

Fig. 2.13 Tear strength of flat sheet with a crack perpendicular
to tension load (infinite sheet width).

Variations of residual strength as a function of material types and crack lengths
were given as test data in Fig. 2.13. Any residual strength analysis of damaged
panels must account for significant variations in other parameters affecting W_e
such as panel geometry, the type and rate of loading, the temperature, and the
sheet thickness. Provided that the ranges of important parameters or combinations
are not exceeded, good agreement can be realized between predicted stresses and
measured stresses.

The local dynamic effects of the partial failure in a fail-safe structure have
been the subject of some concern in the calculation of residual strength. It is
thought that these effects consist of two parts:

1. The sudden release of the stored (elastic strain) energy in the failed element
    may impose appreciable overloading on the adjacent elements.

2. The stiffness of the subassembly (as a wing) would be appreciably reduced as
    a result of element failure; the wing would then attempt to assume a new
    static deflection but in so doing would experience oscillations about the new
    position.

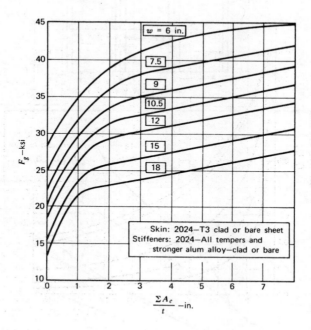

Fig. 2.14   Ultimate strength of flat, stiffened panels and
slightly curved panels with a circumferential
crack.   From W. T. Shuler [374], Figs. 20-22.

As a result of extensive dynamic tests, Spaulding [25] has said: "Whatever the
failure and deflection dynamics may be, they are not additive. Failure occurs so
rapidly that its dynamic effects have completely damped out before the wing will
have appreciably changed its deflected position...." In all cases measured, the
dynamic stress increment was less than 5% of the average stress, but the strain
increment was as much as 30% of the total average strain. This high strain near
failure is associated with the plastic deformations that permit only a small
stress change.

Technique of Fail-Safe Design. Tension surfaces, as wings and tails, form a good
example of structural subassemblies to which the fail-safe approach may be applied.
One type of construction is shown in Fig. 2.17. The spanwise aluminum "boards" or
machined extrusions are joined by steel attachments. When a board fails, its load
is transferred through the attachments to those adjacent. These spanwise rivet
rows must then be able to perform this transfer anywhere along the span, in addition
to performing their normal shear transfer. This condition requires that a greater
joint strength be present in all splices than is characteristic of safe-life

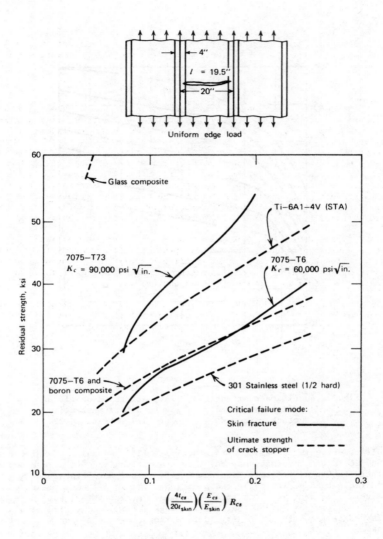

Fig. 2.15 Residual strength curves for aluminum alloy
panels stiffened with 4-in. wide adhesive
bonded crackstoppers of different materials.
From D. Y. Wang [375], Fig. 1.

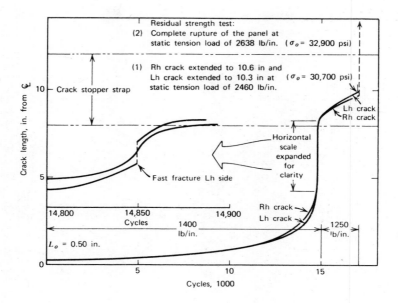

Fig. 2.16   Test results for panel of Fig. 2.15.   From
D. Y. Wang [375], Fig. 2.

designs. The needed joint strength can be obtained with the fewest fasteners and
a minimum of overstress in the adjacent members if the attachments can withstand
considerable plastic deformation before failure.   Fastener deformation distributes
the load to be transferred over a greater number of attachments and reduces the
load to be taken by each. This "plastic" transfer of load can be obtained by
using steel attachments made critical in bearing in the "board."

The board type of construction is advantageous when the use of a high strength
alloy is dictated by compression requirements. Alloys such as 7075-T6 are less
resistant to crack propagation than are the lower strength types such as 2024-T3.
Cracks that may develop within a board will find a natural barrier at each spanwise
joint, which temporarily arrests its development and allows some time for its
discovery. Machined board construction has the added advantage of permitting the
use of raised pads to minimize points of stress concentration and to provide
integral reinforcement around discontinuities.

A truly fail-safe structure must not fail when a shear beam is damaged for any
reason. Here, multiple beam construction has the advatage of supplying alternate
load paths for tension, in case any single beam web should fail. Assuming that a
fatigue crack would start as usual at the lower tension surface, it would probably

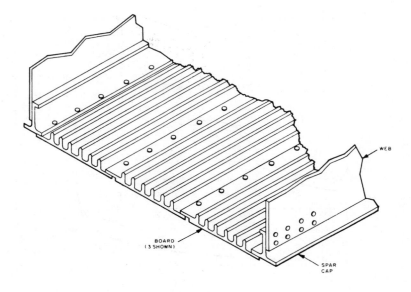

Fig. 2.17   Typical fail-safe construction for tension
surfaces.   From E. H. Spaulding [25].

run upward through the web unless stopped in some manner.   Figure 2.18 shows one
type of "crackstopper" that is formed basically by the creation of an artificial
web splice and stiffener located about 1/3 of the distance above the lower surface.
The remaining 2/3 of the shear web would be able to sustain considerable load
should the first 1/3 be broken.   Even though the scheme has satisfactorily stopped
fatigue cracks, its use involves considerable extra weight, parts, and assembly
labor.

A more attractive design, which avoids the artificial splice, is shown in Fig. 2.19.
Here the web is thickened at the attachment to the spar cap (tee extrusion).
Steel attachments are made equally critical in bearing, in the web and cap; vertical
stiffeners are added as required on the web.   Upon the fracture of a spar cap, the
load to be transferred divides between the thickened web and the adjacent surface
structure.   Bearing deformation in the holes will allow the load to be distributed
over a number of rivets, minimizing the load in each.   The thickened web is sized
to carry the additional sudden load.   If, instead, the shear web should fail
first, it could crack completely from bottom to top.   As above, sizes and attachments are chosen to prevent the failure of the cap, the deeper spar caps with the
additional flanges being proportioned to take fail-safe shear loads as caps of a
framed-bent structure.   The integral stiffeners on the machined web in conjunction
with steel web-to-cap rivets complete the structure.

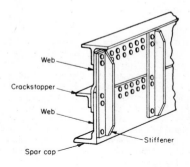

Fig. 2.18 Shear web crackstopper.

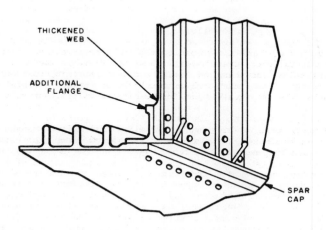

Fig. 2.19 Fail-safe shear web design.

The criterion of criticality in bearing is particularly important in the design of the joint elements. Sufficient yielding to permit redistribution of the load upon failure of a major member can be achieved by using an acceptable tension-shear-bearing (T-S-B) ratio and accommodating the stress-concentration factors at the holes. Not incidentally, the tension-shear-bearing ratio here will be quite different from that for a separation-type joint where the bolts are essentially in tension only.

Inspections and Inspectable Areas. Reference has been made above to the relationship between fatigue life and the frequency and severity of inspection. Upon consideration of any but the simplest of structures, a very practical limitation appears in the form of blind areas, and it follows that the fail-safe approach to design can be applied only to those areas that are inspectable. It is thus mandatory that blind or noninspectable areas must be kept to a minimum; when they are unavoidable the design must be such that the growth of the maximum initial crack during the structure's life leaves an adequate margin against its critical length. This situation can be controlled by having data on the maximum crack size to be expected of the material, its shape and processing; and/or by an occasional teardown inspection of one unit in the group. For inspectable regions, the determination of inspection intervals as a function of initial size of defect and of its rate of growth is discussed at length in Section 3.2.4.

It is, then, nearly impossible to implement the fail-safe design approach in all elements of a complex structure or machine, but the potential gain is so great that many successful efforts have been made. Figure 2.20 [55] portrays several methods of designing for inspectability in conformance with the requirements of a fail-safe design.

2.5   FACTORS OF SAFETY VS. RELIABILITY

Discussion of a safety factor is necessarily included in any approach to fatigue design. In so doing, one immediately finds a severe problem in misunderstanding, not only among engineers and designers but also among researchers. This condition seems to originate chiefly in the attempt, made implicitly or explicitly, to carry over from static design the idea of a limit load and its associated stress, and the setting of a safely higher allowable fatigue stress through the use of a safety factor. Such a simple procedure is usually adequate for static design in which the limit load, normally expected to occur only once or at most a few times during operational life, is multiplied by one factor to account for a number of uncertainties and various kinds of scatter; the procedure ignores the differences in material behavior when subjected to static or to dynamic (fatigue) loads. Probably the most important of these "scatters" are the uncertainties in the values of the applied load, and in the strength properties. The factor used must be sufficiently large to reduce the probability of failure to an acceptable level in the subjective sense; it is often obtained by traditional methods (1.15 on yield or 1.50 on ultimate), or by consideration of the statistical properties of the data, or as a last resort, by estimation.

A somewhat obscure thought that the fatigue problem can be adequately treated by applying only one such factor has appeared in many discussions, which often seem misdirected to the point of whether the factor should be applied on the load (stress) or on the life. Considering this last point, a discerning look at the relative ΔS's and ΔN's obtained by taking an increment along the curve (at any point short of the endurance limit) readily indicates that the scatter in life is controlling. As to the use of two or more factors Lundberg [22] suggests $K_{tot} = K_n J$ where K_n, the limit fatigue factor, is the increase of area necessary to provide a fatigue life with an n% probability of failure (area above that to

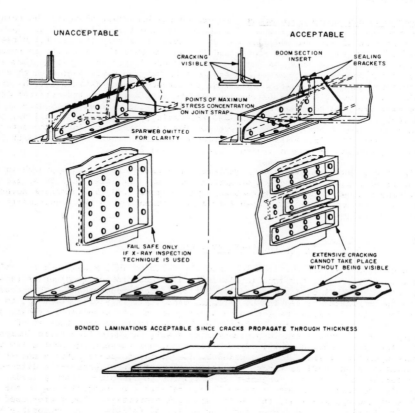

Fig. 2.20 Design for inspectability. From
A. J. Troughton and J. McStay [54].

carry the static ultimate load), and J is the fatigue safety factor intended to
cover the scatter in fatigue properties. It is to be noted that the need for K_n
arises from the adoption of a static strength as the reference point in fatigue
design. The only apparent advantage of such referencing is the attendant knowledge
of any increase in area or weight necessary to determine a safe fatigue life.
Other factors have been suggested to cover specific variances: machined and
assembly dimensions; complexity, that is, the difficulty in identifying and analyzing all load paths; and different types of service, as well as the expected
variations in load magnitudes and sequences.

Albrecht [50] has considered the factor-of-safety approach based upon the average fatigue stress from a very limited number of full-scale test specimens. The factor of safety is defined as: $FS = S_F/S_D$, where S_F is the average fatigue stress at failure and S_D is the design level stress. By the theory developed in the reference, a value of S_D can be determined for a 90% confidence level on the mean and for a 99.9% probability of survival. Since most designers are more familiar with ultimate or limit load analysis techniques, it was considered desirable to form the relationship between the design stress and the experimental fatigue limit by means of a required factor of safety that is a function of the number of specimens tested. The parameters that have been tabulated include: the factor of safety, S_D, the mean stress, the number of test specimens, the confidence level, distributions of S_F (normal and log-normal), and the variance $V = \sigma/S_F$. From data accumulated on full-scale fatigue tests of aircraft and helicopter parts, it appears that values of V between 0.10 and 0.20 are realistic. For example, if $V = 0.20$ the factor of safety required using a normal distribution and a single test specimen is 5.3. This may seem severe but is statistically correct for the required probability of survival. For a log-normal distribution, however, a factor of safety of only 3.3 is required, indicating the importance of the proper distribution.

A factor of safety approach is also a current requirement of the Civil Aeronautics Manual [51]: the minimum number of (fatigue) test specimens is dependent on the oscillatory test level as listed in Table 2.4. For the above requirement the lowest failure stress level is used as a design fatigue limit instead of a mean fatigue limit reduced for a given confidence and survival rate, as was proposed by Albrecht.

TABLE 2.4

Minimum Number of Test Specimens	(Applied Stress Level) (Critical Stress Level)
4	1.1
3	1.25
2	1.5
1	2.0

The contrasts between "fatigue" design and "static" design may be outlined as in Table 2.5, from which it follows that, in static design, if $FS = 1$, then $P_f = 0$. For the common design case wherein a constant, or at least not decreasing, degree of safety is desired during operational life, a uniform value cannot be given for the safety factor because the scatter varies with life; and, more fundamentally, the different types of designs, as "safe-life" or "fail-safe," would require different, and often unknown, values for the safety.

Structural design is a probabilistic problem in which the desired structural "integrity" or "reliability" must be considered; as a first premise it is sometimes assumed that the overall reliability should be distributed among the components so that the probability of failure of each component is proportional to its weight. After a favorable distribution of the probability of failure has been established, it is then necessary to design each component with the proper factor of safety to attain this goal. The factor of safety, defined in this manner, is the ratio of the _probable_ strength to the _probable_ critical force, and is a function of the desired reliability and the variations in the loads and strengths. Such a factor is no guarantee of safety but only a statistical quantity which will, if properly chosen, result in an efficient structure. It should not be invariant or arbitrary but selected with careful deliberation. If the applied and failing loads are sharply defined (insignificant variability), then a factor of safety as low as 1.01 could result in an extremely high reliability. On the other hand, if the variability is high, a large factor of safety would be required to provide a small probability of failure.

Present safety factors are based upon past experience and engineering judgment, and are inevitably affected by the subjectivity of the designer. There are at present two major conditions in the engineering community that are acting to increase the difficulty of assigning a proper factor of safety, and to increase the penalty for an improper choice. The structural requirements of modern vehicles are becoming more stringent, requiring the designer to consider new failure criteria, new materials and constructions, and more complex loading conditions, for example,

TABLE 2.5   Comparison of Static and Fatigue Design Parameters

	Static Design	Fatigue Design
Lifetime	Infinite assumed, but no value positively known.	Required to be known, and frequently less than infinite.
Load level	Constant.	Constant or varying, or both, in some alternating sequence(s).
Load period	Infinite (one period duration), or a few very long periods.	Cyclic, of known periods, and of known total duration.
Load frequency, or strain rate	Zero.	Constant or varying, or both, in some alternating sequence(s).
Probability of failure	Zero assumed, actually an unknown finite value.	Finite.
Factor of safety	Finite, small number, that is, 1.25 for airframes, 4.0 for bridges.	Not applicable in probabilistic approaches. Limited applicability in deterministic approach.

fatigue of a fibrous weave with a variable temperature-load history. These conditions have resulted in greater variability in the applied and failing loads than had been encountered in the past, and the effect has tended to increase the probability of failure for structures designed with the arbitrary, standardized factors of safety.

The second condition is the growth and extension of calculation procedures for numerical reliability, usually of electronic circuitry. The applications of these procedures to structural analysis have, however, been severely limited by the inadequacy and lack of definition of the methods, and particularly by the lack of statistical data on loads and strengths. The psychological pressure to generate numerical reliability numbers, instead of a factor of safety, as representing structural integrity, has been very high but without the apparent recognition of the inapplicability of such numbers or of the false illusion of security that high R values often represent. An industry survey recently concluded that, while statistical procedures were being used in structural design and analysis, little credence could be given to (high) structural reliability magnitudes determined by state-of-the-art procedures. Vastly more data, and a new methodology incorporating the effects of fail-safe, safe-life, and maintenance provisions would have to be developed before structural reliabilities could be considered as valuable end results. Since the data and methodology are accruing at glacially slow rates, obtaining a general structural reliability is effectively impossible, but in those simple cases for which "reliable" data and distributions are known, Section 3.2.3 indicates the future possibilities.

A nomographic treatment for factors of safety is given in Section 3.2.3.

2.6   MINIMUM WEIGHT DESIGN

This term has three widely differing interpretations: first, that a minimum weight design has been achieved when the weight of an item meets or comes in under a budgeted figure. Such a budget derives from the breakdown of the all-up capacity of a launching vehicle, airframe, or transport device and really has no relation to the weight of any of the individual items, only that their sum shall not exceed the given limit. Second, there is the prosaic engineering approach, with at least a trace of cut-and-try, wherein the results of stress analyses and vibration tests are fed back to obtain a "trimmed" weight, somewhat lower than that of the initial paper design. This practice is basically the attempt to redistribute the material in a qualitatively more efficient manner, thus reducing the weight. Since the relationships among all the parameters may be rather complex, especially in a buckling failure mode, and since in general the weight/strength factor is not the reciprocal of the strength/weight factor, a very well-controlled procedure is needed. The third interpretation of the term implies the use of such relatively extensive procedures as the use of structural indices [14], or of nondimensional analysis [38] or of mathematical synthesis [40]. These have all been well worked out and generally provide very acceptable results, but they do not necessarily yield "the minimum" weight as a single-valued function.

In considering this problem it is essential that the relationships between the variables of, and the constraints on, the design be fully recognized. It is often assumed that a minimum weight design results when the instability or failure, in all modes acting independently, occurs simultaneously. Intuition and experience both indicate that an instability in any one mode tends to increase the possibility of failure by decreasing the stiffness and stability in all other modes. If the design can be made such that all modes are equally stable, as web- or flange-buckling in a channel, then there is no interaction and the stability of each mode can be defined by an independent stability equation. The attainment of instability in any mode will then result in instability in all modes and a collapse of the structure. If the number $\underline{n}$ of unspecified dimensions is equal to, or less than, the number of possible buckling modes, then the so-called minimum weight design occurs when the stability stresses of the $\underline{n}$ lowest modes are made equal. If there are more unspecified dimensions than possible failure modes, additional restrictions must be imposed, in the form of equations defining an optimum stress level or an area distribution.

This frequently made assumption--that a design for which all modes of failure are equally probable will also exhibit the minimum weight--can in general be shown to be untrue, that the resulting weight is not necessarily the absolute minimum. Nevertheless, the assumption has proved to be very useful, in that designs based on it appear to be of lower weight than those made under other constraints.

A number of schemes, for either analysis or synthesis, have been developed with this assumption as a base. While the details of the schemes may vary widely, essentially all of them make use of expressions for stability or failure that involve the loading conditions, the material properties, and a limiting value of some chosen parameter. For computerized treatments this limiting value is often called an "alarm." It is usually defined with the dimensions of a stress, frequently the yield or ultimate; of course, a fatigue allowable value may be taken.

Correlation of this value with that given by the solution is easy by direct comparison, but the real problem still remains: the establishment of a proper fatigue allowable. The ease of the powerful techniques of synthesis and of nondimensional analysis are really restricted to the handling of complex load and stress geometries, and especially when computerized, to providing that necessary insight into the effects of changing sizes and properties by performing many iterations in a short time. The latter might often be thought to be no restriction, but it does not address itself directly to the problem of setting the fatigue stress. This latter, and major, problem is treated at some length in Chapter 3.

2.7   MACHINES CONTRASTED TO STRUCTURES

The interest in the fatigue behavior of machinery apparently predated that of structures, due no doubt to the more frequent failures. Wohler's experiments with railway components were based on the need to account for the effects of live-loads, but it is only relatively recently that recognition has been made of the presence of live-loads in structures, to a significant degree in terms of fatigue. "Numbers of cycles" is a parameter easily recognized for rotating or reciprocating engines, and, together with the tacitly assumed intent to build for an infinite lifetime, lead to the early view of fatigue as an "endurance limit" of some sort, usually stress. The development of ideas that strain might be a more applicable parameter than stress in some cases, [287], the growing need for weight control, and general economics have expanded the view of fatigue behavior to include the whole time spectrum, not just infinity. The lifetime of machinery can usually be based on its operating modes and its general functional nature: the duty factor for a central station turbine is quite different from that for a bulldozer. The lifetime of a structure is, of course, related to its function--a truck frame vs. a bridge truss--but the geographic location and natural environment may also be highly effective. Wind and seismic loads and their durations, as well as the live-loads, dictate a finite lifetime, unless all of the actual stresses are inefficiently low. Solar thermal gradients are of considerable significance in large structures and will generate tens of thousands of cycles in "a few years." There have recently been a number of bridge disasters, at least one* of which seems, on cursory analysis, to be the victim of simple fatigue of the bars due to vibration loading excited by the wind, and accelerated by high stress concentration factors and stress corrosion.

In the past, the term "cycles" has generally meant the number of counts of the maximum stress, but evidence now indicates that the life-controlling parameter is the number of traverses of a stress range from some value to the maximum. This condition is especially important for structures subjected to random loading, where the determination of a number of "cycles" may not be straightforward, see Section 3. For designs to finite lives, the stress range is fundamentally important, for the life is proportional to $1/(\Delta S)^5$, approximately, the exponent 5 being the sum of 2 plus the exponent m in the crack growth rate relation (Sec. 3.2.4).

---

*The Silver Bridge on the Ohio river, an eye-bar truss type, which collapsed in December 1967.

CHAPTER 3

PREDICTION OF FATIGUE LIFE

The objective of all fatigue analyses, calculations, testing, and second-guessing is an acceptable combination of load and life. The designer's chief task appears in two parts: (1) to find a proper expression of the load and its relationship with time or frequency in order to establish the allowable stress level, and (2) to control the hardware design such that geometry and material properties do not force the actual stress at some location above the previously set allowable value. At the present state of the art, the prediction activity includes both of these tasks as prerequisite to the application of a damage rule, usually the linear cumulative rule of Palmgren-Miner. Both linear and nonlinear rules are discussed below, but emphasis is placed on the former because of its practicality and simplicity, plus the difficulty and occasional impossibility of applying a nonlinear rule. Irrespective of which rule is to be used, a major problem arises: that of the nature of the load-time relation. Does it result from a true random process (gaussian, etc.) or from a single-amplitude sine wave, or a set of discrete-amplitude sine waves? For the latter two cases the definition of a cycle is fairly evident and the number of cycles is determined by simple counting. The event to be counted, however, is not universally agreed upon; it may be the peaks, level crossings, ranges, range-pairs, and so forth. For the random response the definition of a cycle is not at all evident; thus the counting problem becomes even more complex. While the evaluations of event-counting methods are voluminous and varied (for example: [368]), there seems to be a growing body of opinion that the most appropriate counting method is that called "Rainflow", [Ref. 398]. A number of the Life Prediction Methods discussed below deal with data generated by sine-wave excitation or constant-amplitude loading. Despite the growing recognition of the discrepancies between this sine and the real-life random excitation, it has not been until recently that test equipment capable of true random output was available. The problem has been solved by the use of servo-controlled, hydraulic machines wherein the demand signal, on magnetic tape, is compared with the actual applied load or strain signal from a transducer on the specimen. The error is then used to reduce the difference to zero, the frequency response of the loop readily encompassing normal test frequencies - up to 60 Hz, approximately.

With such machines producing much data, the dependence on the older sine data should eventually become reduced. But the bulk of available data is still sine-generated, even though randomizing has also been applied to the sequencing of the discrete values in the multiple-amplitude sine tests. This latter procedure is sometimes called block programming; of course, the sequence may be ordered random blocks or steps, simple repetitive or another kind, see Fig. 3.1. The effects of these sequences are included in the discussions, particularly in Section 3.2.1, but in general it is the true random responses with which the designer must deal, for it results from natural forces, not the synthetic, mathematically simple sine wave. As an aid to the designer's efforts, emphasis is placed on Section 3.2.1, which portrays two methods of handling random waveforms. The first is to be preferred for its rigorousness, simplicity, and lack of synthesis; the second is not less rigorous but is more useful in extrapolations or modifications of existing design for which values of a characteristic frequency and of the transmissibility are already known.

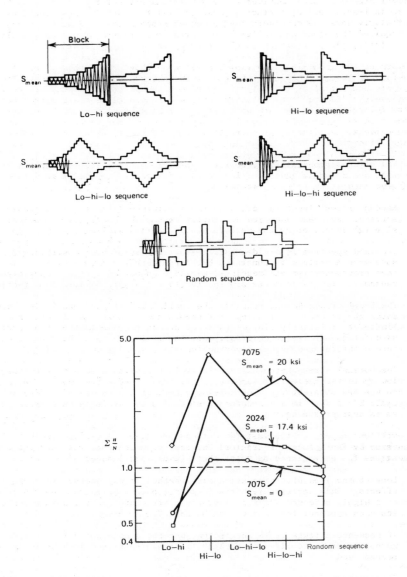

Fig. 3.1 Effect of the sequence of block loadings on fatigue life.

Another equally important facet of the life prediction process is treatment of the remaining or residual strength of the damaged structure (as distinguished from residual stress). Such considerations are intimately related to Fracture Mechanics and to fail-safe design; further discussion will be found in those Sections. Particularly have the considerations of crack propagation by Fracture Mechanics now acquired the power and validity conferred by successful numerical results, [275], [312], [382] et al are excellent examples of its application.

The form of the input load function has been the chief topic above, and it has been tacitly assumed that the life prediction methods utilized the conventional, monotonic properties of the materials. However, prediction methods are evolving which use the "fatigue" or cyclic properties for much added realism (see Sec. 4.1.2). In the following brief summary of the types of life prediction methods, the dependence of these types on both the load function and on the different properties is quite evident. All the methods are followed by the application of a damage rule, usually linear.

1. Use of an appropriate fatigue curve, S-N or Goodman, for the applied stresses that are assumed known or calculable. Limited by the available data to sine wave input (constant amplitude).

2. Random process input, usually constant amplitude in test, but variable in service. Response analyzed by Fourier method and converted to equivalent sine (g) level, or by the methods of Section 3.2.1.

3. The load spectrum is divided into a small number of blocks within which the sine-wave amplitude is constant, but is different for each block. Blocks are then programmed for the required, variable number of cycles, repetitively or randomly. Limitations are similar to (1). Analysis by method of Section 3.2.1.

4. The Local Strain Method requires the cyclic or "fatigue" properties; accomodates the sequence of loading, pre-strain and mean stress effects, but is limited to relatively simple states of stress (as are most of the currently used fatigue damage procedures). It also easily accepts the results of the more sophisticated events-counting schemes, as Rainflow.

5. The Nominal Stress Method is a relatively approximate scheme, adaptable to low cycle fatigue. It assumes that the smooth σ-N curve and the notched S-N curves are straight lines on log-log paper; at 10^6 cycles $S = \sigma/K_t$, and at 1 cycle $S = \sigma K_t$. It accomodates plastic flow but neglects mean stress effects, as is appropriate in LCF.

Ten excellent papers on the last two methods are given in [435], together with a discussion by Prof. Fuchs of Stanford University, which ends with several cogent suggestions for an engineering approach to Cumulative Damage:

a. Look at the load histories and decide whether or not one should expect sequence effects. This decision requires judgment, but no more than the choice of the most highly stressed areas in a proposed structure, or the assumption of the loads on a member that failed in the field.

b. If sequence effects can be neglected, use the simple nominal stress range method to estimate life to crack initiation. Condensed load histories are recommended.

c. If sequence effects must be expected, use a local strain method to estimate life to crack initiation. Condensed load histories are recommended.

d.   If the crack initiation forecast shows danger, check whether crack propagation
     will be dangerous or not. Note that the stress levels, including self-stresses
     or residual stresses are very important for crack propagation. Condensed
     load histories are less advisable at this stage because crack propagation
     damage is proportional to a much lower power of the stress range than crack
     initiation damage, (about third power vs. about eight power).

In addition to those methods discussed below, there are a number of other [41],
most of which have had a very limited use. Some are based on arbitrary conditions
of loading, type of specimen, and special factors of safety; others on a combination of component tests at a constant load level and the calculated life of a
standard specimen for a given spectrum. Since these methods lack general applicability, they will not be discussed further here, except to mention the new,
sophisticated schemes for summing the damage cycle-by-cycle, see [399] et al.
Descriptions of the two most widely accepted events-counting schemes will be found
in Section 4.2.1.

3.1   METHODS OF PREDICTION

There are numerous identifiable methods for the prediction of fatigue life; fundamentally, they all involve the concept of the gradual accumulation of damage
during the process of loading. The differences among the methods appear from the
emphasis placed on some particular aspect or formula for the representation of
either or both the applied loading spectra or the allowable S-N data. The methods
fall naturally into three categories according to these differences:

I.   LINEAR CUMULATIVE DAMAGE, based on specific S-N data for each specimen type.

   1.   MINER'S RULE [30], the well-known linear summation of the fractions of
        fatigue damage expressed in terms of the cycle ratio (n_1/N_1) where
        failure is hypothesized when $\Sigma \, (n_1/N_1)=1$. The rule may be used directly
        with standard fatigue curves; it is discussed at some length in Section
        3.2.1.

   2.   A new method of rigorous analysis of random responses (Sec. 3.2.1.) pro-
        vides the displacement and its characteristic frequency directly from
        empirical traces of power spectral density vs. the frequency. The
        method does not require any prior knowledge of fundamental frequency,
        transmissibility, or other parameters unique to the object under analy-
        sis.

   3.   LUNDBERG'S FFA METHOD [72], based on Miner's rule, establishes equations
        to represent the applied loading spectra and the allowable S-N data,
        obtains a closed form solution for the total damage and thus for the
        corresponding predicted fatigue life.

   4.   SHANLEYS "IX" method [69] is based on a concept of the rate of formation
        of slip bands. An equation is derived for the fatigue damage utilizing
        a mathematical expression for the S-N curve, which results in Miner's
        summation.

   5.   LANGER'S METHOD [29] and GROVER'S METHOD [70] are essentially the same
        and require S-N data which separate the stages of crack initiation and
        crack growth. Then, utilizing Miner's hypothesis, they derived a life
        prediction. The required data is generally not available.

6.   Smith's residual stress method [71] uses Miner's rule and includes the
     residual stresses from plastic yielding at higher loads. It is a very
     elegant method of stress analysis but not practical for the complexity
     of loading history met in actual service.

Ia.   LINEAR CUMULATIVE DAMAGE based on ε-N data and the Cyclic Properties

7.   The Local Strain Method [431] places emphasis on the local cyclic
     stresses and strains at the root of the notch where failure actually
     originates. The local values are derived from the nominal via Neuber's
     Rule, and the cyclic properties are used instead of the usual monotonic
     values. This combination of conditions makes the approach very realistic
     and it is coming into widespread use. Discussion and examples will be
     found in Section 3.2.1.

II.   NONLINEAR CUMULATIVE DAMAGE, based on specific S-N data for each specimen
type.

8.   The original base for the modified CORTEN-DOLAN method [36] was a rela-
     tion between fatigue damage and the number of cracks formed as a function
     of the largest varying load in the sequence, with the growth of cracks
     to occur at all load levels. The result was an expression of nonlinear
     cumulative damage in terms of stress ratios and included damage from
     stresses below the endurance limit. In modification, the equation was
     converted to a cycle ratio basis that assumed no contribution from
     loadings below the endurance limit. Evaluation of the nonlinearity
     coefficient indicated results very closely equivalent to those of the
     linear rule, see comment in Section 3.2.2.

9.   SHANLEY'S "2X" METHOD was based on essentially the same reasoning as his
     "1X" method with the additional assumption that one of the coefficients
     in the rate equation was stress-dependent. This condition strongly
     increased the rate of crack growth, resulting in a nonlinear form for
     the damage summation.

10. HENRY'S METHOD [73] is a procedure for reducing the allowable S-N curve
     step-by-step, accounting for the damage of prior loads. The order of
     load application must be known. The method may be applied only to a
     somewhat limited class of materials because of the form of the mathe-
     matical expression of the S-N data.

III.   CUMULATIVE DAMAGE, LINEAR OR NONLINEAR, from Damage Boundaries or Modified
S-N curves.

11. THE STRESS CONCENTRATION METHOD is, in practice, a procedure of refined
     stress analysis for evaluating the ratio of the peak stress in a discon-
     tinuity to the nominal stress. It must be coupled with a damage theory
     in order to make a life prediction, Miner's hypothesis often being
     applied. The stress concentration factor may be used to: (1) define
     specifically the peak stress, the allowable stress or number of cycles
     then being determined from the appropriate S-N curve for the new
     unnotched material; or (2) select an appropriate S-N curve from a graded
     set of curves for notched specimens. This S-N curve may be considered
     as the damage boundary for the design for which no specific or identical
     test data exist. The choice of an S-N curve is then an arbitrary speci-
     fication of the damage boundary shape before this boundary has been
     confirmed by tests.

12.   IN THE FATIGUE QUALITY INDEX METHOD, a fatigue test of a completely
detailed specimen is conducted under the full spectrum of service loads.
The results are analyzed with a set of standardized S-N data fixed for
the purpose of providing a scale of measurement of fatigue quality. A
stress concentration factor is determined which, by interpolation,
makes $\Sigma n/N = 1$ exactly, the factor being defined as the fatigue quality
index. It may be compared with an acceptance standard, and, as a life
prediction method, the establishment of the index provides an arbitrary
damage boundary.

13.   THE FREUDENTHAL AND HELLER METHOD [32] is to construct a fictitious S-N
curve by use of a stress interaction factor derived from a statistical
analysis of tests in which blocks of sinusoidal loading were applied in
random order. The probability distribution function of the stress
amplitude (or so-called "load spectrum") was the commonly assumed exponential "gust spectrum"; the resulting damage law contains a parameter
which is expressed not only in terms of the material properties, but
also in terms of this particular load distribution. Care would be
required in its application to fatigue loads having other probability
distributions; unfortunately, a lack of both the type and quantity of
the necessary data tend to preclude its practical application. (However,
Freudenthal's efforts have provided much other highly useful information;
see Sec. 3.2.1).

14.   THE MARCO AND STARKEY METHOD [33] defines damage boundaries through a
power relation of the cycle ratio, the exponent being made either stress
or load-dependent. (See also Item 16 below.)

15.   THE KOMMERS NONLINEAR DAMAGE HYPOTHESIS [76] is based on the non-
linearity of the damage boundaries determined from two-step load tests.
The boundaries were functions of both the load levels and the cycle
ratios in each load level. (See also Item 16.)

16.   THE RICHART AND NEWMARK METHOD [77] formalized and added experimental
verification to Kommer's hypothesis. Methods 13, 14, and 15 are essentially all one and seem to be overly complex for wide usage.

3.1.1   CONSERVATISM EVALUATED

The several methods of prediction listed above vary over a very wide range of
comprehensiveness, conservatism, level of mathematics, and inherent error. Such
wide variation makes it difficult for the individual designer to judge the usefulness and particularly the conservatism of any one method; for a rational design,
some means of judgment is required. An evaluation of several of the more useful
methods [78] has been made in two forms, depending on the scale on which the
comparisons were measured.

1.   COMPARISONS OF LIFE CYCLES. Each of the methods evaluated was required to
predict the fatigue life of the test specimen from the unit applied load
spectrum, using whichever form of the allowable data that was specifically
applicable to the method. The degree of conservatism was defined by the
ratio N_t/N_p, where N_t is the test life and N_p, the predicted life. Quite
apparently, if $N_t/N_p < 1$, the prediction is unconservative, and the reverse is
true if $N_t/N_p > 1$.

2. COMPARISON OF THE CHANGE IN STRESS LEVEL REQUIRED TO PREDICT EXACTLY THE TEST
    LIFE. The stress scale may be used to evaluate the various prediction methods
    by the determination of a proportionality, k, by which all stress levels of
    the test spectrums are raised or lowered to arrive at the exact prediction of
    the specific test life. Again, the degree of conservatism is defined by the
    comparison of the predicted to test results in the ratio form:

$$\text{degree of conservatism} = \frac{S_{vtest}}{S_{vadj}} = \frac{S_{vtest}}{S_{vadj}k} = \frac{1}{k}$$

Shanley's "2X" Method is quite conservative, but there is some lack of agreement
regarding the stress concentration method when using the concept of linear cumulative damage with standardized S-N curves. The other methods lie closely within a
band containing Miner's prediction, an agreement to be expected, since all these
methods except the tangent intercept method are variations of the use of the
linear rule, some of which vary only in the details of curve fitting. Stress
curves of the same methods lie in the same order, with relatively minor differences
in the degree of conservatism. This scale of comparison is informative to the
designer with respect to the weight of material necessary to achieve a given
fatigue life, and in the assessment of any factor of safety on loads as related to
the required reliability. In general then, it may be concluded that:

1. THE MINER LINEAR CUMULATIVE DAMAGE METHOD is an eminently practical and
    versatile way of approaching the fatigue problem. Its accuracy is commen-
    surate with the uncertainties of the loading history and of the available S-N
    data; its maximum inaccuracy in predicting specimen life needs to be no more
    than about one order of magnitude. (See Sec. 3.2.1.)

2. THE STRESS CONCENTRATION FACTOR METHOD, when used alone, is frequently less
    than adequate, mostly because of the uncertainties in the calculation of the
    individual factors. However, its use in preliminary design, when backed up
    by full-scale test of potentially critical members and joints, forms a pro-
    cedure of generally acceptable accuracy.

3.2   TREATMENTS OF CUMULATIVE DAMAGE

Considerable progress has been made in the understanding of the fatigue mechanism
during the past two decades, and plausible theories have been proposed to explain
qualitatively a number of the phenomena accompanying cyclic straining. Nevertheless,
the fundamental research has not, as yet, made any significant impact on quantitative methods for the design of structures subject to random loads, so that in
general the engineer must rely on "laws" of very limited applicability.

Such laws are based on the results of the latest research, but during their evolution a mathematical formulation must be introduced, inevitably containing empirical
parameters that can be obtained only from tests of a large number of specimens.
The prediction of the resulting law can thus be used with complete confidence only
for those cases in which the parameters were applicable; the only way to avoid the
proliferation of laws to cover numerous special circumstances is a greater emphasis
on basic research in the physics of the process. However, the numerous conditions
that a fatigue law is expected to cover is, to some extent, artificially generated
by the fact that the majority of fatigue testing equipment has, at least until
recently, been capable of applying only sinusoidal loading to the specimen.

Fatigue laws have been formulated, therefore, in an effort to describe the results
of tests made with a type of loading that is probably the exception rather than
the rule in a large number of practical applications. It is probable that there
is considerably more hope of deriving a "law" of fairly general applicability for

loadings that, like many service loadings, are truly random in character. By definition, there can be no effect because of the order in which the loads are applied, and it is felt that a number of concepts common to previous analyses may be used with perhaps more success in this case.

One such concept is that of "fatigue damage." This term is used to describe the gradual "deterioration" of a metal during cyclic straining, and a damage function D is postulated, the value of which increases from zero to unity during the life of the part. The function D may be looked upon as describing, in a rather vague way, the many complicated physical changes that take place in a metal during the initial formation of striations on the surface, the initiation of microcracks, and their propagation to final critical size.

Many attempts have been made to describe the degree of cumulative damage by use of the general expression for the transfer of energy or mass. Damage may be not too indirectly related to the change in energy states between that of a fatigued piece (whether fractured or not) and that of the same piece in a pristine, uncycled state. Damage is often described by a "crack" parameter that, in the form of number, size, shape, distribution, etc. of the cracks, may be interpreted as a change in the state of energy in the immediately adjacent volumes. This ΔE is composed in varying proportions of the strain energy, both elastic and plastic, the loss due to heating by internal friction or damping, and a term for the change in shape of the microvolumes. The general expression usually takes the form:

$$\Delta E \text{ or } \Delta m \text{ or } \Delta D \backsim CF^k$$

where F, the intensity factor, has the dimensions of temperature, voltage, pressure, stress, etc. The constants, C and k, relate to time in the form of period or rate and other appropriate dimensions, k being usually the rate factor, as the slope of a characteristic curve.

Griffith [46] balanced the elastic energy released by the propagation of the crack with the energy required to form the new surface in brittle material, and Irwin [47] has applied the idea in the analysis of the "crack toughness" of engineering alloys. The latter approach is becoming known as "Fracture Mechanics" and by empirical modification can now accommodate material of ductile behavior. In general, this analysis involves a quantity K_c as a fracture-toughness parameter, (the stress-intensity), the nominal stress and a crack length. Their relationship, for a simple geometry, is:

$$K_c \backsim S \sqrt{a}$$

where a is the crack length at the instant the crack becomes unstable and produces rapid failure of the part. In terms of strain, the first expression may be rewritten as $c=N^k(\Delta e_p)$ in which the exponent is applied to the number of cycles and is the slope of the log e_p vs. log N curve; e_p is the plastic strain and c is a material constant. Coffin [88] has shown that e is e_f at fracture, ($N_f=1/4$ cycle), and $c=e_f/2$. Thus the fracture ductility can be utilized to determine fatigue behavior, at least at low N. In the more familiar stress terms,

$$C=NS^b$$

where the exponent b, the slope of the log S – log N curve is applied directly to the intensity factor, stress. Both the nonlinear and damage boundary methods described below for predicting fatigue life may utilize this expression, with the slope or its reciprocal as the exponent on stress. The major difficulty in implementation is that the "constants" are at best approximations, and the prediction methods seem to be either straightforward but limited by simplifying assumptions, or cumbersome in the attempt to improve the accuracy.

The simplest rule based on concepts of fatigue damage is the linear one of Palmgren and of Miner [28, 30] and this can be shown to be derivable from any assumption regarding the rate of damage accumulation, provided that it is assumed to be continuous and to depend only on the instantaneous value of the stress. In spite of the known tendency of the Palmgren-Miner rule to overestimate life, it is still used extensively, first, on account of its simplicity and, secondly, because more sophisticated rules have yet to be shown to be generally applicable to all types of service loading. Moreover, a body of experience has been built up which enables an estimate to be made of the error to be expected in any set of circumstances.

3.2.1   THE LINEAR, OR MINER, RULE

It appears that the Miner rule arose from efforts to reduce available information and data on fatigue behavior to design practice. In 1945 Miner commented that progress with methods of fatigue analysis had been slow because of lack of basic information on material, and more important, because no method of handling any but the simplest problem had been available (i.e., the direct use of an S-N curve). Any solution requires several tools before one can make other than a simple comparison based on tests; even in comparative tests, the experiment may be easily oversimplified to the extent that significant effects are lost. These tools must provide the following:

1.   Information on the loading conditions: level, frequency of occurrence, number of cycles, and so forth.

2.   A method of relating various loading cycles; that is, cycles of stress that have different ratios of minimum to maximum stress, as well as differences in mean stress level.

3.   Information on the number of cycles to failure of the material or part at various stresses: S-N, or constant-life curves.

4.   Means of evaluating the cumulative effects of cycles of stress at various levels.

There was already a considerable background for these conclusions [29, 30, 43, 44]; it is to be noted that the concept of an accumulation of damage, however defined, during the period(s) of cyclic stressing was widely accepted. The argument arose over the rate of accumulation and the proper mathematical expression of a damage law. Miner's approach was to formulate an expression that would meet the requirements and accommodate the information listed above in the simplest possible manner. He assumed that the phenomenon of cumulative damage under cyclic stressing was related to the net work absorbed by the specimen. The number of stress cycles applied, expressed as a percentage of the number to failure at the given stress level, would be the proportion of useful life expended. When the total damage, as defined by this concept, reached 100%, the fatigue specimen should fail. Failure was defined as the visible appearance of a crack, since such a condition prohibits further use or operation of most structural or machine elements. A failure limit of 10^7 cycles was established and only applied maximum stresses above the endurance stress (for the aluminum alloy used, at 10^7 cycles) were considered.

If W represents the net work absorbed at failure, then:

$$\frac{w_1}{W_1} = \frac{n_1}{N_1}; \quad \frac{w_2}{W_2} = \frac{n_2}{N_2}; \quad \frac{w_3}{W_3} = \frac{n_3}{N_3}; \text{ etc.} \tag{1}$$

Since $w_1 + w_2 + w_3 + \ldots w_n = W$, hence

$$\frac{w_1}{W_1} + \frac{w_2}{W_2} + \frac{w_3}{W_3} + \ldots \frac{w_n}{W} = 1 \tag{2}$$

Substitution from Eq. (1) yields

$$\frac{n_1}{N_1} + \frac{n_2}{N_2} + \frac{n_3}{N_3} + \ldots \frac{n_n}{N_n} = 1 \tag{3}$$

or

$$\Sigma \frac{n}{N} = 1 \tag{4}$$

To check the validity of Eq. (4), Miner performed 11 tests at two or more stress levels and 11 tests in which a percentage of damage was imposed at one stress ratio; then the ratio and load were changed and the life was run out to failure. Alclad 24S-T sheet was loaded axially at stress ratios of R = +0.50, +0.20, and −0.20. An assumption of sinusoidal loading was made and the photos of the stress-time traces from the oscillograph appear very close to true sinusoids. One of Miner's curves is shown in Fig. 3.2a for the specimens at two stress levels; Fig. 3.2b is an expansion for three typical specimens to illustrate the loading cycles employed and the agreement of the final failure with the cumulative damage concept. These curves are based on the ratios n/N. At any one value, n_1/N_1, a percentage of useful life has been consumed, and this proportion is used to determine the starting point of n_2 at a different stress.

Thus specimen S19 was run at 33,300 psi for 135,000 cycles. Since N_1 = 243,000 cycles, n/N_1 = 135/243 = 0.55, and at the lower stress of 29,600 psi this was equivalent to 0.55 N_2 = 0.55 (330,000) = 181,000 cycles, which is the point from which n_2 starts in Fig. 3.2b. Now, since n_2 = 141,000 cycles at failure, n_2/N_2 = 141/330 = 0.43 and $\Sigma n/N = n_1/N_1 + n_2/N_2$ = 0.55 + 0.43 = 0.98. For the 11 tests at two or more stress levels, the average value of $\Sigma n/N$ is 0.98, and the maximum and minimum are 1.45 and 0.61, respectively. If only the last level of loading is plotted for these tests (assuming that only this final level of stressing caused the failure), the average value of $\Sigma n/N$ is 0.37, thus indicating that the damage from the other loading levels cannot be ignored. For the second set of 11 tests at various R and load, the average value of $\Sigma n/N$ was 1.05, with maximum and minimum values of 1.49 and 0.80. The average $\Sigma n/N$ for all tests was 1.015.

By the use of a modified Goodman diagram, as in Fig. 3.3, the fatigue characteristics of a material or part can be compared, even though the information available has been obtained for varying ratios of minimum to maximum stress. In general, test data from Alclad 24S-T specimens do not result in such straight lines, but the deviation is not great. In Fig. 3.4 the lower limit curve bounding the shaded area has been derived from the straight line diagram of Fig. 3.3. It is useful in presenting a cross-plot of stress (expressed as a percentage of the ultimate tensile strength) against stress ratio R, for a given number of cycles. The upper limit curve represents the maximum deviation of test values of various experiments, including some unpublished data on Alclad 24S-T.

Miner's own conclusions indicated that, from the then available evidence, the concept of cumulative damage:

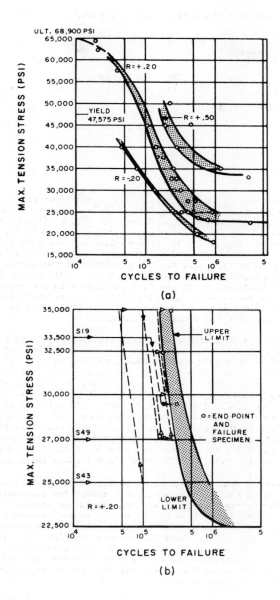

Fig. 3.2 S-N curves representing Miner's cumulative damage rule:
(a) for specimens at two stress levels; (b) an expansion
for three typical specimens. From M. A. Miner [30].

1. Holds true for Alclad 24S-T aluminum alloy and probably for the other high strength aluminum alloys.

2. Provides a simple, practical (and conservative) means of analyzing fatigue problems.

3. Should be experimentally investigated for steels and other materials to determine its range of usefulness. (This has been done, see especially [11], [56], and [57].)

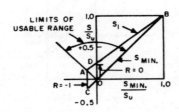

Fig. 3.3 Modified Goodman diagram for the Miner rule.
Line CB represents the minimum stress line,
while AB represents the maximum stress line.
The stress ratio R equals the ordinate of CB
divided by the ordinate of AB.

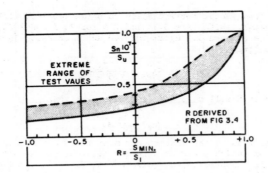

Fig. 3.4 Correlation between test results and
straight line from Fig. 3.3.

Since Miner published his hypothesis, much work has been done on fatigue damage, with the rather well-accepted conclusion that the rate of damage accumulation is not linear. Thus, despite its great simplicity and usefulness, the Miner rule is open to some comment if taken literally. Particularly, the parameters of mean stress, loading sequence, notch effect, and static loading prior to fatigue testing are not included in the treatment. Other experiments [69] under special conditions have yielded data from which summations as high as 300 have been obtained; except for the comment on the average n/N of the last load level, Miner makes no accommodation of the effects of loading sequence, that is, high-low or low-high. The effect of mean stress appears to be buried in the several values of stress ratio used, with the implied conclusion that it has no effect.

The two discussions immediately following are particularly appropriate in delineating the advantages and shortcomings of the linear rule and in comparing predictions of life under various conditions of loading. Miner [30], Corten and Dolan [36], and Gassner [69] had used essentially the same type of loading, sequences of a very few (2 to 8) discrete amplitudes; while, more recently, Schijve [65] explored the effects of full program loading (many amplitudes) and Kowalewski [68] went on to random excitation (see Sec. 3.2.3).

In the program or spectrum tests for the validity of Miner's rule, the complete load spectrum is applied to the component or structure in periodic sequences, each period containing the load amplitudes in quantities corresponding to the spectrum. A certain sequence of load cycles with different amplitudes is repeated periodically; in one period, the cycles with the same amplitude are grouped together. It is further common practice to arrange the load cycles in each period in an order of increasing and/or decreasing amplitudes.

Many tests involved an approximation of the Taylor gust spectrum [66] sequence, which implies a linear relation between the gust velocity and the logarithm of the number of exceedances, expressed as the analytical function:

$$U = -5.625 \log m + 43.75 \tag{5}$$

where U is the gust velocity, and m is the number of times that U is exceeded in 10^7 miles of flying. The mean stress is derived from the ultimate stress by

$$S_m = \frac{S_U/1.5}{LF} \tag{6}$$

where LF is the load factor corresponding to a (maximum) 50 ft/sec gust, and 1.5 is the safety factor. With LF = 3.33 and 2.5 for 7075 and 2024, respectively, S_m turns out to be 6.3 and 9.0 kg/mm or 9.0 and 12.8 ksi. Thus a 50 ft/sec gust corresponds to a stress

$$S_{a_{50}} = \frac{S_U}{1.5 - S_m} \tag{7}$$

and a gust velocity U will result in a stress amplitude

$$S_{av}\left(\frac{U}{50}\right) S_{a_{50}} \tag{8}$$

By the use of Eq. 7 and 8 the gust spectrum of Eq. 5 can be converted to a stress amplitude spectrum, which plots as a series of steps on the S-N axes. Such a series of steps is supposed to be equivalent to Taylor's straight line spectrum. The spectra and the effect of the block sequences are shown as Fig. 3.1, the general conclusions being that the sequence of loading and the level of mean stress have a pronounced effect on the value of Σ n/N. Tests in which S_m = 0 tend to result in $\Sigma n/N < 1$, while those in which $S_m > 0$ give $\Sigma n/N > 1$.

In a comparison of the results of program and random inputs, gaussian excitation was applied such that the small, notched (K_t = 1.77) aluminum specimens were subjected to simple bending, with a mean stress of zero. Figure 3.5 shows the life functions as drawn for the 50% values of the scatterband, the 90% lines having the same shape and lying about 10% in life below the 50% values. For the one-load-level tests, about 15 tests were performed at each level; for the program tests, the length of the cycle was chosen so as to result in a range of 5-25 program cycles to failure. The life function line for random loading and the number of zero crossings at failure is plotted at the 5-sigma stress level. As

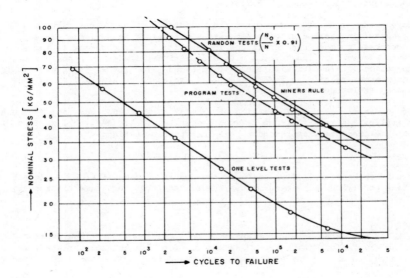

Fig. 3.5   Life functions for various loadings.   From J. Kowalewski [68].

this is the highest level in corresponding program loading, the two functions are directly comparable. The horizontal distance between the two curves is nearly constant, which means that, on a log-log plot, a constant factor exists between the lines at every stress level, found for this example to be about 2. Compare Freudenthal's results [89] and Fig. 3.6.

The life function line computed by Miner's rule should coincide with that from program loading, but it can be seen that Miner overestimates the life--in this range, up to a factor of 3. Although, in this life range, Miner's rule and the

random test results are not very far apart, it may not be concluded that this will
also be the case at longer lives. It is to be expected that the difference will
become even greater at lives of 10^7 and 10^8 cycles because more and more stress
levels will fall below the endurance limit and consequently will do no damage,
according to the linear cumulative damage rule. Therefore Miner's rule (derived
on a basis of sine input) may give fatigue lives even longer than the corresponding
lives from random loading.

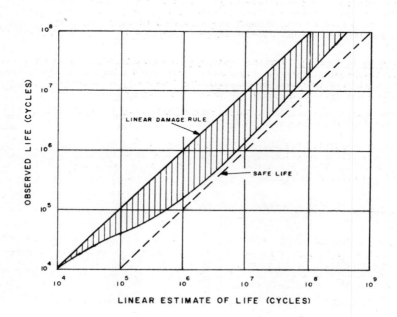

Fig. 3.6   Observed fatigue lives vs. linear estimates.
From A. Freudenthal and R. Heller [32].

These results all tend to support the lack of conservatism in literal application
of the Miner rule; even for specimens having a theoretical stress 2.4 times the
nominal, the value of the average summation was significantly greater than one.
Although the term "conservative" was used in Miner's conclusions and appears
justified by his data on a numerical basis, the extension of this justification by
the use of Fig. 3.4 is hardly safe. The deviation in this figure is from the
Goodman relation, which is itself an approximation. The truth of this matter
appears much better illustrated by Freudenthal's results [32], Fig. 3.6 showing
that, at least in terms of life the linear rule is unconservative by about one
order of magnitude (at a reasonably high number of cycles, $N > 10^4$). The position
of the safe lifeline could perhaps be considered as overly conservative, since the
band between it and the linear damage line contain all data points from a very
large number of specimens. Thus the deviation in life shown is no more than one

order from that of the linear rule. An almost identical situation was found in similar tests on 2024 aluminum and another comparison in terms of predicted vs. achieved lives of aircraft structural components, Fig. 2.10, shows very similar trends.

Damage Calculation for Constant Amplitude Loading. A simple, straightforward method exists for the fatigue analysis of elements subjected to sine-wave excitation, permitted chiefly by the presence of a true resonant frequency in the response, as contrasted to the absence of such in a random response. Also, some simplification is possible because many types of structures under sine excitation (or constant amplitude loading) show a predominant single-peak response; thus the frequency band to be considered as damaging is relatively short. However, the very practical limitation that natural processes produce random rather than sine excitation severely restricts the usefulness of such analysis. For those particular structures that happen to produce a predominant single peak in their response to random excitation, a sine-based analysis might not produce significantly different results (this topic is discussed in the next subsection).

Consider, as an example, the determination of the portion of fatigue life consumed in the vibration test of a solar cell panel. The test produced the dynamic response curve in Fig. 3.7 for the first model of the honeycomb panel, which was loaded uniformly on one face with bonded cell dummies. The panel was 14 in. x 34 in. x ¼ in. thick with 7075-T6 faces 0.005 in. thick and an adhesively bonded core of 0.0008 in. x ¼ in. foil. It was framed with a riveted zee extrusion, which also carried the hard-point mounting hinges and latches at the quarter-points of the long dimension. Thus the fundamental mode of vibration was that of a wide beam with maximum deflection at the center and maximum moment at the supports. The following conditions and assumptions applied:

1. The 1-g sine wave input, normal to the plane of the panel, was swept at a linear rate of 10 Hz.

2. The bending deflection and stress are related to the input amplitude and to the transmissibility or amplification factor as in Fig. 3.8; the maximum stress at resonance was 50 ksi; the mean stress was zero (R = -1.0).

3. No significant stress concentrations were present in the faces.

4. The resonant frequency in the fundamental mode was 200 Hz.

5. Damping was such as to limit the transmissibility or amplification factor to 15.

6. No failure occurred.

7. The Miner criterion was used as a simple summation, that is, not set equal to 1, since no failure occurred.

The response curve is squared off in equal time increments i as allowed by the linear sweep rate; then a graphical integration is performed for n_i, as in Fig. 3.8. This operation presumes that the structure vibrates at the center frequency of the bandwidth i for the time interval, which is taken here as 0.5 sec. The fatigue life corresponding to S_i is found from such data as that in [1], Fig. 3.3.1(g) or (j), taking the median line. The Miner summation is then formed, yielding about 1.5% for this sweep. Should the values at the lower limit of the S-N band be taken to represent the allowable life, the damage accumulation would be roughly 3%. The curve in Fig. 3.8 is simply the resonance portion of that in Fig. 3.7 on an expanded frequency scale for convenience.

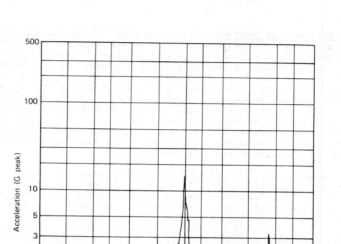

Fig. 3.7 Sine response of solar panel, first design. Input of 1 g.

Considerable approximation is evident in this method, particularly that due to the use of small-scale S-N curves. But the major objective is achieved, that of finding the order of magnitude of the damage accumulation. The value happened to be satisfactorily low, probably any value below about 10% would be so considered.

However, the peak stress was sufficiently high to draw attention to the potential danger in the assumption that K_t is indeed one or less everywhere, and the beneficial results of added damping are seen in Fig. 3.9. Nylon sleeves were installed at the hinges, and the frame was attached by adhesive bonding instead of riveting. The major (200 Hz) response was halved, at the small expense of an increase in the low frequency (80-90 Hz) response, the latter having no significant effect on the damage accumulation. It should be noted that only five terms make any contribution to the summation, and only one of these is really significant. It is often adequate

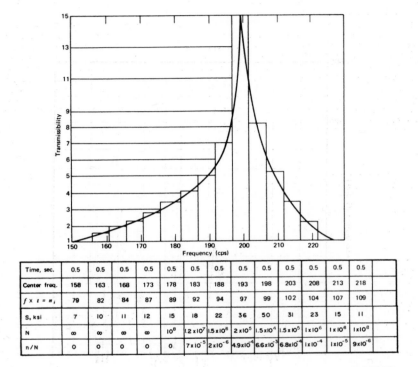

Time, sec.	0.5	0.5	0.5	0.5	0.5	0.5	0.5	0.5	0.5	0.5	0.5	0.5	0.5	
Center freq.	158	163	168	173	178	183	188	193	198	203	208	213	218	
$f \times t = n_i$	79	82	84	87	89	92	94	97	99	102	104	107	109	
S, ksi	7	10	11	12	15	18	22	36	50	31	23	15	11	
N	∞	∞	∞	∞	10^8	1.2×10^7	1.5×10^6	2×10^5	1.5×10^4	1.5×10^5	1×10^6	1×10^6	1×10^6	
n/N	0	0	0	0	0	7×10^{-5}	2×10^{-6}	4.9×10^{-4}	6.6×10^{-3}	6.8×10^{-4}	1×10^{-4}	1×10^{-5}	9×10^{-6}	

$\Sigma\, n/N = 0.008026$

Fig. 3.8   Calculation scheme for damage accumulation
under sine-wave excitation.

to confine the summation to the time or frequency band between the half-power
points, that is, the verticals intersecting the rising and falling curves at 0.707
times the peak amplitude.

For tests in which the sweep rate is proportional to frequency, the expression
below is used to calculate the number of cycles:

$$n = \frac{t_f'(f_2 - f_1)}{\log_e(f_2/f_1)}$$

where t_f = the time width of the frequency band, in sec

    f_1 = the lower frequency of the band, Hz

    f_2 = the upper frequency of the band, Hz

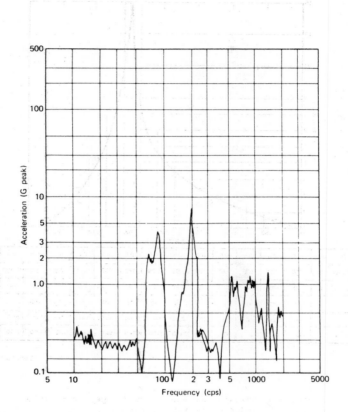

Fig. 3.9   Sine response of solar panel, final model.

Rework of the example for this sweep rate yields a summation of about 2.4%, as contrasted to 1.5% for the linear rate. While there is a large proportionate difference between these results, the net effect of the different rates is small because the majority of the damage is done at the lower frequencies, near resonance. For logarithmic sweep rates, the number of cycles at the amplitude of the half-power point is given as

$$n = \left[\frac{t f_n}{\ln 2}\right] \ln \left[\frac{f_n + 0.5\Delta f}{f_n - 0.5\Delta f}\right]$$

where t = the sweep rate in octaves/sec
f_n = the resonant frequency
Δf = the bandwidth at resonance and is equal of f_n/t

Damage Calculation for Random Loading. The objective of this analysis of responses to random inputs is to find the actual stresses and the corresponding number of cycles, or equivalently, to determine the magnitude and characteristic frequency of the displacement. One popular approach is rather limited in application: treatment of the response waveform as an equivalent sine wave is possible only for that circumstance in which a single predominant peak appears, so that a frequency becomes available by observation. Consideration of many actual plots indicates that such circumstance does not always exist, Figs. 3.10a, 3.11a and 3.12a from the radar pedestal of the Lunar Module are typical. The approach herein is directed to finding a displacement, then strain, from which the applied stress can be derived, and to find the corresponding frequency which, with the time of excitation, provides the number of cycles. The method is capable of handling response curves of any graphical form, the single-peak form appearing only as a special case.

It appears that a common, practical method of data-gathering results in plots of acceleration density vs. frequency, g²/Hz vs. f, and with the accelerometer usually located at other than the critical section of the member. If, in contrast, the data were in the form of a time history of the strain from a gage at the critical section, the damage calculation would be straightforward, once the event and its counting scheme were chosen. But for the popular form above, (PSD vs. f), the general case is attacked by expressing the rms value of acceleration, a, velocity v, and displacement s as functions of frequency:

$$\sigma_a{}^2 = \int \Phi(\omega)d\omega = 2\pi \int \Phi(2\pi f)df = \int \phi(f)df \tag{1}$$

$$\sigma_v{}^2 = \int \frac{\phi(f)}{\omega^2}df = \frac{1}{4\pi^2} \frac{\phi(f)}{f^2} df \tag{2}$$

$$\sigma_s{}^2 = \frac{1}{16\pi^4} \int \frac{\phi(f)}{f^4} df \tag{3}$$

where Φ is the PSD function of ω, and ϕ the corresponding function of f. We assume that the "random" input is of Gaussian form, and that only the rms values are known. The joint probability density is p (a,v,s), and the variables are statistically independent, i.e., the probability that v will be in the interval (v, v+dv) is not dependent on the values of a or s. This density may be expressed as:

$$p(a,v,s) = \frac{1}{(2\pi)^{3/2}\,\sigma_a\sigma_v\sigma_s} \exp\left(-\frac{a^2}{2\sigma_a{}^2}\right) \exp\left(-\frac{v^2}{2\sigma_v{}^2}\right)\exp\left(-\frac{s^2}{2\sigma_s{}^2}\right) \tag{4}$$

In this analysis, the frequency is defined as one-half the number of reversals of stress or the number of times the velocity becomes zero, per unit time. The period is the time taken by the velocity to change by the amount dv, or $T = dv/|a|$. The frequency, $|a|/dv$, is the number of such periods that can be fitted into unit time, here taken as one second. Multiplying Eq. 4 by $|a|/dv$ and integrating over all a and s from minus to plus infinity yields the frequency of velocity zero-crossings or stress reversals:

$$f_v = \frac{\sigma_a}{2\pi\sigma_v} = \frac{\sqrt{a^2/v^2}}{2\pi} \tag{5}$$

Similarly, the displacement frequency becomes:

$$f_s = \frac{\sqrt{v^2/s^2}}{2\pi} \tag{6}$$

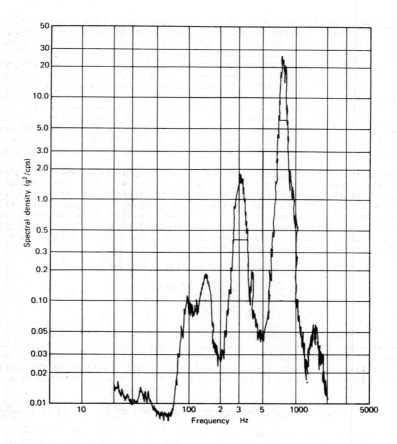

Fig. 3.10a   Random response with single peak, (log-log).

and the rms displacement amplitude is obtained by multiplying Eq. (3) by g, the gravity constant, (386 in/sec^2):

$$S = g \sqrt{\frac{1}{16\pi^4} \int \frac{\phi(f)}{f^4} \, df}$$

and rearranging to yield:

$$S = g \sqrt{(a^2/f^4)/16\pi^4} \tag{7}$$

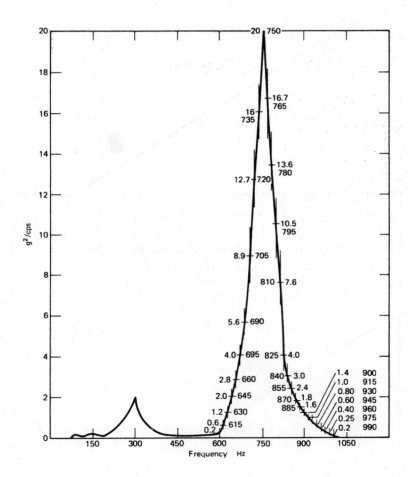

Fig. 3.10b   Linear replot of Fig. 3.10a.

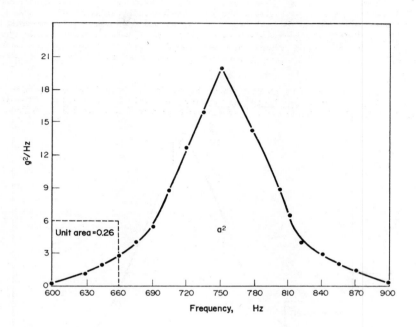

Fig. 3.10c   Peak plot, acceleration.   Plot area = 1.913;
scale factor = 6 x 60.

The gravity constant is also present in Eq. (5) and (6) but falls out by virtue of the ratio form. The limits of integration were infinity, but if one were interested in the frequency of zero-velocities at some displacement above an arbitrary thresh-old, s_o, Eq. (5) would take the form:

$$f_v = \frac{\sigma a}{2\pi \sigma v}\left[\int_{s_o}^{\infty}\left(\frac{1}{\sqrt{2\pi}\sigma s}\right)\exp\left(-\frac{S^2}{2\sigma_s{}^2}\right)ds\right] \tag{8}$$

The expressions above are valid for any bandwidth but there is little point to including the entire (nominal) range of 2000 Hz. Energy considerations support this comment: the area under the peaks of the PSD curves may be taken as a general indication of that portion of the spectrum where the response energy is concen-trated, thus the choice of a bandwidth for analysis in the first two examples below is fairly obvious. But for the frequently-found noisy response of Fig. 3.12a, several linear replots had to be made to assure the location of that bandwidth having significant energy, really, to locate the upper end of the band.

The displacement, from Eq. (7), may be transposed along the structural element to the critical section via the appropriate functions of Mechanics-of-Materials.

OFD - C'

Then the strain and finally a stress are calculated. The number of cycles is simply f_st where t is the time of excitation. The method of counting cycles and what should be counted as a "cycle" is the subject of much discussion (typically [367,368]), probably because of differences of opinion as to the mechanism of fatigue damage. The use of this method is supported by the data of Clevenson and Steiner [369], their plots showing that, within certain limits, the fatigue life (of aluminum alloys) is not much affected by the ratio of peaks to zero-crossings, nor by several different loading spectra, $\phi(\omega) = K_1\omega^{-2}$; $K_2\omega^0$; $K_3\omega^{+2}$; which include many of the spectra specified for transport vehicles.

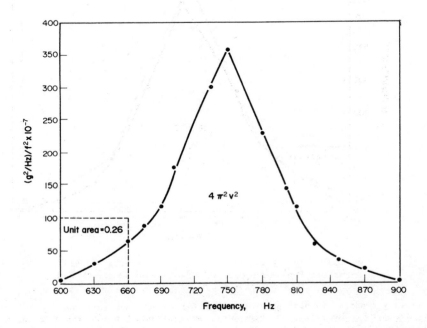

Fig. 3.10d   Peak plot, velocity.   Plot area = 2.031;
scale factor = $100 \times 60 \times 10^{-7}$.

The method is now applied to the three types of response wave forms commonly found in practice. Consider the frequency and displacement of the element whose response is shown in Fig. 3.10a. This plot was chosen for the first example because it is typical of many structures in that the great majority of the energy (area) is contained in a single peak, that at about 750 Hz. Since the analytical relation-ships above are not known, the integrations will have to be done numerically, requiring that the log-log plot be redrawn on linear scales as in Fig. 3.10b. For accuracy in the integration this peak is expanded as in Fig. 3.10c, the effect of the 300 Hz peak being regarded as negligible. The area as (acceleration)^2 is

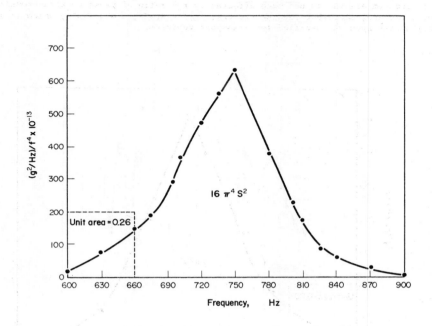

Fig. 3.10e   Peak plot, displacement.   Plot area = 1.806;
scale factor = 200 x 60 x 10^{-13}.

measured by planimeter, and the scale factor established.   Then for v^2 a calculation scheme is set up as in Table 3.1.   A plot is made of v^2 vs. f, as in
Fig. 3.10d; the above steps are repeated for s^2, resulting in Fig. 3.10e.   The
frequencies and the displacement are determined from Eqs. 5, 6 and 7 by introducing the ratios of the scale factors, and of the scale factor to unit area,
yielding

$$f_v = \sqrt{\frac{1.425}{1.535} \left[\frac{60 \times 6}{60 \times 100 \times 10^{-7}}\right]} = 760 \text{ Hz}$$

$$f_s = \sqrt{\frac{1.535}{1.408} \left[\frac{6000 \times 10^{-7}}{12000 \times 10^{-13}}\right]} = 735 \text{ Hz}$$

$$s = 386 \sqrt{\left[1.408 \left(\frac{12000 \times 10^{-13}}{0.26}\right)\right]} / 1550 = 0.00069 \text{ inch}$$

This displacement is then used in deriving a stress value at the required (critical)
section, and this stress occurs for a number of cycles equal to the product of the
frequency and the exposure time, $f_s t$.

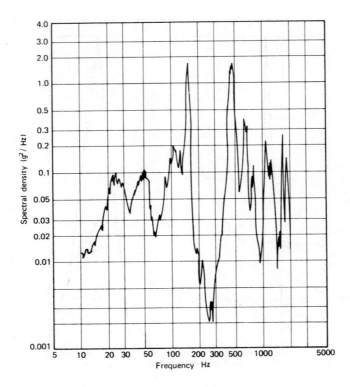

Fig.3.11a   Random response curve with double peaks (log-log).

The response of Figs. 3.10a or 3.10b is sufficiently single-toned that the characteristic frequency, or so-called resonant frequency is easily estimated, indeed
the difference between the observed 750 Hz of the acceleration peak and the calculated 735 Hz of the displacement zero-crossing is negligible.   But now consider
the response shown in Fig. 3.11a, taken at a slightly different location on the
same device with the same input.   How does one estimate a frequency here?   The two
obvious peaks and their immediate vicinity are replotted on linear scales, as in
Fig. 3.11b, and the above calculation scheme repeated to produce the velocity and
displacement plots, Figs. 3.11c and 3.11d.   The frequency of the velocity zero
crossings is found to be 272 Hz and that of the displacement to be 141 Hz, a
significant difference at least in percentage terms; the RMS displacement is
0.003 in.   Beyond the general intuitive feeling that both f_v and f_s should fall
between those of the peaks on the power-spectral-density plot, and in the vicinity
of the lower frequency peak because of the suppression of the effects of the

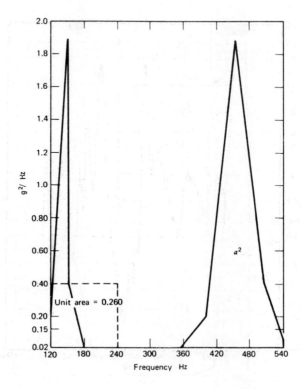

Fig. 3.11b   Peak plot, acceleration.   Plot area, 0.863;
scale factor, 0.4 x 120.

higher frequencies by the division by f^4, no estimation of values seems feasible.
Note should be taken of this effect of applying the analytical relationships
(i.e., dividing the power-spectral-density by f^2 and by f^4) in suppressing the
peaks, especially the higher frequency peak: compare the relative magnitudes of
the pairs of peaks in Figs. 3.11b, 3.11c, and 3.11d, accounting for the scale
factor.

Another very frequently observed type of random response, Fig. 3.12a shows no
obvious peaks, or some large number of peaks depending on the viewpoint.   Tentative
replotting on linear scales indicated the dominance of the lower frequencies, that
is, below about 30 Hz, and Figs. 3.12b, 3.12c, and 3.12d provide values from which
the f_v was calculated as 16.2 Hz; f_s as 10.1 Hz; and the displacement, s, as
0.097 in.   The use of a reduced bandwidth, 5 to 30 Hz as contrasted to 0 to 30 Hz

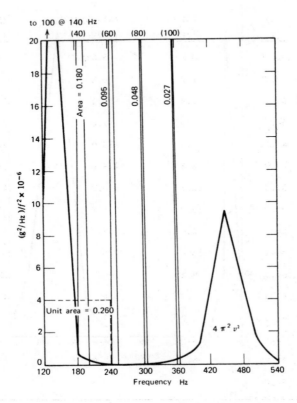

Fig. 3.11c   Peak plot, velocity.   Plot area, 1.163 total;
scale factor, 4×10^{-6} (120).

above, gave values of 16 Hz, 12.3 Hz, and 0.076 in., respectively, the large
percentage differences supporting the dominance of the lower frequencies.   The
effective response was zero above 30 Hz.   It will be noticed that the recorded
trace did not start until the frequency reached 10 Hz; therefore the values below
10 Hz had to be synthesized.   Since, in test work, it is impossible to keep the
acceleration constant at very low frequencies, a shift is usually made to constant
displacement.   Assuming that this had been done, the displacement plot is completed
from 10 Hz to 0, constant at the 10 Hz level.   Then the corresponding velocity and
acceleration values are calculated at several frequencies in this band, and those
plots completed, from which the above values of f_v, f_s, and s were determined.

TABLE 3.1   Calculation Scheme for Random Responses

(1) f	(2) f^2	a^2 (3) g^2/Hz	v^2 (4)=(3)/(2) $(g^2/Hz)/f^2$	s^2 (5)=(4)/(2) $(g^2/Hz)/f^4$
600	36×10^4	0.2	5.6×10^{-7}	15.4×10^{-13}
630	39.7	1.2	30.1	76
660	43.7	2.8	64.4	148
675	46.0	4.0	87.0	189
690	47.6±	5.6	118.0	248
700	49.0	8.8	179.0	368
720	51.8	12.7	246.0	475
735	54.0	16.0	298	554
750	56.2	20.0	355	632
780	60.1	13.6	226	378
800	64.0	9.2	144	225
810	65.5	7.6	116	177
825	67.6	4.0	59	88
840	70.8	3.0	42.3	60
870	75.8	1.8	23.8	32
900	81.0	0.2	2.5	3

All of these values are RMS as calculated; no weighting factor was used in the
illustrative calculations.   Since true resonance can occur only in sine wave
action, the frequencies must not be regarded as resonances: it is only in the
very strongly single-toned responses, as in Fig. 3.10a that the differences between
random and sine behavior, and in the values of the acceleration and displacement
frequencies, are small.

To illustrate a damage fraction calculation by this method, consider the response
of Fig. 3.12a as produced by an accelerometer on a truss member in the pedestal
base.   The member was considered as a beam (6061-T6 aluminum weldment) with clamped
ends subject to both lateral displacement and rotation at the center because of
the attachment of a heavy black box.   The response then represents the combined
displacements; but, for simplicity here, it is treated as lateral only.   The
displacement calculated above as 0.097 in. is transposed via the standard equations

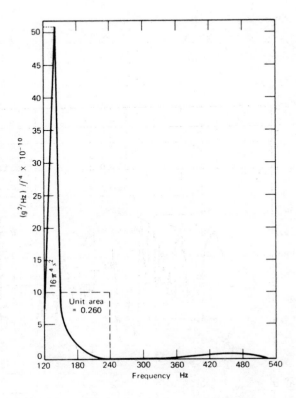

Fig. 3.11d Peak plot, displacement. Plot area, 0.229;
scale factor, $10 \times 10^{-10}(120)$.

to a load, and then to the moment at the accelerometer station. From this value
the moment at the critical section (through the attach holes) is derived, resulting
in a nominal, net section stress of 8100 psi. The mean stress calculated from the
box load was only 1200 psi, but the residual stress from the possibly incomplete
stress relief treatment was estimated at 5000 psi for a total mean of 6200 psi.

The K_t for the section was taken at 2.5; thus from [1], Fig. 3.3.1(k) the fatigue
limit is about 3.5×10^5 cycles. The time of excitation was 4 min per axis per
run; then the number of cycles in the z axis (the direction parallel to the hole
axis) for z excitation is $4 \times 60 \times 10.1 = 2640$. For the x and y excitation directions the coupling factors were 0.8 and 0.4, respectively. These conditions are
listed in Table 3.2 which illustrates the summation of the damage fractions.

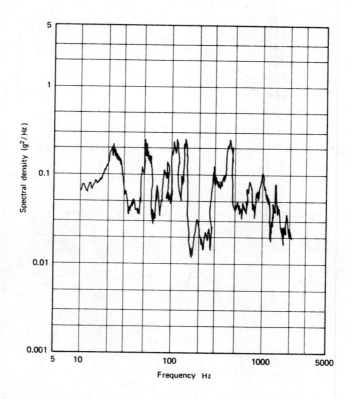

Fig. 3.12a   Random response curve with multiple peaks,(log-log).

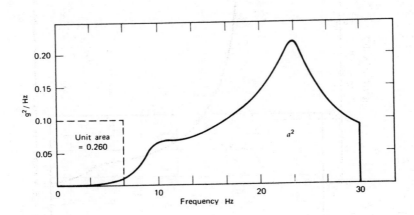

Fig. 3.12b Peak plot, acceleration. Plot area, 0.981; scale factor, 0.1 x 6.67.

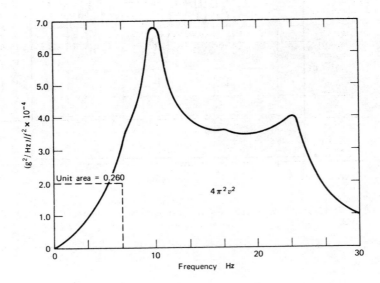

Fig. 3.12c Peak plot, velocity. Plot area, 1.870; scale factor, $(2.0 \times 6.67) \, 10^{-4}$.

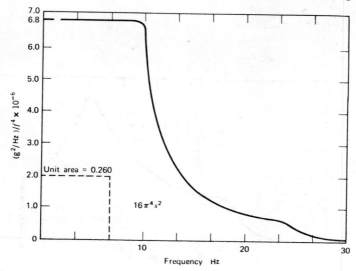

Fig. 3.12d   Peak plot, displacement. Plot area, 1.852;
scale factor, $(2.0 \times 6.67) \, 10^{-6}$.

Thus, very approximately, 1% of the life was used up in one test run. The several
factors contributing to the inaccuracy include: (1) the assumption that the
coupled frequencies are equal to f_v, (2) that the K_t's are equal in all modes,
(3) the estimation of the residual stress level, (4) the small-scale S-N curves
used, etc. However, the major objective of illustrating the method of finding the
effective frequency is achieved.

TABLE 3.2

Axis	S	K_t	Coupling Factor	S_{max}	n	N	n/N
x	8100	2.5	0.8	16,200	2640	5.0×10^6	0.00053
y	8100	2.5	0.4	8,100	1640	$1.0 + 10^8$	0.00004
z	8100	2.5	1.0	20,200	2640	3.5×10^5	0.00765
						$\Sigma n/N =$	0.00822

The Local Strain Method*

This method is based on knowledge of the local stresses and strains at the notch root as derived by the Neuber Rule from data on smooth specimens, Figs. 3.13 and 3.14, or from strain gages reading nominal values on the part, (as it is frequently

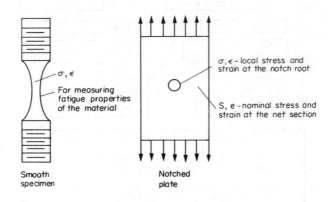

Fig. 3.13 Nominal and local stress and strain.

impossible to measure directly at the root). Service data is, of course, preferred because analytical results must be based on a model for which there is adequate confidence. This condition then reduces the applicability to very simple states of stress, essentially to uniaxial. The method uses a linear cumulative damage rule:

$$\Sigma \ \frac{L_R}{L_f} \ < 1.0 \tag{1}$$

where:

L_R = life requirement set by the designer for each machine function

L_f = fatigue life of the part for the related machine function

to predict time or cycles to initiation of a "committee-sized" crack.

The term "committee-size" was first jokingly suggested in a talk by W. T. Bean, who defined a committee-size crack as one of sufficient size that a committee, without the use of optical aids, could agree it existed. Upon more serious consideration, it appears to be a useful definition in that it defines the point where an owner might claim a part failed. In most steel parts, the defined crack will be on the order of 1/16 to 1/4 in. in length. Thus, the method is especially

---

*This material on the Local Strain method was made available thru the courtesy of Lee Tucker, Deere & Co., and the SAE.

well adapted for application to machine parts wherein the time and distance for stable crack propagation to the committee-size crack is relatively short.

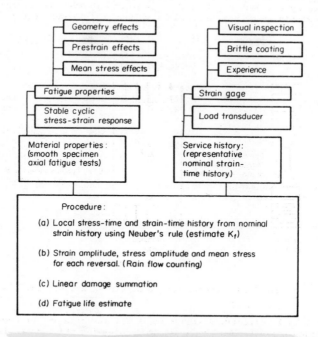

Fig. 3.14   Schematic of the Local Strain Procedure

The cyclic or "fatigue" properties of materials are utilized so that the effects of work hardening/softening may be included. However, most alloys soon stabilize, so that the greater portion of their fatigue life is spent under stable stress-strain response conditions. This condition can be depicted by the cyclic stress-strain curve in Fig. 3.15 and the equation:

$$\frac{\Delta\varepsilon}{2} = \frac{\Delta\sigma}{2E} + \varepsilon'_f (\Delta\sigma/2\sigma'_f)^{1/n'} \tag{2}$$

where:

$\Delta\varepsilon/2$ = strain amplitude

$\Delta\sigma/2$ = stress amplitude

n' = cyclic strain-hardening exponent

ε'_f = fatigue ductility coeffecient (see Table 3.3)

σ'_f = fatigue strength coefficient (see Table 3.3)

E = modulus of elasticity

TABLE 3.3 – Fatigue Properties, Definitions and Approximations

Fatigue Property	Definition	Approximation
ε'_f Fatigue ductility coefficient	Intercept of the log $(\Delta\varepsilon_p/2)$ – log $2N_f$ plot at $2N_f = 1$	$\varepsilon'_f \simeq$ monotonic fracture ductility, ε_f
c Fatigue ductility exponent	Slope of the log $(\Delta\varepsilon_p/2)$ –log $2N_f$ plot	$c = \dfrac{-1}{1+5n'}$ *
σ'_f Fatigue strength coefficient	Intercept of the log $(\Delta\sigma/2)$ – log $2N_f$ plot at $2N_f = 1$	$\sigma'_f \simeq$ monotonic true fracture strength, σ_f
b Fatigue strength exponent (Basquin's exponent	Slope of the log $(\Delta\sigma/2)$ –log $2N_f$ plot	$b \simeq -1/6$ log $2\sigma_f/S_u$

*n' = cyclic strain hardening exponent taken as the slope of the log $\Delta\sigma/2$ – log $\Delta\varepsilon_p/2$ plot.

NOTE: SEE TABLE 4.2b FOR VALUES

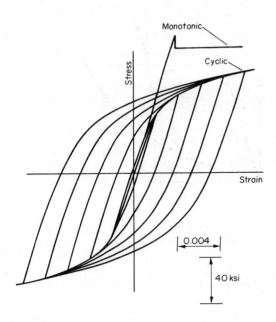

Fig. 3.15   Monotonic and cyclic stress–strain curve for
AISI 4140, Rb 92–97.

Further, the hysteresis loop shape, shown in Fig. 3.15 can be approximated by a similar equation, if the coordinate system is moved to the loop tips:

$$\frac{\varepsilon_1}{2} = \frac{\sigma_1}{2E} + \varepsilon'_f(\sigma_1/2\sigma'_f)^{1/n'} \tag{3}$$

where:

ε_1 = strain measured from loop tip

σ_1 = stress measured from loop tip

These equations and their corresponding plots are illustrated by Fig. 3.16. Note that the outer loop shape is nearly the same for all of the hysteresis loops.

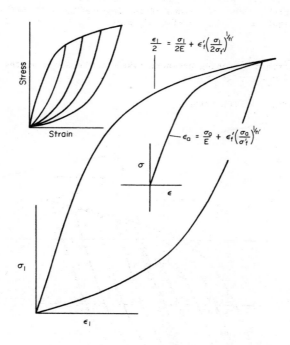

Fig. 3.16   Illustration of cyclic stress–strain curve and hysteresis loop shape.

In this procedure, the local stress-strain behavior of the notch root material must be estimated. Plastic strains of considerable magnitude take place, thereby requiring the non-linear stress-strain response of the material to be followed.

In addition to the cyclic stress-strain response, the fatigue strength and ductility must be measured or estimated.

<u>Fatigue Properties</u>. Historically, fatigue resistance has been characterized by plots of stress or strain versus life obtained from cyclic bending, torsion, or axial tests. Except for fatigue strength at very long lives, bending fatigue results are difficult to interpret because of stress gradient effects and lack of control of a basic material parameter, such as stress, strain, or plastic strain. Torsional tests include biaxial stress effects and sometimes exhibit different failure modes. For example, in low cycle torsional tests, low-ductility metals normally fail in the direction of maximum tensile stress, while high-ductility metals fail in the direction of maximum shear stress. The axial test avoids most of these difficulties, and thereby provides the better test for defining fatigue strength.

A set of fatigue properties which describes the empirical relationships amongst stress and strain and life are listed and defined in Table 3.3 and illustrated in Figs. 3.17, 3.18, 3.19. The resulting relationships for stress amplitude versus

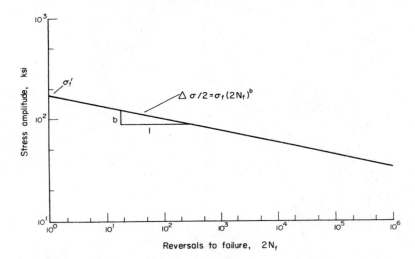

Fig. 3.17 Stress amplitude vs. reversals to failure
for Man-Ten steel, (A440 steel).

reversals, plastic strain amplitude versus reversals, and total strain amplitude versus reversals are given in Eqs. 4-6:

$$\Delta\sigma/2 = \sigma'_f (2N_f)^b \tag{4}$$

$$\Delta\varepsilon_p/2 = \,'_f (2N_f)^c \tag{5}$$

$$\Delta\varepsilon/2 = \Delta\varepsilon_e/2 + \Delta\varepsilon_p/2 = \frac{\sigma'_f}{E}(2N_f)^b + \varepsilon'_f (2N_f)^c \tag{6}$$

where:

$2N_f$ = reversals to failure

$\Delta\varepsilon_p/2$ = plastic strain amplitude

$\Delta\varepsilon_e/2$ = elastic strain amplitude

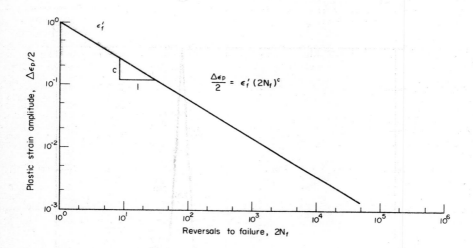

Fig. 3.18   Plastic strain amplitude vs reversals to failure for Man-Ten steel.

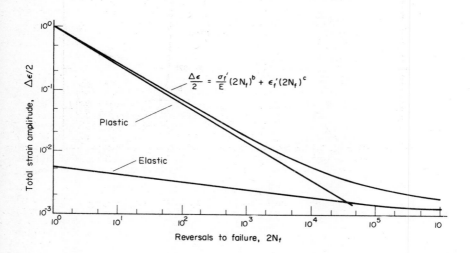

Fig. 3.19   Total strain amplitude vs reversals to failure for Man-Ten steel.

The properties just discussed are determined from testing smooth specimens under completely reversed, constant amplitude cycling, but in actual service the material is seldom, if ever, used under those conditions. Three influential factors: prestrain, mean stress, and geometry, have been studied in an effort to evaluate their effect upon fatigue resistance of metals.

<u>Prestrain Effects</u>. A few cycles of plastic strain applied to a test specimen at the beginning of a test may significantly affect the fatigue resistance of the specimen when cycled at a lower strain range. In steel, prestraining usually reduces the life and this effect is shown graphically for Man-Ten* steel in Fig. 3.20.

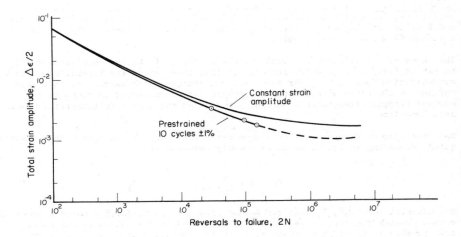

Fig. 3.20   Effect of prestrain on total strain-life behavior of Man-Ten steel.

In most service situations, prestrain will have occurred. Assuming this, a conservative but realistic approach would be to prestrain all smooth specimens prior to cycling to determine fatigue properties. Strain-controlling at 2% strain range for 5-10 cycles, then decreasing the range at a uniform rate to zero, appears to be a reasonable method for prestraining a specimen. For cases where prestrain in known not to occur, a second set of specimens which are not prestrained should be tested to measure quantitatively prestrain effects, as well as analyze the parts which are definitely not prestrained.

---

*Man-Ten is a U.S. Steel Corp. tradename for ASTM A440-63T steel.

Mean Stress Effects. For a given strain range, tensile mean stresses generally decrease fatigue life, while compressive mean stresses generally increase it when compared to the life at zero mean stress, the Goodman diagram being traditionally used to account for these effects. It has been shown that the mean stress effect in hard steels can be accounted for by assuming a change in the fatigue strength coefficient equal to the mean stress, resulting in:

$$\sigma_a = (\sigma_f - \sigma_o)(2N_f)^b \tag{7}$$

This equation can be shown to be similar to the Goodman relationship, except ultimate strength is replaced by the true fracture strength.

Geometry Effects. The fatigue resistance of metals is affected by geometric conditions. Most of this influence is believed to be due to size effect in that a small volume of (highly stressed or strained) material is less likely to contain a critical flaw than a larger volume.

This same effect is noted for small notches. Tests of notched specimens usually result in fatigue lives which are greater than those using the theoretical stress concentration factor, K_t. For this reason a fatigue notch factor, K_f, has been defined as the ratio of the fatigue strength for an unnotched specimen to the nominal fatigue strength of a notched specimen at the same life (generally measured at a long life).

Determination of K_f by means other than extensive fatigue testing has been investigated, resulting in Eq. 8, which is usually conservative:

$$K_f = 1 + \frac{K_t - 1}{1 + a/r} \tag{8}$$

The material constant, "a", is determined from fatigue tests of a notched specimen versus unnotched specimens at long life. For steels, the following empirical relationship in terms of ultimate strength has been suggested, where a is in inches and S_u in ksi:

$$a = 10^{-3} \left(\frac{300}{S_u} \right)^{1.8} \tag{9}$$

A more detailed discussion concerning the fatigue notch factor is presented below.

Analytical Procedure. The fatigue life estimate is based on the assumption that the local damage done to the structure by the service history is equal to that done by the same stresses and strains imposed on an axial test specimen of the same metal. Also, a linear damage summation and Neuber's rule (432) are assumed valid.

Nondamaging Strain Cycles. To minimize the analysis, strain excursions which produce insignificant damage should be excluded before any analysis is performed. A conservative estimate of the maximum nondamaging strain range can be derived from the following assumptions:

1.   The material behaves elastically:

$$\Delta\sigma = E\Delta\varepsilon \tag{10}$$

$$\Delta S = E\Delta e \tag{11}$$

where:

$\Delta\sigma, \Delta\varepsilon$ = local stress and strain range

$\Delta S, \Delta e$ = nominal stress and strain range (see Fig. 3.13)

2.   The notch stress is equal to the fatigue notch factor times the nominal stress:

$$\Delta\sigma = K_f \Delta S = K_f E \Delta e \tag{12}$$

3.   The mean stress effect can be accounted for using a form of Eq. 7:

$$\Delta\sigma_{cr} = \frac{\Delta\sigma}{1 - \sigma_o/\sigma'_f} \tag{13}$$

$$\sigma_o = \sigma_1 - \Delta\sigma/2 \tag{14}$$

where:

σ_1 = maximum likely local stress to be encountered in the service history

$\Delta\sigma_{cr}$ = completely reversed stress range from the material's stress-life plot at long life, say 10^7 reversals

Substituting Eqs. 12 and 14 into 13 and solving for the nominal strain range results in an estimate of the maximum strain range which can be excluded from the strain-time history without significantly affecting the life estimate:

$$e = \frac{\Delta\sigma_{cr}(\sigma'_f - \sigma_1)}{K_f E(\sigma'_f - \Delta\sigma_{cr}/2)} \tag{15}$$

Fatigue Notch Factor. An accurate estimate of the fatigue notch factor is an essential part of the life analysis since it can significantly affect the predicted life. A constant amplitude test of the component or machine will usually provide the best estimate of this factor. In this test, the life and nominal strains are measured; then, using smooth-specimen fatigue data, a trial-and-error procedure is followed until a value is found which predicts the measured life. This value of K_f is then used in the estimate of life under variable amplitude strain histories.

An example of this approach is given in the Applications section below, but for economic reasons, this is not generally a practical solution. Usually an estimate will have to be made based on an analysis involving notch geometry and material properties. As a first approximation, the fatigue notch factor can be assumed equal to the theoretical elastic stress concentration factor; this is conservative for small notches since K_f is always less than K_t.

For a less conservative estimate, an empirical relationship between the notch factor and the theoretical factor may be used. Peterson's relationship, Eq. 8, is a typical example of this approach and others which are generally based on long life fatigue test results. Conventionally, the fatigue notch factor is defined as the ratio of stress amplitude of a smooth specimen to the net section stress amplitude of a notched specimen at equal fatigue life. This way of defining K_f results in a fatigue notch factor which is variable over the life range. This variability of K_f is generally attributed to a combination of plasticity at the notch root and size effects. These two effects can be separated by using Neuber's rule which proposed that the theoretical elastic stress concentration factor is equal to the geometric mean of the actual stress and strain concentration factors:

$$K_t = (K_\sigma K_\epsilon)^{1/2} \tag{16}$$

Eq. 16 may be altered by replacing K_t with K_f and defining K_σ and K_ϵ as the ratio of the stress or strain of a smooth specimen to the net section stress and strain of a notched specimen at the same life which results in:

$$K_f = \left(\frac{\Delta\sigma}{\Delta S} \frac{\Delta\epsilon}{\Delta e}\right)^{1/2} \tag{17}$$

When K_f is defined by Eq. 17, the effects of plasticity are essentially removed. The fatigue notch factor becomes approximately a constant over the life range, and its variation from K_t is thought to be primarily due to size effect.

Note that at long lives K_ϵ and K_σ are equal, and Eq. 17 then yields a resulting K_f which is equal to that defined in the conventional manner. After K_f has been estimated, the nominal strain-time history must be converted to local stress and strain-time histories.

Nominal to Local Stresses and Strains. Local stresses and strains are calculated from nominal strains using Neuber's Rule in the form:

$$\Delta\sigma\Delta\epsilon = K_f^2(\Delta S\Delta e) \tag{18}$$

With the right-hand side constant, a rectangular hyperbola is represented. To use this equation, the nominal stress as well as the field-measured nominal strain must be known. When the nominal strains are elastic, Eq. 11 may be combined with Eq. 18:

$$\Delta\sigma\Delta\epsilon = (K_f\Delta e)^2 E \tag{19}$$

If the nominal strains are inelastic, nominal stresses must be determined from the cyclic stress-strain response of the metal, Eqs. 2 and 3 provide approximations to the stable cyclic response condition for estimating the fatigue life.

Stress-strain response may be followed graphically from a set of experimentally obtained hysteresis loops such as are shown in Fig. 3.15, but for a complex strain history the graphical approach is extremely time consuming. Since most large-scale data acquisition systems are designed for computerized data reduction and analysis, an appealing approach is to simulate the stress–strain response with a computer [433]. A solution for the local stress and strain peaks is found by determining the intersection of the stress–strain response curve and the Neuber hyperbola, Eq. 18. Note that the axis of the hyperbola is located at a previous maximum or minimum and not at the initial origin; again, the response may be followed graphically. However, the model [433] may be more convenient to use for most strain-time histories. The technique of following the stress-strain response to the hyperbola requires care in choosing the stress and the strain range to be used. Subtracting sequential maximum (minimum) from minimum (maximum), generally called "range-counting" will usually result in an incorrect solution. A new method called "rainflow" counting [398] will determine the correct stress and strain ranges to be used in Eq. 18. This method divides the strain-time history into reversals which do not form closed hysteresis loops; a more detailed description is given below. Once the Neuber hyperbolas have been determined, their intersection with the stress-strain response defines the local stresses and strains at the peaks. An example of the local stress-strain response intersecting the Neuber hyperbolas is shown in Fig. 3.21. Having determined the local stresses and strains, the stress and strain ranges and the mean stress for each reversal are defined. Again, the rain flow method must be used to define reversals for damage analysis.

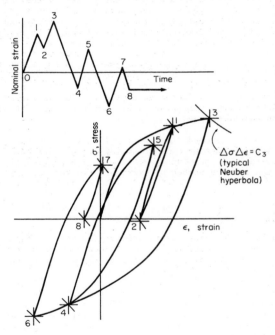

Fig. 3.21 Local stress–strain response intersecting the Neuber hyperbolas.

Throughout this discussion, the term <u>reversal</u> has been used in preference to <u>cycle</u>, since it is an easier term to define. A reversal is counted when the measured signal changes direction. The term cycle, which becomes somewhat ambiguous when applied to a random amplitude history, is equivalent to two reversals for constant amplitude cycling. However, the use of reversals or half cycles is noncritical in the actual implementation of this procedure, as long as one is consistent.

Cumulative Damage Summation. In order to maintain simplicity, and because more complicated damage summations have not proved to be more accurate, a linear damage rule is used to estimate life. A visible crack is assumed to be present when the damage summation expressed in Eq. 20 approaches:

$$D = \Sigma \ \frac{r}{2N_f} \tag{20}$$

Where:

  r = number of reversals of a given magnitude in strain
        or stress-time history

  $2N_f$ = number of reversals to failure for the same given
          magnitude

Fatigue life for any counted reversal, $2N_f$, is determined from either the material's stress-life relationship, Eq. 4, or the total strain-life relationship, Eq. 6. These empirical equations are defined by constant amplitude, completely reversed fatigue tests, therefore, prestrain and mean stress effects must be taken into account before using the relationships. Prestrain effects are treated by using fatigue data from prestrained smooth specimens to define the material's fatigue properties. As has been suggested, one practical approach is to assume prestraining has occurred and use only prestrained fatigue data. Mean stress effects can be accounted for by using Eq. 13 which results in a completely reversed stress range which has a life equal to that of the measured nonreversed cycle.

The choice between using the stress-life curve or the total strain-life curve is dependent upon the life range. In the low cycle region, the total strain-life curve is more sensitive than the stress-life curve. Also, mean stresses do not exist, and prestrain is not a factor to consider. In the intermediate to high cycle region, mean stresses can exist, and, therefore, the stress-life plot is probably more convenient to use.

Once the damage has been summed, the fatigue life for the design function described by the measured service history can be estimated. A prediction will be in terms of either time-to-failure or events-to-failure. The fatigue life estimate is simply equal to the time or number of events in the service history divided by the damage summed.

SUMMARY

The final goal of this procedure is to predict the service fatigue life of a part for it's combined design functions using a linear summation of damage, Eq. 1. The fatigue life, as calculated by this procedure, represents an experimentally measured service history.

This approach has the advantage of being able to include the effects of loading sequence into the service life summation. For example, it may be desirable to know the effect of a "once-in-a-lifetime" severe load on the service life. The damage done by the single high load generally does not significantly contribute to the damage summation itself; however, the residual stresses set up by that one severe load may so contribute. It has been shown that a Neuber analysis will correctly account for the effects of an over-load. Depending on whether the high load sets up a compressive or tensile residual stress, the fatigue life of the severely loaded part may be more or less than for the part which was not severely loaded. The technique for performing the analysis is first to measure the high load to determine the local residual stress and strain which it sets up, and then use that residual stress and strain as the starting point for the analysis of other service histories.

APPLICATIONS

To illustrate the analytical method, fatigue life for a representative part was estimated and compared to test results. The example is a notched beam subjected to a pure bending moment; fatigue lives for both constant amplitude loading and two-level block loading were estimated.

Notched Beam in Bending. For this example, a notched Man-Ten steel bar was loaded in a 4-point bending fixture. Eight different constant amplitude levels and thirteen different two-level blocks were applied to the notched beam. Failure of the beam was defined as a fatigue crack 0.2-0.3 in. long

Since the nominal net section remained essentially elastic, nominal strains were calculated from the flexure formula, Mc/IE.

The fatigue properties for the Man-Ten steel were given in Figs. 3.17, 3.18, 3.19 as a log-log linear relationship approximated over the life range of interest. Eq. 21, which describes the stress-life relationship, results from substituting the fatigue strength coefficient and exponent into Eq. 4.

$$\Delta\sigma/2 = 170(2N_f)^{-0.12} \qquad (21)$$

where: $\Delta\sigma$ is in ksi.

For the block loading tests, the one large cycle in the block was assumed to prestrain the material. It was, therefore, necessary to modify Eq. 21 to include the prestrain effect, so three additional smooth specimens were tested, as recommended previously, to measure the effect on the life plots. The data are reported in Table 3.4 and plotted on the strain-life plot in Fig. 3.20. The stress-life relationship not illustrated is approximated for the life range above 10^4 reversals by Eq. 22.

$$\Delta\sigma/2 = 260(2N_f)^{-0.165} \qquad (22)$$

where: $\Delta\sigma$ is in ksi.

The local stresses and strains were found by first following the monotonic stress-strain response to the Neuber hyperbola for the largest maximum in the history. From that point on, the stable stress-strain response was followed to all future hyperbolas. The fatigue notch factor required for Neuber's rule was estimated by four different approaches, as will be explained later.

Since the histories were simple (Fig. 3.22), computer simulation was not necessary, rather, a graphical routine, using an experimentally-obtained set of hysteresis loops shown in Fig. 3.23, was employed in the analysis. Hysteresis loops for each

TABLE 3.4 Fatigue Results for Strain Controlled Smooth Specimen Tests
of Man-Ten Steel After Prestraining 10 Cycles at ± 1%

Specimen Number*	Stress Amplitude,** $\Delta\sigma/2$, ksi	Total Strain Amplitude, $\Delta\varepsilon/2$	Plastic Strain Amplitude, ** $\Delta\varepsilon_p/2$	Reversals to Failure, $2N_f$
1	44	0.003	0.0015	32,800
2	38	0.002	0.0013	107,000
3	36	0.00175	0.0005	161,300

*Specimens were 1/4 in diameter, 3/4 in gage length.
**Half-life value, essentially stable.

block history were illustrated in Fig. 3.22   The local stress and strain ranges
and local mean stress for each reversal were determined from the graphical plot
using the rain-flow counting method. Mean stress was accounted for by calculating
an equivalent completely reversed stress, using Eq. 13. For the constant amplitude
tests, where prestrain did not occur, Eq. 21 was used to calculate the life of
each reversal, and in the programmed loading tests, where prestrain did occur,
Eq. 22 was used. Fatigue life for each test in terms of cycles or blocks is
simply the inverse of the damage for a single cycle or block as calculated by
Eq. 20.

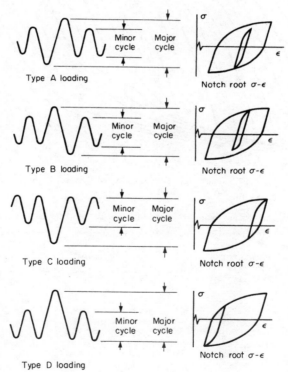

Fig. 3.22 Loading cycle and notch root stress-strain hysteresis loops.

The actual median fatigue life from the constant amplitude tests is compared to the estimated median life in Table 3.5 and Fig. 3.24. Four different fatigue notch factors were introduced into the analysis to illustrate the effect of K_f on the life estimate. The following list gives each value and its origin:

1. $K_f = 2.62$ $K_f = K_t$

2. $K_f = 2.49$ Eqs. 8 and 9

3. $K_f = 2.15$ a trial and error approximation based on constant amplitude data of the beam

4. $K_f = 1.90$ Eq. 8 with "a" value from Ref. 434

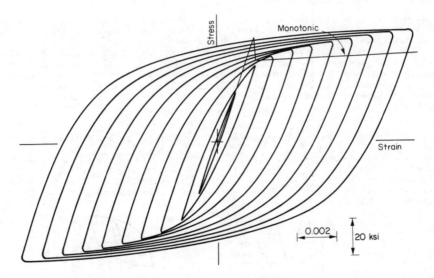

Fig. 3.23 Stable stress-strain hysteresis loops for
Man-Ten steel.

TABLE 3.5 Comparison of Fatigue Life Data with Estimated Life
for Constant Amplitude Loading

	Load-P, kips		Reversals to Failure, $2N_f$	Reversals to Failure, $2N_f$ Estimated			
No.	Max.	Min.	Median Value	$K_f = 2.62$	$K_f = 2.49$	$K_f = 2.15$	$K_f = 1.90$
1	16.0	1.0	3.6×10^4	2.8×10^4	3.3×10^4	4.1×10^4	5.1×10^4
2	17.5	2.5	3.7×10^4	2.8×10^4	3.3×10^4	4.1×10^4	5.1×10^4
3	16.7	3.3	5.3×10^4	3.1×10^4	3.6×10^4	4.4×10^4	7.0×10^4
4	15.0	0.6	5.8×10^4	3.1×10^4	3.4×10^4	4.1×10^5	6.6×10^5
5	15.0	5.0	1.0×10^5	6.0×10^4	7.6×10^4	1.2×10^5	1.9×10^5
6	15.0	5.0	1.6×10^5	6.0×10^4	7.6×10^4	1.2×10^5	1.9×10^5
7	11.0	1.0	1.8×10^5	1.1×10^5	1.6×10^5	2.5×10^5	5.9×10^5
8	13.8	6.3	4.9×10^5	2.0×10^5	2.4×10^5	4.1×10^5	7.6×10^5
9	8.5	1.0	1.1×10^6	5.0×10^5	6.4×10^5	1.4×10^5	2.3×10^6

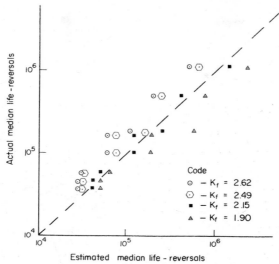

Fig. 3.24   Actual vs estimated life, constant amplitude loading.

Fig. 3.25 presents the comparison of actual median life to estimated life for the program loading tests. Only the first and third estimates for K_f were calculated for this comparison. As expected, setting $K_f = K_t$ resulted in a conservative estimate. Estimating K_f from the constant amplitude data gave the closest estimate in most cases.

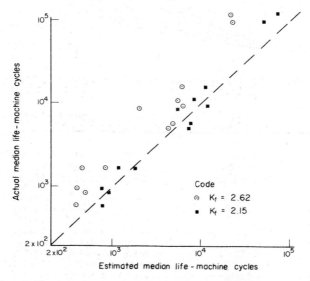

Fig. 3.25   Actual vs. estimated life, programmed loading.

DESCRIPTION OF RAIN-FLOW COUNTING METHOD

The rain-flow counting method proposed by Matsuiski and Endo [398] is designed to count reversals in accordance with the material's stress-strain response. Since the analytic procedure is based on the comparison of stress and strain from constant amplitude tests with variable amplitude stresses and strains from service history, it is essential that the counting routine defines reversals in a comparable manner. The definition of reversals or pairs of reversals must be consistent with the constant amplitude hysteresis loop. Previously proposed methods, such as range counting, do not recognize this requirement.

An illustration of this point can be seen in Fig. 3.26. Material response clearly indicates a reversal should be counted from point 3 to 6, and a pair of reversals, a closed loop, should be counted for the interruption from point 4 to 5. Rain-flow counting defines the reversals exactly in this manner.

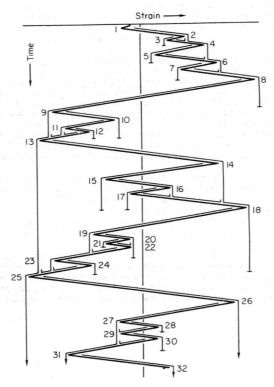

Fig. 3.26 Example of Rain-flow Cycle Counting Method.

Note that range counting would have defined one reversal from point 3 to 4, a second from point 4 to 5, and a third from point 5 to 6. Particularly, the reversal counted from point 5 to 6 is not representative of any hysteresis loop which would be duplicated in constant amplitude cycling.

A detailed description of the counting routine follows:

"The rain-flow cycle counting method is illustrated in Fig. 3.26   The strain-time history is plotted so that the time axis is vertically downward, and the lines connecting the strain peaks are imagined to be a series of pagoda roofs. Several rules are imposed on rain dripping down these roofs so that cycles and half cycles are defined. Rain flow begins successively at the inside of each strain peak. The rain flow initiating at each peak is allowed to drip down and continue except that, if it initiates at a minimum, it must stop when it comes opposite a minimum more negative than the minimum from which it initiated. For example, in Fig. 3.26 begin at peak no. 1 and stop opposite peak no. 9, peak no. 9 being more negative than peak no. 1. A half cycle is thus counted between peak nos. 1 and 8. Similarly, if the rain-flow initiates at a maximum, it must stop when it comes opposite a maximum more positive than the maximum more positive than the maximum from which it initiated. For example, in Fig. 3.26 begin at peak no. 2 and stop opposite peak no. 4, thus counting a half cycle between peak nos. 2 and 3. A flow must also stop if it meets the rain from a roof above. For example, in Fig. 3.26 the half cycle beginning at peak no. 3 ends beneath peak no. 2. Note that every part of the strain-time history is counted once and only once.

"When this procedure is applied to a strain history, a half cycle is counted between the most positive maximum and the most negative minimum. Assume that of these two, the most positive maximum occurs first. Half cycles are also counted between the most positive maximum and the most negative minimum that occurs before it in the history, between this minimum and the most positive maximum occurring previous to it, and so on to the beginning of the history. After the most negative minimum in the history, half cycles are counted which terminate at the most positive maximum occurring subsequently in the history, the most negative minimum occurring after this maximum, and so on to the end of the history. The strain ranges counted as half cycles, therefore, increase in magnitude to the maximum and then decrease.

"All other strainings are counted as interruptions of these half cycles, or as interruptions of the interruptions, etc., and will always occur in pairs of equal magnitude to form full cycles. The rain-flow counting method corresponds to the stable cyclic stress-strain behavior of a metal in that all strain ranges counted as cycles will form closed stress-strain hysteresis loops, and those counted as half cycles will not. This is illustrated in Fig. 3.27."

If this rain-flow method is followed, a variable amplitude history can be reduced to a series of reversals whose maximum and minimum values are defined. Mean stress is then known for each reversal, and a completely reversed stress amplitude can be calculated by Eq. 13 which is directly comparable to constant amplitude material properties. Fatigue life can then be calculated using Eq. 4.

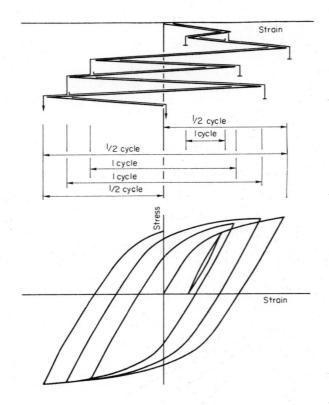

Fig. 3.27   Rain-flow Counting Method applied to a strain
            vs time sequence.

DAMAGE CALCULATIONS WITH LOAD SPECTRA

The term "load spectrum" is widely used throughout the literature of fatigue and is generally taken to mean a plot of the number of occurrences or exceedances of a given load level per unit time or distance. Such a curve is established by summing the number of events in a load-time trace by any of the common counting methods: peak, level-crossing, Rain-flow, etc., for either periodic or random functions; it thus turns out to be a "probability" curve, as in Fig. 3.60. Since the time interval under consideration can, in the practical sense, be only a small portion of the total time or life, the spectrum is only a set of the total population. The results of operating on the set can, then, be extrapolated to the total life only by means of some appropriate distribution function. If the function is known beforehand, it becomes possible to establish a spectrum and predict the life without test data on the structure or models of it. But herein lies a major question in the consideration of fatigue life prediction: "What is the proper distribution function of loads arising from natural phenomena such as atmospheric gusts, road surfaces, etc.?"

The data to the present indicate a strong predominance of the normal or gaussian distribution, examples being aircraft pressurization and maneuver loads, and the loads on automobiles under many conditions. The straight-line distribution also is found for some automotive components, and the log-normal for gust loads. Spectra based on these distributions plus the rectangular are sketched in Fig. 3.28 the graphical similarity with the conventional S-N is immediately evident - in fact, the latter may be regarded as a limit spectrum. It is possible to construct load or stress spectra from mission profiles, with results as in Fig. 3.29 if the transfer functions relating vehicle weight, speed and so on to stress are known, and more directly, from service or laboratory data [368]. Adjustment or reduction factors are applied according to the relevant certification specification, the factor depending chiefly on the number of specimens involved. A typical automotive stress spectra is shown in Fig. 3.30.

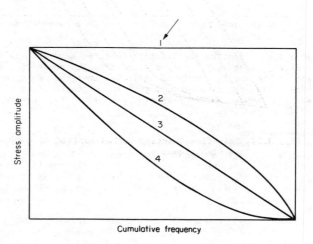

Fig. 3.28   Typical stress spectra for various distributions.
1-rectangular; 2-normal; 3-straight line; and
4-log normal.

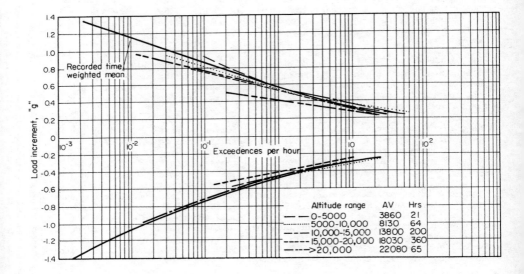

Fig. 3.29   Flight-load spectra for pressurized, twin-engine aircraft.

Characteristic response spectra for static structures have been investigated, typically for short, girder-type bridges, resulting in frequencies of occurrence as shown in Fig. 3.31a and b. Analysis indicated that the Rayleigh distribution gave the best fit, with the final result that the histogram form of loading, Fig. 3.31c, was recommended as properly simulated input for testing the steel for such bridges. (See also Sec. 4.2.1).

The shape of the spectrum is important in design and it depends on numerous items: the variation in mean stress; the factor by which the structural area is increased above that to satisfy static strength requirements; and on design parameters such as the slope of a wing-lift curve. The slope of a spectrum is commonly used to describe it for design operations such as matching a given limit spectrum, and in controlling the Fatigue Quality Index. Further discussion of the latter will be found in Sec. 3.2.2.

Applications of the linear damage rule may be made for: (1) combinations of a given load spectrum and an S-D curve for direct calculation of the percent damage in a single load-carrying member, and (2) computerized versions of essentially the same operations expanded to accommodate many members, and multiple load spectra and S-N curves. As an example of the direct method, the design of a spar joint to withstand gust loading is outlined in Fig. 3.32 [142]. The fluctuating stresses, expressed as a percentage of the static ultimate stress have been derived from

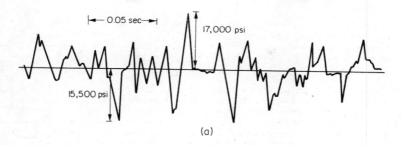

(a)

Fig. 3.30a   Typical road-load recording: stress fluctuations
with time in steering tie rod over cobblestone
test route.

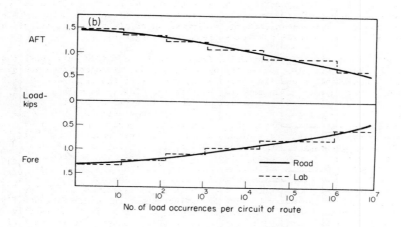

Fig. 3.30b   Typical road-load histogram and
laboratory step program.

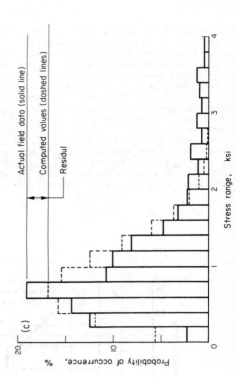

Fig. 3.31   Actual vs Computed
Frequency of Occurrence.

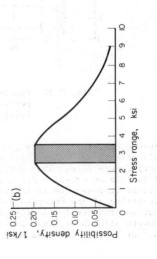

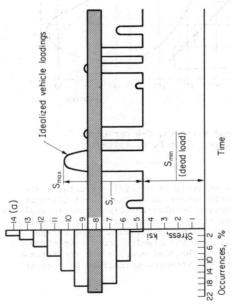

measurements of gust velocity. To simplify the procedure, these stresses have
been grouped into intervals covering a range of 4% of the ultimate stress, and the
number of stress fluctuations within each interval is plotted; thus 10 stress
fluctuations occurred with a magnitude between 10 and 14% of the ultimate stress.
The damage caused by each group of stress fluctuations, according to the linear
damage law, is equal to the ratio of the number of cycles applied to the number
required to cause failure in the joint at that stress range, and this ratio is
plotted as percentage damage in the right-hand curve. The cumulative damage is
then equal to the sum of the damage caused by each group. In the example, this
amounts to about 30%, giving an estimate of the life of the joint subjected to the
gust loading shown of a little over 30,000 hours. It can be seen from Fig. 3.32
that most of the damage is caused by gusts within the relatively narrow band of
4 to 14% of the ultimate stress, the maximum at about 8%, corresponding to a gust
velocity of about 10 ft/sec. This stress range is often chosen, therefore, for
constant amplitude laboratory tests on aircraft components.

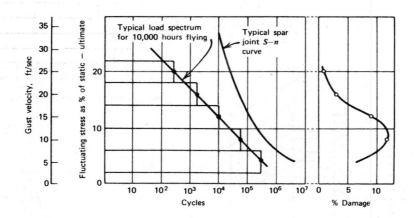

Fig. 3.32   Combination of load spectrum, S-N curve and
            percent damage curve. From P. G. Forrest
            [142], Fig. 147.

Numerous computer programs for the fatigue analysis of aircraft structures are
applicable to any general fatigue damage calculation using the linear (Palmgren-
Miner) damage rule. They may be made conducive to mission-flight profile analysis
and include an equation for a gust loads spectrum, calculation of rigid-airplane
vertical response due to turbulence, and the automatic definition and damage
calculation of the Ground-Air-Ground cycles from the mission profile spectrum.
Input information is usually given the form of the load spectra and the fatigue
strength allowables, that is, S-N data with the stresses normalized on a chosen
limit or failure criteria such as S_{yp} or S_{ult}.

A most interesting example of the inverse situation, i.e., that wherein the operator wholly controls the input load levels and numbers of cycles, is illustrated by a device for experimentation in nuclear fusion. Here, the required spectrum is given, Fig. 3.33, and several calculated S-N curves are noted in their relation to it. The problem is solved in Fracture Mechanics, using values of critical toughness, crack growth rate and initial crack size characteristic of the alloys, their processing and the sensitivity of the NDE procedure. The critical region was identified by stress analysis as the ligament between a hole and a free edge (in tension) in the side plate on the case for the major magnetic coils. The calculation scheme for producing the S-N curves is given in Table 3.6. The requirement for acceptance of an alloy for this service is that its S-N curve be always above and to the right of the spectrum – thus, several alloys were rejected.

This general situation gives rise to the concept of allowable and non-allowable sequences of operation. In performing the function of the device, the coils may be programmed to produce field or load pulses in any combination of load levels and numbers of cycles. Thus, the accumulation of fatigue damage in the form of crack length must be monitored, even for those alloys that show acceptance on Fig. 3.33. The reason behind this condition is that the critical crack length varies inversely with (stress)^2, so it may be possible to find a sequence starting with many low-stress cycles which accumulates sufficient crack length to later override the (now shorter) critical length for cycles at some higher stress. Examples of such sequences are shown in Table 3.7.

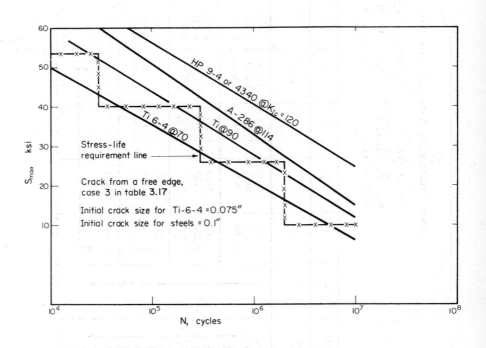

Fig. 3.33 A required spectrum and calculated S-N curves for several alloys.

TABLE 3.6 Typical Worksheet in Crack Propagation

CASE: Finite-width Strip in Tension, with Single-edge Notch

$K = SQ \sqrt{\pi a}$

Mat'l.: A-286 $m = 3.25$

$a_{cr} = [K_{Ic}/S_{max}]^2/\pi Q^2$ $K_{Ic} = 114$ ksi $\sqrt{inch}$ $S_{max} = 20$ ksi.

$da/dN = C(\Delta K)^m$ $C = 3.0 \times 10^{-10}$

1	2	3	4	5	6	7	8	9	10	11	12
	b	a_i	a_i/b	Q	a_{cr}	$\sqrt{a_i}$	ΔK	ΔK^m	da/dN $\times 10^{-10}$	$a_{cr}-a_i$	$N_{rem.}$
W	14	.050	.0035	1.0001	10.34	.224	7.92	833	2500	10.290	41×10^6
		.200	.0142	1.021	10.12	.448	16.14	8427	25280	9.920	3.84
		1.500	.1070	1.160	8.913	1.224	50.26	345500	103600	7.413	0.072
		2.000	.1428	1.171	8.834	1.414	58.56	555500	1,666000	6.837	0.041
		3.000	.2142	1.214	8.517	1.732	78.2	1,422000	4,266000	5.517	0.013
		5.000	.3571	1.335	7.745	2.236	105.6	3,705000	11,115000	2.745	0.0025
28		6.000	.4285	1.410	7.333	2.449	121.3	5,875000	17,627000	1.333	0.00075
		6.5	.4642	1.460	7.082	2.549	131.7	7746	23240	0.582	0.00025
		6.8	.4857	1.480	6.986	2.607	136.6	6711	26133	0.186	0.00007
		6.850	.4892	1.498	6.940	2.617	138.7	9,173000	27,520000	0	0

TABLE   3.7   Allowable and Non-Allowable Operational Sequences

BLOCK NO.	1	2	3	4	5	6
Stress, ksi	1.5	6	13.3	24	40	53.2
Applied cycles,n	870×10^3	6×10^3	14×10^3	5×10^3	42×10^3	4×10^3
Critical Crack,a_{cr}	VERY LARGE		~ 10.0	5.000	2.176	1.176

SEQUENCE A$_1$

	c	f	e	b	a	d
a_i	0.303	0.721	0.651	0.168	0.080	0.303
da/dn	0	0	5×10^{-6}	2.7×10^{-4}	2.1×10^{-6}	8.7×10^{-5}
Δa	0	0	70×10^{-3}	13.5×10^{-3}	88.2×10^{-3}	34.8×10^{-2}
	.303		.721	.303	.168	.651

$a_{final} = 0.721$ inch

SEQUENCE B$_1$

	f	e	d	c	a	b
a_i	.645	.645	.579	0.436	0.080	0.1682
da/dn	0	0	4.7×10^{-6}	2.85×10^{-5}	2.1×10^{-6}	6.7×10^{-5}
Δa	0	0	.066	.143	88.2×10^{-3}	2.68×10^{-1}
	$a_f = 0.645$ inch	.645	.579	.168	.436	

SEQUENCE B$_2$

	f	e	d	c	a	b
a_i	1.8765	1.876	1.096	.866	0.080	0.710
da/dn	0	12×10^{-9}	8×10^{-6}	6.7×10^{-6}	2.1×10^{-6}	5.2×10^{-4}
Δa	0	.0005	.780	.231	.630	.156
$a_f =$	1.8765	1.8765	1.876	1.096	.710	.866

Sequence B$_2$ will run because at no time does an a exceed an a_{cr} - i.e.,

B$_2$ Block (a) at 40 ksi accumulates .710 inches; .710 < 2.176
Block (b) at 53.2 ksi starts with .710 and adds .156 for a total
    of .866;   .866 <1.176
Block (c) at 24 ksi starts with .866, adds .230; 1.096< 5.0, etc.

SEQUENCE C$_1$

	a	b	c	d	e	f
a_i	0.080	0.080	0.080	0.094	0.136	0.556
da/dn	0	0	10×10^{-7}	8.3×10^{-6}	1×10^{-5}	3.1×10^{-4}
Δa	0	0	.014	.042	0.420	1.240
	0.080	0.080	.094	0.136	.556	$a_f = 1.796$

But Sequence C$_1$ will not run because the total growth does exceed an a_{cr}.

C$_1$ Block (a) a = 0
Block (b) a = 0
Block (c) a = 0.094; 0.094 <<<10.0
Block (d) 0.094 + 0.042 = 0.136; 0.136<<5.00
Block (e) 0.136 + 0.420 = 0.556; 0.555 <2.176
Block (f) 0.556 + 1.240 = 1.796; 1.755 >1.176

Truncating block (f) to 2×10^3 would allow C to run, by making a = a_{cr}.

3.2.2   NONLINEAR AND DAMAGE BOUNDARY RULES

The linear rule does not accommodate the effect of prior stress history in assuming that it has no effect on the response of the material to subsequent cycles of repeated loading and that stresses below the original fatigue limit develop no damage. This condition with its resulting lack of accuracy and conservatism has prompted many efforts to formulate a more comprehensive and correct rule. Such efforts have generally yielded expressions of considerable complexity, at least in comparison to the Miner summation, occasionally requiring data that is not usually available. The situation is similar to that in the study of Fracture Mechanics, in which the Larson-Miller parameter may, in the sense of basic simplicity, be likened to the Miner Rule; other parameters have been developed to give some improvement in accuracy at the expense of added complexity. However, the application of a nonlinear rule may be dictated by accuracy requirements or other circumstances, and, of the the more important rules identified in Section 3.1:

1.   Freudenthal and Heller's interaction method [32] is difficult to apply due to lack of appropriate data.

2.   Corten and Dolan's attempt [36] to accommodate the effect of prior stress history is often limited by the inability to predict actual stresses in a complex structure.

3.   Results of the Stress Concentration Factor treatment are strongly dependent on the handling of the local geometric shapes.

4.   The Fatigue Quality Index requires a very comprehensive view of the necessary vs. the possible value of K, and of the penalties of forcing all details to have the same value.

The last two methods are discussed in more detail below.

The Stress Concentration Factor Method. Such a method exists in stress and fatigue analyses because practical machine and structural members are rarely of uniform section. Changes in section, commonly called notches, interrupt the stress pattern and tend to make it non-uniform, resulting always in an increase in local stress level. Analysis that involves stress concentrations may be regarded as treatment of exceptions to Saint-Venant's principle - in simplifying the given shape for purposes of stress analysis, it is not permissible to apply the assumptions usually invoked. When dealing with conventional shapes and loadings, one usually considers that plane sections remain plane, that end effects disappear in a short length, etc. But this type of assumption is satisfactory only when Saint-Venant's principle can be applied, that is, when the sections being analyzed were not close to a point of local load application, or to a sharp discontinuity in the force flow (abrupt change in cross section). Recent reconsideration [98] of the principle of vanishing end effects has not altered this conclusion.

The results of the stress concentration factor (SCF) method are, then, the corrected local stresses, rather than the nominal values, and these may be used with essentially any of the rules for calculating damage. Graded sets of S-N or constant-life curves are often utilized with these corrected stress values and the mode of calculation may be any of those possible, from the simple Miner summation to the probabilistic considerations, depending on the type and accuracy of available information. The dominant factor in stress concentration factor treatments seems to be the emphasis on the effect that local geometry is having on the stress distribution; thus, when properly applied, the method produces a more realistic and accurate evaluation of the actual stress levels. To neglect consideration of the notch and shape effects in any but the most primitive of shapes is generally to produce an unacceptable solution.

Notches. In geometric terms, the distribution of stress in a loaded element may be regarded as analogous to the flow of fluid in a closed conduit. Even though there is no velocity term in the former, the concept of streamlines or force lines is very helpful in visualizing the effect of notches on stress distribution. This condition is illustrated by Fig. 3.34a in which a flat plate with edge notches is subjected to a uniform tension, P. Since the spacing of the flow lines may be regarded as inversely proportional to the intensity of stress, it is evident that the latter must be greater at the notched or reduced section--all the "flow" must pass through the restricted area. Thus, the theoretical stress concentration factor is defined as the ratio S_{max}/S_{nom}, where S_{nom} is the P/A stress. Now, consider the geometry of the plate as an entity in relation to that of the notch: if the plate were very long compared to its width, the angle ϕ_1 would approach 180° and the plate profile would tend toward that of the force lines. If ϕ_2 were increased, so would the effective value of r; but h would decrease and the shaded wedges would tend to disappear, again bringing the plate profile nearer that of the force lines.

In a notched plate under tensile elongation, tensile and shear forces exist on the interface between the wedges and the inner section. Should the wedges be machined off, these forces would disappear and thus could not provide the added stress intensity at the notch. The presence of the wedges is considered detrimental to the stress conditions within the plate, and it then follows that the radius r, depth h and angle ϕ_2 are the important parameters in determining the severity of the notch effect: the value of K_t. The angle, and primarily the depth, control the additional stress contribution from the wedges; the notch radius defines both the intensity and the gradient of the stress over a very small area at the root. The concentration effect increases as the radius and angle decrease, and as the depth increases. A fine saw cut forms the classic example of a severe notch.

The stress system internal to a notched member is always that of combined stresses, even though the applied force may be only unidirectional. In a grooved cylindrical bar under uniform axial tension, radial stress arises at the root of the notch because of the curvature of the force lines. From the equilibrium conditions for the axial loading of curved beams, the radial stress at any radius may be calculated, provided only that the axial stress and the force line curvature are known. Since this is an integrated effect, the radial stress increases from zero at the surface to a maximum at the center. The triaxial tension results in a decrease of the shear stress in the interior of the bar. This condition explains why the tensile failure of ductile materials, after necking down, appears to be of a brittle, tensile nature near the center but of a shearing nature nearer the surface: the cup and cone. There should be no sharp intersection between the cup and the cone; they are distinguished only by that point on a radius at which, going from the surface to the center, the shear component becomes less than the tension component. The latter is then resolved on the slip planes of each grain to a shear stress, with failure occurring when the tension rises sufficiently to produce the critical resolved shear stress of the material. Tests of specimens with machined grooves of various depths have shown an apparent increase in the ultimate tensile strength. This condition results from the radial tension; a higher average axial stress over the notched region is required to produce slip.

Stress distributions are also affected by holes, in either tensile or compressive loading. One example is a wide plate with a small central hole, under edgewise compression. The curvature of the force lines causes a lateral tension on those edges of the hole at which the tangent is normal to the force line. For the ideal case of an infinitely wide plate, this lateral tension is equal to the nominal compression, and may cause longitudinal cracking in a brittle material. There is also a concentration effect at the ends of a horizontal diameter where, for the infinitely wide plate, the longitudinal compression is three times the nominal

compression. The lateral tension is particularly bad because any crack, perpendicular to the force field, represents a severe notch and tends to propagate
itself by tearing. In general, the dimension of the hole perpendicular to the
compressive force field is the most important parameter, having the same effect as
the depth of an external notch. This argument also holds for longitudinal tension,
the local compression and tension interchanging positions at the ends of the
horizontal and vertical diameters.

The stress concentration factor is defined as the ratio of the maximum local
stress to the nominal stress. Figure 3.34b illustrates the general relationships
for both tension and bending as typical cases of loading. The nominal tensile
stress is P/A, where A is the area of the minimum cross section, and similarly,
the nominal bending stress is calculated as if the bar were of uniform depth d.

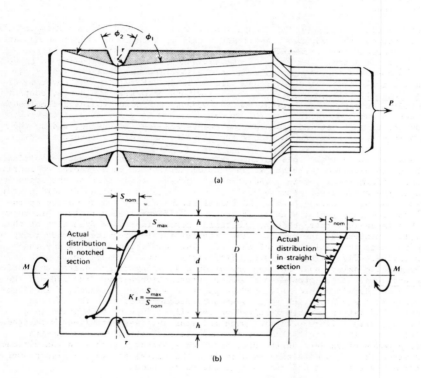

Fig. 3.34   Force lines and the stress concentration factor.

Neuber [95] has shown that K_t may be calculated from the geometry of the notch
alone--it is not a function of material properties--and that its value lies between
one and infinity. The form of such equations is:

$$K_t = 1 + f(a, W)$$

where a is the length of the notch or crack and W is a dimension measured in the same direction. For a simple case, as a bar in bending, f(a,W) is $(\sqrt{a})$x f(crack length/width). This $\sqrt{a}$ term also shows up as a fundamental parameter in Fracture Mechanics considerations, see Section 3.2.4. Although calculation of K is possible from purely analytical considerations, the empirical boundary conditions make it sufficiently difficult so as to be impractical for all but the most demanding or unique design jobs. Adequately accurate values of K_t have been determined by experiment for a great variety of notch sizes and shapes, and for many loading conditions. Tabulations and plots, as typically in Fig. 3.35, for essentially all conditions likely to be encountered, exist in many references of which [93] is

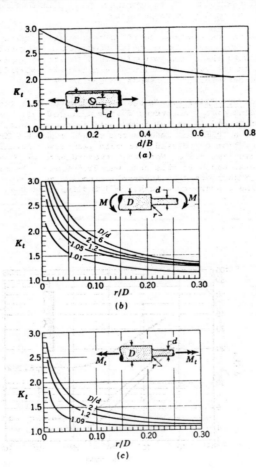

Fig. 3.35 Values of stress concentration factors: (a) flat plate and central hole, in tension; (b) round shaft with shoulder fillet, in bending; and (c) round shaft with shoulder fillet, in torsion. From R. E. Peterson [93].

perhaps the most comprehensive. There is rarely any need for guessing at the stress concentration factor, particularly since the weight of a member is usually related very closely to the value of the stress concentration factor used in sizing its cross section.

It is interesting to note the existence of many practical situations, as in Fig. 3.35, wherein the maximum value of K_t is about 3.0. But this condition is distinctly not to be regarded as a rule of thumb; there are many other geometries and loading conditions leading to K_t values of the order of 20.0, that is, for an elliptical hole in a plate under tension normal to the major axis. The fact that such values can exist should give the designer pause, indicating the possible reconsideration of the need of such geometry.

The accuracy of the values assigned to the stress concentration factors, or K_t's, has recently been re-evaluated [363, 364]. Comparisons of results of theoretical relaxation methods with those from experiments with strain gages and with birefringent coatings lead to the conclusion that the generally available data [93] are quite adequate, and occasionally somewhat conservative.

<u>Plasticity and Notch Sensitivity</u>. It has been shown by many tests that use of the full theoretical value of K_t in fatigue design will generally give results on the safe side, but, for materials of any reasonable degree of ductility, plastic strain occurs in the notches and reduces the stress concentration effect to a value somewhat below that calculated from pure geometry. Actual data showing this effect are plotted in Fig. 3.36. Note that the ordinate is ΔS, the stress increment above the mean stress, which in this case was 10,000 psi. Examination of the curves shows that ΔS does not vary inversely with K_t, as might be expected. Increasing K_t by using a sharper notch has a dwindling effect on the decrease

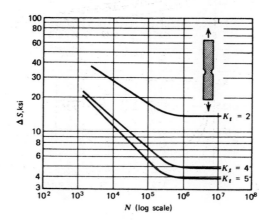

Fig. 3.36   Neuber effect, in 7075–T6 aluminum, axial
loading, S_m = 10 ksi. From H. J. Grover
[365].

in ΔS; it has been found experimentally that there is a limiting small value of the notch radius below which there is no additional stress concentration effect: the Neuber effect. The ratio between the apparent increase in local stress in fatigue and the increase predicted by the elastic theory of stress concentration has been defined [93] as notch sensitivity. The notch sensitivity factor is given as:

$$q = \frac{K_f - 1}{K_t - 1}$$

where K_t is the theoretical elastic, and K_f is the fatigue strength reduction factor:

$$K_f = \frac{\text{fatigue strength of unnotched specimens, at N cycles}}{\text{fatigue strength of notched specimens, at N cycles}}$$

The scale of notch sensitivity for a material then varies between no notch effect, $q = 0$, and full theoretical notch effect at $q = 1$. Measured values of the fatigue notch factor and of notch sensitivity have been observed to vary with (1) the material, (2) the type and severity of notch, and (3) the type and severity of loading. Table 3.8 shows typical values of K_f and of q for notched specimens of several materials at $N = 5 \times 10$ cycles in rotating bending tests. For a number of reasons, these values should not be considered representative of the materials to an extent suitable for quantitative design. The values are tabulated here only to illustrate the variation of K_f and q with differing materials. It will be observed that some alloys show a K_f equal to 1, indicating no reduction in fatigue due to the notch, while others show a value as high as 1.6, indicating a strength reduction about equal to the value of K_f for the notch used. The last two items were included to show that, although q is not a true material constant, the trends are evident despite widely differing notches; that is, both aluminum alloys show low q's, even though their K_t's varied by 4 to 1. Specific scatter bands for K_f's for 2024 and 7075 are given in Fig. 4.69 as functions of K_t and the maximum nominal stress. For the K_f's for 4130 as a typical alloy steel, see Fig. 4.44.

TABLE 3.8   Stress Concentration Factors, K_t; Fatigue Notch Factors, K_f; and Notch Sensitivity Indices q, for Various Alloys in Rotating Bending.

Alloy	K_t	K_f	q
Aluminum 2024-0	1.6	1.0	0
Magnesium AZ80-A	1.6	1.1	1.6
Stainless steel, type 18-8	1.6	1.0	0
Structural steel (BHN = 120)	1.6	1.3	0.5
Hardened steel (BHN = 200)	1.6	1.6	1.0
Gray cast iron	1.6	1.0	0
Bronze forging	1.6	1.0	0
Aluminum 7075-T73	6.7	1.8	0.13
Titanium 6Al-4V	3.5	2.8	0.72

Since the notch sensitivity of all alloys generally rises with increased strength and hardness, the steels form a most interesting and practical group for study because of the great range of strengths and hardness possible. Figure 3.37 illustrates the relatively greater sensitivity of the hard steels; Fig. 3.38 shows the effect of this sensitivity by the rapidly diminishing returns in trying to increase

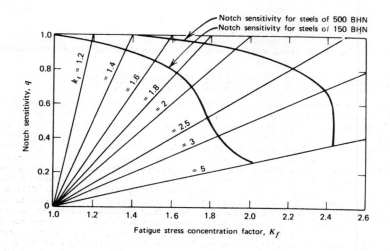

Fig. 3.37   Relations among notch-sensitivity, strength, and
the stress concentration factor for steels.

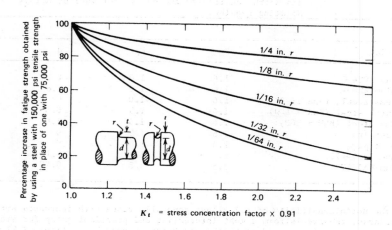

Fig. 3.38   Fatigue strength vs the stress concentration factor for steel.
From Climax Molybdenum Co. (brochure), New York, 1961.

fatigue strength by choosing a steel of higher tensile strength and corresponding higher hardness. For example, for a part whose geometry gave K_t = 2.0 and had a radius value of 1/64 in., doubling the ultimate tensile strength (UTS) from 75 to 150 ksi would increase the fatigue strength by only 25%. If it were required to increase the fatigue strength by 75% or more, then the K_t must be reduced to about 1.2 max and the radius increased to about 3/16 in., these design modifications being, or course, not independent.

Another important aspect of plastic deformation near a notch is the residual stress remaining after unloading from some high value. Figure 3.39 illustrates

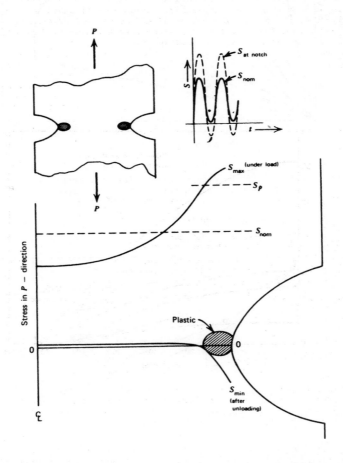

Fig. 3.39 Residual stresses after unloading of a notched specimen. From H. J. Grover [348], Fig. 46; by permission of the Naval Air Systems Command.

this for a notched sheet of an aluminum alloy under zero-to-tension loading.   The curve marked S_{max} shows the longitudinal stress along a transverse section through the center.   Near the notch, S_{max} exceeds the proportional limit, S_p, and there is a region (cross-hatched) of yielding.   Upon unloading, the edge of the notch goes to a compressive value.   Hence, not only the stress amplitude, but also the mean stress at the notch was affected by the combination of stress concentration (diminished from K_t by plastic deformation) and residual stresses introduced by the local deformation.   Because of these conditions, the usual (conservative) practice of applying the stress concentration factor only to the stress amplitude or alternating component may be open to question.   But there are partly compensating effects: (1) at higher mean stress the maximum stress is high enough so that local plasticity makes the actual stress concentration factor less than K_t, and (2) at low mean stress the observed fatigue strength is not as sensitive to mean stress as to stress amplitude.   The situation is well-portrayed by Fig. 3.40.   In practice, it appears that application of an effective fatigue-strength-reduction factor to just the stress amplitude often provides a reasonable prediction of the allowable stress amplitude for a notched specimen.

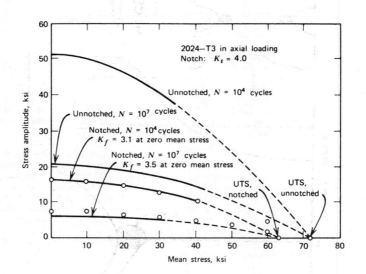

Fig. 3.40   Constant life diagram for notched specimens.   From
H. J. Grover [348], Fig. 53; by permission of the
Naval Air Systems Command.

K_f and the Criterion of Failure.   If engineering materials conformed to elastic theory, and if fatigue strength depended on the value of the maximum principal stress, K_f would be equal to K_t.   But K_f will be less than K_t if failure does not depend on the criterion of the maximum principal stress.   For essentially all notches, the maximum stress occurs at the surface, where one of the principal stresses is zero. As long as the other two principal stresses are of the same sign, the criterion of failure by maximum shear stress gives the same result as that of the maximum principal stress. If failure depends on the maximum shear strain energy, a slightly different result obtains. Letting K_t' be the stress concentration factor, and with one principal stress equaling zero, we find that

$$K_t' = K_t \ \sqrt{(1 - C + C^2)}$$

where C is the ratio of the second principal stress to the maximum principal stress. While there is some evidence favoring the shear-strain-energy criterion, the dependence of fatigue failure on any of the possible criteria is not rigorously known; in any event, the difference between K_t and K_t' is less than 15%.

The Shape Effect. The geometric characteristics of notches, especially the root radius, are of course, the most effective shape factors controlling fatigue strength. But the term "shape factor" or effect is more comprehensive and includes the configuration of the entire part and particularly the profile of its cross section. There is not complete agreement on the effect of cross-sectional shape on flexual fatigue strength; early experiments [313] rated round, diamond, and square sections in that order for decreasing fatigue limit, and later work [314] put the diamond first and the round second. The general explanation seems to be that the diamond would have a greater percentage of its volume in the highly stressed corners, providing both a high stress gradient and the "corner effect." The latter is considered as the relatively greater restraint on the grains in the corners compared to those in the interior and would be larger for the diamond than the round.

Apparently, the only difference between a "diamond" and a "square" cross section is that, in the former, the neutral axis in bending is a corner diagonal, while in the latter it is a central line bisecting a pair of sides. In any event, the differences in fatigue limit are small; for a structural steel (0.24%C), the S_e for a diamond was 9 to 13% higher than for a round, in the range of 10^5 to 10^7 cycles to fracture. A far more theoretical treatment of the effects of nonuniform stress distribution may be found in Cochardt's discussion of his stress distribution function [315] and in Lazan's volume-stress function [316].

Treatments of the stress concentration factor usually consider only those changes in section that reduce the load-carrying area, or inward-projecting notches. While this condition is usually controlling, neglect of the effect of external projections may lead to an incomplete analysis. A classic example exists in the common eye-bar, the stress concentrations due to a specific geometry being given in Fig. 3.41. The stresses are those found by photo-elastic measurements and are represented by the polar diagram gbiej with the hole edge as a base line. The greatest stress is not in the section ck, where it could be expected, but is in the section di at point d, and is about 4.2 times as great as the nominal stress s_o in the eye section ck. Curve mno gives the stress variation in the section di from the inner edge to the outer edge, and it shows that the outer third of the eye section is under compression instead of tension. The maximum compressive stress normal to the inner surface of the hole (pqr) is not at the centerline, but is located about 45° downward because of deflection of the eye. Consequently, primitive design for taking the tensile load on the eye section and tacitly assuming $K_t = 1$ would be inadequate. This matter is further discussed in the part of Section 3.4.1 on Lugs and Clevises; and Fig. 3.150 provides stress concentration factors for various profiles.

With reference to the effect of three-dimensional profiles of the whole part, the connecting rod of Fig. 3.42 is typical of the improvement possible in fatigue life. Note particularly that the changes between the old and new designs are not

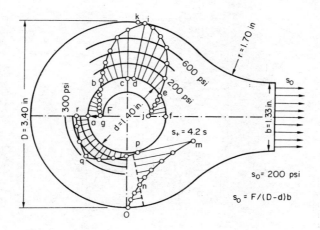

Fig. 3.41   Stress distribution in an eye-bar.
(From Ref. 3.60).

drastic in the layman's sense but that they are always in the direction of a
larger radius blend between sections, to reduce the discontinuity in the stress
flow. Crankshafts are also fine examples of the effect of shape on service life.
Figure 3.43 shows several of the recognizable stages of evolution: straight-
through boring of the pins and journals for roughly double the life of the initial,
primitive shape; finishing the ends of the bore in a barrel shape and increasing
the width of the web for another doubling of life; and finally, adding two stages
of profile adjustment and stress-relieving grooves. Such practice raised the
fatigue limit stress from 6200 psi to 22,000 psi for a production shaft in nodular
iron. It should be noted that some of the latter shapes are not possible by
classical forging, the time-honored method, and probably not economic by any
variant of forging, but that these improvements become possible both technically
and economically by casting. Although the method is essentially unlimited as to
shape, considerable metallurgical effort was required, along with production
control, to produce the required soundness. Casting also makes feasible such
other details as internal bosses (locally thickened sections) to reduce the stress
concentration factors associated with the oil holes; further consideration is
illustrated by Fig. 3.44.

The effect of external geometry on the fatigue behavior of welded joints has been
rather extensively explored (317); particularly, the stress concentration factors
at the transverse transition of the bead and parent metal are discussed in
Sec. 3.4.1. Correct and incorrect profiles for deep, stepped holes are shown in
Fig. 3.45.

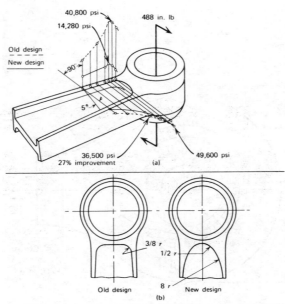

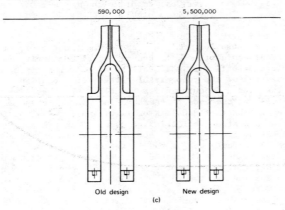

Life in cycles		
25,000 lb compression + 12,500 lb tension	728,000	over 10,000,000
33,200 lb compression + 16,600 lb tension	146,000	400,000

Fig. 3.42 Connecting rod profiles. (a) Effect of torque load on stress at eye end; (b) design change of eye end to avoid stress concentration due to torque; (c) V and circular crotch forks. From Climax Molybdenum Co. (brochure), New York, 1962.

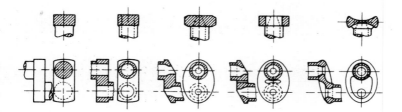

Fig. 3.43 Evolution of the section profiles of a crankshaft.
From R. Cazaud [386], Fig. 213.

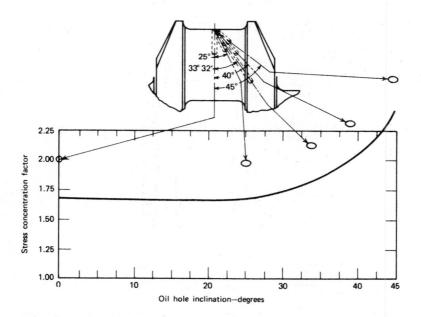

Fig. 3.44 Effect of orientation of holes on stress concentration
factor. From Climax Molybdenum Co. (brochure), New York,
1961.

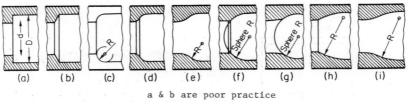

a & b are poor practice
c - i are good practice

Fig. 3.45 Shapes for stepped holes in shafts.

Very acute intersections, as shown in Fig. 3.46, also have a highly detrimental
effect on fatigue strength. Such edges may arise by inadvertent accumulation of
other angles or by the fabrication process, particularly shearing. Fatigue data
from such samples indicate a comparatively great reduction of, and increased scatter
in mean life.

Designing with Stress Concentration Factors. Perhaps foremost in the designers'
mind as he approaches a task involving stress concentrations should be to accept
the fact that they exist and may be ignored only at his peril. Once this condition is
recognized, two approaches seem possible:

1.   The semianalytical, wherein the equations of Section 3.2.2 and 4.2.2,
     together with an empirical tensile strength, may be used to calculate
     a notch fatigue strength.

2.   The wholly empirical, wherein the results are taken directly from an
     S-N or constant life plot, the data for which were, hopefully, derived
     under the same conditions as those of the problem.

The first method is limited essentially to the steels because of the relatively
low scatter in their fatigue data, but for most of the other common alloys,
particularly the aluminums, the degree of scatter makes it rather imprecise. The
use of published data and curves, especially as given in (1) is certainly the
most direct and popular method, but one very serious condition is that the degree
of identicality between the test conditions and those of the problem must be
known. At least the following must be accounted for:

1.   Alloy composition, treatment, and surface finish,

2.   Part geometry and size,

3.   Loading geometry,

4.   Ambient conditions of temperature, and corrosion, and there is no
     limit to the degree of departure that the designer might make.

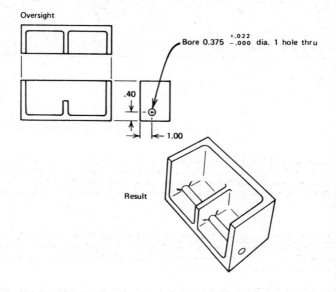

Fig. 3.46 Feather edges as stress concentrations. In drafting
drafting, the hole was shown in the end view only;
other details were omitted. In service, fatigue
cracks originated at the feather edges. From H. J.
Grover, S. A. Gordon, and L. R. Jackson [9], by permission of the Naval Air Systems Command.

Some inherent discrepancies among parameters are known to exist, that is, many
parts are triaxially stressed in use, but most fatigue data is biaxial at best. Many
S-N and constant life diagrams exist for both the smooth and notched conditions; the
major problem is to find the nominal stress, and then the stress concentration
factor.

As noted earlier, this volume makes no attempt to include stress analysis as a topic;
coverage is given in the problems below only for coherency, and some of the
shapes were chosen with the simplicity of analysis in mind.

Example 1. The simple little bracket in Fig. 3.47 illustrates that situation so
often met by the fatigue designer: after the shape, size, material, loads, and
method of fabrication are fixed, then the question is asked, "Will it work, or how
long will it last?" In this analysis, the effect of the residual stress from the
forming operation and of varying the natural frequency are also considered.

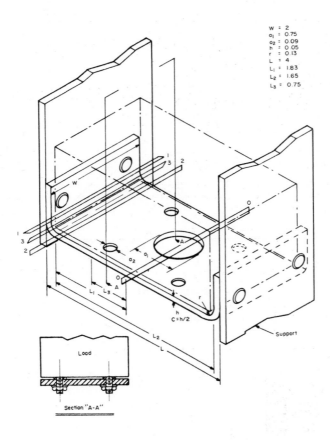

W = 2
a_1 = 0.75
a_2 = 0.09
h = 0.05
r = 0.13
L = 4
L_1 = 1.83
L_2 = 1.65
L_3 = 0.75

Fig. 3.47 A simple bracket.

The bracket is formed of 2024-T3 aluminum alloy by a conventional brake operation: cold bending with no subsequent stress relief. The load consists of a boxed, solid state switch weighing 6 lbs. secured to the bracket by studs through four small holes. The nuts are torqued sufficiently to prevent relative motion; the load does not strengthen or stiffen the bracket.

In application, the bracket and load are part of an airborne electronic chassis. The chief source of vibratory loading is the residual unbalance in a 12,000 rpm rotor; flight test data indicate that this will not exceed 1.3g at the bracket supports. Maneuvering loads up to 3g may be expected. The design life of the chassis is 1000 flights of 5.0 hours duration, plus the time for testing, run-up, idling, and so forth. The maximum design temperature for this equipment is 130°F.

Static Stresses. From a first-round visual examination it appears that the critical sections are 0-0, 1-1 and 2-2. It turns out that the ultimately critical section is not one of these, but rather section 3-3, whose analysis is given by Eq. 40, below. However, to continue in the conventional design procedure, consider section 0-0 first: the nominal static beam stress based on the net section is given by Eq. 33 below taken from [93], Fig. 86:

$$S_{0nom} = \frac{6M}{(w - a_1)h^2} \tag{33}$$

Where M = the maximum moment
       W = the beam or sheet width
       a_1 = the hole diameter
       h = the beam or sheet thickness

The determination of the moment, M, becomes a problem of finding the proper length of span, a consideration forced by the presence of the curved ends.

Assuming the conditions and geometry as in Fig. 3.48, we can show that the maximum moment, M_0max, for the general case, in terms of the two important parameters, r/L and L_3/L_1, is:

$$M = \frac{PL_1[\tfrac{1}{2}(1 - L_3/L_1)^2 + \tfrac{\pi}{2}(1 - L_3/L_1)r/L_1 + (r/L_1)^2]}{1 + \tfrac{\pi}{2}(r/L)}$$

$$= \frac{3(1.83)[\tfrac{1}{2}(0.59)^2 + \tfrac{\pi}{2}(0.59)(0.068) + (0.068)^2]}{1 + \tfrac{\pi}{2}(0.068)}$$

$$= 1.2 \text{ in. lb.} \tag{34}$$

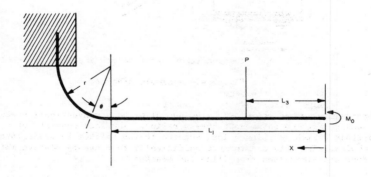

Fig. 3.48 Nomenclature for curved beam.

Note that the load P in Eq. 34 is one-half the total load. Now, from Eq. 33

$$S_{0nom} = \frac{6(1.2)}{(2 - 0.75)(0.05)^2} = 2300 \text{ psi} \qquad (35)$$

To account for the stress concentration due to the holes, the section at the large hole only needs to be treated because, by inspection, S_{nom} is much larger here than at a section through the small holes for the term $(W - a_1)$ is much smaller than $(W - 2a_2)$, that is, the minimum net section occurs at 0-0. So, from [93], Fig. 86, bending of a finite width plate with a transverse hole, $K_t = 1.4$, $(a/h = 0.75/0.05 = 15$, $a/w = 0.75/2 = 0.375)$; thus, at section 0-0,

$$S_{max} = K_t S_{nom} = 1.4(2300) = 3220 \text{ psi}$$

We now consider section 1-1:

$$M_{1max} = M_{0max} - P(L_1 - L_3 + r)$$
$$= 1.2 - 3(1.83 - 0.76 + 0.13)$$
$$= 2.4 \text{ in. lb.} \qquad (36)$$

and the nominal stress at 1-1, disregarding the sign, is:

$$S_{1nom} = \frac{M_{1max}C}{I}$$
$$= \frac{24(0.05/2)}{2 \times 10^{-5}}$$
$$= 3000 \text{ psi} \qquad (37)$$

The static tensile stress here is only: $P/A = 3(2 \times 0.050) = 120 \text{ lb/in.}^2$ and is neglected. At Section 2.2,

$$M_{2max} = M_0 - P(L_1 - L_3)$$
$$= 1.2 - 3(1.075)$$
$$= 0.375 \text{ in. lb.} \qquad (38)$$

and

$$S_{2nom} = \frac{M_{2max}C}{I}$$
$$= 2550 \text{ psi} \qquad (39)$$

In the region of the bend, the conditions that act to increase the nominal stresses are:

1. Curvature of the beam
2. Residual stress due to permanent deformation caused by the bending operation.

At sections 1-1 and 2-2, the curvature is theoretically zero and neither condition applies. However, it is reasonable to consider that the maximum stress in the bend region is that due to some maximum moment plus the effects of (1) and (2) above, that is, S_{max} will occur at a section very close to section 1-1, where θ approaches 90°, designated as 3-3. In the form of an equation,

$$S_{max} = \frac{Mc}{I} \qquad (40)$$

where $M = M_{0max} - P(L_1 - L_3 + r \sin \theta)$

For all practical purposes, $\sin \theta$ can be set equal to one; then Eq. 40 becomes identical with Eq. 36 and,

$$M_{3max} = M_1 = 2.4 \text{ in. lb.}$$

then

$$S_{3nom} = S_{1nom} = 3000 \text{ psi}$$

The mode of loading and the curvature of the beam act to increase the nominal stress by a factor K_t, which can be determined from Fig. 3.49. These curves have been calculated from the following equation:

$$K_t = 1 + B\left(\frac{c}{96}\right)\left[\frac{1}{(r-c)} + \frac{1}{r}\right] \tag{41}$$

rewritten to this more convenient form for rectangular cross sections from [93, Eq. 86], and in which B is a constant equal to 0.5. Then, for an r/c value of $0.125/0.025 = 5$ and a thickness h = 0.05, $K_t = 1.85$, to be applied to S_{3nom}.

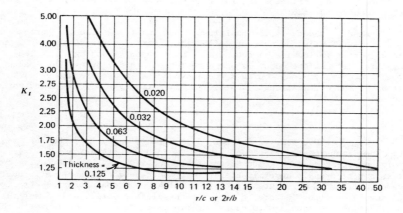

Fig. 3.49   Stress concentration in curved sheet subject to bending. $K_t = 1.00 + 0.5 \ (I/bc^2)[1/(r - c) + (1/r)]$; K_t = stress concentration factor; b = width of sheet; c = half-thickness of sheet; r = radius of curvature; and I = moment of inertia.

The residual stress S_r from the plastic strain during bending can be shown to be a function of yield strength, the modulus, and r/c ratio, never to exceed half the yield. Its value is given by:

$$S_r = \frac{S_y}{2}\left[1 - \left(\frac{S_y}{E}\right)^2\left(\frac{r}{c}\right)^2\right] \tag{42}$$

and since for most structural alloys $(S_y/E)^2$ is very small (<3%), Eq. 42 may be reduced to:

$$S_r = \frac{S_y}{2} \qquad (42a)$$

or for 2024-T3, $S_r = 48,000/2 = 24,000$ psi. Now collecting the static stresses at the critical section 3-3, there results:

$$S_3 = (S_{3nom})K_t + S_{3r}$$

$$= (3000)1.85 + 24,000$$

$$= 29,500 \text{ psi, static} \qquad (43)$$

The maneuvering load is considered as a multiplier on the S_{3nom} term only; from Eq. 43:

$$S_{3max} = 3(3000)1.85 + 24,000$$

$$= 40,600 \text{ psi}$$

By comparison with the values of S_{tu} of 65,000 psi and S_{ty} of 48,000 psi, a comfortable margin of safety appears. Even if the stock for the bracket were taken in the transverse direction of the rolled sheet, the values of 64,000 and 42,000 psi still compare favorably.

Dynamic Stresses. Under static conditions, the symmetry of the load and the supports permitted only one mode of deflection, that of vertical bending. Under dynamic or vibratory loading, a second mode will appear, rocking about some longitudinal axis, because the mount is not a center-of-gravity type. The loading and geometrical conditions under which a coupled mode may arise are discussed in [39]. The stress on the critical section 3-3, resulting from the combined bending and torsion is:

$$S = \sqrt{S_b^2 + S_s^2}$$

where $S_b = S_{3nom} = 3000$ psi, from above; $S_s = 3T/Wh^2$, where T is the applied torque. The dimensions of Fig. 3.47 give $S_s = 3600$ psi, then S = 4700 psi. The value of the stress concentration factor to be applied to this alternating component is not rigorously determinable, but it would seem conservative to apply the same value as for the curvature effect in bending. The concentration effects on the rectangular section in torsion should be negligible until, and if sufficient torsional deflection develops to cause the flange to lose parallelism with the support, concentrating the force on one side of one rivet. Thus, the actual stress is taken as $K_t S = 1.85 (4700) = 8700$ psi.

Consideration of the dynamic stress includes the vibration input amplitude and frequency, the transmissibilities from the rotor to the bracket and of the bracket itself, as well as its natural frequency. Starting with the 8700 psi as the one g, live-load stress, one must determine the vibratory stress levels and the number of cycles at these levels. The input vibration level from the residual unbalance of the rotor is given by the parabolic curve in Fig. 3.50a as seen at the bracket supports. The transmissibility of the bracket proper is the curve peaking at a resonant frequency of 1100 cps, best obtained directly by test. If such experimental results are not available, a "standard" resonance curve may be assumed, with a maximum T value of about 10 to 12, with the resonant frequency estimated from acoustic tests or by calculation from the static deflection.

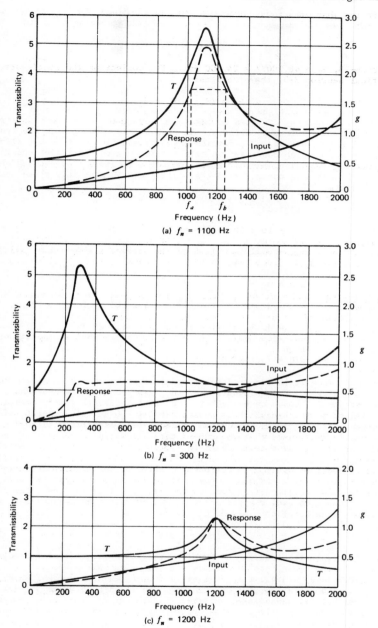

Fig. 3.50 Bracket design curves.

The response-g loading on the bracket is then the product, T x (input g), shown by the dashed curve. The stress at any frequency is this loading times the one-g stress, or:

$$(S_{act}) = S(loading)_f$$

Thus, at say 800 Hz,

$$S_{act} = 8700(0.78) = 6800 \text{ psi}$$

Now, to determine the fatigue life, one needs to compare the number of cycles, n, in a (narrow) bandwidth, centered on any frequency at the corresponding stress with the number of cycles, N, allowable at that stress, summed over the frequency spectrum. The time dimension is furnished by the duration of the several periods of rotor operation which constitute a flight. Suppose that a flight consists of a 10-min. taxiing and idling period before and after each 300-min. cruise, and that service and maintenance accumulate 3 hrs. idling and 30 min. full speed operation every 50 flights.

Consider the several periods of operation:

Taxi. $n = 10 \times 60 \times \frac{2000}{60} = 2 \times 10^5$ cycles

$S_{3a} = 0.21 (8700)$, from Fig. 3.56a, = 1800 psi at 333 Hz
N, (from (Fig. 4.66) = ∞ cycles

$n/N = 0$

Cruise. $n = 300 \times 60 \times \frac{12000}{60} = 3.6 \times 10^6$ cycles

$S_{3a} = 1.15 (8700) = 11,000$ psi at 2000 Hz
N, (from Fig. 4.66) = 5×10^8 cycles by extrapolation;
then
$n/N = (3.6 \times 10^6)/(5 \times 10^8) = 0.0072$

Run-up. Since it is required to integrate for the number of cycles at each frequency and stress, and the equation of the response curve is not known, the numerical method is chosen as being both practical and sufficiently accurate. A convenient incremental bandwidth is chosen here as 200 Hz. In general, of course, the accuracy increases as the bandwidth is decreased, especially in the regions of high slope. It is assumed that the run-up rate allows full response of the bracket to be developed, and the rate is taken for simplicity as 200 Hz/sec. See Table 3.9.

Since run-up and run-down each occur once per flight, the life fraction due to this exposure is ~0.006. As usual, the great majority of the life fraction accrues at resonance (1100 Hz); the treatment in this region, that is, the calculation of the time or number of cycles applied during the sweep may be conservatively approximated, as was done here; or n may be determined from:

$$n = \frac{f_b - f_a}{0.69R}$$

where f_b and f_a are the frequencies bounding the resonance band at the half-power point and R is the sweep rate in octaves per sec. The half-power point is 0.7 (peak T) = 0.7 (4.9) = 3.43; then from Fig.3.50a, f_a is seen to be 1020 Hz and f_b is 1240 Hz. In a vibration test on a shaker, the sweep rate would be known or measured; here it is assumed as 2 oct/sec. Then, from the expression above, n = 164 cycles, and the life fraction for run-up and run-down is 0.005 instead of the 0.006 above.

TABLE 3.9

f_1	f_2	n	S_{3a}	N	n/N
0	200	200	750	∞	0
200	400	200	1,800	∞	0
400	600	200	3,700	∞	0
600	800	200	6,700	1×10^8	0.000002
800	1,000	200	14,100	5×10^7	0.000010
1,000	1,100	200	21,000	7×10^5	0.002860
1,100	1,200	200	16,000	1×10^7	0.000020
1,200	1,400	200	10,000	1×10^8	0.000020
1,400	1,600	200	10,000	1×10^8	0.000020
1,600	1,800	200	10,000	1×10^8	0.000020
1,800	2,000	200	10,000	1×10^8	0.000020

$$\Sigma \, n/N = \overline{0.002972} \simeq 0.003$$

Service.

$$\text{Idling:} \quad n = 3 \times 60 \times 60 \times \frac{2000}{60} = 3.6 \times 10^5$$

$$S_{3a} = 0.2 \, (8700) = 1740 \text{ psi at 333 hz}$$
$$N, \quad (\text{from Fig. 4.66}) = \infty$$

$$n/N = 0$$

$$\text{Full-speed:} \quad n = 30 \times 60 \times \frac{12,000}{60} = 3.6 \times 10^5$$

$$S_{3a} = 0.9 \, (8700) = 7830 \text{ psi at 2000 hz}$$
$$N, \quad (\text{from Fig. 4.66}) = 5 \times 10^8 \text{ cycles}$$

by extrapolation; then

$$n/N = (0.0036 \times 10^8)(5 \times 10^8) = 0.00072$$

plus an estimated number of run-ups, say 10, which would add 10 (0.006) = 0.06. Thus, the total life fraction, Σ_T consumed per flight, for the bracket resonant at 1100 hz, is:

$$\Sigma_T = \Sigma \left(\frac{n}{N}\right)_{\text{taxi}} + \Sigma \left(\frac{n}{N}\right)_{\text{cruise}} + \Sigma \left(\frac{n}{N}\right)_{\text{run-up}} + \Sigma \left(\frac{n}{N}\right)_{\text{service}}$$

$$= 0 + 0.0072 + 0.006 + \frac{0.00072 + 0.06}{50}$$

$$= 0.0144$$

a value that permits only about 70 flights to fracture of the bracket, obviously an untenable design.

In an attempt to increase the bracket life, it was redesigned, retaining the same proportions, such that is now was resonant at 300 cps; see Fig. 3.50b. The resonant frequency was deliberately lowered to detune the bracket further from the frequencies of high input. The one-g, live-load stress at 300 cps would be:

$$\frac{S_2}{S_1} = \sqrt{\frac{f_1}{f_2}}$$

$$S_2 = 8700 \sqrt{\frac{1100}{300}} = 16,800 \text{ psi}$$

The summation for this new condition follows:

Taxi.

$$n = 10 \times 60 \times \frac{2000}{60} = 2 \times 10^5 \text{ cycles}$$

$$S_{3a} = 0.65 \ (16{,}800), \text{ from Fig. 3.50b} = 10{,}400 \text{ psi}$$

(No additional term for time at resonance is put on here as was done in the 1100 hz run-up calculations, because the response curve shows no indication of following the transmissibility curve, that is, the bracket is thoroughly detuned)

$$N, \text{ from Fig. 4.66} = 5 \times 10^8$$
$$n/N = (0.002 \times 10^8)(1 \times 10^8) = 0.0004$$

Cruise.

$$n = 300 \times 60 \times \frac{12{,}000}{60} = 3.6 \times 10^6$$

$$S_{3a} = 0.85 (16{,}800) = 14{,}200 \text{ psi}$$
$$N, \text{ from Fig. 4.66,} = 2 \times 10^8$$

$$\frac{n}{N} = \frac{(0.036 \times 10^8)}{(2 \times 10^8)} = 0.018$$

The summation need not be carried further, for the results already indicate a bracket life of only some 50 flights, a comparatively poorer design than the first.

These results lead to the conclusion that this is not a proper design for the bracket; deflection and transmissibility are too high, leading to excessive stress levels from the fatigue viewpoint, although these stresses were acceptable statistically and dynamically (at one cycle). A second design change by adding flanges along the span greatly reduced deflection, transmissibility, and stresses, and yielded a resonance at 1200 hz. Although the response curve did reach the full level of the transmissibility curve, Fig. 3.50c, the total life fraction was essentially zero. The difference in action of the three bracket designs is particularly well illustrated by two comparisons:

1.   Although the response of the 300 hz bracket is much lower than that of the 1100 hz bracket, the former still did not yield an acceptable fatigue life, due to high deflection and stress.

2.   Although the responses of both the 300 hz and 1200 hz brackets were relatively low, and although the latter had the higher peak response, it provided a useful fatigue life because of a higher natural frequency, lower deflection, and lower stress.

The residual stress of 24,000 psi is considered fairly high for this alloy, but as may be seen from the slopes of the constant life lines in Fig. 4.23, this value turned out not to be predominant in the design. For shorter lives at higher alternating stresses, the constant lifelines take on greater slopes, increasing the importance of the mean or residual stress.

Example 2   Determination of Lifetime under Repeated Axial Loading. Consider the piston shown in Fig. 3.51 as part of a hydraulic system. The mode of loading is derived from the pressure-time curve for this loop of the system and, with some simplification, can be stated as changing from a compression value to a tension value once each pressure cycle, that is, R is negative. The loads are such as to produce a nominal compression of 25 ksi and tension of 50 ksi in the stem. The static or holding pressure stresses the stem to a nominal mean value of 12.5 ksi in tension, and the working pressure range provides a stress amplitude of 37.5 ksi. ([50-(-25)]/2 = 37.5), and (37.5 - 25 = 12.5).

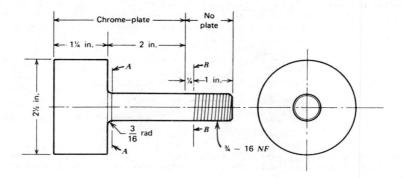

Fig. 3.51 A hydraulic piston. Material: AISI 4140; S_u = 150 ksi.
Loading: Axial, cyclic; 50 ksi tension; 25 ksi compression;
R = -0.50.

Critical regions to be examined are readily chosen as the fillet section A–A and
the end of the threaded section B–B. The critical mode of loading is taken as the
axial tension. The stress concentration factor K_t at A–A is calculated by considering the piston as a T-head bolt, and using Fig. 108 of Ref. 93:

$$r/d = .25; \qquad m/d = 2.0; \qquad 1d/m^2 = .208$$

Since .208 is less than 1/3 (the dividing line in Fig. 108) use K_{tA}. Then D/d = 3.25
and K_{tA} = 3.3. Now the fatigue notch factor is required. Ref. 93 gives the notch
sensitivity for this steel as 0.90, leading to:

$$K_f = 1 + q(K_t - 1)$$

$$= 1 + .9(3.3 - 1) = 2.07$$

If we start with standard S–N data for the given material (i.e., axial load,
non-plated, K_t = 1) as curve 1 in Fig. 3.52, the effect of plating is accounted
for by reducing all ordinates by about 35%, resulting in curve 2. This value was
derived from data on this particular alloy, 4140 steel heat-treated to 150 ksi
ultimate strength, and chrome-plated with no specific bake-out to eliminate the
effects of hydrogen embrittlement. Effects of plating on other steels are given
in Section 4.4.1. Now the effect of notches and mean stress are presented in the
constant-life diagram, Fig. 3.53. From curve 2 of Fig. 3.35 (unnotched, plated
bars) the S–N data are plotted directly for each level of mean stress. Then the
notch effect is entered by dividing these ordinates by K_t = 2.07, resulting in
another set of lifelines, designated by the dashed lines. The mean and alternating
stresses define a point X, whose location at about 5 x 10^4 cycles indicates a
relatively short life. This general procedure applies the stress concentration
factor to the alternating stress component only, static fatigue not being considered

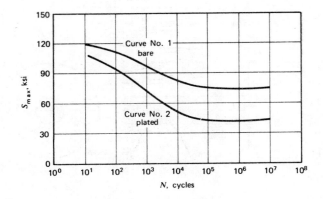

Fig. 3.52 S-N curves for bare and chrome-plated 4140 steel.
$K_t = 1.0$; $R = -1.0$.

as a significant factor in this type of part and loading. If a numerical factor
of safety is to be used, it should be applied directly to the value of K_f and will
act to adjust the ordinates of the dashed lines in the constant life diagram. As
the standard S-N curves are considered to represent a failure probability of 50%,
a minimum factor of safety of 2 is recommended.

A similar calculation must be made at the end of the threaded section B-B. The
relevant thread dimensions are: major diameter 0.750 in., minor diameter 0.673 in.,
and the theoretical flat at the root is 0.0078 in. If we assume the best machining
practice for cut threads, the root radius cannot be over 0.0039 in. Then, from
Fig. 29 of Ref. 93, and the ratios:

$$r/d = .0039/.673 = .0058$$
$$d/D = .673/.750 = .893$$

K_t is derived as 5.3, and $K_f = .9(5.3 - 1) = 3.88$. Now the constant lifelines,
Fig. 3.54, for this condition are established, as above, by dividing the ordinates
of the standard curves by 3.74. For the same mean and alternating stresses,
Point X, the part will now have a lifetime of less than 10^3 cycles, quite apparently the controlling and intolerable condition.

Examination of such parts having more than one SCF is highly informative as to a
means of equalizing the probabilities of failure arising from each notch, or as
modifications of the design details to obtain roughly equal lifetimes at each
notch. Study of the K_t plots for thread roots indicates that to obtain a K_t value
equal to that at the first fillet, the minimum radius at the root would have to be
about 0.053 in. This requirement immediately leads to a question as to the validity
of this design, a _threaded_ stem, for this piston under the given loading cycle.
While it may be possible to obtain satisfactory fatigue life at this section by

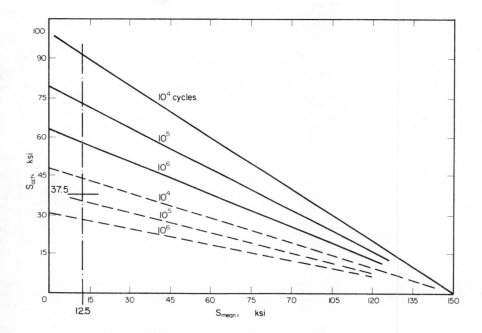

Fig. 3.53 Constant-life plot for section A-A of hydraulic piston.
Full lines from curve 2 of Fig. 3.52; Dashed lines with
a K_f of 3.08 applied.

increasing the diameter of the threaded portion and the use of rolled threads, the
designer may be forced to consider other means of attachment, as possibly, locked
pins or wedges. This example emphasizes that conventional design and fabrication
processes, however high their quality, may not result in parts possessing adequate
lifetimes. It may also be observed that Fig. 3.51, as drawn, exhibits some of the
shortcomings in drafting practice described in Section 3.3, especially as to the
omission of the detail on the thread run-out.

The Fatigue Quality Index. A means of measuring the quality or excellence of the
design with regard to its life has been established as the Fatigue Quality Index,
which is basically the stress concentration factor, K_t, with certain limits on its
application. Fig. 3.55 shows S-N curves collected from tests of variously notched
specimens; the dotted curve represents that interpolated K_t value which just
makes $\Sigma(n/N) \simeq 1.0$. Then, for a required life, the design stress limit is defined,
and often expressed as:

$$S_a = (EC/N) + S_e$$

OFD - E*

(see Sec. 4.1 for deriviation, and Sec. 3.4.5 for similar forms). A replot of
these parameters, as in Fig. 3.56 outlines the permissible design region for a
given load spectrum and immediately indicates the penalty of departing from the
design point. The designer's chief task then is to insure that the geometry of
the details adheres to this value of K_t.

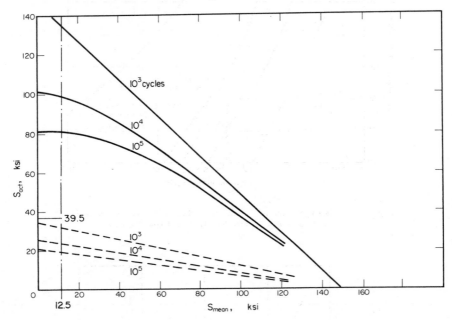

Fig. 3.54 Constant-life plot for section B-B of hydraulic piston.
Full lines from curve 1 of Fig. 3.52; Dashed lines with
a K_f of 4.2 applied.

The Index varies with the load spectrum, generally increasing with decreasing
slope of the spectrum. For unnotched 7075-T6 Alclad sheet, a slope decrease from
3 to 2 yielded little change in K_t, about 1.9 to 2.1, but in riveted butt joints a
decrease 2.5 to 1 let K_t go from 3.7 to 4.3; for 2024-T3 joints, a decrease of
3 to 1 let K_t go from 4 to about 7, for the same life in each case.

The use of the Index, or common value of K_t, for all details is being more highly
regarded, particularly from the viewpoints of of consistency and economy: once
the value is set for the assembly there is no benefit from improving the quality
(lower K_t) of one or a few items, and the upper limit of K_t (lower quality) describes
the quality level to be achieved by all items; otherwise the assembly is not built
to specification. Statistical treatment of the results of tests on a large number
of aircraft components indicates a peak of the (somewhat skewed) distribution
curve at $K_t = 3.5$, with a range of about 1.5 to 9.

Langer has also successfully applied this approach to heavy pressure vessels; see
Section 3.4.3.

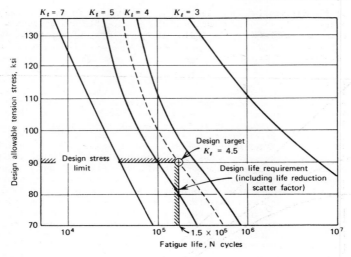

Fig. 3.55   Design Stress level vs. life and fatigue quality
of the part.   From W. J. Crichlow [312].

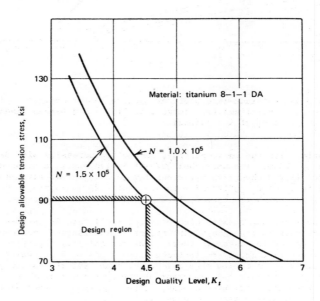

Fig. 3.56   Design stress level vs. detail quality level, K_t.
From W. J. Crichlow [312, Figs. 28, 28].

3.23 CONFIDENCE AND STATISTICS

The numerical reliability of structures and machines is being handled somewhat gingerly, a major reason being that, while the extreme-value (worst case) approach is being more widely recognized as often producing inefficient and overweight structures, it also enjoys a significantly high confidence. Many designers, as well as their supervisors and customers, seem reluctant to give up the tried-and-true approach of using factors of safety for the unfamiliar dangers of the statistical method. But it is becoming increasingly apparent that designing for the worst-case combination of extreme values, however improbable, cannot be tolerated if any claim of minimized weight is to be made. The inconsistency with the truth of not admitting to any probability of failure (P = 0 is a highly special case in the general theory), together with the rising use of statistics and reliability in the electrical design disciplines, has led to re-examination of the traditional approach. Two conditions have acted to restrain the universal application of statistical methods to structural design: (1) a general unfamiliarity with these methods among mechanical and structural designers, and (2) a lack of data in statistical terms such as tabulated standard deviations of strengths and dimensions. While a number of organizations such as the National Aeronautics and Space Administration (NASA), the Air Force, and the Aerospace Industries Association are understandably concerned over both conditions, and progressive actions are being promoted, other than elementary results have not yet been achieved. From the sparse data available, a few typical values are given later in this section.

Confidence in Design. The probabilistic approach to design is based upon the distributions of the applied and failing stress indices, as obtained from a limited sampling of the populations. The only way by which a distribution could be completely and positively known would be to test the entire population, but this is hardly practical and would defeat the need for satistical data. Values for design are based upon sampling, and there is always the finite probability that a sampling would result in an optimistic representation of the actual population. To obtain a high degree of confidence, it may occasionally be necessary to reduce the experimentally determined estimates of the mean and standard deviation values. The adequacy of the design is evidenced by the resulting strength of the specimens built to the design specifications. A failing strength equal to the strength predicted by analysis would indicate a confidence of 0.50 that the structure was designed with the correct failure distribution and resulting probability of failure. A higher or lower strength would indicate correspondingly higher or lower confidence in the failure population utilized in the design. If confidence in the accuracy of the failure population is not deemed sufficiently adequate, then either the structure must be redesigned with a more conservative estimate of the failure population that has the required confidence, or the present design must be acceptable with the probability of failure that would result from this more conservative estimate.

Tests are used to increase knowledge of the population or of a subset characterized by the design. Increased confidence in design values can be obtained by increasing the number of tests (n), although the effect becomes smaller for large n, or by conservatively estimating the design values. Design values in [1] are usually presented with a high degree of confidence; the A and B values represent a 0.95 confidence that these values will be exceeded 99 and 90% of the time, respectively.

Example 3

Consider the design and testing of a fatigue-resistant structure utilizing an arbitrary cumulative damage theory. The following was ascertained:

1.   Analysis of experimental data on similar structures, together with the cumulative damage theory, resulted in an estimate of the failure index with a mean value $\bar{D}_f = 0.800$ and a coefficient of variability $\gamma_f = 0.10$, values used in designing the structure.

2.   The loading history and variability of the material and elemental fatigue data indicated a coefficient of variability of $\gamma_a = 0.05$.

3.   The structure was designed for a limit stress level, which resulted in $\bar{D}_a = 0.490$ and the factor of safety, $r = 1.63$. This corresponded to the desired probability of failure of $\phi(u_f) = 0.0001$.

4.   Three specimens were fabricated and tested in fatigue.

5.   A representative loading history approximating a percentage (5%) of the total design life of the structure was repeated until the specimens failed. The failures corresponded to $R_i \bar{D}_a$, where $R_i = 1.55$, 1.65, and 1.75. This resulted in an average test factor of:

$$\bar{R} = \left(\frac{1}{n}\right) \Sigma R_i = 1.65$$

Now the following parameters are required, assuming that the coefficient of variability is satisfactorily defined from previous experience.

1.   With what confidence, can it be said that the structure has a probability of failure of (1), = 0.0001, (2), = 0.001, and (3), = 0.01?

2.   What should the value of test factor $\bar{R} = (1/n) \Sigma R_i$ have been in order to predict a probability of failure of 0.0001 with a confidence of a, = 0.75, b, = 0.90, and c, = 0.99?

3.   How would the confidence change if the number of specimens increased from 3 to a = 5 specimens, b = 9 specimens, and c = 12 specimens, while $\bar{R}$ remained at 1.65?

The results of the analysis of the test data are presented in Table 3.10; they indicate that very little can be done to increase the confidence in this design, as evidenced by the test results, unless one is prepared to strengthen the structure, reduce the loadings, or accept a higher probability of failure.

Table 3.10   Analysis of Test Data

Probability of Failure, $\phi(u_f)$	Failure Deviations, u_f	Mean Design Ratio, r	Mean Test Ratio, $\bar{R}$	Number of Specimens, n	Confidence Deviations, u_c	Minimum Confidence, C
Confidence for probability of failure, $[(\bar{R}/r) - 1]n^{1/2}/\gamma_f$						
0.0001	−3.715	1.63	1.65	3	0.213	0.584
0.001	−3.090	1.48	1.65	3	1.989	0.9766
0.01	−2.326	1.33	1.65	3	4.167	0.99998
Scatter factor for given confidence, $r[(u_c\gamma_f/n^{1/2}) + 1]$						
0.0001	−3.715	1.63	1.650	3	0.213	0.584
0.0001	−3.715	1.63	1.694	3	0.675	0.75
0.0001	−3.715	1.63	1.751	3	1.282	0.90
0.0001	−3.715	1.63	1.849	3	2.326	0.99
0.0001	−3.715	1.63	1.63	3	0	0.50
Effect of number of specimens on confidence, $[(\bar{R}/r) - 1]n^{1/2}/\gamma_f$						
0.0001	−3.715	1.63	1.65	3	0.213	0.584
0.0001	−3.715	1.63	1.65	6	0.3048	0.618
0.0001	−3.715	1.63	1.65	9	0.3681	0.644
0.0001	−3.715	1.63	1.65	12	0.4250	0.665

SOURCE: H. Switzky [62].

The following points can now be made.

1.   Statistical techniques can be employed to convert a structural reliability requirement to an equivalent factor of safety problem. Conversely a given design or test data can be analyzed for confidence.

2.   Statistical techniques offer logical procedures for extrapolating to the limits of available test data. Empirical relationships can be assumed and the available data analyzed to obtain satisfactory approximations to the solution.

3.   The designer should be cognizant of the design philosophies and capable of translating them into a logical design procedure to secure a desired structural reliability, while retaining recognition of the limitations of these procedures.

Structural Reliability. "Part-failure-rate" provides the fundamental term with which electronic system reliability is expressed but failure rate is not applicable to structural systems. There is no direct analog to "part" nor, hence, to "parts-failure-rates" in structures; but for sub-and component-assemblies information covering these dimensions is beginning to appear [304-308]. Small pressure tanks of Al or Ti are quoted as having failure rates of 0.05 to 0.50 failures per million hours, or 0.0004 per pressurization cycle. It is possible to specify numerical reliability for some simple shapes as a hemispherical tank shell (a value of 0.9999 has been noted), but general part failure rates cannot be available for custom structures and their individual elements. The general difficulties and restrictions arising from the lack of completely adequate data and methods of calculation were discussed in Section 2.5; the examples below should be considered with these restrictions in mind.

Investigation of the reliability of a structure requires study of the relationship between structural strengths and system loads--reliability being defined here as the probability that strength is greater than stress. This relationship is shown schematically in a Warner diagram, Fig. 3.57, where (M_S, σ_S) represents the composite strength function including the principal strength (stress) parameters, and (M_L, σ_L)

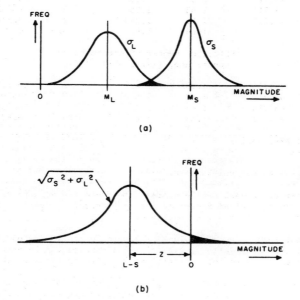

Fig. 3.57 Warner diagram. (Figs. 3.63-3.67 from
E. B. Haugen [64]).

is the combined stress function including the principal load parameters. The
symbols M and σ represent the two defining characteristics of the normal function:
the mean (M) and the standard (σ), as given by the following equations:

$$M = \frac{\sum\limits_{i=1}^{n}(m_i)}{n} = \frac{m_1 + m_2 + \cdots + m_n}{n}$$

$$\sigma = \frac{1}{n}\sqrt{\sum\limits_{i=1}^{n}(M - m_i)^2}$$

$$\sigma = \frac{\sqrt{(M - m_1)^2 + (M - m_2)^2 + \cdots + (M - m_n)^2}}{n}$$

where $m_1, m_2, \ldots, m_n$ represents the population of values in a normal distribution of
parameters. Each of the curves in Fig. 3.57a may be represented by an equation of
the form:

$$f(x) = \frac{1}{\sqrt{2\pi}\sigma} \cdot \exp\left[-\frac{(x - M)^2}{2\sigma^2}\right]$$

The function is symmetrical about M and extends mathematically to ±∞. Thus,
regardless of strength and stress levels, an overlap area of strength and stress
curves will exist, however small. The area of overlap can be interpreted to
represent instances when when stress exceeds strength, that is, instances of
structural failure. By writing the difference function $(L - S, \sqrt{\sigma_L^2 + \sigma_S^2})$, the
incidence and/or probability of L>S may be calculated, with typical results as
shown in Fig. 3.57b. The "reliability," or probability, that strength is greater
than stress or load is the area under the curve to the left of zero, R is determined
by taking a number of standard deviations, z, from zero to the ordinate, z being
defined as:

$$z = \frac{|M_L - M_S|}{\sqrt{\sigma_L^2 + \sigma_S^2}}$$

The value of z, and thus of R, may be found in the standard tables of the normal
function. The numerical value of the reliability is sharply dependent on z, and
the number of z's is occasionally termed a safety index, thus taking on the connotation of a factor of safety. Here the designer is particularly cautioned against
any feeling of security by calculating R to many decimal places, at whatever
magnitude (as in the examples below) on an assumed z. The making of such an
assumption may be the only avenue of procedure open to the designer in the absence
of specific information on the value of z, (the correct number of σ's) for his
material and form in a particular condition of heat treat and loading, but the R
value so determined usually contains this element of assumption. A short table of
R for various values of z or nσ is given in Table 3.11. Three-σ is a value often

Table 3.11

z	R
0.000	0.5000
0.500	0.6915
1.000	0.8413
2.000	0.9772
2.327	0.9900
3.000	0.9987
3.080	0.9990
3.610	0.9999

used and, as noted, yields a usefully high reliability but not necessarily an adequate level.

The load function (M_L, σ_L) will in many instances be some combination of (1) axial, (2) moments, and (3) shear loads; and may be represented by:

$$M_{(L_1, L_2, \ldots, L_m)}, \qquad \sigma_{(L_1, L_2, \ldots L_m)}$$

For example, if

$$\begin{aligned}
\text{thrust} &= (T, \sigma_T) \\
\text{aerodynamic drag} &= (A, \sigma_A) \\
\text{dynamic pressure} &= (D, \sigma_D)
\end{aligned}$$

and

$$\text{force due to moments} = (M, \sigma_M),$$

the load function will be of the form

1.   Mean = algebraic sum of the component mean values, $M_L = T + A + D + M$ (in consistent units)

2.   Standard deviation, $(\sigma_L) = \sqrt{\sigma^2_T + \sigma^2_A + \sigma^2_D + \sigma^2_M}$

The strength function (M_S, σ_S) will commonly be a composite due to the inclusion of such strength parameters as allowable stress, structural geometry, section area, moment of inertia, modulus of elasticity, and fatigue and creep allowables, represented by:

$$M_{(S_1, S_2, \ldots, S_m)}, \qquad \sigma_{(S_1, S_2, \ldots, S_m)}$$

A simple example will show how strength parameters are combined to form the strength function.

$$\begin{aligned}
S &= \text{average stress (allowable), lb/in.}^2 \\
P &= \text{strength, lb, and} \\
A &= \text{area, in.}^2
\end{aligned}$$

where the material strength is described by (S, σ_S), the area by (A, σ_A), and the total strength by (P, σ_P). Then (P, σ_P) will equal:

$$P = S \cdot A \qquad \text{and} \qquad \sigma_P = \sqrt{S^2 \sigma^2_A + A^2 \sigma^2_S + \sigma^2_A \sigma^2_S}$$

and the strength function will be:

$$\begin{aligned}
(P, \sigma_P) &= (S \cdot A, \ \sqrt{S^2 \sigma^2_A + A^2 \sigma^2_S + \sigma^2_A \sigma^2_S}) \\
&= (M_S, \sigma_S)
\end{aligned}$$

The term $A^2 \cdot \sigma^2_S$ is sometimes negligible and can be dropped.

From the strength and stress distribution functions above, the difference function is written:

$$(|M_L - M_S|, \sqrt{\sigma_L{}^2 + \sigma_S{}^2})$$

which, with z expressed as above, leads to $R = f(z)$.

Some parameters are complex distributed functions that include more than one distributed parameter. It has been stated that moment of inertia is normally distributed because of variation of geometrical values about mean values. An example of such a statistical description is given by taking the moment of inertia about the z axis, Fig. 3.58, to obtain:

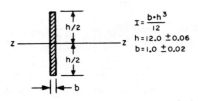

Fig. 3.58   Beam inertia.

$$3\sigma_h = 0.06, \quad \sigma_h = 0.02$$

$$3\sigma_b = 0.02, \quad \sigma_b = 0.0066$$

(Tolerances are assumed equal to 3σ.)   Then:

$$M_I = \frac{1 \cdot 12^3}{12} = 144 \text{ in.}^4$$

$$\sigma_{h \cdot b} = \sqrt{1^2 \cdot (0.02)^2 + 12^2 \cdot (0.006)^2} = 0.0816$$

$$\sigma_h^2 = \sqrt{2} \cdot h \cdot \sigma_h = 0.339$$

$$\sigma_I = \frac{1}{12} \cdot \sqrt{(144)^2 \cdot (0.0816)^2 + (12)^2 \cdot (0.339)^2}$$

$$= 1.062 \text{ in.}^4$$

$$(M_I, \sigma_I) = (144, 1.062) \text{ in.}^4$$

There will be instances when reliability requirements are the basis of design. Consider a problem in which reliability R is specified and a load (M_L, σ_L) is known. Then, since $z = f(R)$,

$$\sigma = M_S^2 - 2M_S M_L + M_L^2 - z^2(\sigma_L^2 + \sigma_S^2)$$

$$M_S = M_L - z \cdot \sqrt{\sigma_L^2 + \sigma_S^2}$$

and

$$\sigma_S = \sigma(S_1, S_2, \ldots, S_n)$$

Thus

$$M_S = f(\sigma_S^2)$$

Here $S_1, S_2, \ldots, S_n$ are the strength parameters, some of which may be known. For instance, if the material is known, M_S will be a function of tolerance on geometry. If material and geometry are known, modification of thermal environment may control M_S. Thus, what is sometimes an indeterminate problem may be resolved. It is evident that the control of tolerances will influence reliability and often permit reduction of weight.

As another example, consider a beam required to support a combined load (L, σ_L), where $L = T + W + D$, (T, W, and D are distributed load components),

$$\sigma_L = \sqrt{\sigma_T^2 + \sigma_W^2 + \sigma_D^2}$$

and this load produces the effective moment (M_L, σ_L). Taking the standard moment expression for a beam, $M = SI/c$, where:

> (S, σ_S) describes the material strength
> (I, σ_I) describes the moment of inertia
> (c, σ_c) describes the distance from the neutral axis to the extreme fiber

then we obtain

$$\sigma_{I/c} = \frac{I}{c} \sqrt{\frac{\sigma_c^2 I^2 - \sigma_I^2 c^2}{\sigma_I^2 + \sigma_c^2}}$$

and

$$\sigma_S = \sigma_{(I/c \cdot S)} = \sqrt{S^2 \cdot \sigma^2_{I/c} + I/c \cdot \sigma^2_S}$$

$$\sigma_S = \frac{I}{c} \sqrt{\frac{\sigma_c^2 I^2 S^2 - \sigma_I^2 c^2 I^2 + \sigma_S^2 c^2 I^2 + \sigma_c^2 S^2 I^2}{c^2 + \sigma_c^2}}$$

Nomograms for Factors of Safety

The outline below establishes a nomographic method of solving the frequent problem of determining the required factor of safety for a given probability of failure, or the inverse. The initial assumption is that both the applied load and the available strength of a structural element may be represented with sufficient accuracy by the normal distribution curve, as a Warner diagram: Fig. 3.57   The load L and the strength S are each specified by its mean m and its standard deviation σ. The factor of safety is defined by the ratio of the mean strength to the mean load, M_S/m_L, with failure occurring in the overlap region of the distributions where the strength is less than the load. The greater the scatter in the variates, the greater must be the factor of safety to assure a given reliability. The difference between the strength and the load (the difference between two normal functions) is the reserve R, also normally distributed and having the characteristics:

$$m_R = m_S - m_L \tag{44}$$

$$\sigma_R = \sqrt{\sigma_L^2 + \sigma_s^2} \tag{45}$$

$$f_R(x) = \frac{1}{\sigma_R \sqrt{2\pi}} \exp\left[-\frac{1}{2}\left(\frac{x - m_R}{\sigma_R}\right)\right]^2 \tag{46}$$

Failure occurs for negative values of R, so that the probability of failure is found by integration of the frequency function over the region for which R is negative

$$P = \int_{-\infty}^{0} f_R(x)dx \qquad (47)$$

Only the major steps in the integration are shown:

introduction of a new independent variable:

$$t = \frac{x - m_R}{\sigma_R}, \qquad (48)$$

and use of the notation:

$$k = \frac{m_R}{\sigma_R} \qquad (49)$$

together with the fact that $dt = dx/\sigma_R$ leads to the intermediate form:

$$P = \frac{1}{\sqrt{2\pi}} \int_{k}^{\infty} \exp\left(-\frac{t^2}{2}\right) dt \qquad (50)$$

The transformation of Eq. 50 by means of the property that the normal probability integral is an even function, and that, with $k = \infty$, its value is unity, results finally in

$$P = \frac{1}{2}\left[1 - \frac{1}{\sqrt{2\pi}} \int_{-k}^{+k} \exp\left(-\frac{t^2}{2}\right) dt\right] \qquad (51)$$

Numerical values of the integral are obtainable from tables, as in [91], and a few points are listed in Table 3.12 in the event that a P vs. k or P vs. k^2 plot is desirable.

TABLE 3.12

P	k	k^2
0.1	1.282	1.642
0.01	2.236	5.412
0.001	3.090	9.549
0.0001	3.719	13.830
0.00001	4.265	18.190
0.000001	4.754	22.600

A highly useful relationship has been established [92] among the factor of safety, SF; the reserve (in the form of $k = m_R/\sigma_R$); and the coefficients of the variation of load and strength, $V = \sigma_L/m_R$ and $V_S = \sigma_S/m_S$, resulting in:

$$SF = \frac{1}{1 - k^2 V_S^2} + \sqrt{\frac{1}{(1 - k^2 V_S^2)^2} - \frac{(1 - k^2 V_L^2)}{(1 - k^2 V_S^2)}} \qquad (52)$$

with k as given above, or by

$$k = \frac{F - 1}{\sqrt{V_L^2 + F^2 V_S^2}} \qquad (53)$$

A graphical solution of Eqs. 51 and 53 is shown as Fig. 3.59; although satisfactory for many conditions, it is approximate. More accurate values can be calculated from Eqs. 51 and 52, using tabular values of k.

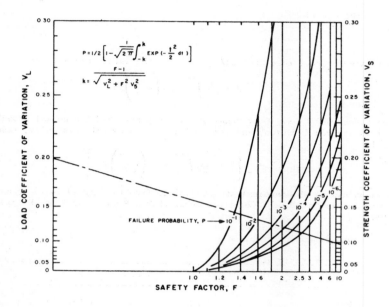

Fig. 3.59 Factor of safety nomogram. From R. B. McCalley [92].

Consider a simplified example in which the load characteristics are m_L = 15,000 lb. and σ_L = 3000 lb; the strength coefficient of variation V_S = 0.10, and it is required to find the factor of safety for a failure probability of less than 1 in 1000. The load coefficient of variation is, then,

$$V = \frac{\sigma_L}{m_L} = \frac{3000}{15,000} = 0.20$$

With the numerical values of these three parameters in hand, the solution is demonstrated by drawing a straight line on the nomogram between V_L = 0.20 and V_S = 0.10; at the intersection with the P = 10^{-3} line, read down to a value of safety factor of F = 1.85 approximately. The value calculated from Eq. 52 is 1.84. It should be noted that the value of the factor of safety is related to the terms by which the strength is specified. Here the mean strength was used; some specifications quote minimum strength, which is about 2 standard deviations lower than the mean. Thus, a factor of safety on minimum strength would be lower than one based on mean strength.

Distributions of Loads and Strengths

In a properly executed fatigue analysis, serious consideration must be given to the type of dispersion or distribution of the fundamental parameters of loads, strengths, and physical dimensions. Both a mean value and a variation around it are naturally occurring events, thus a realistic description can be established on the mean and a standard deviation. As examples, consider:

1.   The tensile yield of 2024-T3 can be written as $(M_{F_{ty}}, \sigma_{F_{ty}}) = (56.3, 3.2)$ ksi, or the fatigue allowable for -T4 at 10^6 cycles $(R = -1.0, K_t = 1.0$, axial load) as $(M_F, \sigma_F) = (26.0, 4.2)$ ksi.

2.   The load from a rocket motor thrust might be $(M_T, \sigma_T) = (1.5 \times 10^6$ lb., ±3 percent), or the load on a freight car spring as $(L, \sigma) = (10,100, 8200)$ lb.

3.   A machined parallelopiped could be dimensioned with a 3σ tolerance as

$$\text{width} = 0.374 \pm 0.015 \text{ in.,} \qquad \sigma_w = 0.005 \text{ in.}$$

$$\text{height} = 1.250 \pm 0.030 \text{ in.,} \qquad \sigma_h = 0.010 \text{ in.}$$

$$\text{length} = 15.0 \pm 0.12 \text{ in.,} \qquad \sigma_l = 0.04 \text{ in.}$$

A meaningful statistical description must include the principal stresses (loads) and strengths as they produce a resultant or combined effect. Such a description is possible because the algebra of normal functions is closed; that is, the sum, difference, product, or quotient of two normal functions is a normal function. Typically, an expression for combined loads is a single normal function.

Distributions are reasonably assumed to be normal and the variables statistically independent, because many populations encountered in engineering seem to have a normal distribution to a good degree of approximation. If a complete theory of statistical inference is developed based on the normal distribution alone, one has in reality a system that may be employed quite generally because other distributions can be transformed to the normal form. The Rayleigh distribution of the peak response amplitudes of random vibration is a major exception to the usual normal distribution and was discussed further in Section 3.2.1. Data on the mean and standard deviation for allowable fatigue stresses does not seem to be generally available, but the applied stress, normalized on response acceleration, can be compared to an allowable fatigue stress as determined by test, or from the appropriate S-N or constant-life curves.

Considering the static strengths of the many aluminum and titanium alloys, we find that the scatter is quite small, in terms of the coefficient of variation $\sigma/\bar{x}$, only about 5 to 7 percent. The strengths tend to follow a log-normal distribution, but, strictly speaking, material properties cannot be normally distributed, since the normal function is continuous from $-\infty$ to $+\infty$. In conventional aluminums, zero strength occurs about 20 to 30σ below the mean, but values of more than 6 or 7σ are usually of little more than academic concern. Thus, the normal distribution is a very acceptable approximation.

Some typical variations in aluminum alloys related to thickness, direction within an extrusion, and production parameters are listed in Table 3.13.

TABLE 3.13 Variations in Yield Strength of Aluminum Alloys

Thickness	2014-T6 Bare Sheet Mean, $\bar{X}$	Standard Deviation, σ
0.020-0.039	74 ksi	3.7 ksi
0.040-0.499	75	3.0

F_{ty} or S_{ty}	2014 Extrusions Mean, $\bar{X}$	Standard Deviation, σ
Longitudinal	64.1 ksi	4.3 ksi
Transverse	56.3	3.2

2024-T3 Alclad Sheet, 0.064

	σ, Within a Lot	σ, Across All Lots in 6-Months Production
F_{tu} or S_{tu}	1240 psi	1300 psi
F_{ty} or S_{ty}	1120	1500

The effects of the manufacturing process in terms of tolerances on dimensions and on alighnment at assembly are of concern to the designer. A mathematical model has been constructed [303], expressing the probability of no-failure as a design characteristic, using as inputs the applied factor of safety and the coefficients of variation of the applied load and of the appropriate strength of the material. Although a factor of safety is involved, it appears that, if Material Review Board action is required on material degradation resulting from manufacturing discrepancy, the decision can be based on at least the semiquantitative changes in reliability.

Random Loading Statistics

Because of the widening recognition that the effects of natural/random loading sequences may be effectively treated only in statistical terms, it is useful to examine their statistical properties. Much of the investigation and analytical work is being done in aircraft design, and terms descriptive of that discipline are used in the following discussion.

Stationary, Ergodic, Narrow Band Processes. The rather overwhelming title to this subsection is the statistician's description of the response of a linear system to loading by random vibration. For a rigorous treatment of random vibration, a knowledge of probability theory would be necessary; but the material below presents a useful approach in engineering terms and algebra with no significant losses, except possibly for the purist.

A few definitions are given to base the discussion properly, and it is helpful to recall that a major characteristic of random vibration is that its amplitude is not predictable from one moment to the next, differing therein from harmonic vibration.

1.   The ensemble is the entire history in time and amplitude of the random signal, incorporating an infinite number of samples and ranging in time from plus to minus infinity.

2.   A stationary random process is one in which the ensemble does not vary its probabilistic properties with time.

3.   An ergodic process is a stationary one in which one sample is representative of the ensemble.

4.   A narrow band process is a stationary random process whose mean square spectral density $S(\omega)$ has significant values only in a band or range of frequencies whose width is small compared to the magnitude of the center frequency of the band. Narrow band processes are typically encountered as response variables in strongly resonant systems when the excitation variables are wide band processes, that is, the response is a narrow band process if it exhibits a single distinct peak.   The term "resonance" is restricted to response to sine excitation, but the difference is small for sharply peaked responses.   This distinction was illustrated in Section 3.2.1.

Since the most important source of fatigue loading has been defined as random vibration (or nondeterministic vibration in that prediction of the amplitudes is not possible), a probabilistic description is required.   A complete description would call for an infinite set of probability distributions, but fortunately for most applications much less is needed, and a great deal can be accomplished by consideration of only the "first order" probability distributions.   Such a distribution for values of $x(t_1)$ at a fixed value of t_1 may be described by a graph such as Fig. 3.60, which shows the probability density function $p[x(t_1)]$ or just $p(x)$ if there is no ambiguity concerning the value of t involved.   This function has the property that the fraction of ensemble members for which $x(t_1)$ lies between x and x + dx is $p(x)dx$.

With regard to "probability," a useful, but in this context heuristic, concept leads to the definition of probability as the fraction of favorable, or successful, or otherwise defined events out of all possible events or occurrences.

Crandall and Mark [13] point out that probabilities are inherently non-negative, they can be only zero or positive, and that probabilities of mutually exclusive events are additive.   Thus, the probability that a sample lies between any two verticals in Fig. 3.60 is the sum of the probabilities that a sample lies in each of the dx intervals that make up the integral:

$$\int_{x_1}^{x_2} p(x)dx$$

The probability that the value of x lies between plus infinity and minus infinity, is unity, that is, there is 100 percent certainty that x is somewhere in this interval (with the implicit understanding that x is a real number).   This condition further implies that the area under the curve is unity or that

$$\int_{-\infty}^{+\infty} p(x)dx = 1 \tag{54}$$

It should be noted that the probability density $p(x)$, which may be interpreted as giving the fraction of successes per unit of x, has the dimensions of $1/x$.

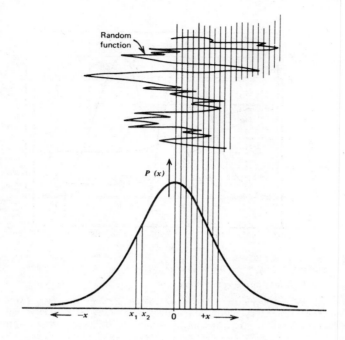

Fig. 3.60   Gaussian distribution of a random function.

Most structures and machines must not only survive but operate while subjected to random vibration. The amplitudes and frequencies of occurrence of the peak forces from such excitation are among the most useful functions in fatigue calculations. In the attempt to characterize such a signal, varying randomly with time, one might suggest the measurement of some average or mean value of the amplitude. But it is soon recognized that random vibration must have a zero average amplitude (as does a sine wave), or else the vibration generator would be accelerated away from the structure. If, however, the purely mathematical operation of squaring the amplitudes at each of many increments along the time axis is applied, the negative signs are eliminated; if, further, these squared values are summed and divided by the number of samples, the mean square value of the random signal results:

$$\bar{x}^2 = \frac{x_1{}^2 + x_2{}^2 + x_3{}^2 \cdots + x_n{}^2}{n} \tag{55}$$

An analysis may be made of the frequency of occurrence of each of these amplitudes as a function of time; the results would appear as a curve showing the relative frequency and the probability density, as in Fig. 3.61. This curve represents the normal or gaussian distribution characteristic of natural, and of artificially induced, random loadings. The total area under the curve was noted as being normalized at unity, and the area of any vertical slice is the probability that the signal would be at that amplitude at any time. The usefulness of the curve is its ability to predict the probability that the amplitude will be between certain

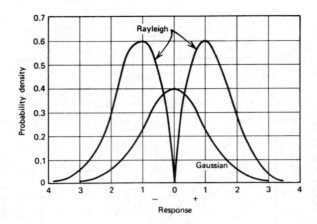

Fig. 3.61   Probability distributions.   Rayleigh: $P(x) = (x/\sigma^2)[\exp(-x^2/2\sigma^2)]$;
gaussian: $\lambda = (1/\sigma\sqrt{2\pi})[\exp(-g_r^2/2\sigma^2)]$.

limiting values or below certain limiting values at any instant, for it is not possible to predict exactly the amplitude at any time. The method not only provides an educated guess as to the instantaneous amplitude, but it also enables a measure of the intensity of the process. The intensity is usually specified by the mean square, σ^2, sometimes called the variance, or by the root-mean-square, σ, or standard deviation, of the signal.

As has been mentioned, the gaussian (RMS) distribution is characteristic of natural loadings; but the _peak_ loads, and therefore the peak stresses induced in the structure, are given by the Rayleigh distribution, which is compared with the gaussian in Fig. 3.61. It should be noted that the maximum number of peaks, or the maximum density of peaks, occurs at the RMS (one-sigma) value of the random process. It is frequently necessary to know the probability of exceeding a given amplitude (load or stress). The curves in Fig. 3.62 are the integrals of those in Fig. 3.61 and provide this information for both distributions. At the three-sigma cut-off it is seen that the probability of exceeding the peak amplitude is 1.0% while that of exceeding the RMS value is only about 0.4%. A gaussian process remains invariant with a linear system; that is, if random vibration is applied to a structure whose response may be assumed linear, the response is gaussian also, irrespective of where it is measured, although the RMS value or standard deviation may change. Also, the sum of two or more gaussian processes is gaussian; if a structure is simultaneously subjected to random vibration from more than one source, the response amplitudes will have a gaussian distribution.

In addition to the above discussion, it is necessary to consider the _rapidity_ of the action of a random process and of the differences in response of a structure to variations in rapidity. The latter is sometimes described in terms of the degree of mismatch in terms of mechanical impedance [39]. Or one might think qualitatively of the structural response as related to the sharpness of the spikes in the input curve. It is quite possible that a peak input of relatively long duration could cause a failure, whereas a peak of short duration might not because

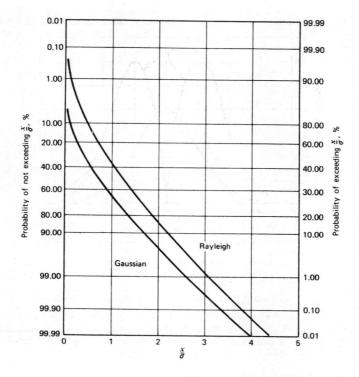

Fig. 3.62 Comparison of Rayleigh and gaussian excedances.

the structure was incapable of responding to it, that is, there was an impedance mismatch. It should be recalled that stress is time-invariant but that strain is time-dependent, at least for metals and the millisecond range. In order to describe the random input signal completely and its action upon the structure, one must determine its frequency spectrum.

The method is that of resolution of the complex random signal into a combination individual sine wave, which varies in amplitude and phase, by means of the Fourier integral:

$$S(\omega) = F(jw) = \int_{-\infty}^{+\infty} f(t)\exp(-jwt)dt \qquad (56*)$$

The difference between this integral and the series is that the latter is composed of a fundamental plus an infinite number of harmonics, whereas the integral is composed of an infinite number of frequencies that bear no harmonic relationship to

*Equations 56-60 from "Random Vibration," by permission of Academic Press and the authors: S. H. Crandall and W. D. Mark.

each other. As in the case of the series, there is a relationship for changing
from a function of time to a function of frequency, thus permitting the random
wave form to be converted to a continuous spectrum of frequencies, typically as in
regions 1 and 3 of Fig. 3.63.

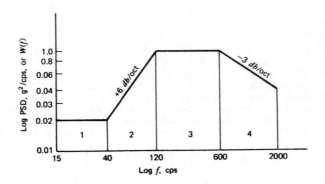

Fig. 3.63   A typical power-spectral-density envelope.

Earlier in the discussion, it was shown that the square of the instantaneous amplitude of the signal was a meaningful representation of the intensity; similarly, it
can be shown that the square of the frequency is also useful.

The conversion from time to frequency is:

$$\int_{-\infty}^{+\infty} f(t)dt = \frac{1}{2} \int_{-\infty}^{+\infty} |F(\omega)|^2 d\omega = \int_{-\infty}^{+\infty} |F(f)|^2 df \qquad (57)$$

$$\overline{x^2(t)} = \int_0^{\infty} \lim_{t \to o} \left[\frac{2}{T}|F(f)|^2 \right] df = \int_0^{\infty} W(f)df \qquad (58)$$

This function, W(f), is called the spectral density and when applied to vibration,
it is indicative of the amount of energy in a frequency band (which approaches zero
as a limit). To obtain the total energy in the random vibration process, this
spectral density is integrated over the entire frequency spectrum, the energy being
equivalent to the mean square of the process. In adding the contributions of the
different sinusoids to obtain the total energy, it is necessary to get the sum of
the mean squares of each of the sinusoids. When the spectral density is given in
terms of the mean square, the total energy is obtained by merely adding the spectral
density for each frequency increment along the spectrum.

Many vibration specifications now give the spectral density in mean square terms,
as g^2/Hz; which strictly speaking, should be written g^2/Hz. In many specifications,
and technical literature, the term __power__-spectral-density is used, when the meaning
really is __acceleration__-spectral-density. The specification envelopes are usually
plotted as f vs. g^2/Hz. Another descriptive term often used is that of "white
noise," indicating that the spectral density is uniform over the spectrum or some

frequency band within it. It is considered that all frequencies contribute to the process in a manner similar to the contribution of all colors to white light. "Band-limited" white noise is a specific designation for a specific random process consisting of uniform spectral density from one frequency to another, as from f_1 to f_2 and from f_3 to f_4 in Fig. 3.63.

From the above, then, it follows that the mean square of a constant-spectral-density random process is simply:

$$g_{rms}^2 = W(f_2 - f_1); \qquad f_2 > f_1 \tag{59}$$

which in the form

$$g_{rms} = \sqrt{\frac{\pi W(f) f_n}{2T} \left(1 + \frac{1}{T^2}\right)} \tag{59a}$$

with T as the force transmissibility, provides for direct conversion of a simple power-spectral-density envelope to a single frequency sine wave of equivalent energy. But, when dealing with power-spectral-density curves of nonconstant ordinates, one must recall that the (usual) straight line on the log-log scale does not define a triangle on the linear scale. The area under such power-spectral-density curves is given by:

$$g_{rms}^2 = \frac{3W}{3 + m} \left[f_2 - f_1 \left(\frac{f_1}{f_2}\right)^{m/3} \right] \tag{60}$$

where m is the ordinate slope of the region in db/octave, plus or minus, and

f_1 = the initial frequency

f_2 = the final frequency, $f_2 > f_1$

W = the acceleration density a f_2

A special case appears for the slope of -3 db/oct; a solution to this indeterminate form is:

$$g_{rms}^2 = f_2 W \ln \left(\frac{f_2}{f_1}\right) \tag{61*}$$

The summing of an entire power-spectral-density envelope may be done by:

$$G_{rms} = \sqrt{g_1^2 + g_2^2 + g_3^2 \cdots + g_n^2} \tag{62}$$

As an example, take the envelope of Fig. 3.63 as a possible random vibration specification. For region 1 use Eq. 59:

$$g_1^2 = 0.02(40 - 15) = 0.50.$$

For region 2 use Eq. 60:

$$g_2^2 = \frac{3(0.01)}{3 + 6} \left[120 - 40 \left(\frac{40}{120}\right)^{6/3} \right] = 3.8.$$

*Equations 61-63 from [302], by permission AIAA.

For region 3 use Eq. 59:

$$g_3^2 = 0.1(600 - 120) = 48.0.$$

For region 4 use Eq. 61:

$$g_4^2 = 2000(0.04) \ln \left(\frac{2000}{600}\right) = 96.0.$$

Then by Eq. 62, the total equivalent G_{rms} is:

$$G = \sqrt{0.50 + 3.8 + 48 + 96} = 12.2$$

Thus, a sine wave of amplitude 12.2 g_{rms} applied at the natural frequency of the structure would be expected to provide the same damage potential as the random wave of the given power-spectral-density over the entire bandwidth.

In general, the response of a loaded structure will be equal to the mean square of the input times the square of the transfer function. The general form of Eq. 62 is then:

$$W(f)_{out} = T^2(f)W(f)_{in} \tag{63}$$

or

$$G_{rms} = \left[\int_{f_1}^{f_2} T^2(f)W(f)_{in} df\right]^{1/2} \tag{64}$$

meaning that, at a given frequency, the product of T^2 and $(PSD)_{in}$ yields the $(PSD)_{out}$. To obtain the response over the complete spectrum, this procedure is followed at each increment of frequency. Solutions for several single-degree-of-freedom systems are given in Fig. 3.64.

From the statisticians point of view, fatigue is a stationary, narrow band process which turns out to have three criteria of failure:

1.   Failure due to the first excursion up to a certain level, a condition easily recognized as static failure in 1/4 cycle, or the first cycle in which the applied stress exceeded the ultimate strength, after a number of nonfailure cycles. In the theory of probability, this type of problem is known as a "first-passage problem" and is sufficiently difficult that an exact solution is unknown. In regard to practical design for fatigue, the first condition is irrelevant and the second is at best marginal; thus, the criterion is of little further interest here.

2.   Failure due to the response above a certain level for too great a fraction of the time. Such a condition might arise in excessive displacements in supports for sensitive circuitry, as in on-board computers and camera-tube guns, resulting not necessarily in fracture but in loss of information because of changes in capacitance, and so forth. While the case is statistically interesting, it would again represent a trivial condition for the design of primary structural or machine elements.

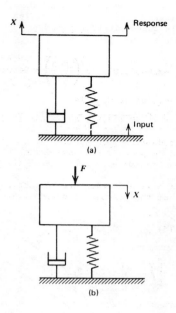

Fig. 3.64   Response of a single-degree-of-freedom system to a random function of acceleration.

(a) Input on base in g^2/Hz. Response in in., rms:

$$x_{rms} = 1/[2(386)]\sqrt{W(f)T/8\pi^3 f_n^3};$$

response in G_{rms}:

$$G_{rms} = \sqrt{(\pi/2)W(f)f_n[T + (1/T)]}.$$

(b) Input on mass in lb^2/Hz. Response in G_{rms}:

$$G_{rms} = 1/386\sqrt{(\pi/2)W(f)Tf_n}.$$

3.   Accumulation of damage. By a rigorous extension of the method leading to Eqs. 62 and 63, Crandall and Mark [13] have established an expression for the total expected damage at time $T\{E[D(T)]\}$ as:

$$E[D(T)] = \frac{\nu T}{c}(\sqrt{2}\sigma)^b \Gamma\left(1 + \frac{b}{2}\right) \tag{65}$$

where ν_o = the expected frequency of the stress history, that is, the natural frequency, Hz.

σ = the standard deviation of the stress, psi

Γ = the gamma function

T = the time interval, sec

and where b and c are as given by the conventional damage law: $c = NS^b$, with
values taken from an experimental plot of log S vs. log N for the material. Miner's
hypothesis is used as the failure criterion, and the stress and strain functions of
time are supposed to be narrow band processes.

An example of the use of Eq. 65 is chosen from the design of an electronics box to
be carried in the Lunar Module; see Fig. 3.65. The vibration environment was as
defined in Table 3.14. According to the linear damage criteria, the total damage
is the sum of that in each mode:

$$E[D(T)] = E[D_1(T)] + E[D_2(T)] + E[D_3(T)] \qquad (66)$$

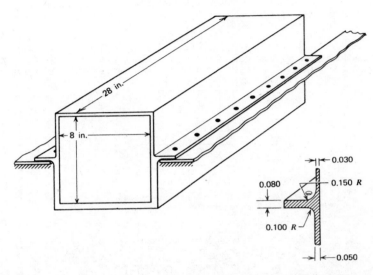

Fig. 3.65   Lunar Module box. All dimensions are approximate.
Load $\approx$ 30 lb; 6061-T6 aluminum with electron-beam
welded corners.

TABLE 3.14   LM Vibration Environment

Mode	$W_0, g^2/Hz$	T_1, sec	ω_1, rps	ω_2, rps
Launch	0.0375	1800	30π	4000π
Space flight	0.02	600	100π	4000π
Lunar excursion	0.05	2400	40π	4000π

The box was carried by the two integrally-machined flanges, which were supported by the LM secondary structure. The critically stressed section was that through the flange, the fillets and the adjacent wall. The unsymmetrical fillets and the unequal thicknesses of the flange and the wall made the calculations of the moments, the deflections, the spring constants, and the natural frequency too tedious to be reproduced here. For the purpose of this example, the result is given as $\nu_0 = 132$ Hz. The constants b and c were evaluated from a log S-log N plot of 6061-T6 as $b = 7.65$ and $c = 10^5(30 \times 10^3)^{7.65}$, and the gamma function value is taken from standard tables.

Since the time between zero crossings varies, the only frequency that can be determined is the statistical average frequency, ν_0. Actually, the mean deflection and mean stress are not zero but, since the mean stress is not extremely high, it will not significantly affect the fatigue life. As an additional factor supporting this statement, it is noted that the constant lifelines for this alloy are quite flat. Since the mean value of the damage varies as the b power of the standard deviation of the stress, a value for the latter is taken as

$$\sigma_y = \tfrac{1}{3}S_{bmax} = \frac{1}{3}\left[\frac{2.24E_e\delta_d tK_t}{1_1{}^2}\right] \qquad (67)$$

where K_t is the stress concentration factor and the rest of the term is the maximum moment. This assumption will slightly overestimate the stress during the upward deflection of the flange relative to the support and provides an additional safety factor. Then,

$$S_\sigma = \frac{0.75E_e\delta_d tK_t}{1_1{}^2} \qquad (68)$$

The expected number of crossings of the level "a" per unit time is given by:

$$\nu_a = \int_0^\infty \dot{y}_2 p(a,\dot{y}_2)d\dot{y}_2 \qquad (69)$$

where $p(a,\dot{y}_2)$ is the joint probability density function given by:

$$p(a,\dot{y}) = \frac{1}{2\pi\sigma_{y_2}\sigma_{\dot{y}_2}} \exp\left[-\frac{1}{2}\left(\frac{a^2}{\sigma_{y_2}^2} + \frac{\dot{y}_2{}^2}{\sigma_{\dot{y}_2}^2}\right)\right] \qquad (70)$$

Using Eq. 70 to evaluate the integral in Eq. 69, we obtain the result:

$$\nu_a = \frac{1}{2\pi}\frac{\sigma_{\dot{y}_2}}{\sigma_{y_2}}\exp\left[-\frac{a^2}{2\sigma_{\dot{y}_2}^2}\right] \qquad (71)$$

The expected number of zero crossings per unit time then becomes:

$$\nu_0 = \frac{1}{2\pi}\frac{\sigma_{\dot{y}_2}}{\sigma_{y_2}} \qquad (72)$$

The value of σ_{y_2} can be obtained from:

$$\sigma^2 \cong E(y_2{}^2) \qquad (73)$$

Since

$$y_2 = H_{y_2}(\omega)e^{j\omega t}$$

then,

$$\dot{y}_2 = j\omega H_{y_2}(\omega) e^{j\omega t}$$

and,

$$H_{\dot{y}_2}(\omega) = \left.\frac{\dot{y}_2}{\dot{x}_0}\right|_{e^{j\omega t}} = j\omega H_{y2}(\omega) \tag{74}$$

The mean square deviation of $\dot{y}$ $E[\dot{y}_2{}^2]$ can now be related to the input power spectral density by:

$$E[\dot{y}_2{}^2] = \int_{-\infty}^{+\infty} W_{\ddot{x}_0}(\omega) |H_{\dot{y}_2}(\omega)|^2 d\omega \tag{75}$$

Since the input PSD is approximately white (i.e. of uniform magnitude), and the mean value of $\dot{y}_2$ is zero:

$$\sigma_{\dot{y}_2}^2 = \int_{-\infty}^{+\infty} W_0 |H_{\dot{y}_2}(\omega)|^2 d\omega \tag{76}$$

The value of the square of the absolute magnitude of the frequency response has been evaluated as:

$$\int_{-\infty}^{+\infty} |H_{\dot{y}_2}(\omega)|^2 d\omega = \pi \frac{\omega_{n_1}^3 + 4h_1^2 \omega_{n_1} \omega_{n_2}^2}{2(m_2/m_1)\omega_{n_2}^4} \tag{77}$$

where the subscripts 1 and 2 refer to the supports and to the box flange, respectively; and $h_1 = 1/Q$.

Now, values of ν_0, W_0 and $E[D,(T)]$ are calculated for each of the operational modes and summarized in Table 3.15.

TABLE 3.15   LM Fatigue Parameters

Mode	ν_0, Hz	S_σ, psi	$E[D_i(T)]$
Launch	132	6570	$3.12 \times 10^{-6} T$
Space flight	132	4740	$2.36 \times 10^{-7} T$
Lunar excursion	132	7500	$8.64 \times 10^{-6} T$

Using the elapsed time in each mode as given in Table 3.15, together with Eqs. 65 and 66, we find that the total fatigue damage is $E[D(T)] = 0.027$. Strictly speaking, the variance of the damage from its expected value should also be considered when dealing with such an ensemble of time histories, but in this case the damage itself turned out to be negligible, so that there is little point in calculating a variance.

The Miner criterion of failure did not appear in its conventional form, $\Sigma n/N$, but in terms of time. For example, the period of the Lunar Excursion mode was 2400 sec, which generated a life fraction, or $E[D(T)]$ of 0.0208, about 2 percent. One may then say that the vibration of this mode alone could be applied for a time equal to $50 \times 2400 = 120,000$ sec for failure, or indeed t/T (instead of n/N) equaled 2400/120,000, or 0.02.

<u>Effect of Number of Test Specimens on Mean Life</u>. The difficulties of obtaining a qualitatively sufficient number of fatigue specimens has been discussed extensively elsewhere, and the dangerously inaccurate results of using too few specimens are indicated below.

In test efforts to obtain numerical values of this parameter [366], 100 cantilever beams arranged individually as a single-degree-of-freedom system, and 100 beams as two-degrees-of-freedom systems were excited to failure. The RMS stress and the average zero crossing frequency were maintained constant for each set of beams. The results are plotted in Fig. 3.66 as the mean time to failure vs. the number of specimens in each sample. Several distinctive observations may be made:

    1. The variation in the means is tremendous (i.e., >2/1) among the smaller sample sizes, from 1 to about 50.

    2. It is not until the sample size becomes rather large, above about 80 that the curves begin to show the tendency toward random variations about the central means (which should theoretically degenerate into straight lines as the sample size approaches infinity).

    3. Since the difference between the two means is greater than six standard deviations of the mean, the difference in endurance times between the two types of specimens must be highly significant in the statistical sense. A possible explanation of this significance follows in the next section.

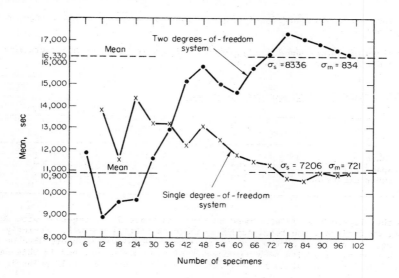

Fig. 3.66 Mean life vs. number of specimens (from Ref. 366).

3.2.4 APPLIED FRACTURE MECHANICS

INTRODUCTION

Within the general discipline of Fatigue Analysis and Design, the concept of Finite Life has evolved as a parameter more useful than Infinite Life or the Endurance Limit. But the measure has remained in terms of the abstract "damage" or cycle ratio. A design was adjusted until the Miner or other Summation gave a result comfortably less than unity, and that effectively ended the effort. Especially lacking in this approach was a parameter or device which the operator or inspector could use directly to monitor life, unless possibly some sort of cycle counter and strain gage. And, more significantly, the approach ignores the presence of the flaws inevitably and observably present in all engineering materials.

The Fracture Mechanics approach to design, selection of material and life prediction is based on direct consideration of the cracklike flaws or defects. Historically, it was first concerned with the high-strength, low-toughness alloys that frequently fractured in a brittle manner at stresses well below yield, even though such alloys were reasonably ductile under static loads. The explanation for such failures has now been found in terms of the material's fracture toughness; the size of the initial defect and its rate of growth under fatigue loading. More recently, much interest has been focused on the lower-strength, higher-toughness alloys and the explanation of failure is generally the same, with the added condition that when the part dimension into which the crack was growing was sufficiently great (i.e., $a_{cr} > W$) and the toughness high enough, the part could spend its entire service life, or life to fracture, in slow stable crack propagation. Some specifications for alloys are beginning to include a maximum size of initial defect.

TRANSITION-TEMPERATURE CONSIDERATIONS

Carbon and low-alloy steels exhibit a transition from ductile to brittle fracture behavior upon lowering the temperature over some short range called the transition-temperature, typically as in Fig. 3.67. Since brittle fracture is even less controllable than ductile fracture, much effort has been expended, by Pellini and others, to correlate the influencing factors. The major results appear as plots of the stress vs. transition-temperature, with the strength limits shown, Fig. 3.68; and as the Fracture Analysis Diagram (FAD), Fig. 3.69. Considering the first: it is generally true that both the yield and ultimate strengths of steel increase with decreasing temperature, coinciding at some very low temperature where the plastic flow properties (elongation and reduction of area) are essentially zero. For a hypothetical steel with no flaws, this nil-ductility-transition temperature would be called the NDT no-flaw. But for commercial steels with small flaws, a decreased level of fracture stress is obtained in the transition range as indicated by the dashed curve labeled "fracture stress decrease due to small, sharp flaw". The highest temperature at which the decreasing stress for initiation due to the small flaw becomes contiguous with the yield strength curve is defined as the nil-ductility-transition (NDT) temperature. Below NDT, the fracture stress curve for the small flaw follows the course of the yield strength curve, as indicated by the continuation of the dashed curve to lower temperatures. The arrows pointing down from the NDT point indicate that increases in the size of the flaw result in progressive lowering of the fracture stress curve to lower levels of nominal stress. As an approximation, the fracture stress is inversely proportional to the square root of the flaw size. The resulting family of fracture stress curves is characterized by a common temperature effect, involving a marked increase in the stress required for fracture as the temperature is increased above the NDT temperature.

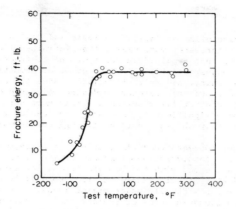

Fig. 3.67   Typical Impact Energy Transition Curve,
for Carbon Steel A212B.

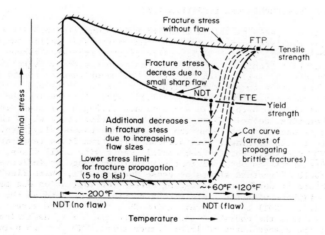

Fig. 3.68   Transition-Temperature features of
steels (from NRL Rept. 5920).

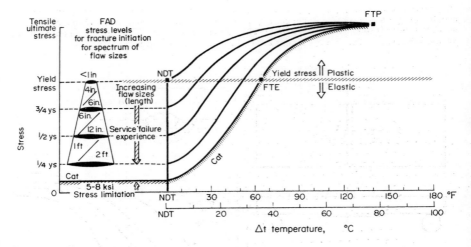

Fig. 3.69 Fracture Analysis Diagram, FAD (from Ref. 42).

The curve noted as CAT (crack arrest temperature) represents the temperature of
arrest of a propagating brittle fracture for various levels of applied nominal
stress. The CAT temperature for a stress level equal to yield strength has been
defined as the "fracture transition elastic" (FTE) temperature and marks the highest
temperature of fracture propagation for purely elastic loads. Similarly, the
"fracture transition plastic" (FTP) temperature has been defined as the temperature
above which the fracture is entirely ductile (i.e., no center cleavage), and the
stress required approximates the tensile strength. The lower shelf in the range
of 5 to 8 ksi, labeled as "the lower limit for propagation", represents that stress
level below which propagation is not possible because the minimum amount of strain-
energy-release required for continued propagation of brittle fracture is not attained.

The Fracture Analysis Diagram provides a generalized definition of the relationships
amongst flaw size, relative stress and temperature by means of a "temperature increment" reference to the NDT temperature. The positioning of the generalized diagram
to specific locations on the temperature scale requires the experimental determination of a single parameter - the NDT temperature, which may be done by a Drop-Weight
Test (DWT), or a Charpy impact test, (CVN).

The essence of the FAD, Fig. 3.69 is the set of four reference points:

    1.   NDT. Restricting the service temperature to slightly above the NDT
provides protection for the initiation of cracks. Service stress may be as high
as yield. No arrest of the propagation of any cracks initially present.

    2.   NDT to FTE. Restricting the service temperature to a level above NDT +
30°F, (i.e., the mid-range of the NDT to FTE region), provides fracture arrest
protection if the service stress does not exceed about one-half yield.

    3.   FTE. As in (2) for service stresses up to yield.

    4.   FTP. Keeping the service temperature above FTP insures that any fracture
which occurs is fully ductile.

The reference points, or temperatures, may be used as design criteria in the roughing-in of allowable stresses, service temperatures and flaw sizes during material selection. Adaptations of the method are useful, as in the consideration of weldability, Fig. 3.70 illustrates the general procedure.

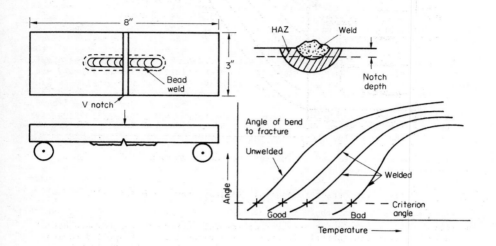

Fig. 3.70 Weldability tests for assessing characteristics
of the heat affected zone.

It should be noted that the cost of steels rises sharply as the NDT is lowered, for this condition is obtained by additional alloy content, special heat treatments for control of grain size, etc.

The relative usefulness of several of the test methods and specimens, as the Charpy V-notch, Drop Weight, Dynamic Teat, etc. is compared in the Section on Fracture Testing.

The Transition-Temperature Approach described above appears eminently applicable in the screening of materials, especially plate stock and forgings for pressure vessels and other weldments. But it is weak in that the test data are not directly usable in design; it does not include treatment of flaws less than about one inch in size, and the stress increments used are very coarse. The latter is particularly important in life prediction for, as will be shown below, life is generally proportional to the reciprocal of stress to the _fifth_ power, approximately. Chiefly for these reasons, this approach is giving way to that of Fracture Mechanics, or Crack Propagation.

THE FRACTURE TOUGHNESS AND CRACK PROPAGATION APPROACH

This method of design, selection of materials and life prediction has evolved rather rapidly over the past 30 or so years, due chiefly to the efforts of Irwin and his colleagues. The status is now such that it may be used with a fair degree of confidence, although with considerable engineering judgment. It is a very powerful method for it deals numerically with imperfect materials and essentially any type of loading and geometry. The older methods of assuming flaw-free material with a comfortable safety factor on strength, and of the Transition-Temperature Approach were frequently not applicable. Even if successful, the degree of conservatism was not known--and basically, there was a pressing need to explain the spectacular failures of the Polaris air bottles, Liberty ships, etc. The development of this method to its present state is interesting both scientifically and psychologically; also it is sufficiently extensive that this presentation will be confined to illustrating the application of some of the relationships and data in practical situations. The reason for its present popularity and for the tremendous effort being spent in expansion is that it does supply coherent explanations for most cases. The interested reader is strongly urged to look into the references for himself, and to follow the continuing evolution.

One of the two basic premises of Fracture Mechanics is that the part contains a sharp-edged crack of engineering size as the ultimate notch; that this crack grows in a slow stable fashion under either static or cyclic stressing, thus, the parameter of growth rate comes into being; that at some "critical" length, the growth rate becomes unstable, leading to "fast" or brittle fracture on the next cycle. The other premise is that there is a material property called "toughness", which is the resistance to crack growth. Toughness is measured in the somewhat odd dimension of: (stress) x $\sqrt{\text{length}}$, often as: ksi $\sqrt{\text{inch}}$, and has not been specifically defined before. It was described only generally by resilience ($S^2/2E$), ductility, etc.

A numerical description of the elastic stress field near the crack tip is also necessary, and for this the "stress-intensity factor" has been established. This factor, K or J, is a single parameter describing the behaviour of the elastic stress field just beyond the small plastic zone at the tip, and in the presence of small-scale yielding. The influence of the level and mode of loading, as well as that of the geometry is sensed in the crack tip region only thru the stress-intensity factor. It is not a stress-concentration-factor by Neuber's definition, but it is to be noted that the $\sqrt{\text{crack}}$ length term in these K:ΔS expressions is identical with Neuber's $\sqrt{\text{notch}}$ radius. Also, as used in Fracture Mechanics, "stress-intensity" has no relation to the same term in the ASME Boiler Code.

K and J are related both to the applied stress and to crack size, thus for every level of applied stress and crack size, there is a corresponding level of applied stress-intensity. In its limiting (maximum) value, K_c or J_c, the stress-intensity is called the Fracture Toughness of the material, a mechanical property measuring resistance to cracking under load. The critical crack size is that size attained when the applied stress-intensity reaches its critical value, i.e., a goes to a_{cr} as K goes to K_c. The critical value of stress-intensity exists in the same sense as does a critical value of strength, such as yield.

The events during crack growth are displayed in Fig. 3.71 with modifications possible, especially the percentage of a_{cr} and yield may be varied according to judgment. These curves are calculable in several forms as shown later. The discussion of crack initiation continues, but for the constructional alloys of conventionally sized section, say 0.50 sq. inch and up, it is generally considered only prudent to take the time to initiation as zero, especially for weldments. These cracks are latent material defects; they are present initially in the alloy as manufactured, their size and number being characteristic of the chemistry, microstructure, all processing, and of the sensitivity of inspection.

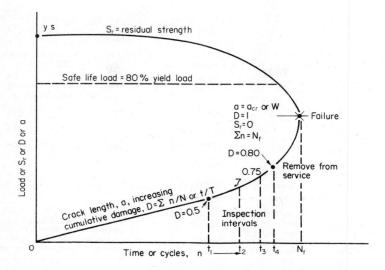

Fig. 3.71   Events during crack growth.

Further, damage may accrue in service due to crack growth, corrosion and abuse.   A
recent investigation of landing gear failures resulted in Table 3.16, where it is
noted that all defects except one are of the surface type.   This condition emphasizes
the importance of surface integrity in highly-stressed parts.   The term "defect" as
used in discussion of Fracture Mechanics does in no way imply "defective" material,
rather, it defines the conventional, engineering-sized discontinuities such as
cracks, voids, cold-shuts, incomplete penetration, etc.   General observation indicates that such defects exist in all alloys all the time after solidification, so
the only question is "how many" and "how big"?

The general relationship between applied stress-intensity and applied stress is
shown in Fig. 3.72a and 3.72b for one classical geometric case: a thru-thickness
crack in an infinite, thin sheet under uniaxial tension.   For practical solutions
to problems with engineering parts and structures, the relationships for many
loading modes and geometric shapes are needed; Refs. 438-40 are atlases of these
equations, and for convenience, a few Cases which occurred frequently are given in
Table 3.17.   The equations or curves for an enormous number of combinations of
shapes and loading modes are available and the problem for the user is to select
the most appropriate one.   If not obvious, the choice is made by engineering
judgment, and if an applicable Case cannot be found rather readily, the design may
be considered for review.

Figure 3.73 is an approximate nomogram relating defect size, stress level and the
fracture toughness of the material.   As such, it provides a "quick-look" at the
variables, but it should not be used in design.

TABLE 3.16   Summary of Typical Initial Flaws in Landing Gear Components

Cause of damage	Dimensions, (in.)	Comments
Processing Operations		
• Localized overtempered martensite	0.10 ⊢⊣ 0.003	Occurs in steel during grinding operations
• Localized untempered martensite	0.10 ⊢⊣ 0.008	Occurs in steel during grinding operations
• Chrome cracking	0.10 ⊢⊣	Crack depth equal to depth of chrome layer
Latent Material Defects		
• Inclusions	0.125 ⊢⊣ 0.005	
• Forging defects	0.10 ⊢⊣ 0.020	Dimensions shown are for forging laps
Mechanical Damage		
• Field inducded damage steel	⊢ 1.0 ⊣ 0.005	
• Field induced damage in aluminum	⊢0.25⊣ 0.02	
• Shop induced tool marks	⊢0.50⊣ 0.003	
Corrosion		
• Corrosion pit as initiation site for stress corrosion cracking in aluminum	0.020 ⊢⊣ 0.010	Depth of crack approximately half that of compressive layer induced by shot peening
• Corrosion pits as initiation site for stress corrosion cracking in steel	0.010 ⊢⊣ 0.005	Depth of crack approximately half that of compression layer induced by shot peening
• Corrosion pit as initiation site for fatigue crack growth	0.01 ⊢⊣ 0.005	Occurs only in fatigue critical regions

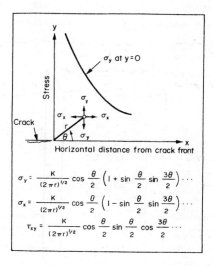

$$\sigma_y = \frac{K}{(2\pi r)^{1/2}} \cos\frac{\theta}{2}\left(1 + \sin\frac{\theta}{2}\sin\frac{3\theta}{2}\right)\cdots$$

$$\sigma_x = \frac{K}{(2\pi r)^{1/2}} \cos\frac{\theta}{2}\left(1 - \sin\frac{\theta}{2}\sin\frac{3\theta}{2}\right)\cdots$$

$$\tau_{xy} = \frac{K}{(2\pi r)^{1/2}} \cos\frac{\theta}{2}\sin\frac{\theta}{2}\cos\frac{3\theta}{2}\cdots$$

Fig. 3.72a   Relationship between the Stress-Intensity Factor, K,
and the Stress Components in the vicinity of a crack
(original equations by Dr. George Irwin).

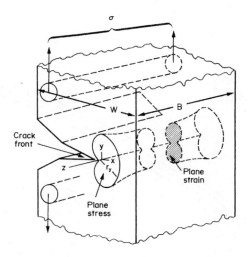

Fig. 3.72b   Schematic of ASTM E399 Compact Tension
Specimen showing the Plastic Zone.

TABLE 3.17   Stress-Intensity Factors

CASE 1   Plate containing a semi-elliptical surface crack

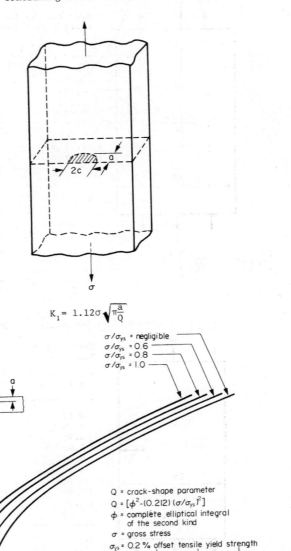

$$K_1 = 1.12\sigma \sqrt{\pi \frac{a}{Q}}$$

σ/σ_{ys} = negligible
σ/σ_{ys} = 0.6
σ/σ_{ys} = 0.8
σ/σ_{ys} = 1.0

Q = crack-shape parameter
$Q = [\phi^2 - (0.212)(\sigma/\sigma_{ys})^2]$
ϕ = complete elliptical integral of the second kind
σ = gross stress
σ_{ys} = 0.2 % offset tensile yield strength

Crack-shape parameter curves

TABLE 3.17 Stress-Intensity Factors (Cont'd.)

CASE 2 Finite-width plate containing a through-thickness crack

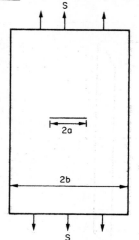

Form Factors for a Finite-Width Plate
Containing a Through-Thickness Crack

a/b	$[2b/\pi a \cdot \tan \pi a/2b]^{1/2}$
0.074	1.00
0.207	1.02
0.275	1.03
0.337	1.05
0.410	1.08
0.466	1.11
0.535	1.15
0.592	1.20

$$K_1 = \sigma \left[\pi a \left(\frac{2b}{\pi a} \tan \frac{\pi a}{2b} \right) \right]^{1/2}$$

CASE 3 Single-edge-notched plate of finite width

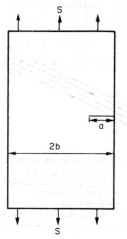

Form Factors for a Single-Edge-Notched
Plate

a/b	f(a/b)
0.10	1.15
0.20	1.20
0.30	1.29
0.40	1.37
0.50	1.51
0.60	1.68
0.70	1.89
0.80	2.14
0.90	2.46
1.00	2.86

$$K_1 = \sigma \sqrt{\pi a} f\left(\frac{a}{b} \right)$$

TABLE 3.17   Stress-Intensity Factors (Cont'd.)

CASE 4   Double-edge-notched plate of finite width

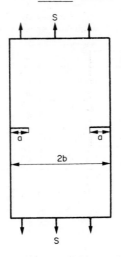

a/b	$\left(\dfrac{2b}{\pi a} \tan \dfrac{\pi a}{2b} \right)^{1/2}$ $L/b \to \infty$
0.1	1.12
0.2	1.12
0.3	1.13
0.4	1.14
0.5	1.15
0.6	1.22
0.7	1.34
0.8	1.57
0.8	2.09

$K_I = 1.12 \, S\sqrt{\pi a}$, for $a/b < 0.5$

$K_I = S\sqrt{\pi a} \left[\dfrac{2b}{\pi a} \tan \dfrac{\pi a}{2b} \right]^{1/2}$, for $a/b > 0.5$

CASE 5   Round bar in tension with circumference crack

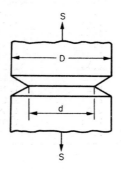

$K_I = S_{max}\sqrt{\pi D} \ f(d/D)$

$K_{II} = K_{III} = 0$

d/D	f(d/D)	d/D	f(d/D)
0	0	0.70	0.240
0.1	0.111	0.75	0.237
0.2	0.155	0.50	0.233
0.3	0.185	0.55	0.225
0.4	0.209	0.90	0.205
0.5	0.227	0.95	0.152
0.6	0.233	0.97	0.130
0.65	0.240	1.00	0

$K_I = S_{max}\sqrt{\pi D} \ f(d/D)$

$K_{II} = K_{III} = 0$

TABLE 3.17   Stress-Intensity Factors (Cont'd.)

CASE 6   Either one or two cracks growing out of a hole in an infinite plate

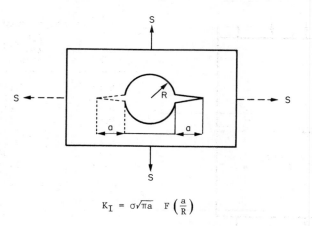

$$K_I = \sigma\sqrt{\pi a} \ \ F\left(\frac{a}{R}\right)$$

a/R	F(a/R) ONE CRACK		F(a/R) TWO CRACKS	
	(UNIAXIAL STRESS)	(BIAXIAL STRESS)	(UNIAXIAL STRESS)	(BIAXIAL STRESS)
0.00	3.39	2.28	3.39	2.28
0.10	2.73	1.98	2.73	1.88
0.20	2.30	1.82	2.41	1.83
0.30	2.04	1.67	2.16	1.70
0.40	1.68	1.58	1.86	1.61
0.50	1.73	1.40	1.83	1.57
0.60	1.64	1.42	1.71	1.52
0.80	1.47	1.32	1.53	1.43
1.0	1.37	1.22	1.45	1.38
1.5	1.13	1.06	1.23	1.28
2.0	1.06	1.01	1.21	1.20
3.0	0.94	0.93	1.14	1.13
5.0	0.81	0.81	1.07	1.08
10.0	0.75	0.75	1.03	1.03
∞	0.707	0.707	1.00	1.00

TABLE 3.17   Stress—Intensity Factors (Cont'd.)

CASE 7a   Notched beam in pure bending

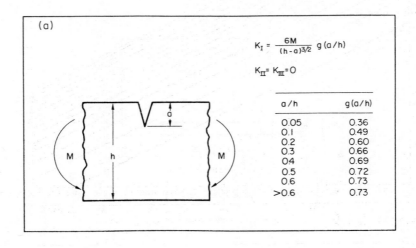

CASE 7b   Three- or Four-point bending

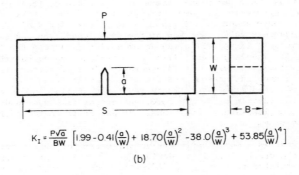

$$K_I = \frac{P\sqrt{a}}{BW} \left[1.99 - 0.41\left(\frac{a}{W}\right) + 18.70\left(\frac{a}{W}\right)^2 - 38.0\left(\frac{a}{W}\right)^3 + 53.85\left(\frac{a}{W}\right)^4 \right]$$

(b)

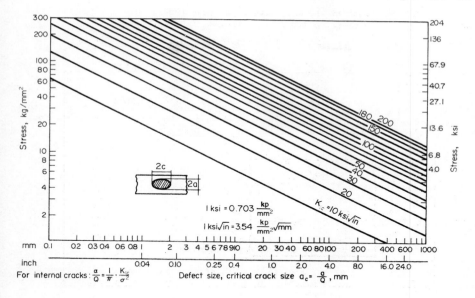

Fig. 3.73   Fracture Mechanics nomogram showing relationships
among stress, defect size and fracture toughness.

Within Linear Elastic Fracture Mechanics, the stress intensity factor is strictly
defined for plane-strain conditions only, by testing Specifications as ASTM E399,
and valid K values relate primarily to thick specimens of high strength and relatively low toughness. But for plane-stress, as evidenced by the thinner plate
(below about 3/4" for steels) of lower strength (below about 70 ksi yield) and
higher toughness (above about 80 ksi √in.), a valid test for K often cannot be performed, so the toughness value is taken at the given condition and called K_c , not
K_{Ic}. The subscripts I, II and III often accompanying K_c denote the fracture mode,
as I: tension; II: in-plane shear; III: out-of-plane shear; but they describe
the plane-strain condition only. It is generally considered that crack growth
occurs only during the tension portion of a stress cycle, thus, the compression
portion is ignored, as are certain effects reputedly originating in crack closure.
These effects: retardation of growth rate, and arrest from overloads have been
confirmed under some laboratory conditions, but use of these "benefits" is considered
imprudent at this time.

J, or the J-integral is a more comprehensive parameter for characterizing the stress-
strain field at the crack tip. It is derived on principles of internal energy
exchange, and is related to K by $J_c = K_c^2/E(1-\nu^2)$. The use of J (or an equivalent K
derived from it) permits treatment of conditions involving considerably more
yielding in tip region (plane stress) than does K, which relates to plane-strain
conditions.

The data for solutions by the Fracture Mechanics Method is given in that Section below, and consists of three items, for several alloy systems:

1.   The most probable size of initial crack, a_i.

2.   The constants $\underline{C}$ and $\underline{m}$ for the crack growth rate equation.

3.   The fracture toughness, K_{Ic} or J_{Ic}; K_c or J_c; as available.

The values have been gleaned from the open literature. While thought to be generally representative, they are not guaranteed in any way, and, except for a few scattered items from Ref. 448, they have no statistical significance.

EQUATIONS

The equations relating the several parameters are given below, written for constant-amplitude stress range.

The range of applied stress-intensity, ΔK, is, in general form, for cracks thru a thickness:

$$\Delta K = \Delta S \sqrt{\pi a} \ F \tag{1}$$

where ΔS is the range of the applied tensile stress, the nominal value remote from the crack tip, calculated as a "P/A" type with no regard for such concentrations as may be present.

a is the instantaneous crack length, in inches, in the direction in which the crack is growing. For a surface crack, a is measured into the thickness. For a through-thickness crack, a is taken along the lateral dimension. See Table 3.17 for illustration and note that what may be the visible dimension of the crack may be either the "length" or "depth" depending on the geometric condition.

F is the form factor, usually given in tabular form as a function of the controlling geometry, such as the ratio of the crack length to the width of a bar; e.g., $F = f(a/W)$. For some Cases, an equation is more convenient, see Case 7b. For surface cracks, an additional factor is necessary to accommodate the crack shape, see Case 1. It is expressed as a function of the ratio of length to depth, and is also a weak function of the ratio of applied to yield stress. Its values are usually given in graphical form, see Case 1. Eq. (1) is then rewritten:

$$\Delta K = F\Delta S\sqrt{\pi a/Q} \tag{1a}$$

Q is the crack-shape parameter for surface cracks, see Case 1. For surface cracks only, Eq. 1 picks up a free-surface correction factor, shown as the numerical coefficient, 1.12.

The critical crack size, a_{cr}, is:

$$a_{cr} = (K_{Ic}/S_{max})^2/\pi \ F \tag{2}$$

where S_{max} is the maximum value of the nominal stress range.

The crack growth rate, da/dN in inches per cycle, is:

$$da/dN = C(\Delta K)^m \tag{3}$$

where C and m are material constants, empirically determined.

For the ferritic-pearlitic structural steels: $C = 3.6 \times 10^{-10}$;
$m = 3.0$.

For those alloys whose growth rates are sensitive to the ratio:
$R = K_{min}/K_{max}$, Eq. 3 takes on the form given by Forman:

$$da/dN = \frac{A(\Delta K)^n}{(1-R)K_c - \Delta K} \qquad (3a)$$

where A and n are different constants. (Generally, the structural steels are not much affected by R, but some aluminums are).

The number of cycles to failure, N_f is, in general:

$$N_f = \int_{a_i}^{a_{cr}} \frac{da}{f(\Delta K, K_{max})} \qquad (4)$$

and, to a first approximation, the solution for Eq. 4 is:

$$N_f = (a_{cr} - a_i)/(da/dN) \qquad (4a)$$

where a_i is the initial size of crack. In the calculation of residual life upon the observation of a crack, a_i becomes the observed length. The method of solution for N_f usually reduces to one of numerical integration due to the form and complexity of Eq. 4. However, Eq. 4 becomes rather simple for some cases, especially when the form factor is unity and the stress cycles is zero-to-a-tension-maximum, so it may be possible to integrate directly for N_f.

If: $F = 1.0$; $Q \sim 1.0$ and $\Delta S = S_{max}$

Then: $K = K_{max}$ and $K = S_{max} \sqrt{\pi a}$;

the number of cycles goes from 0 to N_f as a_i goes to a_{cr}:

$$N_f = \int_o^{N_f} dN = \int_{a_i}^{a_{cr}} \frac{da}{da/dN}$$

$$N_f = \int_{a_i}^{a_{cr}} \frac{da}{C(K_{max})^m}$$

$$N_f = \int_{a_i}^{a_{cr}} \frac{da}{C(S_{max}\sqrt{\pi a})^m}$$

$$N_f = \frac{1}{C} \left[\frac{2}{m-2} \right] \left[\frac{a_{cr}}{[S_{max}\sqrt{\pi a_{cr}}]^m} \right] \left[\left[\frac{a_{cr}}{a_i} \right]^{\frac{m}{2}-1} - 1 \right], \text{ For } m \neq 2$$

For m = 2: $N_f = \frac{1}{CM(\Delta S)^2} \ln\left(\frac{a_{cr}}{a_i}\right)$

where M = flaw-shape parameter = $1.21\pi/Q$, for surface flaws.

The number of runs through a loading or stress spectrum, N_R:

$$N_R = \frac{1}{\dfrac{n_1}{N_1} + \dfrac{n_2}{N_2} + \dfrac{n_3}{N_3} + \cdots}$$

where the subscripts refer to the several levels of stress; n is the number of cycles applied at each level; N is the number of cycles to failure at each level (from Eq. 4). See Fig. 3.79.

The application of these Equations is illustrated in the following section in Procedures and Examples.

PROCEDURES AND EXAMPLES

General Procedure to Calculate Life in the presence of a crack, i.e., to produce Operating Limit Lines, which are plots of crack length vs. cycles to failure, at various stresses, as shown by Fig. 3.74.

1. From the shape of the part under consideration and the direction of maximum tension, find in Table 3.17, or in Ref. 438-40, the Case best representing the situation. While it is possible to calculate the stress-intensity factors, the mathematical manipulation is extensive and has been done for all common Cases. Note the form of the relationships for ΔK, F, and Q.

2. Set up a worksheet, as Table 3.18.

3. Calculate a_{cr} by Eq. 2, to its <u>minimum</u> value, i.e., until $a_{cr} = \underline{a}$.

4. Calculate ΔK for chosen increments of crack length between a_i and a_{cr}, by Eq. 1.

5. Calculate da/dN by Eq. 3.

6. Calculate N_f by Eq. 4a.

Example 1

Consider an A36 steel bar or strip of finite width under tension, with a nick in one edge as from flame-cutting, which will, shortly after introduction into service, develop a crack through the thickness.

a) A reasonable representation of the condition is Case 3, and the only dimension needed is the half-width, <u>b</u>. This assumption holds for thicknesses less than the width.

b) Take values: b = 3 inches; (2b = W)

$$K_{Ic} = 90 \text{ ksi } \sqrt{\text{inch}}$$

$$S_{max} = 20 \text{ ksi}$$

$$C = 3.6 \times 10^{-10}$$

$$m = 3.0$$

and Eq. 1 for Case 3 is: $\Delta K = \Delta SF\sqrt{\pi a}$

c) Choose a_i = 0.050 inch, typically. The initial crack size cannot be zero, and it has been observed that welding and flame-cutting in A36 and A441 steels often produce initial cracks, from thermal shrinkage, of about 0.060 inch. So, any value somewhat below .060 is a proper starting point. The critical length, a_{cr}, will be calculated for a set of increasing a_i's until $a_{cr} = a_i$, which value of a_{cr} will necessarily be the minimum.

d) Set up the first 6 columns in the Worksheet.

e) With the a_i's above, form the ratio a_i/b as required to find the F's.

f) From the table of form factors for Case 3, find F = 1.1.

g) Calculate an a_{cr} by Eq. 2 as 5.860 inch, (vastly greater than a_i).

h) Choose a new greater value of a_i and repeat (e), (f) and (g) until $a_{cr} = a_i$, to within a few percent.

Note that in these trials, Line 4 shows an overrun, i.e., a_{cr} less than a_i, obviously impossible. a_{cr} must lie between 2.253 and 4.882. The last line shows $a_{cr} = a_i$, to about 1%, acceptable.

i) Set up Columns 7, 8 and 9; (7 is only a convenience).

j) Calculate ΔK by Eq. 1 for Case 3; and $(\Delta K)^m$. Omit any overrun, as Line 4.

k) Calculate da/dN by Eq. 3. (x 10 to the appropriate exponent).

l) Calculate $(a_{cr} - a_i)$.

m) Calculate N_f by Eq. 4a.

Columns 3 and 12 now plot as the S = 20 ksi curve in Fig. 3.74.

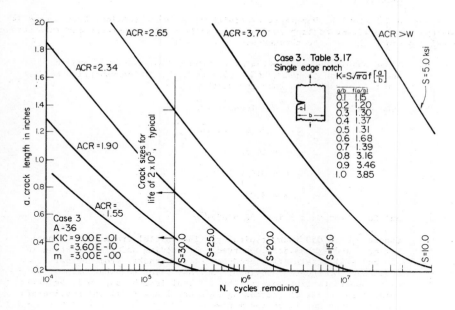

Fig. 3.74 Operating Limit Lines

TABLE 3.18 Typical Worksheet for Operating Limit Lines.

Case 3: Single-Edge Notch, Finite-width Strip in Tension

(1)(2) W b	(3) a_i	(4) a_i/b	(5) F	(6) a_{cr}	(7) $\sqrt{a_i}$	(8) ΔK	(9) $(\Delta K)^m$	(10) $da/dN \times 10^{-10}$	(11) $a_{cr} - a_i$	(12) N_f
6 3	.050	.017	1.10	5.860	.224	8.73	665	2161	5.810	2.688×10^7
	.200	.067	1.14	5.650	.488	18.10	5929	19270	5.450	2.828×10^6
	1.000	.334	1.32	4.882	1.000	46.80	102500	333100	3.882	1.165×10^5
	3.000	1.000	2.86	2.253					0	0
	1.500	.500	1.51	4.260	1.224	65.52	274600	892500	2.760	3.092×10^4
	2.000	.667	1.82	3.540	1.414	91.23	753500	2448900	1.540	6.289×10^3
	2.500	.856	2.25	2.864	1.581	125.10	1953000	6347000	0.364	5.730×10^2
	2.600	.866	2.36	2.730	1.612	134.86	2452700	7971000	0.130	1.630×10^2
	2.650	.883	2.41	2.674	1.628	139.08	2690300	8743500	0	0

Repeat steps (b) through (m) for values of S_{max} of interest.

Notes:

1. When using a hand calculator, it is convenient to do the operations for Columns 8-12 line-by-line.

2. Other convenience columns may be introduced between 7 and 8; $S\sqrt{\pi}$ is useful.

3. A simple computer program may easily be written for these operations, one version already exists.

Operating Limit Lines for a <u>surface</u> crack, part way through the thickness, Case 1, are shown in Fig. 3.76, and Fig. 3.77 gives a variation in format: 2 stresses and 3 flange widths.

Example 2

To calculate life under complex crack-growth conditions; (to produce Operating Limit Lines).

Consider a typical box car body bolster, Fig. 3.75. Such a bolster is one of two lateral beams that collect the car load from the sills and transfers it to a similar bolster in the truck. Bolsters are either box- or wide-flange beams, in bending at small deflection, so the predominant stress in an outer fiber is safely assumed as tension only. Assume that the beam carries a partial-length cover plate and that there is a toe crack in the end fillet weld. Initially, the crack is only part way through the thickness of the under-plate or flange. Case 1 is representative of this situation while the crack is growing through the remaining thickness, then Case 2 takes over for the "thru-the-thickness crack" condition and the growth thereafter. The general procedure is identical to that in Ex. 1, with the following conditions:

Case 1:   a) $\underline{a}$ is the crack depth into the thickness: $\underline{a}$ cannot exceed $\underline{t}$.

b)   The visible length of surface crack is 2c.

c)   The crack grows both ways: into the thickness and along the surface in some fixed ratio of length to depth, (or of the elliptical axes).

d)   The form factor is unity. (F)

e)   The crack-shape parameter, Q, is a constant greater than 1.0.

f) $K = 1.12 \ S\sqrt{\pi a/Q}$.

Case 2:   g) $\underline{a}$ remains equal to $\underline{t}$.

h)   Growth occurs only in the $\underline{c}$ "direction".

i)   The form factor, F, becomes a variable.

j)   The crack-shape parameter, Q, becomes unity.

k) $K = S\sqrt{\pi c}F$.

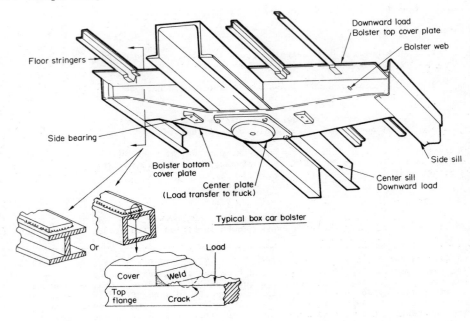

Fig. 3.75 Typical railroad car bolster (main lateral beam).

For the Example take values:

A-441 steel; K_{Ic} = 90 ksi $\sqrt{inch}$; S_{max} = 10 ksi; S_y = 50 ksi;

C = 3.6 x 10^{-10}; m = 3.0; W = 8 inch; b = 4 inch;

$a/2C$ = 0.2; $2c_i$ = 0.060; a = .012; t = .180 inch;

when a = t: $2c$ = .900 and c = .450.

The steps are basically the same as in the first example, but note that on the Worksheet Table 3.19, the operations under Case 1 are limited to the calculation of N_f over the range of a_i from the observed depth to the thickness. The "observed depth" is, of course, derived from the observed length on the surface. When a_i reaches .180, c has reached .450, and a_i is no longer relevant. The crack-shape parameter, Q, is found from the curves given with Case 1 as a function S/S_y (here equal to 10/50), and $a/2c$, (here taken as 0.2, a frequently-found ratio.)

The first Line under Case 2 picks up the c value from the last Line under Case 1. Q becomes unity, and F a variable, then the Procedure is identical to that in Example 1. The results, N_f and c, plot as the dotted line in Fig. 3.76.

The discrepancy between the last value of N_f under Case 1 and the first under Case 2 is due to the highly variable physical conditions at break-thru, and, analytically, to the use of two models (Cases) to describe the same condition, neither model being exact.

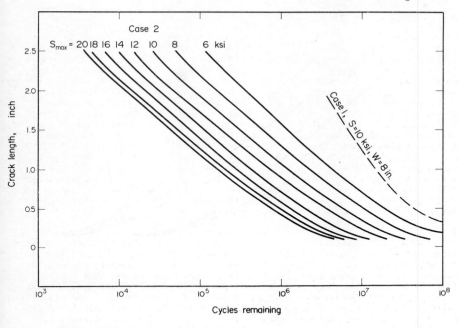

Fig. 3.76   Operating Limit Lines for beams with partial-length
cover plates. Flange width = 8 inch. A441 steel.
K_{Ic} = 90 ksi $\sqrt{inch}$.

Note that the analysis considers the half-width of the strip (flange), and the
half-length of the crack, then proceeds by symmetry.

Note also, that in both Cases the critical crack length is much greater than the
strip width, thus there will be no brittle fracture, the slow, stable growth continuing on through the entire width of the strip. This condition is frequently
found in the low-strength, high-toughness steels.

The Operating Limit Lines resulting from this calculation, Figs. 3.76 and 3.77 are
shown for a variety of conditions. Fig. 3.76 covers Cases 1 and 2 for several
stress levels but only one flange width while 3.77 shows Case 2 only, 2 stress
levels and 3 widths. These conditions were introduced, arbitrarily, to illustrate
the variety which may be encountered, and to indicate some of the directions
which the designer may wish to explore in a search for optimization.

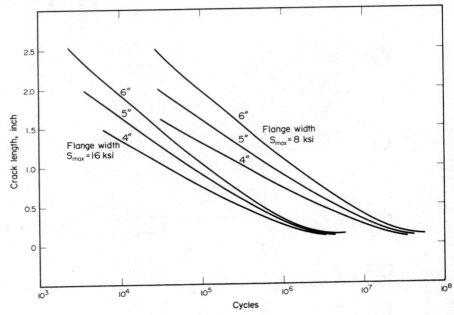

Fig. 3.77 Operating Limit Lines for beams with partial–length cover plates. Flange widths 4", 5" and 6". $\underline{S_{max}}$ = 8 ksi and 16 ksi. A441 steel. K_{Ic} = 90 ksi $\sqrt{inch}$.

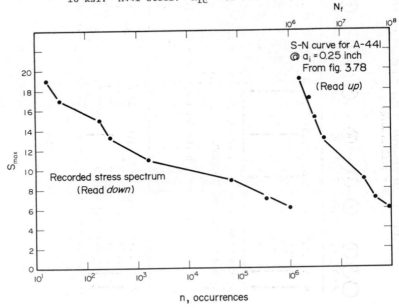

Fig. 3.78 Performance Curve for life under Spectrum loading (Example 3).

TABLE 3.19 Worksheet for Operating Limit Lines Under Complex Crack-Growth Conditions

Case 1 Part-through Surface Crack: Toe crack in fillet Weld.
$K = 1.12 \, S\sqrt{\pi a/Q}$.

① W	② b	③ c_i	④ c_i/b	⑤ F	⑥ Q	⑦ c_{cr}	⑧ a_i	⑨ $\sqrt{a_i}$	⑩ ΔK	⑪ $(\Delta K)^m$	⑫ $(da/dN)10^{-10}$	⑬ $a_{cr} - a_i$	⑭ N_f
8	4	.030	-	1.0	1.3	25.78	.012	.11	1.92	7.03	2107	5.142	5.03×10^8
		.125					.050	.224	3.89	59.00	177.00	5.106	2.88×10^8
		.250					.100	.316	5.50	166.87	500.60	5.056	1.01×10^8
		.375					.150	.387	6.74	306.50	1103.6	5.006	4.53×10^7
		.450					.180	.424	7.38	403.00	1450.0	4.976	3.43×10^7

a_i has become equal to $\underline{t}$, so now the crack is thru-the-thickness, and Case 2 is representative.
$\underline{c_i}$ now becomes the "length".

Case 2 Through-the-Thickness Crack, $K = S\sqrt{\pi c F}$

① W	② b	③ c_i	④ c_i/b	⑤ F	⑥ Q	⑦ c_{cr}	⑧ a_i	⑨ $\sqrt{c}$	⑩ ΔK	⑪ $(\Delta K)^m$	⑫ $(da/dN)10^{-10}$	⑬ $c_{cr} - c_i$	⑭ N_f
8	4	.450	.113	1.007	1.0	25.60		.670	11.90	1689	6080	25.15	4.14×10^7
		1.000	.333	1.05		24.55		1.00	18.13	5966	21478	24.60	1.14×10^6
		2.000	.666	1.20		21.48		1.01	27.41	20608	61824	23.60	3.85×10^6
		4.000	1.000	1.20		21.48		2.00	38.77	58315	174945	21.60	1.23×10^6

Example 3

To calculate Life under Spectrum Loading; (to produce a Performance Curve).

Given:

The box-beam of Ex. 2 with a 0.25 inch toe crack in the end fillet weld of the cover plate on the tension side; through-the-flange-thickness, (Case 2).   The Spectrum is as plotted in Fig. 3.78 and was actually recorded on a loaded 50-ton box car during a 615 mile trip.

Procedure:

From Fig. 3.76, at the given a_i, find N_f at each of several stresses.   Plot on the same axes, see Fig. 3.78.   Note the two abscissae scales: $\underline{n}$ and N_f.   Fig. 3.79 gives a geographical illustration of this Procedure.

Discussion:

The degree of success or failure is the distance on the abscissa between the two curves.   The spectrum plot represents one run, so the ratio N_f/n is the number of runs, N_R.   But since the slopes will, typically, be neither uniform nor equal, no single value of runs can be obtained.   Eq. 5 calculates the average number of runs, and it is observed that roughly equal values will be obtained by taking the ratio of abscissae at the average value of the stress range.   In this Example, the latter gives 6100 runs while Eq. 5 gives 5600.   No single, discrete value of N_R exists because of the random sequence in the natural application of the stress levels.   We resort to averaging as a means of obtaining some "characteristic" value of N_R.   If, in Fig. 3.78, N_R was calculated at each stress level, some N_R's less, and some more than the average would result; this condition appears only under the (invalid) assumption that all cycles of that stress were applied at one time (in one block) before the cycles of any other stress level.   Averaging thus ignores any potential interaction effects, but is generally acceptable for the low-carbon steels which are not particularly susceptible to interaction effects of different load levels.

This Example illustrates a successful design with material of adequate toughness in the presence of this initial crack: N_R is a finite number of considerable magnitude.   Poor designs would be characterized by low N_R's, and disasters by the S-N curve intersecting the spectrum or falling entirely to its left, with N_R then less than one.

For other situations wherein an operator has control of the load or stress sequencing, it is possible to calculate a discrete N_R for different sequences, but this operation requires a check, both analytically and during service, of the accumulating crack length, to insure that it does not overrun the a_{cr} for the stress level in the next block of cycles.   (See the example in the Section of Spectra).

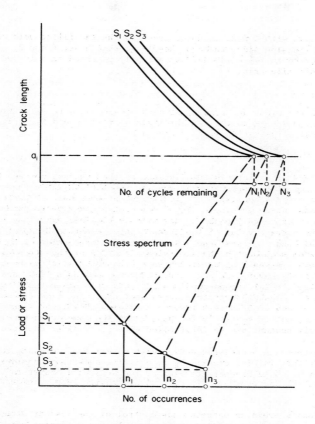

Fig. 3.79 Graphical illustration of procedure to obtain
n, N and S values for plotting Performance Curves.

STRESS CORROSION CRACKING (SCC)

Considerations of SCC are often limited to the static, tensile stressing of smooth specimens. Of these three conditions, only the tension can be extrapolated directly to as-built machine and structural members, for the latter inevitably contain crack-like surface defects and are subject to cyclic loads. The SCC rate as a function of combined static and cyclic stressing can be obtained only by specific testing. The discussion below is confined to the effects of the cracks, and of assembly and heat treat stresses.

Whether the defect extends as a stress corrosion crack depends not only on the nature of the alloy, the environment and the magnitude of stress, but also on the depth of the initial crack. In the presence of a crack, nominal stress is a fiction which can be misleading, and at the present time, the response of a cracked body to loading can be treated quantitatively only by fracture mechanics. Stress corrosion cracks are brittle, that is, for the most part they occur before the onset of general yielding, while the bulk of the metal is still in the elastic state. Therefore, it is appropriate to use the Irwin linear elastic fracture mechanics for SCC even in metals too ductile for the practical measurement of purely mechanical fracture toughness.

The SCC growth rate as a function of applied stress intensity, K, is shown in Fig. 3.80 and the plot resembles the standard da/dN vs. ΔK plots, plus a plateau. In Region II, the SCC rate is high and insensitive to K, but the sensitivity becomes very high at lower levels of K, Region I, and at sufficiently low K there appears to be a threshold which has been designated K_{Iscc}. That such a genuine threshold K exists has been clearly proven for titanium alloys in salt water, and it appears probable in steels; but its existence is in question in some aluminum alloys. The value of such a parameter of K_{Iscc}, where such a genuine threshold exists, lies in one's ability to predict the combination of remote (nominal) stress and size of crack-like surface flaw through the use of the Irwin equation:

$$K^2 = \frac{1.2\pi\sigma^2 a}{\varphi^2 - 0.212\left(\dfrac{\sigma}{\sigma_y}\right)^2},$$

(1)

where a is the depth of the crack, σ is the stress, σ_y is the yield strength, and φ is a factor for the shape of the crack. If one assumes a long, thin flaw and the existence of yield-point stress, then from Eq. (1) stress corrosion would be expected to propagate if the flaw depth exceeded a_{cr}, given by:

$$a_{cr} = 0.2\left(\frac{K_{Iscc}}{\sigma_y}\right)^2$$

(2)

The value of a_{cr} may thus be regarded as a figure of merit which incorporates both the SCC resistance K_{Iscc} and the contribution which yield strength stress levels can make to SCC hazard by virtue of residual or fit-up stresses. These equations are identical in form to those above, and as given with Case 1.

A convenient way of displaying the SCC characteristics of materials is to plot Eq. (2) for various values of a_{cr}, as in Fig. 3.81 . If a given material was found to have a K_{Iscc} as indicated by point x, a surface crack 0.1 in (2.5mm) deep would be deeper than required to propagate a stress corrosion crack in the same environment used to determine the K_{Iscc}. But a crack 0.01 in (0.25mm) deep would not propagate as a stress corrosion crack in the same material in the same environment.

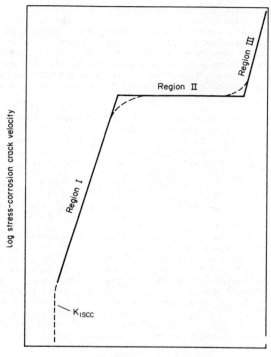

Fig. 3.80   Effect of stress-intensity on SCC kinetics. Region III
is often missing.   Region I may be missing in some
systems.   Regions I & II may be strongly curved.

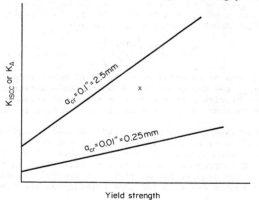

Fig. 3.81   Plot of Eq. 2 for two values of a_{cr}.

A plot of K_{Iscc} data such as Fig. 3.81 can be used in a slightly different way:
If one knows that he cannot detect and remove crack-like flaws less than, say 0.01
in (0.25mm) deep, then if he has yield-point stresses, he must select a material
having a K_{Iscc} above the 0.01 in line.

The K_{Iscc} values for a number of commercial and experimental titanium alloys are
shown in Fig. 3.82 as a function of yield strength. Lines corresponding to
various critical crack depths (following Eq. 2) are included in this plot as an
aid to interpretation of the degree of susceptibility of various alloys. Another
way to regard K_{Iscc} is to view it as linearly proportional to the load-carrying
capability of a specimen or component containing a standard size flaw.

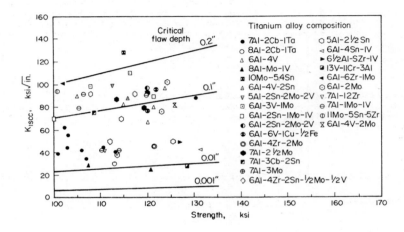

Fig. 3.82 K_{Iscc} data on titanium in sea water.

It is certain folly to design for a sustained working stress equal to a measured
SCC threshold stress, however carefully that determination may have been made, but
rather a sizeable margin must be left. The reason for this margin is that unknown
stresses from heat treatment, fit-up, welding, and thermal expansion are the causes
of a majority of service failures, and one does not want to run the risk that
unknown stresses from such causes may elevate his design stress above the SCC
threshold stress.

Considering first the assembly stresses, (in an aluminum alloy) the sketch of
Fig. 3.83 shows how a component, machined in such a way as to expose the vulnerable
short transverse texture (or end grain), is assembled by bolting to a rigid member
with a slight angular mismatch, so that there is sustained high tensile stress
across the short transverse texture, a sure-fire recipe for SCC failure, if the
alloy is susceptible. The lesson of Fig. 3.83 should be learned thoroughly for
the principles apply to many seemingly different situations. For example, one
common design which causes sustained stress is the tapered pin across the short
transverse texture of a bar or extrusion. The use of forced-fit inserts such as
the tapered pin or bearings in aluminum alloys susceptible to SCC is unwise, and
when such designs must be used, conservative practice dictates that the alloy and
temper be in the "low" or "very low" category as these terms are used in Table 3.20
and Fig. 3.84 .

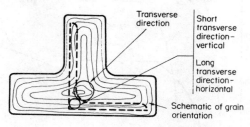

Location of machined angle with respect
to transverse grain flow in thick tee

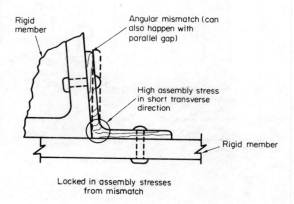

Locked in assembly stresses
from mismatch

Fig. 3.83 One method of generating high sustained
stress across the short transverse direction.

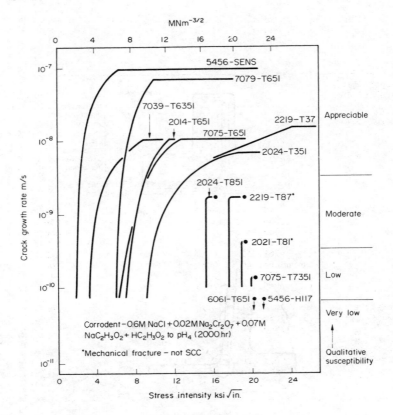

Fig. 3.84   K-SCC rate curves showing various
apparent K_{Iscc} levels.

TABLE 3.20   Categories of aqueous susceptibility of commercial
wrought alloys in plate form--short transverse
orientation.

Susceptibility category	Alloy	Temper
A. Very low	1100	all
	3003, 3004, 3005	all
	5000, 5050, 5052, 5154, 5454, 6063	all
	5086	O, H32, H34
	6061, 6262	O, T6
	Alclad: 2014, 2219, 6061, 7075	all
B. Low	2219	T6, T8
	5086	H36
	5083, 5456	controlled
	6061	T4
	6161, 5351	all
	6066, 6070, 6071	T6
	2021	T8
	7049, 7050, 7075	T73
C. Moderate	2024, 2124	T8
	7050, 7175	T736
	7049, 7075, 7178	T6
D. Appreciable	2024, 2219	T3, T4
	2014, 7075, 7079, 7178	T6
	5083, 5086, 5456	sensitized
	7005, 7039	T5, T6

Other situations involving residual stresses are discussed in Section 4.4.1.

Now consider heat treating stresses. Aluminum and other alloys which are strength-
ened by age hardening are first given a solution treatment at elevated temperature,
quenched, and then are either allowed to age at room temperature or at an elevated
temperature. To establish the right conditions for age hardening, it is necessary
to then cool a given alloy from the solution temperature faster than some minimum
rate, and for many alloys, this requires quenching in cold water. Quenching
usually places the surface in compression, as shown in Fig. 3.85. SCC does not
initiate or propagate when the alloy is in compression, therefore machining off
the compressive layer after quenching is undesirable from the SCC viewpoint. Note
in Fig. 3.85 that if machining should extend to the centerline, stresses there,
if unaltered by the machining operation, would exceed the threshold stress for SCC
for several alloys. If the quenched piece is stretched mechanically, the quenching
stresses are largely removed (Fig. 3.85). This mechanical relief is a standard
mill procedure available for most product forms and converts the -T6 temper to the
-T651 temper.

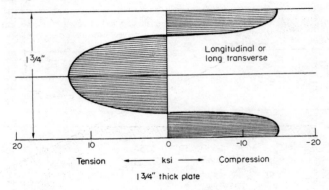

1 3/4" thick plate

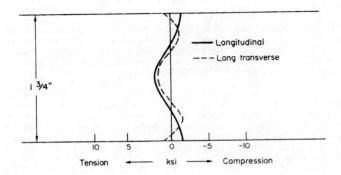

Fig. 3.85   (UPPER) Residual stresses in 7075-T6 plate
             quenched in cold water. (LOWER) Same, after
             2% stretch.

Blind holes in pieces being quenched cause residual tensile stresses in the poorly
quenched regions of the hole unless special fixturing or sprays are used. Quenching
stresses may be reduced by quenching in air or hot water providing the alloy, part
thickness, and mill product form are favorable. But quenching some alloys in
boiling water rather than in cold water, such as 2024-T4, can render the product
highly susceptible to intergranular corrosion without being stressed, perhaps
because of precipitation during the quench. Therefore, one follows faithfully the
heat treating procedure which has been proven for each commercial alloy.

An attempt has been made to correlate the fracture mechanics term K_{Iscc} and its
implications with the behavior of smooth specimens of the same alloy. As an example,
for 7079-T651, K_{Iscc} was estimated at 4 ksi $\sqrt{in.}$, the threshold stress was 8 ksi,
and the yield strength is 67 ksi. It is instructive to compare the implications of
these data graphically, as shown in Fig. 3.86 . Inserting the data above in Eq. 1
and rearranging and taking logarithms of both sides, one finds that the log of the

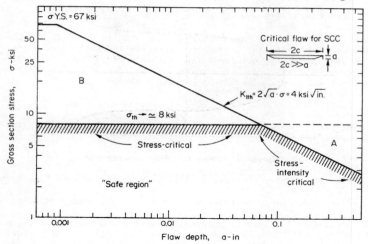

Fig. 3.86   Flaw depth vs. stress for 7079-T651 plate. Sloping
line is plot of Eq. 1. In area "A" smooth specimen
data would erroneously imply safety against SCC. In
area "B" fracture mechanics data would erroneously
fail to predict SCC.

stress in linearly proportional to the log of the flaw depth a, as shown by the
sloping line in Fig. 3.86 . One would suppose that this line should intersect the
threshold stress line at some value of flaw depth approximating either the initial
surface roughness of the specimen or else some value of the flaw size commensurate
with any corrosion pits which may have formed.   In fact, however, as may be seen in
Fig. 3.86 , the intersection is at about 0.06 in (1.5mm).   Either the estimate of
K_{Iscc} is too high, if the concept of K_{Iscc} is valid for this alloy, or some extra-
neous phenomenon is entering.   In any event, as matters now stand, one would have
to be governed by both the smooth specimen threshold data and the fracture mechanics
data.   The threshold data from smooth specimens is incapable of taking account of
the effects of flaws, and clearly from the sloping line one cannot be safe at the
threshold stress level with flaws present deeper than about 0.06 in (1.5mm), assuming the customary surface flaw shape, 10 times as long as deep; likewise, one is
not safe using the fracture mechanics guidance below that flaw size.

It seems to be the present status of SCC technology that one must act upon the most
pessimistic SCC characterization data, whether they be threshold data from smooth
specimens, fracture mechanics data, or data from constant strain rate tests, and in
the absence of positive information to the contrary, it is a prudent rule to assume
that somewhere in the structure there will probably be stresses equal to the yield
strength.

The reader should appreciate that few laboratory SCC characterization tests are of
durations comparable with the desired service life of most structures, and added
conservatism taking this state of affairs into account is warranted in interpreting
laboratory SCC data to engineering practice.

Much relevant comment on the susceptibility of many alloys in various environments
may be found in Ref. 470.

INSPECTION LEVELS AND INTERVALS

The topic of Inspections and Inspectable Regions was discussed in Section 2.4.2, and now the basic question is: "How often must the inspection be made to insure that the growing crack does not reach its critical length, or, conservatively, that the length remains safely below some reasonable portion of the critical length?" The most direct and accurate method to obtain an answer is:

1. Identify the geometrically complex regions of the design.

2. Determine the maximum tension or moment by a spatially detailed stress analysis.

    [(1) and (2) frequently focus on the same region]

3. Calculate the best Operating Limit Line or Performance Line possible.

4. Adjust the Line as experience accumulates by testing and service.

Table 3.21 identifies the type of information needed and the logistics to make the General Procedure work. The cards are to be used directly by the Inspector on the shop floor, or, for the body bolster, on the rip track. The earliest date on a card should be that of fabrication, but if the equipment is already in service, then the best that can be done is to make frequent inspections to enable an estimate of the slope of the growth curve. The Design Office establishes the card; Inspection provides the raw data; Design then evaluates it. A typical drawing of the part showing Inspection Regions is given as Fig. 3.87 . It appears directly on the card to guide the Inspector. Many cards may be necessary to cover all pertinent regions of a large complex structure or machine.

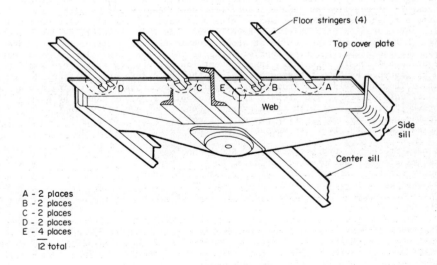

A - 2 places
B - 2 places
C - 2 places
D - 2 places
E - 4 places
12 total

Fig. 3.87   Typical Inspection Regions on Railcar body bolster.

TABLE 3.21   Inspection Outline

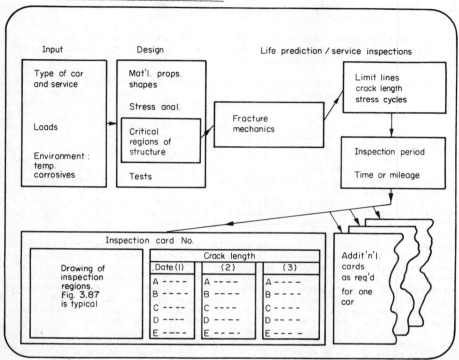

The general procedure being described here can be applied at any crack length; for
large forgings, castings and weldments, visual examination often aided by magnetic
particle or dye penetrant is normally appropriate. But for the smaller sections
and higher confidence, resort is usually made to other Non-Destructive-Evaluation
means (NDE) such as ultrasonic, X-ray and acoustic emission. The basic problem is
not how small a defect can be found, but how big a defect may be missed? At the
present time, the threshold for commercial production inspection is about 0.050 -
0.060 inch, in steels, aluminums and titaniums. Some lab techniques appear capable
of working down to a few thousandths, but unless specific data at the required
confidence level is available, it seems only prudent to take a value of not less
than 0.050 inch for a_i. The Aircraft Structural Integrity Program (ASIP) of the
U.S. Air Force, requires demonstration of <u>safety</u> in the presence of a 0.050 inch
radius corner crack at the worst hole, and of reasonable growth rates from a 0.005
inch corner crack for <u>maintenance</u>, i.e., economically long inspection intervals.

The Design Office files curves such as Figs. 3.76 and 3.77 for the design--suppose
that a card comes in from Inspection with a region noted as cracked to 1.25 inch.
This region normally sees 12 ksi, so from Fig. 3.76 the coordinates indicate about
4×10^5 more <u>cycles</u> to failure. The spectrum curve, Fig. 3.78, shows about 6.5×10^2
occurrences of 12 ksi <u>per run</u> through the spectrum, thus:

$$\text{Runs remaining} = (4.0 \times 10^5)/(6.5 \times 10^2) = 667$$

The earlier discussion of this spectrum and an initial crack of 0.25 inch indicated a total life of 5600 runs, so it has taken (5600 - 667) runs to grow the crack from 0.25 to 1.25 inch, thus:

$$\text{Crack growth per run} = (1.25 - 0.25)/(5600-667) = 2 \times 10^{-4} \text{ in./run}$$

as an average rate.

Now, in Table 3.19, Col. 12, note that the instantaneous rate for a 1.25 inch crack is 26500×10^{-10} in./cycle, by interpolation from Col. 3.   The rate per run at 1.25 inch is:

$$(26500 \times 10^{-10}) \ (6.5 \times 10^2) = 1720 \times 10^{-4} \text{ inch per run}$$

as an instantaneous rate.

Thus, the latter is 860 times the former--such a value points up very sharply the approximation inherent in averaging and that working versions of Operating Limit Lines should have a greatly expanded scale for crack length. Note also, that the cycle scale is for cycles remaining, the inverse of the conventional a vs. N curve such as Fig. 3.71. A plot of such data could be made from Col. 3 vs. Col. 4 in Table 3.19, with Col. 12 giving the instantaneous slopes.

The term "Inspection Level" can mean a number of different things: 1) A minimum defect size; 2) A statistical sample size; 3) A confidence level; or 4) A temperature. The first three items are well known, the fourth represents a correlation between the transition Temperature Approach and Fracture Mechanics Method. This correlation is illustrated by the example below from [465]; it should be compared with Langer's work [178] and that in the section on Pressure Vessels. The objective is to establish the relationship amongst calculated stresses, expected operating temperatures, inspection limits and life criteria for a gas turbine disk.

A quick overview of the steps to be taken is shown schematically in Fig. 3.88. The first is to select a flaw model for study, Fig. 3.88a. Once several assumptions have been made, a plot of cyclic life versus initial flaw size for a constant stress and for various fracture toughness (K_{Ic}) levels is generated, Fig. 3.88b, using a crack growth computer program such as CRACKS [466] or [467]. Plots of this type are required at various stress levels in order to continue with the analysis. The next step is to cross-plot K_{Ic} versus initial flaw size for various stress levels at a suitable constant value of cyclic life, Fig. 3.88c.

Next, an inspection limit must be obtained from nondestructive evaluation for placement on the K_{Ic} versus flaw size plot. Where this inspection limit crosses a stress line defines the K_{Ic} level required for the selected stress and inspection limit (point A, Fig. 3.88c). Use can now be made of K_{Ic} versus excess temperature curves for the materials of interest, Fig. 3.88d. Excess temperature is defined to be the operating temperature (or temperature of a fracture toughness test) minus the 50 percent Fracture-Area-Transition-Temperature (to be labeled simply FATT) for the material. The 50 percent FATT is defined to be that temperature at which the fracture surface of a Charpy bar shows 50 percent fibrous fracture and 50 percent brittle fracture. Reference 468 discusses more details on this measure of fracture resistance. If the K_{Ic} required from point A in Fig. 3.88c is employed, a required excess temperature for the material of interest is obtained. The final step is to calculate combinations of operating temperature and FATT which give the desired excess temperature. The locus of the combinations defines allowable FATT values for a given material, stress, inspection limit and cyclic life as a function of operating temperature, Fig. 3.88e. Attention will now be given to the details of each step starting with the model assumption and ending with a discussion of the use of the final "design" curve of operating temperature versus FATT.

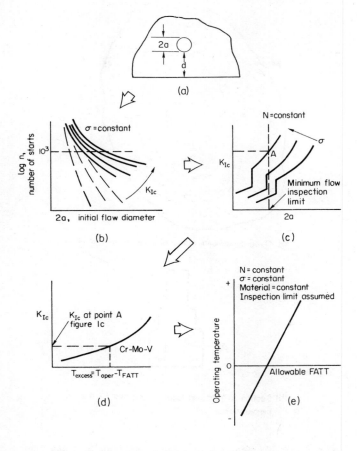

Fig. 3.88   Overview of analytical method.

Any analytical method formulated to cover a complex area like inspection limits for turbine wheels requires some assumptions and simplifications to make the technique manageable. Although any model of flaw could be chosen, only the buried penny-shaped flaw close to a bore and oriented perpendicular to the tangential stress will be considered here, Fig. 3.89a and b.

It is further noted that in this work all flaws will be initially located so that the edge of the flaw is a minimum of 2.54mm (0.100 in.) from the surface of the bore as shown in Fig. 3.89b.

Another problem to be addressed is that of stress distribution in the bore area of a turbine wheel and how to account for various gradients in the model. It was decided to employ only uniform stress fields in model analysis. This technique

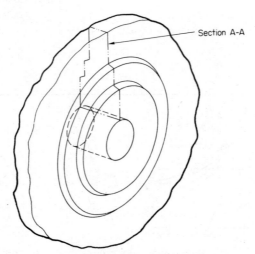

(a) Three-dimensional sketch of a typical turbine wheel

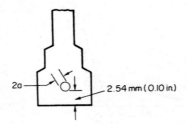

(b) Section A-A showing location assumed for the penny-shaped flaw.

Fig. 3.89   Typical turbine wheel and flaw location used
in illustrative example.

amounts to utilizing a reference stress state which can be adjusted after all calculations are complete to account for gradient cases. This aspect will be considered further after final curves are generated.

Another simplification is selection of a single crack growth curve for all ferritic materials that could be used in turbine wheels. Without this simplification, a different analysis would have to be conducted for each material and operating temperature. However, much literature exists on crack growth properties of ferritic steels and consideration of this data reveals some interesting things. Fig. 3.90 shows some of the available data on both Ni-Cr-Mo-V (open symbols) and Cr-Mo-V (filled-in symbols) at two temperatures. It is noted that the solid line represents an upper bound to all data at intermediate crack growth rates, da/dn and lower bound to the threshold K at very small growth rates. Thus, up to about 10^{-6} m/cycle growth rates, the line represents conservative bounds and is used in the analysis.

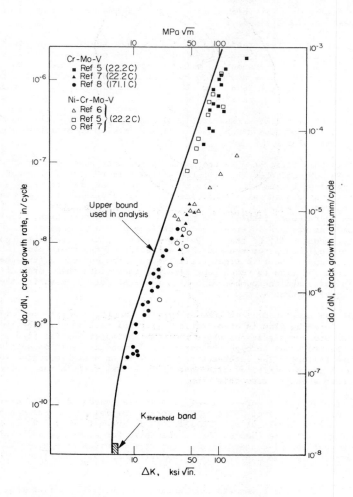

Fig. 3.90   Crack growth data for ferritic steels.

Even though the majority of life is used up in the threshold and linear region of the crack growth curve, the determination of a critical size to cause catastrophic-failure depends on the maximum K achievable in a crack growth test. Although this maximum K value in a crack growth test can sometimes be higher than the plane strain fracture toughness, K_{Ic} of the material as measured in a monotonic test, the convenient conservative assumption will be made here that monotonic K_{Ic} values are equal to the maximum K obtained in a crack growth test. Thus, since it is well known that K_{Ic} varies with temperature in ferritic materials, it follows that the maximum K in a crack growth test will vary with temperature. Because of the above, the K_{Ic} level must be left as a variable in the analysis and in order to insure that all potential toughness values are included, the hypothetical crack growth curve must extend to a very high growth rate. The maximum K or K_{Ic} used here will be 221.4 MPa$\sqrt{m}$, (200 ksi $\sqrt{in.}$).

It will be noted that the transition from the linear region of the crack growth curve to the K_{Ic} level is assumed to be abrupt and growth rates suddenly become infinite, a conservative assumption made to expedite and generalize the analysis. With the crack growth curve now determined, it is possible to pose the question, "How many cycles will it take a flaw with some initial size, 2a, to propagate to a critical size and eventual failure?"

With the analytical model, stress field and crack growth properties established, it is now possible to conduct the basic analysis needed to establish a design or inspection limit plot.

To accomplish this first step, a crack growth computer code such as CRACKS [466] is required. The objective of the computer analysis is to obtain plots of the number of cycles required to reach a certain K_{Ic} assuming some starting flaw size, an imposed cyclic stress and a constant initial flaw location. The computer is typically programmed to solve an equation of the crack growth curve of the form da/dn = f(ΔK) in an incremental fashion, and print out stress intensities, growth rates and flaw size at various intervals. A plot from such a program is shown in Fig. 3.91 for the model. All of the K_{Ic} values for one initial flaw size can be obtained in a single computer run and the generation of the whole plot takes about four or five runs. The investigator simply scans the output to find the number of cycles it took the flaw to reach some K level, then plots the coordinates of the initial flaw size and the number of cycles for that K level assuming the material had a K_{Ic} equal to that K.

Figure 3.91 consists of two families of curves, the lower set of curves is for K_{Ic} levels where the flaw is of a critical size while it is still buried. The upper curves are for situations where the combination of initial flaw size and K_{Ic} allow the flaw to break through the surface and propagate some amount as a surface crack before failure. The intersection represents an unstable region where the buried flaw becomes a critical size just as it breaks through the surface and failure occurs shortly after this event.

Suppose that a series of curves like Fig. 3.91 are generated for several different stress levels. Then a cross plot of K_{Ic} versus initial crack size can be obtained for various stress levels at a fixed life, as is shown in Fig. 3.92 for 7500 cycles. Data from Fig. 3.91 were used to obtain the curve for a stress, of 344.7 MPa (50 ksi). Note that the discontinuities on the stress curves of Fig. 3.92 occur because of the shift from internal critical flaws to surface critical flaws discussed previously.

At this point, it becomes important to establish the smallest flaw that can be found at the location being studied. Suppose that consultation with nondestructive evaluation personnel reveals that a flaw with a diameter of 7.7 mm (0.300 in.) can be found with 99 percent confidence. This limit is shown by the vertical dashed line in Fig. 3.92. Stress, flaw size and K_{Ic} combinations to the left of this line are unsafe for 7500 cycles of life while those to the right of the line are safe.

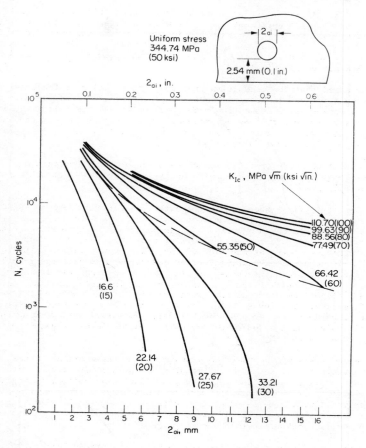

Fig. 3.91   Cycles to attain a given K_{Ic} level vs. various
initial flaw sizes for a uniform stress of 50 ksi.

At this point, plots of K_{Ic} versus the Excess Temperature become important. Plots
of this type for typical ferritic turbine wheel materials can be found in [468] and
Fig. 3.93 shows the lower bound curves for Ni-Cr-Mo-V and Cr-Mo-V materials.

Figure 3.92 can now be converted to allowable operating temperature and FATT limits
through use of Fig. 3.93. To illustrate this step, a stress level of 344.7 MPa
(50 ksi) is selected. Entering Fig. 3.92, this stress requires a K_{Ic} of 56.46 MPa
$\sqrt{m}$, (51 ksi) $\sqrt{in.}$) at the inspection limit of 7.7 mm (0.300 in.). Using this K_{Ic}
to enter the excess temperature plot, Fig. 3.93 and assuming that a Cr-Mo-V material
is to be used, an excess temperature of -45.5°C (-50°F) is required to provide safe
and reliable machine operation. If all combinations of operating temperature are
considered, a plot like the dashed line in Fig. 3.94 showing operating temperature
versus required FATT for a given material at a constant uniform stress, constant

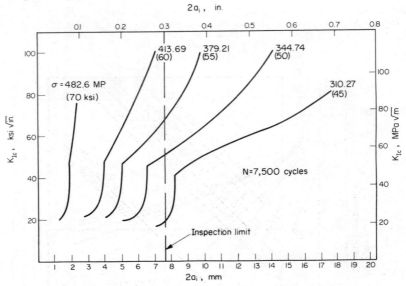

Fig. 3.92 K_{Ic} vs. initial flaw diameter for various
stress levels and constant life.

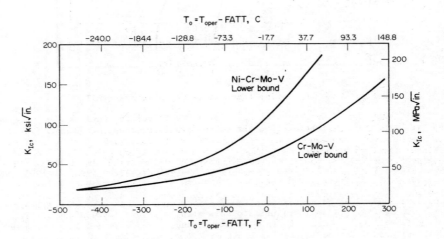

Fig. 3.93 K_{Ic} vs. excess temperature.

life and specific inspection limit is obtained. Other stress and material combinations are shown on Fig. 3.94 and the effects of stress, material selection, and
operating temperature on the inspection level of FATT become obvious.

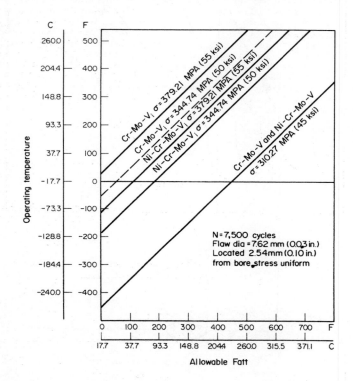

Fig. 3.94   Operating temperature vs. allowable FATT.

Plots like Fig. 3.94 are the end product of the procedure discussed above. These
plots can be used by design and materials engineers to establish a safe acceptance
FATT level for turbine wheels. Further, the plots used to make up Fig. 3.94 can
be employed to make a disposition decision on parts not meeting flaw size limitations. All that is needed to use the plots is a stress level. Either an advanced
strength of materials stress analysis or a more sophisticated finite element
analysis could be used to determine stresses. However, the user should keep in
mind that the plots developed here are already conservative and hence, conservative
stresses should probably not be used. These uses can be illustrated by example.

Suppose it is necessary to establish a FATT limit for material acceptance of
Cr-Mo-V wheels and a desired part life of 7,500 cycles. Nondestructive evaluation
personnel verify that the smallest flaw they can find is 7.7 mm (0.300 in.) at the
most difficult inspection location which is 2.54 mm (0.100 in.) below the bore
surface. The design engineer calculates a stress field for the tangential stress

in the bore area as shown in Fig. 3.95. To be conservative, the peak stress of
413.7 MPa (60 ksi) could be applied uniformly over the section of the part containing the flaw but this would probably be overly conservative. Instead, it is
more reasonable to apply some average stress to the flaw model. Rules of thumb
could be developed analytically or experimentally for determining what stress in
a gradient case should be used when utilizing information from a uniform stress
distribution case. Suppose that it is found that using the average stress between
the surface and the flaw and applying this average stress uniformly over the
flawed section gives the same results as the gradient case. Then, a stress of
344.7 MPa (50 ksi) from Fig. 3.95 could be used in determining allowable FATT.
With the stress established, the design engineer needs to determine the operating
temperature of the wheel when this stress appears. If he finds this to be +4.44°C
(+40°F), he can enter Fig. 3.94 for a stress of 344.7 MPa (50 ksi) to find that
the maximum acceptable FATT is +32.2°C (+90°F).

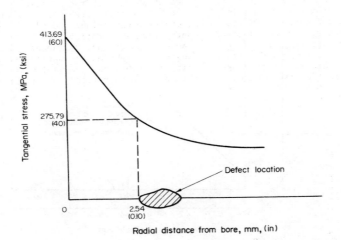

Fig. 3.95 Stress distribution for example.

Suppose a situation arises in which the FATT for a wheel is measured as +10°C (50°F)
with a specified maximum of 32.2°C (90°F) but NDE shows a flaw size of 8.89 mm
(0.350 in.) exists in the part with the edge of the flawed area about 2.54 mm
(0.100 in.) from the bore surface. Production scheduling is critical and the manufacturing personnel request permission to use the wheel in spite of the large defect.
Using some of the same numbers from the previous example, the design engineer establishes that the expected excess temperature (operating temperature--FATT) is -23.3°C
(-10°F). The minimum K_{IC} that can be expected for Cr-Mo-V is obtained from the
excess temperature plot and is 64.2 MPa$\sqrt{m}$ (59 ksi $\sqrt{in.}$). This K_{IC} and the average
stress level of 344.7 MPa (50 ksi) gives an allowable initial flaw size of 8.89 mm
(0.350 in.) from Fig. 3.92 for the design life of 7500 cycles. Therefore, the
design engineer accepts the contention of manufacturing that the wheel can be safely
used. The above are but two examples of the usefulness and versatility of the
methods discussed here.

FRACTURE CONTROL PLANS

It seems almost axiomatic to say that machines and structures of any degree of criticality should be planned, designed, built and operated under a Fracture Control Plan. The creation of most objects has been done under some control, of course, but this is usually evident only in the semi-quantitative sense of a comfortable level of design stress, good design practice to minimize stress concentrations, and use of a tough or "forgiving" material with no defects recognized. While such practice has produced many very successful designs, the lack of numerical correlation amongst these parameters leads to a general inability to predict service life accurately, or to calculate the probability of failure.

Recently the rise of the twin disciplines of Fracture Mechanics and of Non-Destructive Evaluation (NDE) has provided much of the necessary correlation and, thus the base for such Plans. The three original parameters: design stress level; stress concentration; and material toughness remain, with the addition now of the all-important items of initial flaw size and its growth rate. Figs. 3.96, 3.97 and 3.98 show the basic tradeoffs in the exercise of a Plan, and the general conclusion is that the primary parameter controlling sub-critical crack growth is the stress-intensity range, ΔK, raised to a power greater than 2, (about 5 for most steels). Thus, the service life is strongly dependent on the applied stress which is directly related to the stress-intensity range by: $\Delta K \sim \Delta S \sqrt{a}$ (const.). The final factor is the rate of growth of the crack, also related to the applied stress range: $da/dN \sim C(\Delta K)^m$.

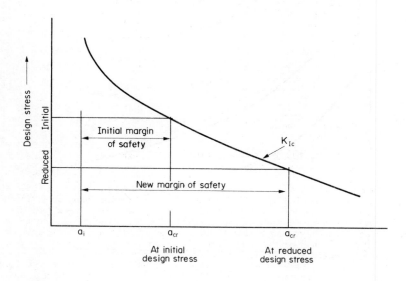

Fig. 3.96   Fracture Control: Increasing Critical
Crack Length by decreasing Design Stress.

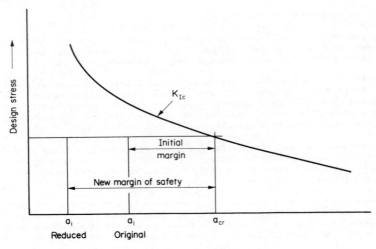

Fig. 3.97   Fracture Control: Increasing Margin of Safety by decreasing size of initial flaw through higher quality, fabrication, and inspection.

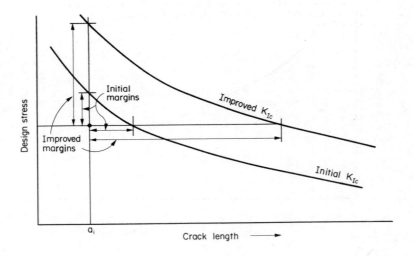

Fig. 3.98   Fracture Control: Increasing Margin of Safety by increasing material toughness.

The range of toughness values available, K_c or K_{Ic} is relatively limited, i.e., to factors of 3 - 5, and the size of the initial flaw is not really subject to much control, considering that the designer must deal with the value of a_i as given by the sensitivity of the NDE equipment. So--the importance of controlling the stress to the lowest reasonable value cannot be over-emphasized.

A very extensive "special case" exists in design for which a true value of the material toughness is of little importance. This situation arises when the K is sufficiently great, and the part width, parallel to crack growth, sufficiently great that the part spends its entire service life in slow, stable crack growth, the controlling parameters being the initial size of flaw and its rate of growth. Another description is that the critical length of crack is greater than the width, and that the applied stress-intensity does not increase significantly toward its critical value, the material toughness. This situation is exemplified by small parts of low-strength, high-toughness material: narrow-flange beams or thick-walled tubing of stainless steel.

The establishment of a Fracture Control Plan is straight-forward; its application requires considerable discipline amongst the many groups within an organization: Design, Analysis, Shop, Inspection, Testing, and especially Marketing. An excellent discussion of Plans is given in [442]; two examples are especially recommended:

> Pressure Vessel Research Committee (PVRC) Recommendations on Toughness Requirements for Ferritic Materials in Nuclear Power Plant Components

and

> American Association of State Highway and Transportation Officials (AASHTO) Material Toughness Requirements for Steel Bridges.

FRACTURE TESTING

Within the concept of Fracture Mechanics, testing is done to gather data on fracture toughness (K_c, J_c, CVN, etc.); to measure crack growth rates; to find Nil-ductility-temperature; and to proof-test assemblies, especially pressure vessels. During the early stage of the overall effort, most of the attention was on fracture toughness, and the reasons for developing the various tests are:

> To determine the probability of catastrophic failure under given conditions.
>
> To choose amongst materials for a given application.
>
> To control material quality at an acceptable level.
>
> To analyze service failures.
>
> To obtain design data, such as maximum working stress, or minimum operating temperature.

See also Table 3.22.

Several types of tests for toughness are noted in the Table, the earliest definitive one being the ASTM Standard Test Method E399-74 for K_{Ic}. In its present form, this test method is for plane-strain conditions only, and the need for broader-based methods has led to consideration of the J-integral, for which no standard method is yet available, although that of Landes and Begley [445] is widely used.

Other ASTM Standard Test Methods are:

- E208-74 Drop-Weight (Nil-Ductility-Transition-Temperature) Test

- E338-68 (1973) Sharp Notch Tension Testing of High-Strength Sheet Materials

- E436-74 Drop Weight Tear Tests of Ferritic Steels

- E602-76T Sharp Notch Tension Testing of Thick High-Strength Aluminum and Magnesium Alloy Products with Cylindrical Specimens

- E604-77 Test for Dynamic Tear Energy of Metallic Materials

and the Recommended Practice:

- E561-75T R-Curve Determination

The emphasis was on the prevention of _fracture_ in the low-toughness, high-strength alloys, and it has been only more recently that the high-toughness, low-strength alloys have come in for consideration. For the latter, it is becoming recognized that much, occasionally all, of the service life may be spent in slow, stable crack propagation, and much effort is being focussed on da/dN test methods, one result being ASTM E647-78T Test for Constant-Load-Amplitude Fatigue Crack Growth Rates above 10^{-8} m/cycle.

The several specimens for these tests are shown in Fig. 3.99.

TABLE 3.22   Fracture Tests

Purpose	Name of Test	Materials	Level of Performance	Applications	Specimen Used
Fracture Toughness	K_{Ic}	All	Plane Strain only	Hi-strength, Lo-toughness. Quantitative Calc. of stress-flaw size.	ASTM E399
Fracture Toughness	J_{Ic} & COD	All	Plane Strain thru to Plane stress	Any strength, any toughness. Quantitative Calc. of Stress-flaw size.	ASTM E399
Crack-Growth-Rate	da/dN	All	----	Stable growth region.	ASTM E647
To insure a crack depth equal to thickness during stable growth	Leak-before-break	Pressure Vessel Steels; Al & Ti	Plane strain thru to Plane stress	Pressure and Vacuum Vessels Quant. calc. of stress-flaw	ASTM E399 for K or Kq; J or JQ
Qualitative selection of Materials; Quality Control	Charpy V-notch, (CVN); Drop-Weight Test, (DWT)	Structural steels to 150 ksi yield	----	Medium toughness steels. Ship steels	ASTM E23
To define NDT, FTE, FTP	FAD	All materials with a characteristic Transition Temperature	Plane strain thru to Plane stress	High toughness materials in the Transition Temperature range.	ASTM E208 for NDT; Mil Std 1601 for DT.

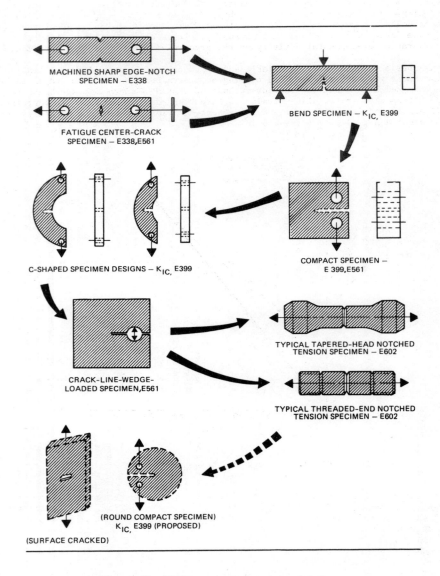

MACHINED SHARP EDGE-NOTCH
SPECIMEN – E338

FATIGUE CENTER-CRACK
SPECIMEN – E338,E561

BEND SPECIMEN – K_{IC}, E399

C-SHAPED SPECIMEN DESIGNS – K_{IC}, E399

COMPACT SPECIMEN –
E 399,E561

CRACK-LINE-WEDGE-
LOADED SPECIMEN,E561

TYPICAL TAPERED-HEAD NOTCHED
TENSION SPECIMEN – E602

TYPICAL THREADED-END NOTCHED
TENSION SPECIMEN – E602

(ROUND COMPACT SPECIMEN)
K_{IC}, E399 (PROPOSED)

(SURFACE CRACKED)

Fig. 3.99 LEFM test specimen progression.

The relatively small size of a <u>Charpy test specimen</u> offers several advantages in terms of material requirements, orientation flexibility, and testing convenience as a simple and inexpensive method of obtaining fracture toughness information. The standard 0.010-in. root radius V-notched Charpy specimen has a wide background of usage, particularly in the steel industry. In recent years, there has been considerable developmental activity on the precracked version of this specimen.

On the basis of information available and the degree of correlation plus testing requirement factors, slow-bending testing of precracked specimens is judged to be the most suitable among the various methods of testing Charpy-size specimens for correlation with static plane-strain fracture toughness, K_{Ic}. Suitable correlation indices derived from this test are the strength ratio (R_{sb}-CV) and the energy per unit area (W/A), Fig. 3.100 and 3.101

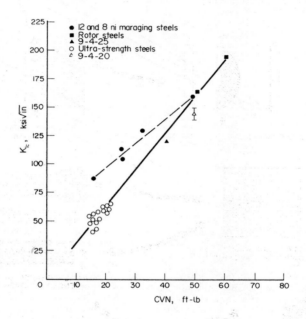

Fig. 3.100 K_{Ic} vs. CVN at CVN upper shelf conditions.

Other methods of testing either the standard or the precracked specimens are of lesser usefulness because of factors such as testing complexity, measurement interpretation, insufficient data for correlation or the specific and limited nature of the correlation.

The <u>dynamic tear test</u> was developed by the U.S. Navy for characterizing the fracture resistance of steels, titanium alloys, and aluminum alloys for the full range of strength level and fracture toughness. Correlations of DT energy values with K_{Ic} results have been established for these materials in the high-strength range; however, these correlations, developed during the late 1960s, are based primarily on K_{Ic} data obtained with single-edge-notch tension specimens that do not conform to current ASTM standard K_{Ic} test methods and 1-in. DT test results. More recent results (of limited number) using ASTM standard K_{Ic} methods generally confirm the accuracy of the earlier results for thicknesses up to 3 inches. The 5/8 inch DT test is covered by MIL-STD-1601 and ASTM E604-77. It appears promising as a rapid, inexpensive quality-control test, Fig. 3.102.

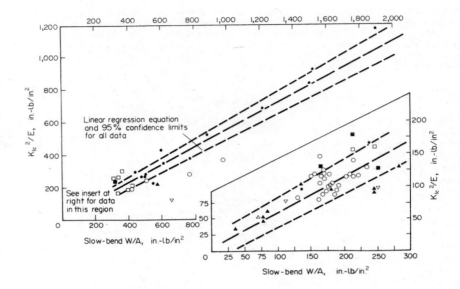

Fig. 3.101 Composite plot of $(K_{Ic})^2/E$ vs. precracked Charpy Slow-bend W/A data for steels, aluminum and titanium.

Because of ease of machining, simplicity of interpretation, and familiarity of industry with tensile testing, the notch-round bar specimen test for quality-control purposes has been rapid and inexpensive. However, problems with respect to notch uniformity and loading eccentricity can produce relatively large scatter bands, and a direct correlation between K_{Ic} and NTS/YS (notch-tensile-to-yield ratio) is still relatively unproven with respect to statistical significance. Hundreds of correlations between K_{Ic} and NTS/YS have been made for a series of aluminum alloys, Fig. 3.103. As such, they provide a reasonable base for using the lower bound of the scatter band for quality control. But, at this time, sufficient resolution of eccentricity problems has not been accomplished for the purpose of using precracked notch rounds for general K_{Ic} testing.

The surface flaw specimen has proven excellent for the development of design data and for use in failure analyses. From the standpoint of quality control testing of fracture-resisting materials, however, high fabrication costs, large specimen size requirements, and inconsistent correlations with ASTM compact specimen fracture toughness data hinder the usefulness of the surface flaw specimen for quality control testing.

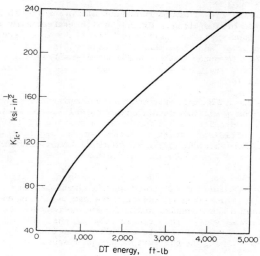

Fig. 3.102   Relationship of K_{Ic} to DT test values for steels.

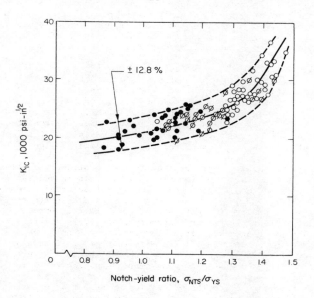

Fig. 3.103   Statistical scatter in K_{Ic}-Notch-Round Correlations
for 2124-T851 based on an arbitrary envelope.

The practice of testing for fracture toughness and crack growth rates is relatively complex, and not free of judgmental factors, to the extent that such testing is not recommended for the small or "part-time" lab. The investment in proper testing machines, instrumentation and specimens is considerable, and the "know-how" of the technicians extremely important. For those situations needing data not available in the open literature, the work should be done by, and in consultation with, an experienced organization.

The general literature does not yet report many statistical descriptions of tough-ness data; much of the available data represents only very small samples, occasion-ally a single point. Critical applications of a particular alloy and form may well require testing of sufficient extent to establish the required level of confidence. Non-destructive testing (or NDE) of parts generally, and particularly for initial crack size, is discussed elsewhere in this Section.

Reference 451 discusses many test methods and specimens from both the viewpoints of correlation of the data, and of promoting the development of rapid, inexpensive tests for quality control. General conclusions and typical correlation plots are given; it will be noted that the correlations are not direct in many cases, and non-existent in a few.

DATA ON FRACTURE TOUGHNESS AND CRACK GROWTH RATE

Such data and associated properties are already quite extensive and are growing rapidly in volume. Table 3.23 provides values for general usage, and it is particularly to be noted that a_i is only that value which Production Inspection would usually find. For any individual application it is strongly recommended that the unique value of a_i be found by direct observation on a "large" number of specimens. The values in Table 3.23 are for engineering sizes, e.g., bars 1/2" in diameter and up; plates 1/2" thick and up. The expected size of defects in foils and fine wires would be far lower, often discernible only by the microscope.

The values of K_{Ic} and J_{Ic} may not be strictly valid according to ASTM Standard Test Method for Fracture Toughness E399 because its conditions, especially specimen thickness, may not have been met, or for some of the high-toughness alloys may have been impossible to meet. However, this situation results in K_c or J_c, and does not invalidate their judicious use. The constants C and m in the crack growth rate equation are usually regression line values on the straight-line portion of the plot of da/dN vs ΔK.

For steels, upper-bound values for C and m have been determined from extensive data as:

	C	m
For austenitic microstructure	3.0×10^{-10}	3.25
For ferrite/pearlite microstructure	3.6×10^{-10}	3.00
For martinsitic microstructure	0.66×10^{-8}	2.25

for ΔK expressed in ksi $\sqrt{\text{inch}}$.

Table 3.24 provides the opportunity for a quick scan of relative growth rates of cracks. Specific data for selected alloys is given in Figs. 3.104 - 3.110 and in Tables 3.25 - 3.27; more extensive data may be found in Refs. 448 and 456. While the values given are thought to be representative, they are not guaranteed in any way, and except for a few scattered items from Ref. 448, they have no statistical significance. The degradation of fracture toughness and the other mechanical properties by radiation is still very incompletely understood and reported; a description of the present situation for Types 304 and 316 SS and 6061 aluminum may be found in Ref. 473.

TABLE 3.23   General Properties For Fracture Mechanics Analyses

Fracture Toughness: K_c or, J_{Ic}, K_c, or J_c, ksi $\sqrt{inch}$

Size of Initial Crack: a_i, inch

Constants in Crack Growth Rate Equation: $dn/dN = C(\Delta K)^m$; (values with * use K or J in psi $\sqrt{in}$., any R; values with ** use K in ksi $\sqrt{in}$., R = 0.0 to 0.25)

At 70°F or as noted, R.H. = 50%; no corrosives

CARBON AND ALLOY STEELS		K	J	a_i	C	m
(1) ASTM A36	Plate	90-120		.050-.080	3.6×10^{-10}	3.0
"	Weldments	90-120		.050-.080	3.6×10^{-10}	3.0
"	Forgings	90-120		.060-.100	3.6×10^{-10}	3.0
A217	Castings	70-75		.050-.080	4.24×10^{-10}	2.9
A296	Castings	85-90		.050-.080	8.49×10^{-11}	3.3
A441	Plate	90-120		.050-.080	3.6×10^{-10}	3.0
"	Weldments	90-120		.050-.080	3.6×10^{-10}	3.0
"	Forgings	90-120		.060-.100	3.6×10^{-10}	3.0
A441	Plate @ -10°F	70		.050-.080	3.6×10^{-3}**	3.73**
A508 Cl 2	Plate, Sub-Surf. flaw in air	120		.050-.080	$.027 \times 10^{-3}$**	3.73**
A508 Cl 3	Plate, Sub-Surf. flaw in air	120		.050-.080	$.027 \times 10^{-3}$**	3.73**
A508 Cl 2	Plate Surf, flaw in H_2O	120		.050-.080	$.038 \times 10^{-3}$**	3.73**
A508 Cl 3	Plate Surf, flaw in H_2O	120		.050-.080	$.038 \times 10^{-3}$**	3.73**
A514B	Plate	100		.050-.080	6.6×10^{-9}	2.25
A517	Plate	150-177		.050-.080	6.6×10^{-9}	2.25
A533B	Plate	120		.050-.080	1.0×10^{-15}*	2.20*
A572 Gr 50	Plate .50 in. @ +70°F	380		.050-.080	3.6×10^{-10}*	3.0*
A572 Gr 50	Plate .50 in. @ +40°F	273		.050-.080	3.6×10^{-10}*	3.0*
A572 Gr 50	Plate .50 in. @ -40°F	150		.050-.080	3.6×10^{-10}*	3.0*

Material	Condition	K	J	a_i	C	m
A572 Gr 50	Plate 1.50 in. @ +70°F	318		.050-.080	3.6×10^{-10}*	3.0*
A572 Gr 50	Plate 1.50 in. @ +40°F	155		.050-.080	3.6×10^{-10}*	3.0*
A572 Gr 50	Plate 1.50 in. @ -40°F	57		.050-.080	3.6×10^{-10}*	3.0*
A645	Plate			.050-.080	6.6×10^{-9}	2.25
AISI 1020	Plate H.R.	120		.050-.080	2.0×10^{-31}*	5.60*
1045	Plate H.R.	120		.050-.080	5.6×10^{-24}*	4.10*
1144	Plate Q&T	56		.050-.080	3.1×10^{-29}*	5.00*
4140	Plate Q&T	56		.050-.080	2.0×10^{-51}*	10.00*
4330M	Plate Q&T	168		.050-.080	6.6×10^{-9}	2.25
4340	Plate Q&T	110-140		.050-.080	6.6×10^{-9}	2.25
HY-80	Plate Q&T	120		.050-.080	2.9×10^{-9}	2.13
HY-130	Plate Q&T	130		.050-.080	2.9×10^{-9}	2.13
HP-9-4	Plate Q&T	110-150		.050-.080	6.6×10^{-9}	2.25
18Ni(190)	Plate Q&T	113		.050-.080	6.6×10^{-9}	2.25
18Ni(250)	Plate	70-90		.050-.080	6.6×10^{-9}	2.25
D6ac	Plate (@Fty210)	90		.050-.080	16.0×10^{-10}	2.60
5Ni-Cr-Mo-V	Plate (@Fty150)	100-120		.050-.080	3.04×10^{-9}	2.26
(2) STAINLESS						
304	Plate	330		.050-.080	3.0×10^{-10}	3.25
304N	Plate	308	3160	.050-.080	1.6×10^{-10}	3.05
304N	Plate @ -328°F	299	2780	.050-.080	2.3×10^{-10}	3.02
304N	Plate @ -452°F	195	1190	.050-.080	9.6×10^{-11}	3.46
308	Plate	300		.050-.080	3.0×10^{-10}	3.25
308	Weld Metal	308		.050-.080	3.0×10^{-10}	3.25
403	Plate	110-140		.050-.080	2.0×10^{-9}	2.35
PH14-8	Mo	160		.050-.080	3.0×10^{-10}	3.25
PH15-7	Mo	60-85		.050-.080	3.0×10^{-10}	3.25
AM350-CRT		120		.050-.080	3.0×10^{-10}	3.25
AM380-SCT		210		.050-.080	3.0×10^{-12}	3.25
Nitronic 33	Plate @ +70°F	342	3800	.050-.080	5.6×10^{-4}	3.82
	Plate @ -328°F	119	440	.050-.080	1.9×10^{-11}	3.98
	Plate @ -452°F	64		.050-.080	9.1×10^{-11}	5.32

TABLE 3.23   General Properties For Fracture Mechanics Analyses (Cont'd.).

		K	J	ai	C	m
Nitronic 50	Plate @ + 70°F	200	1305	.050-.080	5.7×10^{-12}	3.86
	Plate @ - 328°F	128	500	.050-.080	2.3×10^{-12}	3.94
	Plate @ - 452°F	100		.050-.080	2.4×10^{-13}	4.66
NMF3 (18Mn-5Cr)	@ Fty 115	140-165		.050-.080	6×10^{-12}	3.70
	@ Fty 180	100		.050-.080	6×10^{-12}	3.70
A453 (was A286)		114	426	.050-.080	3.0×10^{-10}	3.25
(3) ALUMINUM ALLOYS						
2024-T6	Forged or Rolled	16		.160	3.0×10^{-13}	3.0
2024-T3	Plate (special H.T.)	83		.120-140	3.0×10^{-13}	3.0
2024-T4	Rolled Angle	36		.120-140	3.0×10^{-13}	3.0
6061-T6	Rolled Angle	36		.120-140	3.0×10^{-13}	3.0
7075-T3	Forged or Rolled	20-24		.095-120	5.0×10^{-13}	3.0
7075-T6	Plate (special H.T.)	68		.095-120	5.0×10^{-13}	3.0
7075-T6	Sheet (.090")	42		.095-120	5.0×10^{-13}	3.0
7079-T6	Forged or Rolled	20		.095-120	5.0×10^{-13}	3.0
7178-T6	Rolled	20		.095-120	5.0×10^{-13}	3.0
5083-0	Plate (2")	(est. 40)		.095-120	1.3×10^{-13}	2.0
(4) TITANIUM ALLOYS						
6Al-4V	Forgings,Beta Ann.	70-80		.080	1.96×10^{-9}	3.34
	Forgings, ELI, Beta Anneal	90		.080	1.96×10^{-9}	3.34
	Forgings, E.B. Weld	40		.080	1.96×10^{-9}	3.34
	Forgings, ELI, E.B. or GTA Weld	70		.080	1.96×10^{-9}	3.34
	Sheet, 025", Mill Anneal	117		?		
	Sheet, 040", Spun	60		?		
	Plate	40-70		.080	1.96×10^{-9}	3.34
4Al-3Mo-1V	Sheet, Beta STA	77		?	1.96×10^{-9}	3.34
8Al-1Mo-1V	Sheet, Duplex Anneal	120-150		?	1.96×10^{-9}	3.34

TABLE 3.24 Summary Ranking of Alloy Systems With Respect to Fatigue Crack Growth Rate Characteristics Under Fixed Test Parameters

Test Conditions	Growth Rate Ranking						
	1 (Slowest)	2	3	4	5	6	7 (Fastest)
LMA, R.T., R=.08, RW Dir.	Inconel 718<<	300M ≈ 9-4-.20 9-4-.30	PH13-8Mo <	Ti-6-4[1] <<	All Aluminum		
STW, R.T., R=0.08, RW Dir.	Inconel 718<<	9-4-.20 < 9-4-.30	PH13-8Mo ≈	Ti-6-4 <<	2000 Sers. ≈ Aluminum	300M ≈	7000 Sers. Aluminum
LMA, R.T., R=0.08, WR Dir.	Inconel 718<<	PH13-8Mo <	300M ≈	9-4-.20 < 9-4-.30	Ti-6-4[2] <	All Aluminum	
STW, R.T., R=0.08, WR Dir.	Inconel 718<<	PH13-8Mo << Ti-6-4	2000 Sers. ≈ Aluminum	300M ≈	7000 Sers. Aluminum		
LMA, R.T., R=0.3, RW Dir.	300M < 9-4-.20 9-4-.30 PH13-8Mo	Ti-6-4 <<	All Aluminum				
STW, R.T., R=0.3, RW Dir.	9-4-.30 < PH13-8Mo	Ti-6-4 <<	2000 Sers. ≈ Aluminum	300M ≈	7000 Sers. Aluminum		
LMA, R.T., R=0.5, RW Dir.	Inconel 718<<	9-4-.20 < 9-4-.30 PH13-8Mo 300M	Ti-6-4 <<	All Aluminum			
STW, R.T., R=0.5, RW Dir.	9-4-.30 <<	2000 Sers. ≈ Aluminum	300M ≈	7000 Sers. Aluminum			

(1) at ΔK >11-15 ksi $\sqrt{in.}$
(2) at ΔK >15-20 ksi $\sqrt{in.}$

(From AFML-TR-76-137, TABLE 8.3.10-1)

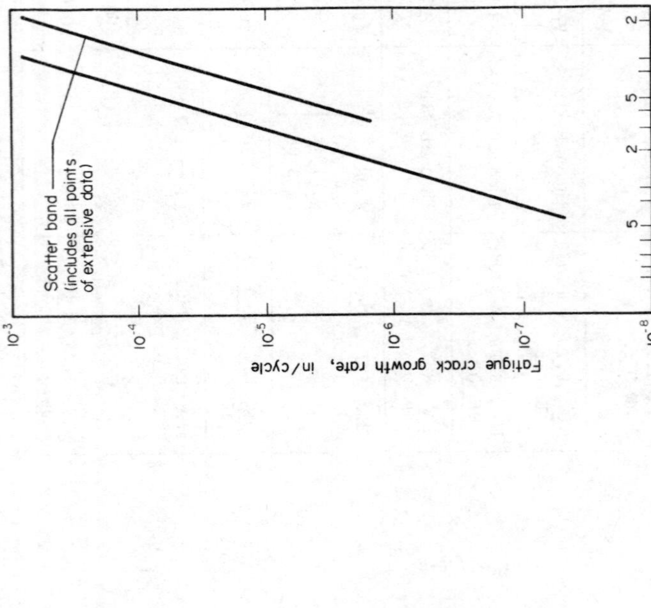

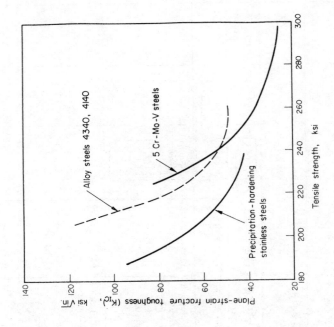

Fig. 3.105 Variation of fracture toughness with strength level for various types of steel.

Fig. 3.104 Fatigue crack growth rate for low-strength steels, Syp < 60 ksi.

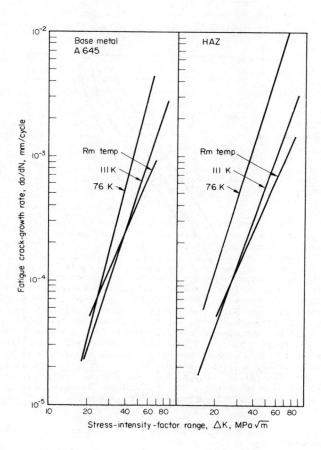

Fig. 3.106   Crack growth rates in 5% Ni steel.

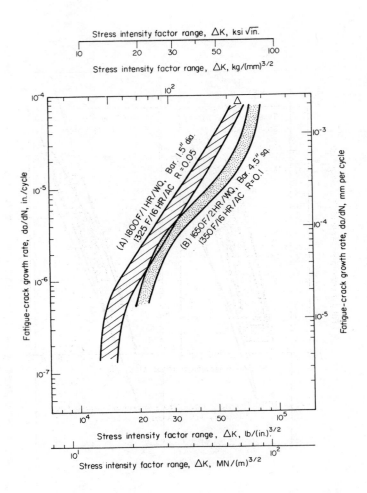

Fig. 3.107   Fatigue Crack Growth Rates for A286 steel.

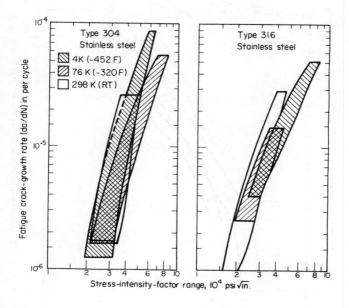

Fig. 3.108   Crack growth rate scatter bands for
304 and 316 stainless steel.

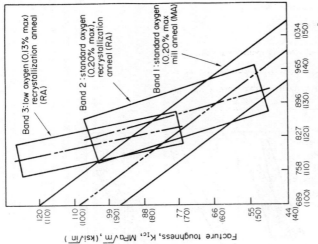

(a) Effects of oxygen content and Thermal Treatment

(b) Bands of 95% Confidence. Least-square lines demonstrate the inverse relationship between K_{Ic} and S_{yp}.

Fig. 3.109 Effects of Oxygen Content and Thermal Treatment on the Fracture Toughness of Ti-6Al-4V (from Ref. 446).

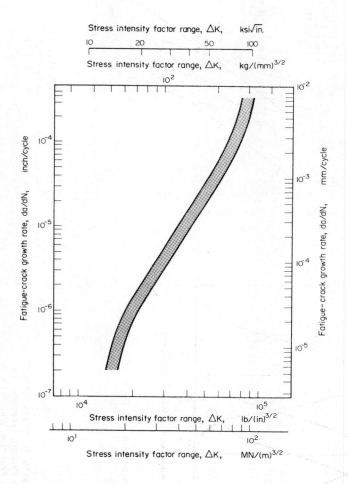

Fig. 3.110   Crack growth rate for fully heat-treated
Inconel 718 at R.T.   One-inch dia. bar.
R = .05.

TABLE 3.25   AVERAGE PLANE-STRAIN FRACTURE-TOUGHNESS DATA FOR TESTS AT LOW TEMPERATURES ON ALUMINUM ALLOYS

Alloy and Condition	Thick., in.	Testing Temp., F	Avg. Yield Strength, ksi	Spec. Design	Spec. Orient.	Spec. Thick., in.	Avg. K_{Ic}, ksi √in.	Min. K_{Ic}, ksi √in.	No. of Spec.	$2.5 \times \left(\dfrac{K_{Ic}}{YS}\right)^2$
2021-T81 Plate	1.0	70	62.3	CT	T-L	1.0	26	–	3	0.44
		-320	73.6	CT	T-L	1.0	36	–	3	0.60
		-423	79.6	CT	T-L	1.0	40	–	3	0.63
2024-T851	3.0	70	63.4	CT	L-T	0.75	27.1	26.9	2	0.46
		0	65.5	CT	L-T	0.75	27.5	25.2	3	0.44
		-65	66.8	CT	L-T	0.75	26.2	24.4	3	0.38
		70	61.0	CT	L-S	0.75	31.3	30.0	3	0.65
		0	63.3	CT	L-S	0.75	31.5	31.4	2	0.62
		-65	64.4	CT	L-S	0.75	30.3	29.3	3	0.55
		70	63.5	CT	T-L	0.75	21.5	19.7	3	0.29
		0	65.5	CT	T-L	0.75	21.8	20.8	2	0.27
		-65	66.5	CT	T-L	0.75	23.3	–	1	0.31
2219-T87 Plate	2.5	72	55(a)	CT	T-S	1.25	26.2	26.1	2	0.57
		-320	67(a)	CT	T-S	1.25	31.4	31.3	2	0.55
		-423	73(a)	CT	T-S	1.25	34.0	33.0	2	0.54
2219-T87 Plate	2.5	72	55(a)	BEND	T-S	1.25	36.3	36.2	2	1.1
		-320	67(a)	BEND	T-S	1.25	42.4	41.6	2	1.0
		-423	73(a)	BEND	T-S	1.25	48.0	47.2	2	1.0
7007-T6 Plate	1.0	70	64.5	CT	T-L	1.0	32.0	–	3	0.62
		-320	80.0	CT	T-L	1.0	23.0	–	3	0.21
		-423	84.8	CT	T-L	1.0	28.0	–	3	0.27
7175-T736 Forging	2.5	70	64.9	CT	S-T	1.0	31.9	29.1	3	0.60
		0	65.9	CT	S-T	1.0	26.6	26.3	3	0.40
		-65	66.6	CT	S-T	1.0	26.3	26.0	3	0.39

(a) Approximate values for yield strength.

TABLE 3.26 Average Fracture Properties of Cryogenic Steels

Steel	Thick., in.	Testing Temp., F	YS, ksi	Charpy-V Energy, ft-lb	K_c, ksi $\sqrt{in.}$
		Base Plate			
A553 Type I	2	−274	121	56	208
(9Ni)		−320	134	48	171
	3	−320	122	81	146
A645	1	−274	96	82	146
(5Ni,0,3Mo)		−320	107	59	106
	1.5	−274	90	30	125
		−320	105	21	82
N−TUF CR−196	1	−274	118	104	198
		−320	133	79	166
		Heat Affected Zone			
A553 Type I SMA[a]	2	−320	−	42	154
CMA[b]	2	−320	−	72	133
A645 SMA	−	−320	−	46	86
N−TUF CR−196 SMA	−	−320	−	61	130
		Weld Metal			
Inco Weld B					
A533 Type I	−	−320	97	39	186
A645	−	−320	100	44	152
Inco 82					
A553 Type I	−	−320	90	98	159

[a] Shielded metal arc.

[b] Gas metal arc.

TABLE 3.27   AVERAGE PLANE-STRAIN FRACTURE-TOUGHNESS DATA FOR TESTS AT LOW TEMPERATURES ON TITANIUM ALLOYS

Alloy and Condition	Testing Temp., F	Avg. Yield Strength, ksi	Spec. Design	Spec. Orient.	Spec. Thick., in.	Avg. K_{Ic}, Ksi $\sqrt{in.}$	Min. K_{Ic}, ksi $\sqrt{in.}$	No. of Spec.	$2.5x\left(\dfrac{K_{Ic}}{YS}\right)^2$
Ti-6Al-4V Plate, Recrystallize Annealed(a)	72	118	CT	L-T	1.5	77.0	74.9	2	1.03
	-65		CT	L-T	1.5	60.4	-	1	0.63
	-65		CT	T-L	1.5	77.0	-	1	1.03
Ti-6Al-6V-2Sn (ELI) Plate, 1600 F, WQ	75	170	BEND	L-S	0.25	38.1	37.5	2	0.13
	-320	258	BEND	L-S	0.25	23.6	22.6	2	0.02
Ti-6Al-6V-2Sn Forging, 1600 F, WQ, 1050 F 1 hr, AC	75	184	BEND	L-C	0.25	31.9	30.4	3	0.07
	75	184	BEND	L-C	0.50	30.5	27.8	3	0.07
	-320	270	BEND	L-C	0.25	22.6	-	1	0.02
	-320	270	BEND	L-C	0.50	24.5	-	1	0.02

(a) Recrystallize Anneal: 1700 F 4 hour, furnace cooled to 1400 F no faster than 100 F per hour, cooling from 1400 F to below 900 F within 45 minutes.

CASE STUDY I: Side Frame of Railroad Car Truck

The General-70 truck shown in Fig. 3.111 is that used under the Metroliner car-bodies; the side frame and bolster are under consideration here.

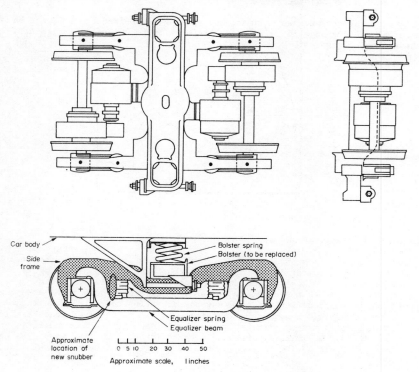

Fig. 3.111   GS1 General-70 Truck.

The following engineering tasks were defined; (1) estimation of material fracture toughness and crack growth rate from handbook data; (2) preparation of a representative service load history from the data available in previous Metroliner studies; (3) fatigue-crack initiation and propagation calculations to estimate economic and safety limits, and; (4) formulation of a fatigue test plan. Comparison of the data on the truck frame alloy with metallurgical reference works indicated that the material closely resembled AISI-2330 cast steel. Hence, the properties of the latter alloy (at 75 ksi UTS, 30 ksi unnotched endurance limit strength) were used in the analysis. Charpy V-notch data for AISI-2330 castings were correlated with K_{ID} values via an empirical relation developed by Barsom and Rolfe:

$$(K_{ID})^2/E = 2(CVN)^{3/2} \tag{1}$$

where K_{ID} is the dynamic fracture toughness (psi $\sqrt{in}$), E is Young's modulus (psi), and CVN is the Charpy impact energy absorption (ft. lb.). These calculations resulted in average K_{ID} values of 124 ksi $\sqrt{in}$ at 75°F and 71 ksi $\sqrt{in}$ at 0°F. (The 0°F value is used to determine critical crack size, under the assumption that fracture occurs during winter operations.)

The crack-propagation rate was estimated by fitting a combined Paris-Forman equation to data available for an alloy with crack-propagation behavior similar to AISI-2330, with the following results:

$$da/dn = 0 \qquad\qquad\qquad \text{for } \Delta K < \Delta K_{TH} \tag{2}$$

$$da/dn \simeq 3.9 \times 10^{-9} (\Delta K)^2 \quad \text{for } \Delta K_{TH} \leq \Delta K \leq \Delta K_{TR} \tag{3}$$

$$da/dn \simeq \frac{3.9 \times 10^{-9} (\Delta K)^2}{(1-R)\frac{K_c}{\Delta K} - 1} \quad \text{for } \Delta K \geq \Delta K_{TR} \tag{4}$$

where da/dn is in units of inch/cycle, and:

$$\Delta K_{TH} \simeq 7(1-R) \text{ ksi}\sqrt{in}. \tag{5}$$

$$K_c \simeq 110 \text{ ksi } \sqrt{in}. \tag{6}$$

$$\Delta K_{TR} \simeq 55(1-R) \text{ ksi } \sqrt{in}. \tag{7}$$

and where $R = s_{min}/s_{max}$ is the stress ratio.

No attempt was made to account for crack-growth retardation due to overloads. Hence, load-history preparation was confined to the development of appropriate exceedance curves. For this purpose, the action of the Metroliner carbody upon its trucks was treated as a stationary Gaussian process driven by the track roughness power spectrum and filtered by the carbody vibration modes. The dominant carbody motions are rigid body bounce and pitch, as far as vertical loads applied to the truck, although minor contributions are also made by the bending modes. Lateral truck loads are similarly influenced by rigid-body sway and yaw. The well-known formulae for calculating the exceedance curve of a stationary Gaussian process were employed:

$$E(x) = \frac{1}{2} \bar{N}_o \exp(-x^2/2\sigma^2) \tag{8}$$

where E(x) is the expected number of exceedances per unit time of the level, x, of the parameter of interest, where:

$$\sigma = \left[\int_o^\infty G_{xx}(f) df \right]^{1/2} \tag{9}$$

$$\bar{N}_o = 2 \left[\frac{1}{\sigma^2} \int_o^\infty f^2 G_{xx}(f)(df) \right]^{1/2} \tag{10}$$

and where $G_{xx}(f)$ is the power spectral density of x as a function of frequency, f, in Hz, (the assumption of a stationary Gaussian process cannot be justified either theoretically or from the available test data in the present case; it was adopted merely as a convenience to obtain first-cut estimates of the load environment within the program schedule).

Exceedance curves were first estimated approximately by adjusting the short-run Metroliner test data to correlate with the vendor estimates. Comparison of the track roughness spectra from the original test with spectra obtained several years later at 16 additional locations on the Metroliner route indicated that the track roughness statistics could reasonably be treated as constants. An example is given in Fig. 3.112 which compares the track alignment spectra from the two data bases. It was then postulated that exceedance levels are influenced primarily by operating speed, which varies considerably from one section of the route to another. Under this hypothesis, the load history was treated as a sequence of stationary Gaussian processes (one for each speed block) which could all be derived

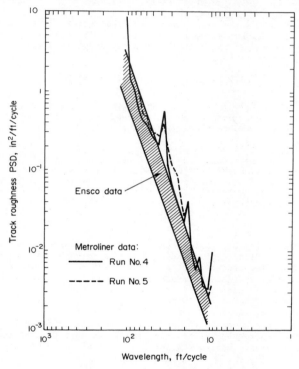

Fig. 3.112 Northeast Corridor track conditions, (1971-75).

from the Metroliner test results as a "baseline" case in the following manner. Let $R(\lambda) = C\lambda^m$ represent the track roughness spectral density as a function of spatial wavelength, λ. (The constants C,m are derived by fitting the upper bound of the data in Fig. 3.112.) Then:

$$\lambda = V/f \qquad (11)$$

$$G_{xx}(f) = G(f) = kR(\lambda) = kC(V/f)^m \qquad (12)$$

where $G(f)$ is the measured baseline acceleration PSD and where V is the operating speed during the test. The parameter, k, is unknown but may be assumed independent of V because the acceleration response is determined primarily by fixed carbody vibration modes. In a similar manner,

$$G_i(f) = kC(V_i/f)^m \qquad (13)$$

is the response PSD in the ith speed block, and both k and c can be eliminated to obtain the block response in terms of the baseline PSD:

$$G_i(f) = (V_i/V)^m G(f) \qquad (14)$$

Block response statistics can now be obtained from Eqs. 9 and 10, and the route exceedance curve is derived by summing the blocks. Alternatively, the zero-crossing rates can be expressed "per route mile" and a service-life exceedance curve can be given:

$$E(x) = 1/2\ \bar{N}_o\{L_s/\sum_i L_i\}\sum_i (L_i/V_i)\exp\{-(V/V_i)^m(x^2/2\sigma^2)\} \qquad (15)$$

where L_s is the service life, L_i the block length, and where $\bar{N}_o$ and σ are the statistics of the baseline exceedance curve. Exceedance curve calculations were based on detailed operating speed-distance profiles for the Metroliner (current and projected) provided by AMTRAK.

Analysis for fatigue-crack initiation was conducted with Miner's linear damage-summation hypothesis. When Miner's hypothesis is applied to Gaussian random-load fatigue, it can be shown that the expected rate of damage accumulation can be calculated from:

$$D = \int_o^\infty \frac{-E'(s)ds}{N(S_a, S_m)} \qquad (16)$$

where $E'(s)$ is the derivative of the stress exceedance curve, and where s is the total stress level consisting of a random alternating component, S_a and a constant mean component, S_m. The denominator of Eq. 16 is simply the number of cycles to failure, taken from a constant-amplitude S-N diagram, for the load condition S_a, S_m. The damage rate D is per unit time if the exceedance curve is given in its usual form, or "per service life" if an expression like Eq. 15 is used. For lateral analysis of the truck frame legs, the correct process is "zero-to-tension," i.e. $S_m = S_a = S/2$ per cycle, and the correct value of N can be found from a Smith diagram (or dimensionless constant-life plot, [452]) for $R = 0$. For vertical loading, S_m is constant and corresponds to the static weight of the Metroliner carbody and passenger load.

Since the retardation effect has been ignored, crack-propagation as described by Eqs. 2 through 7 is also a linear damage-summation theory, and can be treated like crack initiation as far as the mathematical formulation is concerned. This is accomplished by assuming an initial crack size, a_i, and then integrating da/dn from a_i to the critical crack size determined by K_{ID} and a "once per life" or similar stress to determine the number of cycles to fracture, N. In the present case, K_{ID} at $0°F$ and the material UTS were used to make the analysis conservative. This calculation is done at constant amplitude, being repeated for several amplitudes to establish an equivalent "S-N" curve for crack propagation. Equation 16 can then be used to estimate D for propagation damage due to random loading. Equivalent "S-N" curves for several assumed initial crack sizes are compared with the ordinary S-N curve in Fig. 3.113 for AISI-2330 steel. The Tiffany formula for a semi-elliptical surface crack:

$$K_I = 1.1 \ S \ M_K \sqrt{\pi a / Q} \qquad (17)$$

(where Q, M_K are parameters which depend upon crack shape, wall thickness, and yield strength) was used to determine when $K_I = K_{ID}$ and to compute ΔK (s replaced by $s_{max}-s_{min}=2s_a$) for the da/dn integration. Results for the truck frame lateral analysis are shown in Fig. 3.114. The abscissa illustrates the benefit of high 0°F fracture toughness in the existing truck frame alloy (71 ksi $\sqrt{in}$). The figure also illustrates the high risk which would result from combining an alloy with no nickel (K_{ID} = 45 ksi $\sqrt{in}$ at 0°F) with an inspection method not 100 percent reliable for detection of 0.15-inch-deep cracks.

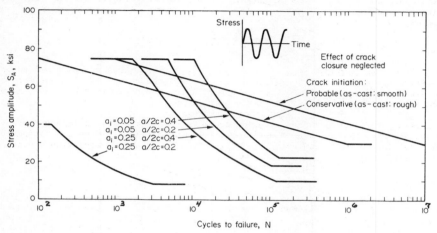

Fig. 3.113   Fatigue endurance vs. damage tolerance
for AISI 2330 steel.

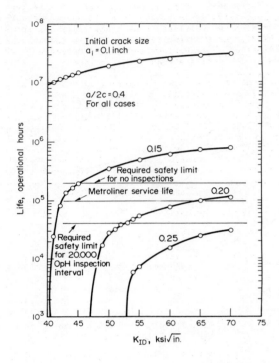

Fig. 3.114   Dependence of Safety Limit on fracture
toughness and inspection reliability.

ASSESSMENT AND FATIGUE TEST PLAN

The results of the fatigue calculations indicated that the General-70 truck frame is a very conservative design. The crack initiation damage rates were negligible, while the propagation rates indicated life limitations only for extremely large initial flaws. The largest assumed flaw (0.25 inch deep, 1.25 inch surface length) had a calculated safety limit of 140 operational hours, but initial damage in this size range could be ruled out as a consequence of the fact that no service cracking has occurred in the average 30,000 to 40,000 operational hours per railcar which have already been accumulated. Also, since the planned modifications would not influence lateral loading, we concluded that a new fatigue test of a complete truck was not necessary, and that smaller-sized initial damage which might affect structural integrity after modification could be effectively controlled by magnetic-particle inspection of the frames at modification time.

The bolster, being a new article, must be fatigue tested to verify the structural design. However, since the load-transfer paths to the bolster were simple and well-defined, it was felt that this article could be tested realistically as a separate component. The constraint to constant-amplitude testing (see Introduction) was of some concern because it is well known that constant-amplitude tests may be either conservative or unconservative when the test article contains cutouts, fastener holes, fillets, etc. Since the bolster design had not been finalized at this point, we chose a conservative approach which assumes that the vertical, lateral and longitudinal loads are in phase at the frequency ratio.

$$\text{Vertical/Lateral/Longitudinal} = 2/1/1 \qquad (18)$$

The longitudinal loading was established deterministically by counting the number of significant braking/traction events in the Metroliner operating profile (19 per one-way run) and by assuming that the wheels would engage at the maximum wheel/rail friction coefficient (0.15) for these events. Over the post-modification service life of the vehicle, these calculations resulted in 3.28×10^5 cycles at 0.15 "g". From Eq. 18, the corresponding figures are 3.28×10^5 lateral and 6.56×10^5 vertical cycles per life, these figures being used to enter the appropriate exceedance curves (Eq. 15) to obtain corresponding lateral and vertical acceleration amplitudes. A "scatter factor" of 4 was then incorporated by extending the test plan to 4 times the number of cycles given above, with the final results:

Vertical: 2.6×10^6 cycles @ 0.30 "g"

Lateral: 1.3×10^6 cycles @ 0.15 "g"

Longitudinal: 1.3×10^6 cycles @ 0.15 "g"

The "scatter factor" concept assumes that the fatigue test article will be average, and seeks to protect the fleet by placing mean fatigue life well above service life.

CASE STUDY II

PART I

Given: A welded-plate bracket, Fig. 3.115 bolted to a rigid wall and loaded downward through the pin, with a remote stress of 52 ksi in Region A. Material is 4340, y.s. = 238 ksi. K_{Ic} = 46 ksi $\sqrt{in}$.

Cracks are observed from both sides of the hole, (A) approximately 0.25 inch, each.

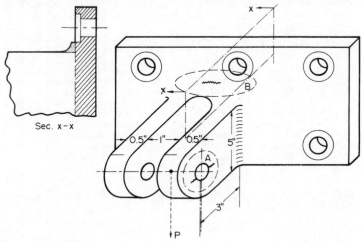

Fig. 3.115   Bracket.   (Holes and 0.5" dia.,
spotfaces are 1.0" dia.).

Question: Was the design a safe one for the region near the pin-hole? (not: Is the bracket safe now, as cracked?)

Solution: The geometry is suitably represented by Case 6 (in Table 3.17). Calculate the critical stress and compare with that applied.

Any stress intensity factor solution for this problem must reflect both the presence of the free surface of the hole and the finite width of the bracket. For cases such as this, an approximate solution can often be developed by applying separate correction factors for the various free surfaces. Let $f_1(a/R)$ be the correction factor for the hole and $f_2(a_e/b)$ be a correction factor for the finite width of the plate as in Case 2. The quantity a_e is an effective crack length for computing the finite width correction. Apply corrections for both effects by writing K_I in the form:

$$K_I = \sigma\sqrt{\pi a}\,f_1(a/R)\,f_2(a_e/b) \tag{1}$$

This is clearly an approximation since it neglects any interaction of the two effects, such a procedure is common practice and is normally sufficiently accurate for design calculations.

Returning to the problem at hand, compute the ratio $\frac{a}{R}$:

$$\frac{a}{R} = \frac{0.25 \text{ inch}}{0.25 \text{ inch}} = 1.0$$

From Bowie's solution (Case 6)

$$f \frac{(a)}{(R)} = 1.45$$

For computing the finite width correction, an effective crack length a_e of $a + R$ will be used. Thus,

$$\frac{a_e}{b} = \frac{a + R}{b} = \frac{0.25 + 0.25}{3} = 0.167 \text{ inch}$$

From Case 2

$$f_2 (a_e/b) = 1.02$$

Solving Equation (1) for the stress and letting critical conditions apply:

$$\sigma_c = \frac{K_{IC}}{\sqrt{\pi a c}\ f_1 (a/R)\ f_2 (a_e/R)}$$

(2)

$$\sigma_c = \frac{46 \text{ ksi } \sqrt{\text{in.}}}{\sqrt{\pi (0.25 \text{ in.})}\ (1.45)\ (1.02)} = 35 \text{ ksi}$$

Since the critical stress is well below the imposed load of 52 ksi, the design will <u>not</u> tolerate an inspectable crack.

It is interesting to investigate the maximum crack length which could be tolerated by this design. Solve Equation (1) for the crack size and assume critical conditions at the imposed stress of 52 ksi.

$$\sigma_c = \frac{1}{\pi} \left[\frac{K_{IC}}{\sigma_{max}\ f_1 (a/R)\ f_2 (a + R/b)} \right]^2$$

(3)

This gives the solution for the maximum crack length. However, the hole correction f_1 and the width correction f_2 are both a function of crack length. For this crack geometry, estimates can be made for f_1 and f_2: As $a + R/b \rightarrow 0$, $f_2 \rightarrow 1.00$. Assume $f_2 = 1.00$. For the hole correction factor, f, assume $f_1 = 3.39$, which corresponds to $a/R = 0$. After calculating a_c one can determine revised estimates for f_1 and f_2 and iterate:

$$a_c = \frac{1}{\pi} \left[\frac{46 \text{ ksi } \sqrt{\text{in.}}}{(52 \text{ ksi})(3.39)(1.00)} \right]^2 = 0.0217 \text{ in.}$$

check

$$a/R = \frac{0.0217}{0.25} = 0.088$$

for which, from Case 6:

$$f_1 = 2.81$$

Also

$$\frac{a + R}{b} = \frac{0.0217 + 0.25}{3} = 0.091$$

for which, from Case 2:

$$f_2 = 1.00$$

Calculating a_c again:

$$a_c = \frac{1}{\pi} \left[\frac{46 \text{ ksi } \sqrt{\text{in.}}}{(52 \text{ ksi})(2.81)(1.00)} \right]^2 = 0.032 \text{ in.}$$

Three more such iterations finally lead to the true critical crack length:

$$a_c = 0.038 \text{ in.}$$

This is a very short crack and could perhaps correspond to a nick caused by a tool at the hole edge; therefore, a re-evaluation of the design is inevitable.

In a situation similar to this, four design modifications are possible:

   a)   Reduce allowable operating stress.

   b)   Reduce hole diameter.

   c)   Use a different material (higher K_{Ic}).

   d)   Reduce plate thickness and still meet stress criteria.

Functional constraints may prohibit reducing the hole diameter. The fourth alternative would be possible for some materials. By reducing the plate thickness (increasing the remote stress) one takes advantage of the material's ability to deform plastically which results in higher values of fracture toughness. For the present material though, it is felt that a moderate reduction in thickness would not significantly improve the fracture toughness, and for a thickness reduction, a wider bracket would be required to meet stress criteria. In this case, probably, the best solution is to select a different material having a higher toughness which also may have a lower operating stress due to a lower yield strength. This attacks the real weakness of the present design; i.e., a low toughness steel ($K_{Ic} = 46$ ksi $\sqrt{\text{in.}}$) with a high yield strength ($\sigma_{ys} = 238$ ksi) which permitted a high operating stress.

Consider a different steel of higher K_{Ic}, say A514B at $K_{Ic} = 100$, and y.s. $= 80$ ksi. Now the critical stress is from Eq. (2):

$$\sigma_c = \frac{100}{\sqrt{\pi(.25)}(1.45)(1.02)} = 76.9 \text{ ksi}$$

and the first cut at the critical crack length is, from Eq. (3):

$$a_{cr} = \frac{1}{\pi} \left[\frac{100}{52 \ (3.39)(1.00)} \right]^2 = .102$$

some 5 times that for the 4340. The interested reader may continue the iteration for the final a_{cr}, which will be a readily inspectable length, and sufficient to provide for reasonable inspection intervals.

CASE STUDY III: A bracket as shown in Fig. 3.115.

PART I   Bracket to be of welded-plate construction.

Given: Material – HY-80 steel y.s. = 80 ksi; K_{Ic} = 120 ksi $\sqrt{in}$.
Remote stress level in region B is zero-to-tension of 50 ksi, normal to wall.
Inspection after fabrication reveals a longitudinal crack in the bead in Region B,
of length 0.150" and depth 0.050", from thermal shrinkage.

Question: Is the bracket design acceptable for the specified service life of
1×10^6 cycles in the presence of such a starting crack?

Procedure: Since the flaw is the surface type, Case I should be representative and
if Q, the crack shape factor, is near unity, it may be possible to solve for N_f
directly by Eq. 6 without the need to use a Worksheet, e.g., Table 3.19.

From Case I:

$$a/2c - .05/.15 - 0.33$$

$$\sigma/\sigma_{ys} = 50/80 = 0.625$$

From the crack-shape parameter curves (with Case I), these coordinates give Q≈1.0,
so Eq. 6 can be used (with 2c = a). The form factor F is only that for the free
surface correction: F = 1.12. Eq. 6 is rewritten:

$$N_f = \frac{1}{C}\left[\frac{2}{m-2}\right]\frac{2c_{cr}}{[S_{max}\sqrt{\pi(2c_{cr})}]^m}\left[\left(\frac{2c_{cr}}{2c_i}\right)^{\frac{m}{2}-1}-1\right]$$

The critical crack size is, from Eq. (2):

$$(2c)_{cr} = (120/50)^2/\pi = 1.83 \text{ inch}$$

C and m, the constants in Eq. 3, are, from the Data Section: C = 2.9 x 10^{-9};
m = 2.13.

Now, substituting in Eq. 6:

$$N_f = \frac{1}{2.9x10^{-9}}\left[\frac{2}{2.13-2}\right]\frac{1.83}{(50\sqrt{\pi 1.83})^{2.13}}\left[\left(\frac{1.83}{.15}\right)^{\frac{2.13}{2}}-1\right]$$

$$= \frac{1}{2.9x10^{-9}}[15.38]\ 6.83x10^{-4}\ [0.17] = 6.16x10^5 \text{ cycles}$$

So the answer is NO, (low by a factor of ∿ 2).

A quick check of the effect of an improved welding procedure to reduce $(2c)_i$
indicates that if $(2c)_i$ were halved to .075:

$$N_f = \frac{1}{(\quad)}\left[\quad\right]\left(\quad\right)[0.23] = 8.33x10^5 \text{ cycles}$$

only about 35% increase. While helpful, this source of improvement is not sufficiently promising to pursue further, especially as the .075" is already down into
the threshold range that production NDE will normally find.

The other parameters which may be considered for change are the toughness and the
applied stress. Changing the stress implies changing the design, perhaps significantly, so try for a higher K_{Ic}. The present value of 120 ksi $\sqrt{in}$. is about as
high as is available in the conventionally weldable alloy steels, but the

nitrogen-bearing stainlesses have very high toughness ($K_{Ic} \sim 350$) and are readily weldable at a yield strength of 50 to 60 ksi. Thus:

$$(2c)_{cr} = (350/50)^2/\pi = 15.6 \text{ inch}$$

$$C = 5.6 \times 10^{-12}; \quad m = 3.82$$

$$N_f = \frac{1}{5.6 \times 10^{-}} \left[\frac{2}{3.82-2} \right] \left[\frac{15.6}{(50\sqrt{\pi}(15.6)^{3.82}} \right] \left[\left(\frac{15.6}{15} \right)^{\frac{3.82}{2} -1} -1 \right]$$

$$= \frac{1}{5.6 \times 10^{-12}} [1.09] \left(\frac{15.6}{6.6 \times 10^7} \right) [67.4] = \frac{1146}{37 \times 10^{-5}}$$

$$= 3.1 \times 10^6$$

While this value is 3 times the requirement, there is no implication of "overkill", first - consider the tolerance on the initial crack size as given by NDE: the probability that other units in the production may have larger cracks, and especially if the public safety should be involved, this safety factor is not too high.

PART 2   Consider the bracket as a ductile-iron casting.

Given: Material is ASTM A439 Type D-2c ($\sim$DIN 1693)

Y.S. = 28 ksi

K_{Ic} = 2700 N/mm$^{-3/2}$; (68 ksi $\sqrt{in.}$)

C = 1.7 $\times$ 10^{-18} mm/cycle; m = 4.58

Remote stress in Region B is zero-to-tension, 15 ksi max. (Dimensions appropriately altered)

Cracked on cooling, as described in Part I

Question: Is the bracket design acceptable in this material for the specified service life of 1 x 10^6 cycles?

Procedure: As in Part I, using Eq. 2 and 6, (metric).

$$\sigma = 15000 \text{ psi} \times 6.896 \times 10^{-3} (\text{N/mm}^2)/\text{psi} = 103 \text{ N/mm}^2$$

$$(2c)_{cr} = [2700/103]^2/\pi = 218 \text{ mm}$$

$$N_f = \frac{1}{1.7 \times 10^{-18}} \left[\frac{2}{4.58-2} \right] \frac{218}{(103\sqrt{\pi}(218)^{4.58}} \left[\left(\frac{218}{3.75} \right)^{\frac{4.58}{2} -1} -1 \right]$$

$$= \frac{1}{1.7 \times 10^{-18}} [.775] \frac{218}{5.16 \times 10^{15}} [187]$$

$$= \frac{31590}{8.77 \times 10^{-3}} = 3600 \times 10^3 = 3.6 \times 10^6$$

Thus the requirement can be met by either material and fabrication.

It should be noted that the a_{cr} is greater than the width of the part, a condition frequently occurring in materials of high toughness at reasonable stresses.

3.3.   DESIGN PRACTICE

The practice of mechanical/structural design divides naturally into two basic categories: geometry and material properties. The designer's first task is usually to select a material, but the choice is often forced by requirements other than fatigue characteristics, as resistance to corrosion or to high temperature. Irrespective of the controlling property, however, the material used in a fatigue resistant design must possess an acceptable combination of many properties:

1. Allowable stress at the required number of cycles

2. Cyclic stress-strain properties

3. Notch sensitivity: q, or K_t/K_f

4. Fracture toughness: K_{Ic} or J

5. Crack propagation rate for slow, stable growth

6. Critical crack size, for fast, unstable fracture

7. Characteristic initial crack size

8. Corrosion rate and compatibility with adjacent materials

9. Sensitivity to fretting

10. Thermal, chemical, and metallurgical stability

11. Size and/or shape sensitivities

12. Sensitivity to surface conditions, as decarburization in steels, micro-cracks in machined beryllium, or the high-oxygen brittle layer on Beta III titanium.

Not all designs necessarily require a value for all possible properties, and data may well not be available for all specific materials, conditions, and forms, but a run-down of the above items is considered as a minimum to validate the choice of materials. Selections are often made on the basis of the first item alone, that is, just picking a point on an S-N curve, a barely permissible procedure that points out the necessity of additional considerations such as:

1. Mode of loading and degree of correlation between available data and design requirements.

2. Applicability of the Miner and Goodman criteria.

3. Availability of the desired form (shape of cross section). Not all alloy and organic compositions are, or can be, produced in all forms.

Further commentary on material choice is given in Section 4.3; foremost in the designer's thoughts during this decision-making period should be the fact that high ultimate strengths are frequently not reflected in high fatigue strengths. In general, the range of fatigue strengths available in any one alloy class (as wrought aluminum) is quite limited, compared to the range of ultimates. For instance, the latter range for the 2XXX, 6XXX, and 7XXX groups is nearly 9 to 1, but the fatigue limit range is only 3 to 1; for the magnesiums, the ranges are 3 to 1 and 2 to 1, and for the titaniums 2 to 1 and about 1.1 to 1. Thus, it seems evident that the fatigue behavior of the final part is much more dependent on the geometry of the part and of its loading scheme than upon the material.

At the present level of knowledge, it is impractical to state quantitatively any limiting rules for fatigue-resistant design because of the great diversity in function, loads, stresses, materials, and environments that may be encountered. Qualitatively, however, a number of useful practices can be defined that, when applied with good engineering judgment, can be expected to reduce potential fatigue problems significantly.

Excellence in detail, weight-saving, and simplicity are all closely related. The result of these combined features is usually low cost, ease of fabrication, longer service life, and increased reliability. This is important from a practical point of view because the scheduled time is often too short to develop and improve a design that just grows out of a non-simple approach. A complex design may look better, more interesting, and more challenging, but more often results in problems during fabrication, test, and service. In general, the same basic rules can be used in any of the industries: automotive, airframe, marine, and materials-processing; also, for the so-called static structures such as bridges, stacks, and tanks, since the wind or seismically induced movements of the "dead" load cause a fatigue effect. Past experience and learning from the one field can be carried over to the other, although some precautions should be taken, particularly since old designs may not be suited to a different environment.

Proposed procedures, whether completely applicable or not, often become fixed rules that govern the design activity. Recommendations or design guides are usually much more appropriate--and about the best that can be done. With this interpretation, some of the more pertinent guides for structural and machine design follow.

1. Keep the design simple.

2. Provide for multiple load paths where feasible (fail-safe).

3. Give extra consideration to tension-loaded fittings and components.

4. Provide generous fillets and radii.

5. Break all sharp edges. Polish critically stressed regions.

6. Make doublers and structural reinforcements result in gradual, rather than abrupt, changes in cross section.

7. Whenever possible, reduce the eccentricity of joints and fittings.

8. Design parts for minimum mismatch at installation for lowest residual and preload tensile strains.

9. Avoid superposition of "notches" in design.

10. Apply suitable factors of safety to net stresses around holes and cutouts.

11. Reduce bearing stresses in riveted and bolted members to design minimums. Design for tension-friction where possible.

12. Protect parts from corrosion.

13. Make a proper selection of materials with cost, strength allowables, fabrica- bility, and environmental effects in mind.

14. Arrange structural elements such that their longitudinal direction (from forming) is parallel to the major load resultant.

15. Laboratory-test all new joints and compare with "time-tried" designs.

16. Pay close attention to fabrication techniques for the optimum forming of components.

17. Establish reliable welding techniques for the reproducibility of joint strengths.

18. Construct rigid and precise tooling for the manufacture of production parts.

19. Provide easy access for the service inspection of structure.

20. Inspection procedures during fabrication and the assembly of the structure should also be provided.

21. Where necessary or desirable, provide vibration and/or shock isolation for fragile items.

22. Select configurations with inherently high structural damping: many joints.

23. Optimize bracket and component resonance frequencies, considering both service environment and equipment fragility.

24. Stagger the resonances for the items within an assembly. (Mismatch impedances of the mounted item and its bracket.)

A summary of these items, and the many more that probably could be written, might be bluntly stated as: "Make it right." Excellence in design results not only from knowledge, experience, and basic talent, but most predominantly from that indispensable ability to see what is needed: to judge the problem and to implement that judgment in the design. To produce adequate hardware, the designer needs the close cooperation of specialists able to perform complex dynamic stress analyses, acousticians, vibration engineers, metallurgists, and specialists in structural testing and reliability analysis as well as those experienced in tooling and manufacturing.

Although a structural designer cannot be expected to be facile in all these disciplines, some knowledge of them is necessary to close the loop, not only to complete the design, but to save time and to improve the assurance of success. As an example of this "outside" knowledge, Fig. 3.116 indicates the level of design experience as derived from the results of vibration testing. The objective, of course, is to keep the transmissibility as low as practical, and the improvement from even "standard" to "good" design is significant. Another means of judging levels of design implementation is shown in Fig. 3.117 in terms of the fatigue life of certain fittings, where both "poor" and "standard" designs become virtually unacceptable above about 10^4 cycles.

A rather long list of design guides or recommendations has been given above; an infinite number of typical cases might be found to illustrate them. The examples in Figs. 3.118 through 3.123 were chosen on the basis of importance and the frequency of occurrence; Fig. 3.124 illustrates the sometimes disastrous results of poor or incomplete drafting practice. The experienced reader may conclude that all of these items are elementary and well-known, but the fact remains that nearly all service failures are fatigue failures caused by the lack of attention to some of these details. Additional examples will be found in the sections on the shape effect and the fail-safe designs.

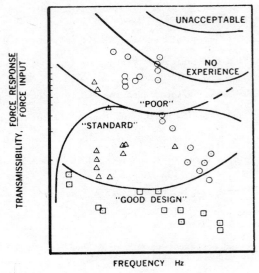

Fig. 3.116 Experience in bracket design; vibration test results.

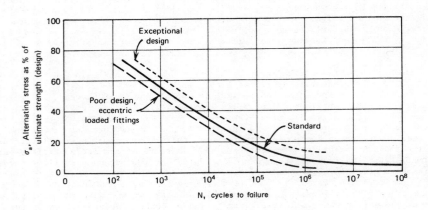

Fig. 3.117 Quality of aerospace structural designs in
aluminum alloys.

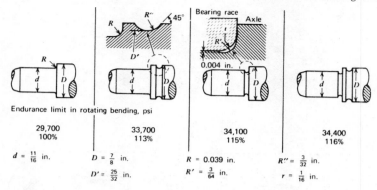

Endurance limit in rotating bending, psi

| 29,700 | 33,700 | 34,100 | 34,400 |
| 100% | 113% | 115% | 116% |

$d = \frac{11}{16}$ in.

$D = \frac{7}{8}$ in. $R = 0.039$ in. $R'' = \frac{3}{32}$ in.

$D' = \frac{25}{32}$ in. $R' = \frac{3}{64}$ in. $r = \frac{1}{16}$ in.

Fig. 3.118 Multiple notches (reduction of the first stress concentration factor by addition of artificial notches).

Method employed		Endurance limit, psi for steel with tensile strength of	
		52,000 psi	71,000 psi
Drilled radially, untreated		22,000	24,000
Unloading notches filed		25,000	27,500
Plain pressed, 0.75 in. diameter, 20 tons pressure		30,000	40,000
Plain pressed, 0.75 in. diameter, 20 tons pressure but with unloading notches also filed		28,000	——
Edge of hole and unloading notches impressed with a profile press		25,000	——
Unloading notches impressed without touching edge of hole		30,000	——
Unloading notches impressed after treatment with plain press		42,500	45,500

Fig. 3.119 The effect of unloading notches around an oil hole on the endurance limit in rotary bending (0.75 in.-diameter test pieces; hole diameter, 0.14 in.).

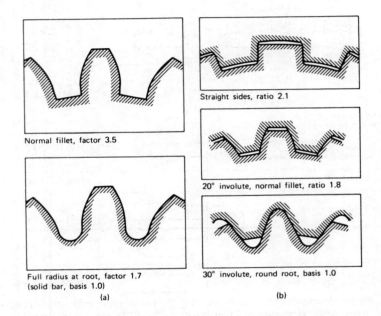

Normal fillet, factor 3.5

Full radius at root, factor 1.7
(solid bar, basis 1.0)

(a)

Straight sides, ratio 2.1

20° involute, normal fillet, ratio 1.8

30° involute, round root, basis 1.0

(b)

Fig. 3.120 Gears and splines. (a) The effect of root radius on the
stress concentration factor of the gear teeth; (b) the
effect of the type of spline on the stress concentration
factor. (Values of K_t for 14-1/2° and 20° pressure
angles are given in [93], Figs. 101 and 102.)

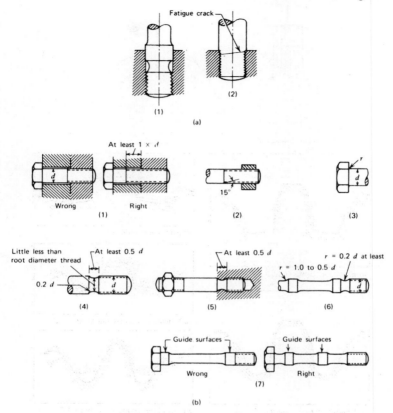

Fig. 3.121   Studs, bolts, and nuts. (a) Improved-type stud with stress-
relieving groove (1) replacing the old-type plain form (2),
which produced fatigue failure. The filet should extend
beyond the first thread. (b) Bolts and nuts: (1) unengaged
thread length between the thread end and the head should be
at least 1 x bolt diameter; (2) run-out angle of the thread
should not exceed 15°, (3) radius under the head should not
be less than 0.08d for small bolts and 0.1d for 3/4-in. bolts
and larger; (4) bolt thread relief proportions for high fatigue
strength, especially when there are bending stresses; (5) this
design of thread relief on studs is recommended over (2) for
greater fatigue strength, especially with incidental bending
or repeated tension impact; (6) proportions for relieved body
bolts shown in (5); and (7) design of bolts used for centering
and locating members where shear and bending forces are present,
minimum body diameter determined by greatest static tension or
torsion force. When the body is reduced to 36% of area of diameter over threads, the repeated impact tension strength is 2-1/2
times as great as a straight bolt having the same diameter as
threads.

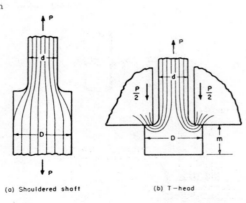

(a) Shouldered shaft   (b) T-head

Fig. 3.122   Shouldered shafts and T-heads: (a) shouldered shaft;
(b) T-head. Note difference in mode of loading between (a) and (b), and between (b) and Fig. 3.124b.
For T-head stress concentration factors, see [93],
Figs. 103–107. Typically, for r/d = 0.1, D/d = 2.0,
and m/d = 0.5, the stress concentration factor is
about 8.4.

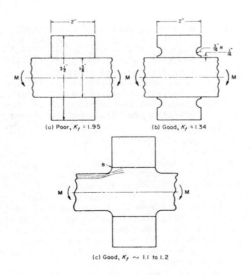

(a) Poor, K_f = 1.95   (b) Good, K_f = 1.34

(c) Good, K_f ~ 1.1 to 1.2

Fig. 3.123   Press- or shrink-fitted assemblies.

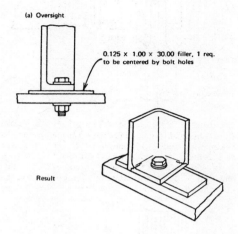

(a) Oversight

0.125 x 1.00 x 30.00 filler, 1 req.
to be centered by bolt holes

Result

Fig. 3.124a Results of poor drafting technique.
(a) In drafting, only one view of
assembly was shown, filler of minimum allowable width for sufficient
edge distance and bearing contact
under nominal loads was called for;
in service, fatigue cracks developed
around bearing web because of lack
of solid practical contact across
the face.

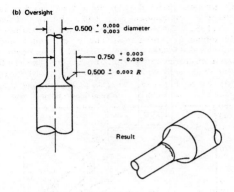

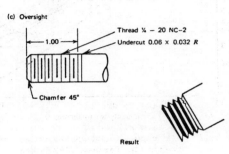

Fig. 3.124b,c   Results of poor drafting technique. (b) In drafting, at least
one possible undesirable combination of tolerances was overlooked;
in service, fatigue cracks developed at sharp edge.   (c) In drafting, no mention of dressing undercut was made; in service, fatigue
crack developed at feather edge.

3.4   SPECIAL TOPICS

3.4.1   JOINTS AND LUGS

The importance of proper joint design cannot be overemphasized. The dependency of
the fatigue behavior of an assembly on that of the joints within it is sharply
outlined in Fig. 3.125. The heavy line is the endurance of the complete wings and
tailplanes, while curve A respresents the mean, and B and C the upper and lower
bounds of the endurance limits of the joints. The conventional approach to the
fatigue design of a multimember structure treats the members as first in importance,
with the joints as inevitable complications, possibly only as necessary evils. In
some cases, it may be more profitable to consider the assembly as a collection of
joints with the members as simple connections among them.

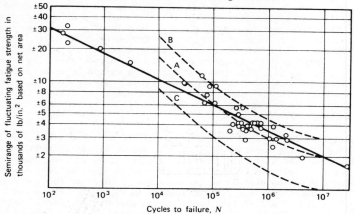

Fig. 3.125   S-N curves for complete wings and tailplanes.
From P. G. Forrest [142], Fig. 143.

A joint is here defined as an assembly of two or more massive pieces of material,
having been mechanically transported to, and being restrained in a position of
contact or of close proximity. Such a definition is intended to exclude the interfaces formed by such methods as plating, painting, vapor deposition, and the like.
It also reduces the scope of the topic to joints made by the four time-proven processes: bolting, riveting, welding, and adhesive bonding. Some of the newer
methods such as joining by explosive forces, by high velocity impact and by the
transfer of momentum to heat are being rapidly developed, but the available data
has not yet been validated in a sufficiently universal manner to warrant their
inclusion here. The four common processes are often thought to present only conventional design problems, but rigorous demands for low weight combined with high
performance require considerable insight regarding fatigue phenomena and the
parameters of shape and material. It is to the latter topic that this section is
chiefly addressed.

It is evident that, in addition to the material, the shape of the parts and the
modes of loading are highly important. The designer may not always have control
of the external envelope, but he must consider the part details and the location
of the joint along the load path. If, for example, the bending component on a bolt
could be eliminated and the loading simplified to shear by changing the joint's

position or orientation, its overall size and weight may be reduced. The shape
of the parts within the joint is, of course, highly significant as it may transform
the external loading to a different mode, and it is inevitably concerned with
stress concentrations. "Shape" here refers to any geometric characteristic such
as the thread form on bolts, the groove profile on weld plate edges or the radius/
thickness ratio at a clevis throat. Since design for resistance to fatigue failure,
at any lifetime, requires close attention to physical detail and since joints are
inherently composed of details, their design for fatigue loading becomes a highly
demanding task.

The overall envelope of the joint may often be dictated by functional, loading,
and displacement requirements, but within the envelope considerable latitude exists
for the arrangement of bolts, rivets, or weld beads. By applying one of the first
principles of design in the form of drawing the load lines, in all three dimensions,
across the joint, it is immediately seen that as usual the straighter the lines,
the greater the strength. Both static and fatigue loading capacities are improved
by the elimination or reduction of transverse components. While the use of lugs
or protruding bosses is often unavoidable, their offset from the load line should
be minimized. Drawing the analogy between load lines and fluid flow results in the
above conclusion, and also that there should be no sudden changes in section of the
connected parts, with the obvious effect on stress concentration factors. If one
considers a shear joint with two or more rows of bolts, it follows that the "flow"
would be improved--made smoother--with the rows placed in tandem along the lines,
rather than staggered.

The least-weight joint or splice is that having the shortest overlap or cover
plate, meaning that the rivet or bolt spacing should be the minimum, practically
and structurally. The minimum structural spacing is, of course, that calculated
on the appropriate criteria of shear, bearing, or tension, while the minimum practical spacing depends on clearances for wrenches or dimpling tools. The latter is
a particularly cogent item for flush riveting; in any event, a practical spacing
must be held to insure well-formed heads and an undistorted sheet. Splices in
structural shapes are becoming less necessary because of the increased sizes of
extrusions and forgings. But, should such a joint be required, the shape may be
divided into strip elements, as the web and flanges, and the splice designed such
that each element has the same stress distribution as in the structural shape.
If it is not possible or practical to splice each element, as in the case of an
angle with a small return flange, the return may be considered to be spliced by
the rivets in a closely adjacent element; the eccentricity so developed is usually
negligible. But the splicing member should be flanged and in the same direction as
the flange of the structural shape for continuity. There appears to be no restriction on the application of this approach to any shape under any combination of
loads.

Many tests have been conducted on bolt and rivet patterns for maximum efficiency
resulting, generally, in the following:

1. The bolts or rivets in the different rows should be located in tandem, that
    is, along the load line, not staggered.

2. The spacing between rows should be the minimum practical.

3. The bolts or rivets in each row should be of the same size.

4. Each row should contain an equal number of bolts or rivets.

An assumption tacit in these conclusions is that each rivet or bolt in a row shares the load equally. Fisher and Struik [396] show the calculated distribution for practical shear joints with 4, 10 and 20 bolts in line, Fig. 3.126a,b,c. The reason for this unequal distribution is the decreasing average shear strength with increasing length--note the negligible effect on joints up to about four fasteners. Also, any local condition that negates the assumption, such as plates of different thickness or stiffness, leads to the possibility of "first" rivet failure and the need for setting aside the recommendations (3) and (4) for equal size and number. For patterns using fasteners of unequal size in shear, it is usually considered that the load on the row will be distributed in proportion to the shear strength of each fastener; when the plate stresses are purely elastic, the load distribution is probably more nearly proportional to the fastener bearing area.

Basically, the envelope of the joint and the fastener pattern need to be considered as an entity in order to determine the effective width of the pattern, and to identify the critical sections. Typical cases are shown in Figs. 3.127a and b.

Bolted Joints. The predominant modes of loading for bolts are tension and shear; bending may be combined with either mode depending on local load paths and fit. Shear joints may be further classified as shear-tension and shear-bearing types.

The general fatigue behavior of bolted joints is given by the expressions:

$$\sigma_{an} = 1.5 + 1500 \left(\frac{13}{N\sigma_{mn}}\right)^{1/2}$$

or

$$N = \frac{13}{\sigma_{mn}} \left(\frac{1500}{\sigma_{an} - 1.5}\right)^2$$

where

σ_{an} = alternating stress for notched material, in kips
σ_{mn} = mean stress for notched material, in kips
N = number of cycles

Both expressions are wholly empirical and, while based on much experimental data [5], may fail for extreme designs.

The fatigue strength of lap joints should be carefully considered by the designer, not only because they may be the only type available, but also because they are basically inefficient due to the high stress concentrations from the offset loading. Investigations using 3/8-in. diameter aircraft bolts in 2024-T3 Alclad (0.100 to 0.375 thick) showed that the fatigue strength at 10^6 cycles is as follows.

1. Increases linearly with the number of bolts in the load line: from 7000 lb/in.^{2} for one bolt to 13,000 lb/in.^{2} for three bolts in line; in plate from 0.100 to 0.125 in. thickness.

2. Increases slowly with the increase in the number of rows of bolts.

3. Increases slowly with the increase in the bolt diameter, from 6500 lb/in.^{2} for 1/4-in. bolts to 8500 lb/in.^{2} for 7/16-in. bolts, at a stress ratio R = +0.25.

4. Decreases with the increase in the plate thickness for any number of bolts in line and for any number of rows; typically for two bolts in line in a single row, it drops from 9700 lb/in.^{2} for a 0.100-in. plate to 5700 lb/in.^{2} for a 0.375-in. plate.

OFD - 1

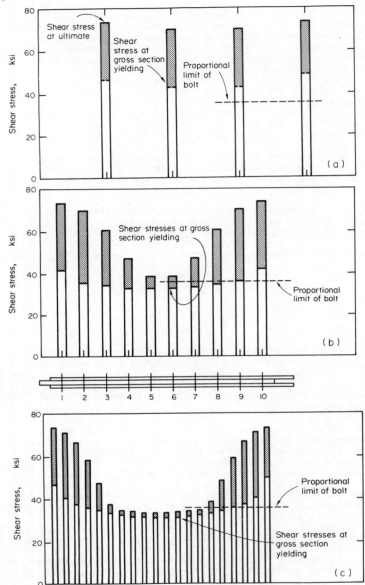

Fig. 3.126   Load distribution in joints with multiple fasteners.   Plate
material S_{yp} = 36 ksi.   Bolts are 7/8" dia. A325.
(a) 4 bolts in line; (b) 10 bolts in line; (c) 20 bolts
in line.

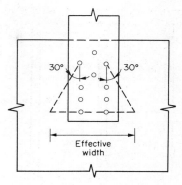

Fig. 3.127a Effective width for a
fastener pattern (from
Ref. 396).

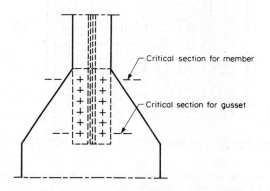

Fig. 3.127b Critical sections for a joint subject to
fatigue loading (from Ref. 396).

The fatigue strength of bolts is, of course, strongly affected by the inherent
stress-raisers, especially those under the head, at the end of the threaded length
and at the end of nut engagement. Major improvements have been accomplished by the
use of elliptical fillets under the head; at thread-runout by rolling the threads,
controlling the runout radii and/or reducing the grip diameter. (More discussion
in paragraph on Stress Concentrations in Bolts, below.) The load distribution
along the thread contact length with the nut is complex, but, generally speaking,
the bolt sees axial tension varying from a maximum value at the first engaged
thread to zero at the first thread outside the nut. The nut sees an axial compression gradient in the same direction. Since the first thread on the nut may be
engaged with one near the end of the threaded bolt section, a serious concentration
of stress may be present. The methods of improving the stress distribution in the
nut, and therefore in the bolt, have been the subject of much study [113, 114,
115, 116]. Some of these attempts have been quite successful, notably that of

thinning the lower threads in the nut, thus allowing them greater local deflection and passing the load further along the engagement length, away from the first thread. Comparison of the performance of such a nut with a standard type on a 220-ksi bolt indicates a 30% increase in allowable fatigue stress at 10^5 cycles, or about 50 times improvement in the fatigue life at 100-ksi stress at the thread root. However, such treatments are rarely cost-effective when compared to the improvements resulting from the treatments listed earlier.

Both bolted and riveted joints inevitably contain holes in the basic sheet or plate; and it has often been observed that a loaded hole may be the most important of all sources of stress concentrations. By the contrasting nature of their geometry, weldments and adhesively bonded joints avoid this particular source of stress concentration. For a single hole in an infinite sheet under tension, the K_t is 3.0 and Fig. 3.128 shows the redistribution, in the sheet adjacent to the hole, of the load that would have been supported by the hole area. It is seen

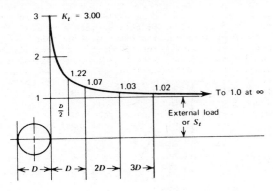

Fig. 3.128 Stress distribution near a hole.

that the peak stress is not much affected by either trimming the sheet to a width of 3 or 4 hole diameters or by adding holes in a row at a pitch of 3 or 4 hole diameters. Thus, for multiple holes at a normal pitch of 3 hole diameters, for example, the accumulative effect of the K_t's for each hole will be much less than their simple product and may be determined by superposition. Considering the original hole plus a new hole on the right and one on the left, the peak stress relative to the gross area stress can be approximated by $3(1.02)(1.02)S_t = 3.13S_t$, an increase of only 4% over that for a single hole, although one-third of the sheet has been removed along this section. Had the stress calculation been based on the net section, the value of K_t would have been 27% greater than that for a single hole.

For designs using a hole pitch less than 3 diameters, the static strength becomes the governing criterion. The fatigue strength is less critical because the low gross-area stress dictated by the static strength greatly reduces the peak stresses. Any bearing stress that may be present will, of course, add to the peak tension stress through a corresponding K_t, but this effect, too, is very highly localized. Thus, for practical design calculation using conventional hole spacings, the peak stress is a function primarily of gross-area stress and the type of hole; the bearing stress/hole spacing ratio has little effect on the peak stress.

The foregoing discussion pertains to joints with straight-shank fasteners. Tapered shank bolts have been recently introduced as a means of increasing fatigue life by the presence of a radial compression in the hole edge. Such loading also includes hoop tension, of course, and some concern had been expressed about the possibility of stress corrosion. There is little evidence of serious corrosion reported as yet but there are numerous reports of greatly increased life, mostly in joints that are rather highly loaded [385], in either constant amplitude or spectrum tests.

Tension Bolts. For bolts in tension, with or without bending, the most suitable criterion for failure is separation of the joint, a condition allowing any number of undesirable results such as loss of pressure, excessive distortion, and impact. Separation of the connected members may occur at loads and bolt stresses lower than those for fracture of the bolt, the latter being considered as ultimate, catastrophic failure. Thus, a major objective in the design of these joints is to show that the bolt can prevent separation while sustaining the maximum design load for the required number of cycles. The first requirement for fatigue, as well as for static design, is the determination of the maximum bolt load, a value that has been the subject of much consideration, resulting generally in the correct but only qualitative conclusion that it is minimized by making the bolt more "elastic" than the bolted parts. Thus, spring constants arise as necessary parameters in the analytical considerations, but unfortunately, the k of the bolted parts is usually calculable for only the simplest shapes (i.e., a pair of plates). For complex shapes, such as connecting rod bosses, resort must be made to direct test. However, a general expression for spring constants may be formed as:

$$k = F/\sigma = AE/L \tag{96}$$

The ratio of the (area) x (modulus) product for the bolt and for the parts:

$$(A_b E_b/L_b)/(A_p E_p/L_p) \tag{96a}$$

may then be taken as a quantitative description of the joint stiffness. The question of the effective compression area of the plates is considered below.

In establishing as expression for the total load on the bolt, the assembly is considered as a pair of springs in parallel with the bolt in tension, and compression in the annular volumes of the plates between the head and nut. To maintain equilibrium, the forces in the bolts and plate must be equal and opposite. The deflections at preload would be equal only if the stiffnesses were equal. As the external or working load is applied via the plates, the total load on the bolt increases, the limit to this action being defined as the condition of impending separation. If the applied load is increased beyond that at separation, then the total bolt load will, of course, continue to increase. This action is shown schematically in Fig. 3.129a. During loading between preload and separation, the changes in bolt and plate deflection are equal.

The load expressions are obtained for a conventional assembly of multiple plates or flanges by writing first the force equality under preload alone:

$$\text{Tensile force in bolt} = \text{Compressive force in plate(s)}$$

$$P_i = C \tag{97}$$

or

$$\frac{P_i L_b}{A_b E_b} = \frac{C L_{p1}}{A_{p1} E_{p1}} + \frac{C L_{p2}}{A_{p2} E_{p2}} + \cdots \cdots \frac{C L_{pn}}{A_{pn} E_{pn}} \tag{98}$$

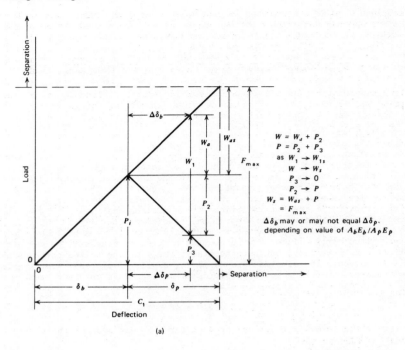

$$W = W_a + P_2$$
$$P = P_2 + P_3$$
as $W_1 \rightarrow W_{1s}$
$$W \rightarrow W_s$$
$$P_3 \rightarrow 0$$
$$P_2 \rightarrow P$$
$$W_s = W_{as} + P$$
$$= F_{max}$$

$\Delta\delta_b$ may or may not equal $\Delta\delta_p$. depending on value of $A_b E_b / A_p E_p$

(a)

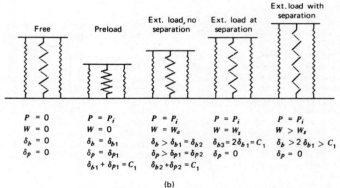

(b)

Fig. 3.129   Force–deflection conditions for a bolted joint.

The relations between the preload and the working load at separation are derived from Eq. 98 and the condition that the compressive preload and deflection become zero. Adding the deflection from the working load to Eq. 98, we obtain:

$$\frac{P_i L_b}{A_b E_b} + \frac{W L_b}{A_b E_b} = \frac{C L_1}{A_1 E_1} - \frac{W_1 L_1}{A_1 E_1} + \frac{C L_2}{A_2 E_2} - \frac{W L_2}{A_2 E_2} + \ldots \frac{C L_n}{A_n E_n} - \frac{W L_n}{A_n E_n} \tag{99}$$

the signs appearing as above because the working (applied) load increases the bolt elongation but decreases the plate compression. Rewriting Eq. 99 through the use of the equalities: 1) the moduli of all plates are equal; and 2) the compression areas in all plates are equal, leads to:

$$\frac{(P_i + W)L_b}{A_b E_b} = \frac{(C-W)L_p}{A_p E_p} \tag{100}$$

where L_p is the total thickness of the plates in the stack.

During application of the working load from zero to value W, the bolt underwent an additional elongation ΔL above that at preload, and the preload compression in the plates decreased an equal amount. The load remaining in the plates is $P_i - \Delta P_i$ where ΔP_i corresponds to ΔL. The new load in the bolt is then equal to the plate load plus W:

$$F_b = P_i - \Delta P_i + W \tag{101}$$

Equating the changes in elongations, we obtain:

$$(\Delta P_i)L \left[\frac{1}{A_1 E_1} + \frac{1}{A_2 E_2} \right] = \frac{(W - \Delta P_i)L}{A_b E_b} \tag{102}$$

which may be reduced to:

$$\Delta P_i = \frac{W}{(A_b E_b / L_b)/(A_p E_p / L_p) + 1} \tag{103}$$

Solving for the total load in the bolt by substituting ΔP_i from Eq. 103 into Eq. 102, we find that:

$$F_{max} = P_i + W \left[\left[\frac{A_b E_b / L_b}{A_p E_p / L_p} \right] \Big/ \left[\frac{A_b E_b / L_b}{A_p E_p / L_p} \right] + 1 \right] \tag{104}$$

which is simplified through the use of $r = (A_b E_b / L_b)/(A_p E_p / L_p)$ to:

$$F_{max} = P_i + Wr/(r + 1) \tag{104a}$$

Similarly, for the total load in the plates, we get:

$$F_p = P_i - W/(r + 1) \tag{105}$$

In view of these relationships, Fig. 3.129b reveals that:

1. Separation occurs only at $W = P_i$ or $W/P_i = 1$ for all values of r. Prior to separation the total bolt load cannot exceed $2P_i$ for any value of r; F_{max} approaches $2P_i$ as r approaches ∞, at $W/P_i = 1$. Although $W_s = P_i$ at separation, $F_{max} \neq 2P_i$ because a portion of W_s is being reacted by compression in the plates. Only when $r = \infty$ (or $E_p = 0$) does $F_{max} = 2P_i$.

2. After separation, F_{max} increases with both W and r, the latter indicating, quantitatively, lower loads in the lower modulus bolts or lower AE joints.

3.   Although the general desirability of "low" bolt loads is not debated, a
     difficulty should be noted with low load, low-r joints in that sufficient
     clamping force, P_i, may be obtainable only with very large bolt elongations.

The above considerations have, through Eq. 104a, related the three prime parameters:
total bolt load, F_{max}; preload, P_i; and working load, W. A further relation would
be highly useful, that between the critical preload, P_c and the alternating component of W, which is W_a. If the initial preload is made equal to P_c, the force
compressing the plates changes between the limits of P_c and 0 during reversal of
the working load between 0 and W. But part of W is used up reacting the preload,
so that the remainder is the alternating component:

$$W_a + P_c = W \tag{106}$$

After some manipulation, we obtain:

$$W_a = W \left(\frac{r}{r + 1} \right) \tag{107}$$

$$P_c = W - W_a$$

and

$$P_c = W \left(\frac{1}{r + 1} \right) \tag{108}$$

For practical reasons, and to provide a margin of safety, the actual preload is
usally made somewhat greater than P_c:

$$P_{act} = c P_c$$

where c is often given a popular engineering factor of safety value between 1.15
and 1.50.

In fatigue design, it is imperative to recognize the effect of the range of alternating loads, and joint design is no exception to the rule. A load ratio is
defined here as:

$$R = \frac{F_{min}}{F_{max}}$$

Then

$$F_{min} = P_c = W \left(\frac{1}{r + 1} \right)$$

$$F_{max} = P_c + W_a$$

$$= W \left(\frac{1}{r + 1} \right) + \left(\frac{r}{r + 1} \right)$$

and Eq. 104a reappears, modified as:

$$F_{max} = c W \left(\frac{1}{r + 1} \right) + W \left(\frac{r}{r + 1} \right) \tag{104b}$$

In fatigue terms the preload, P or P_c, represents the mean stress, and the variable
part of the applied load, W_a, the alternating stress. Thus, the latter is the more
critical and should be decreased, at the expense of increasing P_c whenever possible,
the relationship being dependent on the relative stiffnesses: the r value.

In stress terms, preloaded bolts in a joint under cyclic stressing are subject to that combination of steady and alternating stresses characteristic of fatigue designs. Much has been written on the desirability of protecting the bolt from the effect of the alternating component by increasing the mean stress, or preload. A glance at the master life diagrams shows immediately that the slopes of the lifelines are such that, in general, much life can be gained by boosting the mean stress. However, recalling that the major objective of bolted design is to prevent separation, it follows that the mean stress (in the fatigue sense) from the preload must be sufficient to accommodate the maximum working load. Then the total load on the bolt is, as given by Eq. 104a, composed of the preload mean stress and the working load alternating stress. The suggestion has been made that the effect of the alternating component, S_a could be canceled by making $P_i = F_{max}$ but this is irrelevant, since S_a arises only from W_a and always adds to P_i, leading to a new, higher value of F_{max}. Not only is there no benefit in increasing P_i beyond that value needed to prevent separation, but such action results in a lower fatigue life because of an unnecessarily high mean stress. These points have been obscured in some earlier discussions on this topic [102, 104, 106, 109]; it was apparently thought that, if some preload was good, more was better. The argument above indicates that there is a proper value of preload for each design condition: insufficient preload allows joint separation with its large increase in the alternating component; too high a preload leads to lowered fatigue life because of the high mean stress.

A major result of this deriviation, Eq. 104a, is plotted as Fig. 3.130, for practical values of r, and is normalized on the preload. The related Fig. 3.131 introduces the time dimension (any arbitrary time period) and illustrates very sharply the effect of different joint stiffnesses on the value of the total bolt load. The study of these plots, together with Tables 3.28 and 3.29, yield several useful conclusions. The total bolt load, F_{max}, increases sharply with increasing slope, or r value, a condition that must be controlled in attempts to design for any given F_{max}. Note that, of the two basic parameters in the ratio, E is controllable only by choice of material, while A is adjustable by the details of the joint design. For example r = 2.6 for an AN8 bolt in aluminum plates with no washers, giving an F_{max}/P_i of about 1.7 at separation. The latter value could be reduced to about 1.4 by bringing r into the range of 0.8 to 1.0 through the use of conical steel washers with a large end diameter of 1.15 in. and a half-cone angle of 30°, as in Fig. 3.132. The weight and volume of the washers and longer bolts may preclude their use in highly efficient joints, but one of the often-suggested remedies--that of simply increasing the bolt size--is of little benefit because of the random variation of r with bolt size. The addition of washers at these locations increases the bolt length and decreases its stiffness (k = AE/1), leading possibly to greater fatigue life.

More extreme examples may be noted in Tables 3.28 and 3.29, the maximum r value of 5.4 indicating a total bolt load of nearly twice the preload at separation. As expected, the higher r's are associated with the higher values of E_b/E_p; thus, the higher bolt load conditions appear for steel bolts in magnesium plates. The stiffness, r, varies little with the type of steel in the bolts, but it is, of course, a strong function of the modulus for non-ferrous bolts.

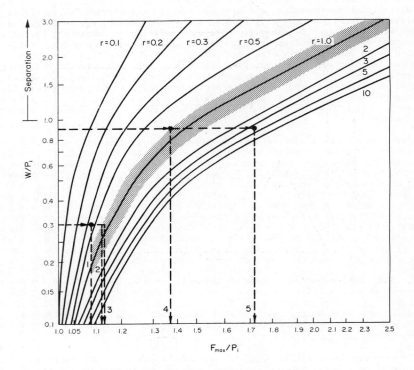

Fig. 3.130   Relationship of Total Bolt Load to the External and Preloads.

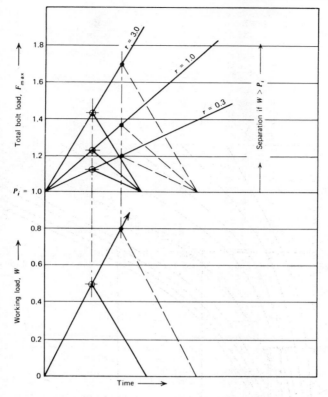

Fig. 3.131   Total bolt load as a function of applied load and time.

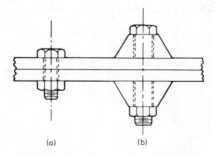

Fig. 3.132   Variation of r and F_{max} for steel bolts
and washers in aluminum alloy plates.
(a) r = 2.5, F_{max}/P_i = 1.7; (b) r = 0.85,
F_{max}/P_i = 1.4.

TABLE 3.28   r Values for NAS Bolts

	Shank Diameter, d_b	Head Diameter, d_p	$\dfrac{(Area)_b}{(Area)_p}$	r in Steel Plate $E_p = 29.5 \times 10^6$	r in Aluminum Plate, $E_p = 10.5 \times 10^6$	r in Magnesium Plate, $E_p = 6.5 \times 10^6$
NAS-3	0.189	0.375	0.25	0.24	0.94	1.51
NAS-4	0.249	0.438	0.47	0.45	1.25	2.03
NAS-5	0.312	0.500	0.64	0.62	1.70	2.76
NAS-6	0.374	0.563	0.81	0.78	2.18	3.50
NAS-7	0.437	0.625	0.97	0.94	2.57	4.18
NAS-8	0.499	0.750	0.80	0.77	2.13	3.45
NAS-9	0.562	0.875	0.69	0.67	1.84	2.98
NAS-10	0.624	0.938	0.80	0.77	2.13	3.45
NAS-12	0.749	1.063	0.98	0.95	2.62	4.23
NAS-14	0.874	1.250	0.95	0.92	2.54	4.10
NAS-16	0.999	1.438	0.93	0.90	2.49	4.00
NAS-18	1.124	1.625	0.92	0.89	2.46	3.95
NAS-20	1.249	1.812	0.92	0.89	2.46	3.95

r Values without washers, 0° cone angle in plates; $r = A_b E_b / A_p E_p$.
Dimensions from NAS-501 Standard Sheet 1, Revision of January 31, 1956. The material is stainless steel type 321, AMS 5645; $E_b = 28 \times 10^6$ psi.

TABLE 3.29   r Values for AN Bolts

	Shank Diameter, d_b	Head Diameter, d_p	$\dfrac{(Area)_b}{(Area)_p}$	r in Steel Plate $E_p = 29.5 \times 10^6$	r in Aluminum Plate $E_p = 10.5 \times 10^6$	r in Magnesium Plate, $E_p = 6.5 \times 10^6$
AN-3	0.189	0.335	0.47	0.45	1.25	2.02
AN-4	0.249	0.398	0.64	0.62	1.70	2.75
AN-5	0.312	0.460	0.87	0.84	2.32	3.74
AN-6	0.374	0.523	1.10	1.06	2.92	4.73
AN-7	0.437	0.585	1.25	1.21	3.33	5.40
AN-8	0.499	0.710	0.98	0.95	2.60	4.20
AN-9	0.562	0.835	0.80	0.77	2.14	3.44
AN-10	0.624	0.898	0.91	0.88	2.42	3.90
AN-12	0.749	1.023	1.16	1.12	3.10	5.00
AN-14	0.874	1.210	1.07	1.04	2.84	4.60
AN-16	0.999	1.398	1.05	1.03	2.81	4.54
AN-18	1.124	1.585	1.00	0.97	2.67	4.30
AN-20	1.249	1.772	0.93	0.90	2.48	4.00

r Values without washers, 0° cone angle in plates; $r = A_b E_b / A_p E_p$.
Dimensions from ANA Standard AN-3 through AN-20, Sheet 1, Revision of April 6, 1964. The material is stainless steel type 321, AMS 5645; $E_b = 28 \times 10^6$ psi.

Tables 3.28 and 3.29 were developed to show the variation in r with the size of standard bolts. The ratio of shank area to annulus area is neither constant nor uniformly varying with bolt size, a condition that leads to wide variation in r and the bolt load. When heavy washers are used to control the bolt load, their outside diameter should be sized to minimize r to the greatest extent practicable. The compression area in the plate, A_p, was calculated as that of a cylindrical annulus (i.e., of 0° cone angle), proportional to $(d_a^2 - d_b^2)$. Thus, the tabulated values of r are conservative to a degree dependent on L, the plate thickness. For L < 1/8-in. the values are essentially as given. A relevant A_p in each design case should be calculated by Eq. 108a from the plate thickness.

Example: Find F_{max} for Tension Bolts

Consider the bolting for a circular steel header plate on a compressed air storage tank, the total force on the plate being 8000 lb, and the design objective being to prevent loss of pressure.

1. Assume a number of bolts and a pattern – here take 10 bolts in an 8" dia. circle.

2. Assume a load distribution to determine the maximum working load – here take a uniform distribution, so W = 800 lb. per bolt.

3. Assume a trial bolt size, and Grade or Class. Here take NAS-4, SAE Grade 2, for which the maximum allowable preload, or clamping load = 1320 lbs (from Table 4 of Ref. 395; or Class T-1 ultimate load = 1837 lbs (from Table 8.1.2(b) of Ref. 1).

4. From Table 3.28 obtain an r value for the trial size as 0.45.

5. Choose a value for W/P_i – here try 0.3.

6. Calculate the preload: $P_i = (1/0.3)W = 3(800) = 2400$ lbs.

7. Enter the curves of Fig. 3.130 on Line 1 at r = 0.45 and $W/P_i = 0.3$ to find $F_{max}/P_i = 1.09$.

8. Calculate the total bolt load: $F_{max} = 1.09(2400) = 2620$ lbs.

It is immediately seen that F_{max} exceeds the capacity of the trial bolt size and grade; recourse to References 395 and 1 indicate that a Grade 8, or a Grade 5 fine thread, or a Class T-4 would be required. While any of these 3 choices form a correct solution, questions of cost and availability may make a further trial with larger, low-grade bolts more interesting.

Consider a NAS-5, Grade bolt.

1. As before.

2. As before.

3. Clamping load = 2160 lbs.
   Ultimate load = 2952 lbs.

4. From Table 3.28, r = 0.62.

5. $W/P_i = 0.3.$

6.   Preload = 3(800) = 2400 lbs., as before.

7. $F_{max}/P_i = 1.12$, (line 2, Fig. 3.130).

8. $F_{max} = 1.12(2400) = 2700$ lbs., which also indicates insufficient capacity.

For a NAS-6, Grade 2 bolt.

1.   As before.

2.   As before.

3.   Clamping load = 3200 lbs.
     Ultimate load = 4530 lbs.

4.   From Table 3.28, r = 0.78.

5. $W/P_i = 0.3.$

6.   As before: Preload = 3(800) = 2400 lbs.

7. $F_{max}/P_i = 1.13$, (from Line 3, Fig. 3.130).

8. $F_{max} = 1.13(2400) = 2725$ lbs.

Comparison of this 2725 lb. load with either the 3200 lb. clamping or the 4530 lb. ultimate capacities indicates a positive margin.

One of the most important factors affecting the total bolt load is the ratio of working to preload, W/P_i. In the example so far, the arbitrary value of 0.3 was used. The general trend is that, as W/P_i increases, the total load comes down, but by definition, so also does the margin between the working (or external) load and the preload (or clamping force). Thus, use of the more efficiently designed bolts with the higher values of W/P_i requires an increasingly accurate knowledge of both W and P_i (Line 4 in Fig. 3.130). This ratio is determined primarily by engineering judgment: in many instances, it would appear desirable to build-in the safety of a wide margin between the external load and the preload (low W/P_i), but this condition is obtainable only at relatively higher total loads.

As noted earlier, joint stiffness has a pronounced effect on total load. Consider the same example again, but with magnesium alloy plates instead of steel. Now, from Table 3.28, the r value is 3.45, leading to higher F_{max} for all W/P_i; for the more efficient joints, the change in plate material raises F_{max} by about 50%, see Line 5 in Fig. 3.130.

In the steel bolt-steel plate combination, the r value increased with increasing bolt size (in the range considered), thus, one would expect the total load to go up, but the bolt capacity fortunately increases faster with size than does the r value. This condition may not hold for all sizes, examination of Tables 3.28 and 3.29 reveals much randomness, especially for the AN group. For joint designs using bolts of other standards, the r values should be calculated for each size and material combination.

The sizing of washers to insure that a sufficient compression area of the plate is brought into action may be done with reference to Fig. 3.133. If we assume that:

$$\text{plate thickness} = L_1$$

$$\text{bolt shank radius} = r_b$$

$$\text{Head bearing radius} = r_a$$

$$\text{hole radius} = r_d$$

and that r_c represents the radius of the effective compression area, then it follows that the washer outer radius, r_w, and thickness, l_w, must be such as to place points B and C on the line OA extended. The compression area is properly taken as the annulus at C-D since the results of studies in photoelasticity show that the volume in compression takes on the profile of a truncated cone with an apex angle between 25° and 33° depending on plate material, thickness and stress level, and that the maximum stress appears near the central plane of the plate. For simplicity in the construction of Fig. 3.133, BF is made equal to $L_1/2$ and the angle is taken as 30°, from which a general expression for the compression area is derived as:

$$A_p = \pi/4[(d_w + 0.50L_1)^2 - d_d^2] \tag{108a}$$

for d_b/L_1 in the range of 0.68 to 1.4. Rigorous derivation for r_c results in a complex logarithmic form, but giving A_p within ±2% of the values from Eq. 108a. A highly useful relationship between thickness and effective radius in flanges or plates is given in Fig. 3.134 for several bolt sizes.

It can also be seen that the washer might be decreased considerably from the proportions shown before any bearing area under the head is lost, but this procedure often leads to failure by local crushing. It is important to develop the full areas in compression at both washer interfaces to avoid this type of failure.

Example

A Class T-8, AN-6 steel bolt is applied to 1/2-in.-thick plates of 2024-T851. From [1], Table 8.1.2(b), the ultimate tensile load on this bolt is 15,100 lbs. Determine the (steel) washer dimensions necessary to develop the ultimate compressive stress in the plates for a factor of safety of 1.5 on the load. The allowable load then is 15,100/1.5 = 10,000 lbs. The ultimate compressive stress, F_c, for 2024-T851 is 60,000 psi from [1], Table 3.2.3.0(b). The minimum compression area to prevent yielding in the plate is then: $A_p = 10,000/60,000 = 0.16$ in.^{2}. First check whether any washer is needed. The area of the bearing annulus under the bolt head is $\pi/4[(0.523)^2 - (0.374)^2] = 0.102$ in.^{2}, dimensions from Table 3.29. Since this area is less than A_p, a washer is needed to prevent local crushing. As a trial washer size, take $l_w = 0.063$, then $r_w - r_a = l_w \tan 30° = 0.036$ in. Now A_p becomes $\pi/4[(0.595)^2 - (0.374)^2] = 0.17$ in.^{2}, sufficiently greater than the 0.16 in.^{2} required.

The nearest standard size plain washer is a Type B, narrow, 0.406 inside diameter; 47/64 outside diameter; 0.963 nominal thickness (from ANSI, B27.2-1958). Applying these dimensions to Fig. 3.133, d_c turns out to be 0.978 in., and d_i is 0.406 in., from which the plate compression area A_c is 0.61 in.^{2}. This relatively large value for A_c means just that the maximum compressive stress in the plate will be 0.16/0.61, or about 1/4 of the plate yield strength, thus allowing for a stress

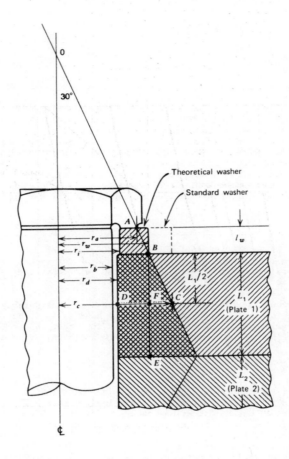

Fig. 3.133 Sizing of washers. <u>Graphical Construction</u>:
1) Point A is located at max radius of bearing
annulus under bolt head.   2) Draw the 30° line
through A, establishing points O, B, and C.   C
is placed at $L_1/2$. 3) Drop the vertical BE.
Then: $r_c = r_w + CF$; $r_c = r_w + 0.5L_1 \tan 30°$;
$r_c = r_w + 0.25L_1$ and the compression area, A_p;
is the annulus at CD, equal to:
$$A_p = (\pi/4)[(d_w + 0.50L_1)^2 - d_d^2].$$

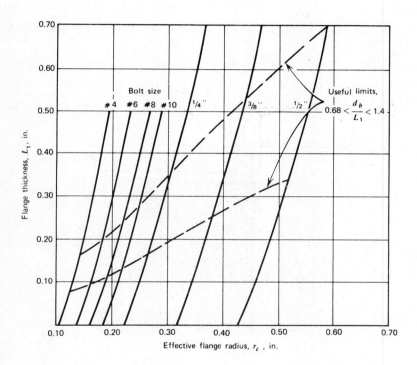

Fig. 3.134 Thickness versus effective radius in bolted flanges.

concentration factor up to 4, a more than ample margin. The fact that the standard washer has a considerably greater outside diameter than that required is neglected--it would be unduly conservative to calculate A_p on a 30° line struck from the standard washer outside diameter.

Checking the effect of this washer on the ratio of F_{max}/P_i: from Table 3.29, r without any washer was 2.92; for these dimensions, r changes to:

$$r = \frac{\pi/4(0.374)^2(29 \times 10^6)}{0.61(10.5 \times 10^6)} = 0.55$$

From Fig. 3.130, for r = 2.92 and a W/P_i somewhat less than 1, say 0.8, F_{max}/P_i = 1.61. At the same W/P_i and r = 0.55, F_{max}/P_i drops to 1.23 indicating that, as usual, the use of a properly sized washer enhances the fatigue life of the bolt by decreasing the r value of the joint, making the bolt "more elastic."

In the derivation above for A_p, (Eq. 108a), the pitch or bolt spacing was assumed to be "proper", i.e., in the approximate range of 2.5 to 5 times the bolt diameter. A brief procedure for finding the most efficient pitch is given in Fig. 3.135a.

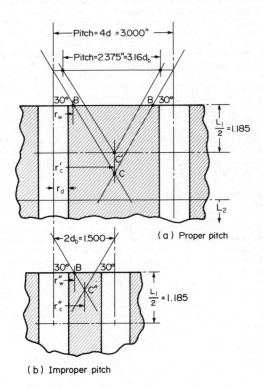

(a) Proper pitch

(b) Improper pitch

Fig. 3.135 Bolt-Pitch Considerations.

Here, the pitch was originally chosen at 4d for 3/4" diameter bolts and, in laying out, the intersection point "C" fell considerably off the mid-plane of the plate. Projection of C to this plane as point C', and thence along the 30° lines results in a spacing of 2.375" (measured) or a pitch of 3.16d. This value could be rounded off to 3d for convenience. The numerical dimensions were arbitrarily chosen.

Fig. 3.135b illustrates the danger of too small a pitch: point C" approaches B with consequent overstressing in the region of the hole edge. Both the short-column and shear-out forces are active in the same region. Comparing the compression stress for a proper pitch of 3.16d with that at a pitch of 2d, indicates an increase by the factor of $(r'_c/r''_c)^2$, or for the dimensions shown: $(1.5/.75)^2 = 4$. Thus, the ever-present problem of shear-out or bearing failure is easily compounded by choosing too low a value for the pitch. Although the compressive and bearing stresses always add, it is probably adequate to use a pitch such that C' does not depart from the mid-plane of the plate by more than $\pm$ 25% of $L_1/2$. Considerations of pitch should also include the ratio of bolt diameter to plate thickness, a practical range being from 1.0 down to about 0.25 min.

For types of threaded fasteners other than conventional, headed bolts, the variation in r due to different design dimensions may be rather wide, sometimes beneficial. For example, a typical blind bolt assembly [100] of nominal diameter 3/8-in. (core bolt diameter 1/2-in.) used in aluminum plates has an r value of only 45% of that for a 1/4-in. AN bolt, leading to a reduction in bolt load of about 20% for the same preload. Recent designs of very high strength bolts have tended to show smaller heads; but some are of the washer type, which means that they have a larger area of the annulus under the head, resulting in a reduction of both r and the bolt load. For proper control of the bolt load, an r value should be found for every head size in combination with the bolt and plate materials.

Another, and serious question, arises from a strict consideration of the loading conditions prior to separation: equal forces and deflections but not necessarily equal spring constants. The forces must be equal for the system to remain in equilibrium, and the changes in length must also be equal to obey the definition of no separation, that is, that all surfaces remain in contact under finite load. In searching for a variable, the spring constants are dissected: the E's may be different but they are invariant; the L's are initially equal; and the ΔL's very small, leaving only the areas as possible variables. The bolt area is known but there always has been some question as to a correct value for the effective area in compression.

Acceptable engineering results have been obtained by the practice of assuming that the compression area was that of an annulus whose outside diameter was that of the median radius of a 45° cone. However, a somewhat smaller angle, 25 to 40°, has recently been derived from data in [99]. Applying the principle of least work, it follows that a volume of the plate near the hole will be raised to a stress level such that equilibrium is always maintained with the work of compression. Actually, the energy will be stored as a function of a double integral of stress and volume. From the practical engineering viewpoint, it appears only that the older practice might be modified to use a slightly smaller compressive area, with no loss in conservatism. Regarding total compression areas and thicknesses and the use of washers and gaskets, it should be noted that the placement of the latter along the load path is of some consequence. The normal location of washers under the bolt head and nut means that their compressive deflection continues as a direct function of the increasing applied load and of the preload. Gaskets or other objects between the plates are subject only to the preload, relaxing as the working load is applied. The introduction of washers or gaskets to a joint increases the bolt length but leaves the flange thickness unchanged. The greater bolt length should lead to an improved fatigue life, other parameters remaining equal.

The problem of finding the maximum bolt load for an actual case is illustrated by a calculation for bolts in the split-line flange of an axial-flow compressor casing, using a method due to Bright [101]. The loading here involves both direct tension and bending. Fig. 3.136 shows typical sections and gives the nomenclature; the term "flange" has the same meaning as the earlier term "plate," $E_f = E_p$, etc.

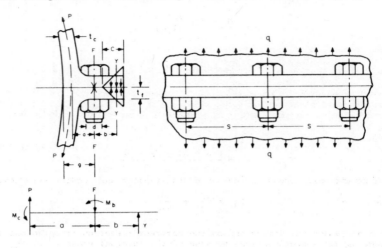

Fig. 3.136   Typical bolted flange.

In the calculation of the moment distribution between the flange and case, it is considered that the flange and bolts act together, that is, undergo the same bending deflection, and the conservative assumption made is that the maximum moment occurs at the bolt. The total moment Pa is distributed between the flange and case in proportion to their moments of inertia, thus:

$$M = Pa = \text{total moment} \tag{109}$$

$$I_c = \frac{St_c^{\,3}}{12} \tag{110}$$

and

$$I_{fb} = I_f + I_b \tag{111}$$

where

$$I_f = \frac{St_f^{\,3}}{12} - \frac{dt_f^{\,3}}{12} \tag{112}$$

$$I_b = 0.049\, d_2^{\,4} \tag{113}$$

then

$$M_{fb} = Pa\left(\frac{I_{fb}}{I_{fb} + I_c}\right) \tag{114}$$

and

$$M_c = Pa - M_{fb} \tag{115}$$

The distribution of M_{fb} between the flange and bolt is given by

$$M_f = M_{fb} \left(\frac{I_f}{I_{fb}}\right) \tag{116}$$

and

$$M_b = M_{fb} - M_f \tag{117}$$

Taking moments about Y, we obtain

$$M_y = P(a + b) - T_b b - M_c - M_b = 0 \tag{118}$$

from which

$$F = \frac{P(a + b) - M_c - M_b}{b} \tag{119}$$

The total static bolt stress is the sum of the tension and bending stresses:

$$f_{bs} = \frac{4F}{\pi d_2^2} + \frac{M_b d_2}{2 I_b} \tag{120}$$

Using AN-5 bolts and the dimensions of the compressor section of a typical aircraft turbine, we obtain a static stress because of the internal pressure of $f_{bs} = 11,500$ psi, compared to a yield strength of 100,000 psi. If detailed calculations involve a factor of safety, it would be applied in the loads analysis on the P term in Eq. 109, a value of 1.15 being commonly used where yielding is the final failure mode.

Before examining the alternating component, the preload and total bolt load are found. Equation 121, a modification of the form given by Dolan and McClow [102], is an expression for W/P_i at separation for this loading condition of tension plus bending:

$$\frac{W}{P_i} = \frac{1/E_f A_f + [(g - a)(0 - a)]/[E_f(I_b + I_f)]}{1/(E_f A_f + E_b A_b) - [(a - t_c/2)(0 - 2a)]/[E_f(I_b + I_f)]} \tag{121}$$

The substitution of typical dimensions for steel bolts in steel flanges yields $W/P_i \sim 0.3$. The working load W corresponding to the static stress of Eq. 120 is 600 lbs; the approximation above indicates that the preload must be 1/0.3 times this value. The total bolt load F_{max} is found by:

1. Obtain r for an AN-5 bolt in steel plates as 0.84, from Table 3.28.

2. Enter the curves, Fig. 3.130, at $W/P_i = 0.3$ and r = 0.84 to find $F_{max}/P_i = 1.2$.

3. $P_i = (1/0.3)W = 3(600) = 1800$ lbs.

4. Then $F_{max} = 1.2(1800) = 2160$ lbs.

The latter discussion might have been left in stress terms and in Fig. 3.130 stress could be read for load.

The working load on the bolts in this example has been treated solely as a static load arising from the internal pressure, its cyclic nature from the turbine start-stop cycle being neglected. The chief source of the alternating component of load comes from the residual unbalance in the rotor at the once-around frequency of 12,000 rpm or 200 hz. The load paths are sufficiently complex so that calculation of this component is impractical; resort has been made to direct experiment. Strain gage data indicate that the peak amplitude of this cyclic force is about 120 lbs., or S_a = 2300 psi. This stress is added to the value of f_{bs} above for a total working stress of 13,800 psi or W = 720 lbs.

Now the appropriate steps in the above sequence are repeated to obtain the total bolt load:

    3. P_i = 3W = 3(720) = 2160 lbs., corresponding to a mean stress of 41,500 psi.

    4. F_{max} = 1.2P_i = 1.2(2160) = 2600 lbs., corresponding to a peak stress of 50,000 psi.

An approximate bolt life may be determined by entering a master diagram for the bolt material with S_a and the new value of P_i as the mean stress. Care should be taken to use a diagram for material with the proper stress concentration factor; if curves for smooth material only are available, the K_t should be calculated and applied to S_a.

A somewhat more accurate, but still not completely rigorous, method is the use of curves for __bolted assemblies__, which comprise at least the bolt and its nut, Fig. 3.137. Here a life of about 1 x 10^5 cycles is indicated; but unfortunately, the curves were obtained with various, low stress ratios, below about (S_{min}/S_{max}) = 0.2, while the date above give a value of 0.8 approximately (41,500/50,000). From the life diagrams of other alloy steels of the same ultimate strength, it is known that an increase in the stress ratio from 0.2 to 0.8 (at the same mean stress) would increase the life by about one order of magnitude, leading to an approximate life of the bolts in this example of 1.0 x 10^6 cycles. Since a practical life requirement is 20 x 10^6, it is concluded that these particular bolts are overstressed, but it is seen that a reduction of only 600 or 700 lbs. in F_{max} would yield this life. Despite the extensive and useful increase in available information on fatigue, very often the designer is still unable to find just what he needs. He should have curves that apply to his problem with their data based on:

    1.   Same material.

    2.   Same strength, ultimate or yield.

    3.   Same shapes, that is, the same stress concentration factors.

    4.   Same stress ratio.

    5.   Same test temperature.

Figure 3.137 represented the best data available for this problem but it requires further information to account for the mismatch in item 4.

Another flange example is given to illustrate the calculation method, using spring constants directly. Consider a pair of standard steel pipe flanges of equal thickness, 0.8 in., with no gasket. The proposed bolting consists of 4NAS-16 steel bolts,

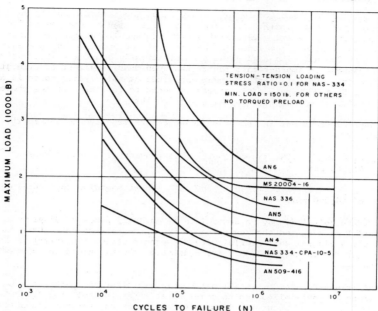

Fig. 3.137 S-N Curves for various bolt-nut assemblies. From
unpublished data, Huck Manufacturing Co.

shank diameter = 0.8 in. = D_s, A_b = 0.502 in.^{2}; basic major thread diameter = 1 in.
= D, minor diameter = 0.846 = D_k; and thread root area = 0.562 in.^{2} = A_k. The
spring constant for the bolt is then:

$$k_b = \frac{E_b A_b}{L} = \frac{(30 \times 10^6)(0.502)}{2(0.8)} = 9.43 \times 10^6 \, \text{lb/in.}$$

The spring constant for the flange is:

$$K_p = \frac{E_p A_p}{L}$$

where A_p is the effective area in compression and is equal to:

$$A_p = \frac{\pi}{4} (d_p^2 - d_h^2)$$

where d_h is the diameter of the bearing surfaces under the nut and head, here equal
to 1.4 in., (no washer) and d_p is the diameter of a 30° cone taken at its median
radius:

$$d_p = 2[2/3(t) \tan 30°] + d_h$$

$$= 2[2/3(0.8(0.577] + 1.00 = 1.610 \text{ in.}$$

Assuming a flange thickness of 0.8 in., the compression area is:

$$A_p = \frac{\pi}{4} [(2.01)^2 - (1.00)^2] = 3.04 \text{ in.}^2$$

(The use of a 45° cone angle would have given about 3.30 in.^{2}) Then:

$$k_p = \frac{(30 \times 10^6)(3.04)}{2(0.8)} = 57.1 \times 10^6 \text{ lb/in.}$$

The steady load P is given as 1000 lbs. and an alternating load W_1 as 8000 lbs. From $W = (P + W_1)/4$ and Eq. 108, the preload to just prevent separation is:

$$P_c = \frac{P + W_1}{4[1/(r + 1)]}$$

$$= \frac{1000 + 8000}{4[1/(0.502/3.04 + 1)]}$$

$$= 2200 \text{ lb. per bolt}$$

The additional alternate force, W_a, is, from Eq. 106.

$$W_a = \frac{W_1}{4[r/(r + 1)]}$$

$$= \frac{8000}{4[0.17/1.17]}$$

$$= 290 \text{ lb. per bolt}$$

and similarly the additional constant force is:

$$P_a = \frac{P}{4[r/(r + 1)]}$$

$$= \frac{1000}{4[0.17/1.17]}$$

$$= 36 \text{ lb. per bolt}$$

The highest stresses will occur at the minimum section or at the thread root, where the area in tension is A_k, and:

$$S_{min} = \frac{P_c + P_a - W_a}{A_k} = \frac{2220 + 36 - 290}{0.562} = 3500 \text{ psi}$$

$$S_{max} = \frac{P_c + P_a + W_a}{A_k} = \frac{2562}{0.562} = 4570 \text{ psi}$$

The stress ratio R is S_{min}/S_{max} = 3500/4570 = 0.755. The mean stress is then $S_{max} + S_{min})/2$ = 4010 psi, and the alternating stress is $(S_{max} - S_{min})/2$ = 560 psi, both in tension.

The shear stress developed by the preload torque is $S_s = 16M/\pi D_k^2$, where $M = \beta P_c D$ and β is the coefficient of friction, taken as 0.15. Then $M = 0.15(2220)(1.00)$ = 332 in. lb., and $S_s = 16(332)/\pi(0.846)^3$ = 2800 psi. The preload shear stress then combines with the S_{max} above to give the maximum resultant tension:

$$S_{tmax} = 1/2[S_{max} + \sqrt{S_{max}^2 + 4S_s^2}]$$

$$= 1/2[4570 + \sqrt{(4570)^2 + 4(2800)^2}] = 5530 \text{ psi}$$

Thus, the benefit of a small back-off, say 5°, after torquing to preload is readily observed. (If S_s tends to zero, S_{tmax} drops about 20% to 4570 psi.)

<u>Shear-Tension and Shear-Bearing Joints</u>. Lap joints and butts with cover plates have long been designed on the basis of shear in the fastener and bearing in the plates as the major modes of loading. Particularly with rivets, however, the axial tension or clamping force is variable from unit to unit and difficult to determine. During recent years the results of observation and of many investigations have been leading to a consensus that the highly clamped joint was superior in both static and fatigue loading. This conclusion is, of course, generally in agreement with the analytical knowledge that the stress concentration factors for holes in plates are fairly high, $K_t \sim 2$ or 3 for round holes, and tending very high as the hole becomes elliptical, a K_t of 25 being possible [93, 99]. It is also fairly common knowledge that even well-driven rivets do not always completely fill the holes, especially if the fit is poor, leading to even further severe concentrations.

The clamping force has been shown to have a dual role [99]: (1) to develop frictional resistance to slip on the contact planes of the joint, and (2) to reduce the stress concentration at the edge of the hole. Any designed increase in the tension/shear ratio is generally beneficial to fatigue life (see the discussion below for Fig. 3.138). However, the effect of clamping force on stress concentrations is complex, a major result being the transfer of the concentration from

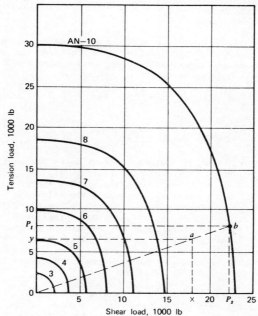

Fig. 3.138 Combined static shear and tension loading for AN steel bolts. (F_{tu} = 125 ksi, F_{su} = 75 ksi.) From E. F. Bruhn [15], Fig. D.1.

the hole edge to the washer edge. In joints designed with high ratios of tension to shear, failures have been observed to start in the plate at the edge of the washer. Thus, the details of washer design take on added importance; particularly, the use of sharp-edged, carburized washers is to be avoided. Properly profiled washers should hold the stress concentration factors in the plates near the theoretical minimums of 2 or 3; and with sufficient clamping force to carry the load by friction, the fatigue life of such joints will be relatively long. Coefficients of friction up to 0.35 have been observed on joint surfaces not specially prepared, which, together with the high strength bolting now available (F_{tu} up to 300 ksi) make a tension-friction joint an achievable design goal. This general benefit from the clamping force has led to more extensive consideration of bolts for joints originally designed as shear-bearing types.

For those joints whose overall restrictions do not permit the elimination of shear bearing, the bolts may be designed for combined loading in tension and shear by an interaction equation of the form:

$$R^3 + R^2 = 1$$

or

$$\frac{x^3}{a^3} + \frac{y^2}{b^2} = 1$$

where x = shear load
y = tensile load
a = shear allowable load
b = tensile allowable load

and the curves of Fig. 3.137. Starting with the known shear, x, and tension, y, loads, point a is located. The extrapolation of a line from the origin through a locates b on the next outer (larger) bolt curve, determining the P_s and P_t allowable loads for that bolt. Then the margin of safety is:

$$MS = \left(\frac{P_s}{x}\right) - 1$$

or

$$MS = \left(\frac{P_t}{y}\right) - 1$$

whichever is the smaller. Plots for other types of bolts are easily constructed; the extrapolation procedure is discussed more fully in [1, 39]. The older elliptical form $x^2/a^2 + y^2/b^2 = 1$ has been superseded by that above; the latter is recommended by [1]. In both expressions, however, the allowable stresses are for static loading, while the requirement is for fatigue allowables. These values are derived, somewhat approximately, by replotting S-N information in the form of Fig. 3.138. The approximation arises because of the absence of corresponding tension-fatigue and shear-fatigue data on specific bolts. It is reasonably conservative, however, to lower the shear allowable as a function of cycles in greater proportion than the tension allowable, because of the much greater increase in stress concentration factors for shear, as mentioned above.

The effect of varying the ratio of areas in tension and in shear is illustrated in Fig. 3.139, using A325 and A354 bolts. This data, typical of the structural steels, follows the interaction equation with the cubic terms very well but it should be noted that, for those slightly less than optimum joints with threads in the shear plane, the curves trend more closely to the unmodified elliptic form. On this plot, the lines for the various ratios of tension/shear area provide information corresponding to those lines for the ratios of S_{min}/S_{max} on the constant life diagrams.

The results of many investigations of joints in the heavier structural steels indicate that, if the stresses are such as to allow survival for approximately 2×10^6
cycles, the life of the joint may be considered as indefinitely long.

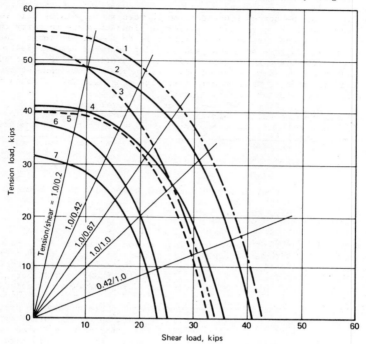

Fig. 3.139   Interaction curves for A325 and A354-BD bolts.
1) A354-BD bolt in Tl plates, shank in shear
plane; 2) A325-HH bolts in A36 plates, s. in
sp; 3) A354-BD, A242 plates, th'ds in s.p.;
4) A325, A36 plates, s. in s.p.; 5) A325 in
Tl plates, s. in s.p.; 6) A325 in A36 plates,
th'ds in s.p.; 7) rivet average. Nominal dia =
3/4" (from Ref. 103).

Upon this basis, a modified Goodman diagram, Fig. 3.140, has been constructed for
the joints of Fig. 3.139. A comparison of these results with design specifications
for A36 steel shows that the bolted joints provided a fatigue life of 2×10^6 or
more at stresses approximately 60% greater than those permitted by the specifications; thus the plot is very conservative.

In regard to the effect of clamping force or bolt tension on the fatigue life of
these joints, measurements of bolt elongation before and at various intervals
during the test showed that the joints providing a life of 2×10^6 cycles or more
had a loss of elongation of less than 20%. For the group of failed joints, the
loss of elongation increased very sharply as the life decreased. These conditions,

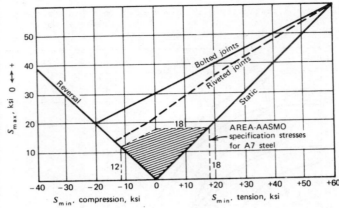

Fig. 3.140 Modified Goodman diagram for A325 and A354 BD bolts at
2 x 10⁶ cycles. From E. Chesson, Jr., and W.H. Munse
[103], Fig. 27.

combined with the fact that the fatigue interaction curves for A325 and A354 bolts
must appear similar to those of Fig. 3.139, sharply illustrate the benefits of using
the highest practical ratio of tension to shear, in terms of force, stress, or area,
(see also Fig. 3.154). The curves of Fig. 3.139 also show the distinct advantage in
fatigue strength of the bolted joints over the equivalent riveted designs.

Many designs for shear joints have been developed: plain and double scarfed,
serrated, keyed, stepped, etc. The results of tests on ten different types of
shear joints are given in [129]; the gross K_t value for each is listed in Table
3.30. It is to be noted that the single shear joint, with its very high K_t of 13
is generally considered unacceptable for high performance applications. Many of
the others are unsuitable, both for their relatively high K_t's and the expense of
their manufacture; the double-shear type at $K_t = 4.1$ is the best all-around choice,
and the double-scarf type at $K_t = 3.2$ is rarely worth the effort.

To summarize a fatigue design approach for bolts in shear-tension loading: if the
criterion of separation applies, the tension component must be treated as in the
previous subsection; in all cases, the bolt should be sized by an interaction
equation or curve using fatigue allowables. The highest practical value of the
tension/shear ratio should also be used, with an exception: for fail-safe structures with redundant load paths, the shear stress in the shear-bearing component
must be sufficiently high to allow the joint to yield slightly upon the loss of a
member, thus permitting the redistribution of the load as required by this design
objective. A general curve of the fatigue behavior of both slip-resistant and
bearing-type joints is given as Fig. 3.141, which shows some concentration of
stresses between 25 and 40 ksi, although the plate stock had yields from 36 to
120 ksi. Approximately the same fatigue strength is obtained by designing slip-
resistant joints on the basis of gross section, and bearing-type joints on net
section. Values taken on the lower-bound line provide a generally conservative
basis for the design of either type of joint subject to cyclic loading.

TABLE 3.30   Stress Concentration Factors for Shear Joints

Type of Joint	Figure Number in [129]	Material	Mean Load (lb)	K_t, Gross
Plain scarf	2c	7075-T6	16,000	8.1
		7075-T6	12,000	5.7
		2024-T4	16,000	4.1
		2014-T6	16,000	6.5
Double-scarf	2j	7075-T6	16,000	3.2
Single-shear	2f	7075-T6	16,000	13.0
Double-shear		7075-T6	16,000	4.1
Stepped double-shear	2g	7075-T6	16,000	4.3
Non-uniform step	2b	7075-T6	16,000	8.1
Uniform step	2h	7075-T6	16,000	6.5
Bolted-keyed	2d	7075-T6	16,000	4.3
Clamped-keyed	2i	7075-T6	16,000	7.3
Serrated	2e	7075-T6	16,000	8.1
Specimen with unloaded holes	2a	7075-T6	16,000	3.1
		7075-T6	20,000	3.2
		2024-T6	16,000	3.2
		2014-T6	20,000	3.0

SOURCE: E. C. Hartman, M. Holt, and I. D. Eaton [129].

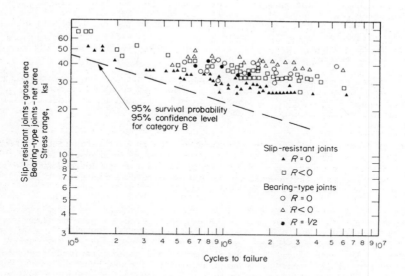

Fig. 3.141   Fatigue behavior of bolted joints (from Ref. 396).

Bolting Materials. The criteria for choosing the bolt material, and the bases
for indexing the comparison of the available materials are as numerous as the
applications. The first criterion often used is that of the tensile strength of
the material, but this approach is quite apt to be insufficient. Many observations
have shown that the only valid strength data come from tests of an assembly of a
fully formed bolt and nut, not from a specimen of the material. Then, too, the
other items in the working environment must be accounted for, such as corrosion
and combined loads, and, particularly in the high strength steels, stress-corrosion
and hydrogen embrittlement. Perhaps a useful base for indexing the materials is
their resistance to the loss of strength or toughness as a function of temperature.

For a reasonable range around room temperature, from, say, about -100 to +250°F,
the designer has a wide choice, so that many parameters must be balanced: strength,
corrosion resistance, the degree of magnetism, cost, and so forth. Outside this
approximate range, the number of material choices decreases rather sharply to the
super alloys and ultra-high strength steels.

On a unit strength basis at normal temperatures, the designer may use the SAE Graded
(0-8) steels, beryllium, titanium, cold-formed aluminum, copper-base alloys, or any
of the three types of stainless steels. On the basis of corrosion resistance, there
are the austenitic and ferritic stainless, copper and nickel alloys, and many non-
metallics such as nylon, Teflon, Delrin, and the polycarbonates. An optimum choice
for both strength and corrosion resistance is often the 316-type stainless in the
5 c condition (min F_{ty} = 100,000 psi), but, of course, the particulars of the
application must be satisfied.

At cryogenic temperatures, down to -423°F, only four bolting materials have been
thoroughly proved: Inconel 718, Unitemp, A-286 and titanium 5Al-2.5Sn, ELI. Other
nickel-base alloys, such as Waspalloy and Rene 41, show promise for this but the
cobalt-base L605 and the maraging steels are not yet completely proven. For the
high temperatures, Inconel 718, Waspalloy, A286 (or A638) and Rene 41 are used to
1600°F; H-11 and AFC 77 to 900°F; and Ti-7Al-12Zr to 750°F. It should be noted
that the alloys useful to the extreme temperatures are also those of high strength,
in the 220 to 300 ksi range. Fig. 3.142 lists approximate ranges of useful temperatures for the several materials.

In the effort to achieve ever-higher strengths and service temperatures, candidate
alloys are continually being brought forth, the most notable one recently being
the Multiphase alloy MP35N. It appears to offer an optimum in combined fatigue
life and resistance to corrosion and temperature, (to 750°F); it's general behavior
is compared to that of 17-7PH in Fig. 3.143.

A special problem in elevated temperature service is bolt relaxation or creep, a
typical case is shown in Fig. 3.144. This phenomenon is characterized by the
Larson-Miller Parameter: $P = T(20 + \log t)$, where T is in degrees Rankine and
t is time in hours. For comparison amongst materials, empirical data is plotted
as P vs. stress at given T's, as in Fig. 3.145.

Figure 3.146 illustrates the general superiority in fatigue of beryllium and
titanium alloys to steel but caution should be expressed regarding their specification. Compared to the steels (at least below a UTS of 220 ksi), they are more
anisotropic and have much higher sensitivities to notches and to the variables in
forging and heat-treating. Both beryllium QMV and three titanium alloys, 6Al-4V,
4Al-4Mn, and 7Al-12Zr, have been satisfactory in high performance joints, but it
is worth noting that the comparatively high endurance of titanium is available
only at a large number of cycles. The 300% margin at 5×10^6 cycles over steels
of comparable strength fades to about 125% at 5×10^4 cycles.

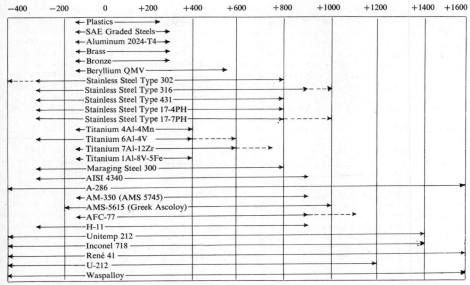

Data from *Metal Progress*, July 1966, by permission ASM.

Fig. 3.142 Working Temperatures for Bolting Materials, °F

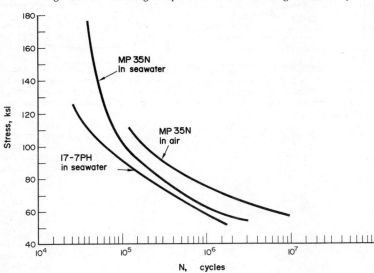

Fig. 3.143 S-N curves for MP35N and 17-7PH bolting alloys (from Ref 402).

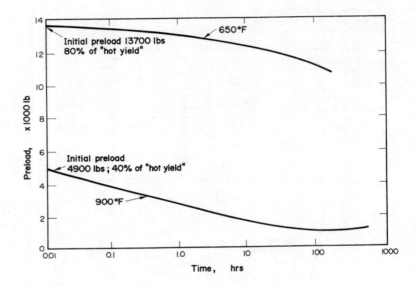

Fig. 3.144   Relaxation of Preload with time and temperature,
for an AISI 8740 bolt and A-286 nut; size M12.

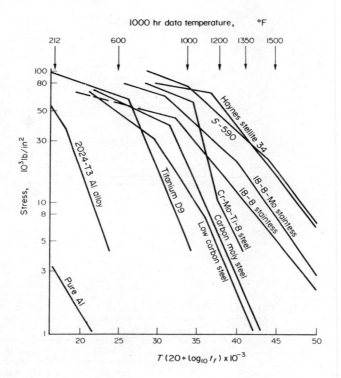

Fig. 3.145   The use of the Larson-Miller parameter
for direct comparison of materials.

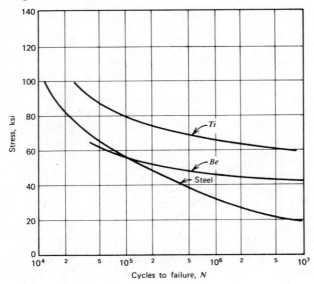

Fig. 3.146   Comparative fatigue of titanium, beryllium and
steel bolts on an S-N basis.   R = 0.1, NAS 1/4-28
(from Ref 111).

In the sizing of bolts for minimum weight design, not only must the strength of
the bolt material be considered but also the strength/weight or strength/density
ratio.   Table 3.31 provides the comparison among some high performance bolting
materials in terms of their strengths and these ratios.   Although beryllium has
the highest ratios, the conclusion is not necessarily that beryllium bolts form
the "best" design.   A design calling for very high strengths may be achievable
only in high strength steels.   Also the use of the high ratio materials leads to
bulky joints, demanding that the overall joint design provide the space.   The
shear-bearing strength of the bolted parts also becomes a design parameter; the
use of the harder, high strength materials may lead to marginal designs in shear
of the parts, not the bolts, particularly if the bolts should not be sufficiently
torqued to provide a friction joint.   Thus, the designer may choose between the
extremes of the high ratio, low strength materials and accept the (larger) fastener
size and lower weight, or choose the low ratio, high strength materials for reduction in size but higher weight.   The size and weight of the nuts and washers form
an important portion of the fastener total.

Preload-Torque Relationships.   The determination of the actual preload in a given
nut and bolt assembly poses a formidable problem.   The most accurate method is by
measurement of the elongation in the bolt with a micrometer, but this procedure
may not always be possible because of inaccessibility, and is clumsy and time-
consuming at best.   Bolts with built-in elongation indicators are available [117];
various deformable or crushable devices, usually in washer form, have been designed
[118, 119] but in most cases an additional gage and measurement are required.
Break-off portions of threaded nuts have been sized such that the driving torque

TABLE 3.31   Strength/Density Comparison of Bolting Materials

Material	Ultimate Tensile Strength, psi, S_u	Shear Strength, psi, S_S	Endurance Limit, psi EL at 10^8 cycles	S_u/Density psi/lb/in^3	S_s/Density psi/lb/in^3	EL/ Density psi/lb/in^3
Low alloy steel	186,000	108,000	40,000	650,000	380,000	140,000
Maraging steel	308,000	190,000	120,000	1,090,000	672,000	425,000
Stainless steel (18-8)	115,000	45,000	34,000	395,000	156,000	120,000
Titanium (6A1-4V)	168,000	103,000	60,000	1,050,000	645,000	375,000
Beryllium	75,000	65,000	45,000	1,136,000	986,000	682,000

develops the correct preload at the ultimate shear stress on this section [120]. Pull-and-swage-type serrated fasteners utilize a tension break-off load to control the preload [121]. This type has some advantage in fatigue loading in that the soft "nut" or collar is swaged around the harder "bolt" or pin serrations, developing a somewhat more uniform distribution of working load along the nut length.

Tightening to preload by means of a torque wrench is perhaps the simplest and cheapest but is apt to be rather inaccurate. The inevitable variation in friction from one set of threads to the next, plus the random presence of lubricant and/or dirt, can result in widely varying preloads. Proper control in manufacturing, handling, and assembly, in addition to frequent calibration of the wrenches, can make this method acceptable but rigid control is necessary to get top performance out of properly designed, top quality fasteners. While the preload may be much smaller than the maximum bolt load, the corresponding stresses may be high depending on the margins of safety used.

Practical programs for establishing and maintaining torque-preload standards consist of three activities:

1.   Frequent calibration of torque wrenches in a master loading fixture.
2.   Large-sample inspection of fasteners for thread condition, performed at the point of assembly.
3.   Continuous monitoring to insure that operators drive into, but not over, the torque range. (The old saying to the effect that "The bolt is at the mercy of the nut on the other end of the wrench" is often true.)

The turn-of-the-nut method of tightening is coming into wider acceptance, especially for field connections where the calibration of wrenches is difficult. In [122], either method for A325 and A490 bolts is permitted. Acceptable torque and preload or clamping force values for the lower and medium strength bolts (up through SAE Grade 8) are provided in [123] but, for the ultra high strength materials, the designer should consult specific data as in [112].

Torquing the bolt into yield has become generally accepted practice to obtain maximum clamping force. But, under fatigue loading, the serious drawback of plastic strain-ratcheting results, which means that the bolt will relax at a rate higher than if torqued to the more conventional 60 - 80% yield; this higher rate obtaining at any temperature. For static, or very low-cycle applications, as buildings, the

practice may not be harmful, and it can be done rapidly and inexpensively by automatic, powered machines.

Loss of preload, or loosening occurs in essentially all bolted joints, at any
temperature, due to seating or embedment, the smoothing of the peaks in the surface
finish and more gross, local yielding at the high spots especially on the threads.
The degree of loss depends on accuracy of alignment and surface finish, and tests
have generally indicated that, for well put-up joints with a finish below about
64 RMS, the loss increases asymptotically to 3-5% of the initial preload during
the first 10 cycles of applied stress. Fig. 3.147 shows typical effects of the
bolt length parameter. Thus, there is considerable advantage to a few cycles of
"shakedown" loading, followed by re-tightening.

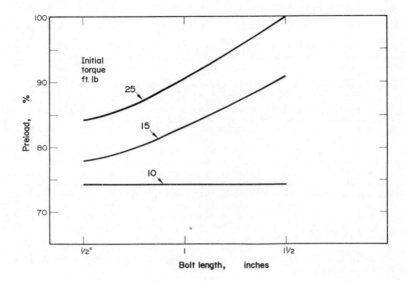

Fig. 3.147   Relaxation of preload as a function of bolt length.
3/8-16, SAE Grade 5.

Effects of Hole Sizing and Interference. In spite of best efforts to design
tension-friction joints with high-strength fastener materials and high clamping
forces, bearing stresses very often exist, and, as discussed above, they can be
high because of extreme values of K_t. Among the methods of reducing these stresses
is that of introducing residual compression on the hole surface. The effects of
"sizing" or "drifting"--the operation of forcing a smooth, oversize pin or ball
through the hole to cause a small plastic deformation--are generally beneficial.
The residual tangential compression absorbs part of the tangential tension from
the active load, thus increasing fatigue life to failure by tear-out. This type
of "preload" is of opposite sign to that in a tension bolt, wherein both the preload and active load add in tension. The results of hole sizing for some specimens
tested in tension-tension with open holes are listed in Table 3.32.

TABLE 3.32

Hole Condition in 7075-T6 Alclad Plate	Fatigue Limit at 10^7 cycles, ksi
1. Hole predrilled; size drilled; edges honed.	8.5
2. Repeat (1), plus sizing 0.001 in. to 0.002 in. with drift pin.	10.5
3. Repeat (1), plus sizing 0.005 in. to 0.006 in. with drift pin.	13.5

SOURCE: Huck Manufacturing Co. [133].

Such results show numerically the expected trend: that the presence of residual compression (below compression yield) improves the fatigue limit in tension. Tests on similar specimens having the holes filled with a self-sizing Huckbolt (interference of 0.002 in. to 0.003 in.) gave a fatigue limit of 21.5 ksi for all three hole conditions. This single value indicates that, although the benefits of the sizing are present, the differences in the three hole conditions were masked by the apparently excessive interference. While the benefits of hole sizing are not debated, the operation is very difficult to control in the shop because of the sharp dependence of the value of residual compression on dimensional tolerances, especially for small diameters.

The relatively large increase, with interference, of radial compression at the expense of a small increase in hoop tension is the basis for the improved fatigue behavior of such joints, particularly those using tapered bolts. The latter provide a far more uniform bearing on the elements of the joint than do the straight-shank interference bolts. Figure 3.148 provides some comparative life data but it should be noted that the benefits of the Taper-lok are available only at the relatively greater expense of hole preparation and higher cost of the fastener itself. The "stress pin" is a new item; it is basically a refined rivet completely filling the hole to the extent of inducing radial compression, plus some plastic compression in the plate edges under the holding collars. Its use forms a joint that is not normally demountable.

It is especially important to distinguish among the states of stress for sized and nonsized holes, with clearance bolt or interference bolt and for open hole conditions. See Table 3.33.

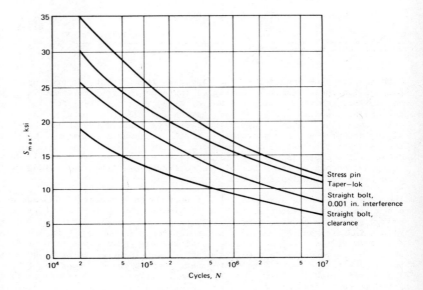

Fig. 3.148 S-N curves for various titanium fasteners and
interferences: No. 10's in two 0.080-in. Alclad
sheets; R = +0.10. From Standard Pressed Steel
Co., Jenkintown, Pa. [111].

TABLE 3.33   Stresses on Hole Surface (Prior to External Loading)

	STRESS	Hole Not Sized	Hole Sized
Open hole	HT[a]	Zero	Small; depends on dimensions
	HC[b]	Zero	Zero
	RT[c]	Zero	Zero
	RC[d]	Zero	Small; depends on Δdiameter
	LT[e]	Zero	Small; depends on Δdiameter
	LC[f]	Zero	Zero
With straight clearance bolt or rivet	HT	Zero, or increased, depending on fit-up	Small; depends on Δdiameter
	HC	Zero	Zero
	RT	Zero	Zero
	RC	Zero, or increased, depending on fit-up	Small; depends on Δdiameter
	LT	Zero	Zero
	LC	Depends on clamping force	Depends on clamping force
With straight interference bolt	HT	Increased, amount depends on fit-up	Small increase, depends on initial compression
	HC	Zero	Decreased from initial value
	RT	Zero	Zero
	RC	Increased, amount depends on fit-up	Considerable increase
	LT	Zero	Zero
	LC	Depends on clamping force	Depends on clamping force
With tapered bolt	HT	Depends on fit-up and pull-in	Not applicable
	HC	Zero	Not applicable
	RT	Zero	Not applicable
	RC	Greatly increased, amount depends on fit-up and pull-in	Not applicable
	LT	Zero	Not applicable
	LC	Depends on clamping force	Not applicable

[a]HT = hoop tension.   [c]RT = radial tension.   [e]LT = longitudinal tension.
[b]HC = hoop compression.   [d]RC = radial compression.[f]LC = longitudinal compression.

Stress Concentrations in Bolts and Nuts.   The fatigue behavior of bolts and nuts
is completely dependent on the control of stress concentrations.   The general
geometry of a bolt-nut assembly provides three distinct locations where the stress
concentration factor or K_t is considerably greater than one: (1) the intersection
of the head and the shank, (2) the thread runout section, and (3) the thread root
on the shank at the nut face.   Typical values of the stress concentration factor
for bolts with cut threads and of tensile strength under 100 ksi (Grade 5 or lower)
are given in Table 3.34.

TABLE 3.34   S.C.F. for Bolts

Type of Thread	Under Head	Thread Runout	Thread Root Radius	K_t
UNC or UNF	2.5	2.4	0.001	6.0
			0.002	4.8
			0.003	3.9
			0.004	3.4
			0.005	3.1
			0.006	2.8
			0.007	2.7
			0.008	2.5
			0.009	2.4
			0.010	2.3
Whitworth	2.5	1.6	---	3.3

The thread form, particularly the root radius, is thus a strong factor in determining K_t, a condition that tends to eliminate such threads as the Acme, Sellers,
and Whitworth thread from consideration for long life bolting.   The American Uni-
fied 60° thread is the base upon which high strength, long life bolting has been
developed.   The basic strength of the bolt is obtained by the combination of composition and heat treatment, and the fatigue life by attention to the details of
form, that is, to the stress concentration factors.   Conventional cut threads have
a stress concentration factor at the root of 3.5 to 5 approximately, tending toward
the higher value if heat-treated after cutting.   Bolts with threads so manufactured
should not be considered for high fatigue applications.   For threads rolled after
heat-treating, the stress concentration factor, or root radius, indeed controls
the life.   A descriptive comparison is given by means of the military specifications for different radii.

MIL-S-7742:   average life 28,000 cycles

MIL-B-7838A: average life 56,000 cycles

MIL-S-8879:   average life 109,000 cycles

Recent work [183] has determined the fatigue life of bolting as a function of the
root radius and the tensile strength, thus contributing the much-needed numerical
values to support the intuitive feeling that "the larger the radius the better."
Results of tests are given in Table 3.35 from which the great advantages of this
form and of rolling the threads after heat treatment are readily observed.

The 75% thread form (MIL-S-8879) has shown high fatigue life for steel bolts but a
refinement of this form has further increased life, or allowed higher loading at
a given life.   The thread has been made asymmetric by removing 5° from the load-

bearing flank, the included angle becoming 55°. The lead or pitch is also reduced a nominal 0.0025 in./in. These changes are applied to the bolt only; they will still fit properly with nuts and holes tapped to MIL-S-8879. Results of S-N tests on 260 UTS bolts with this special thread indicate an endurance limit of about 140 ksi at 10^6 cycles, the maximum available combination of load and life. Corresponding values for the next best form, the 75% symmetrical thread, are lower by about 30%.

TABLE 3.35   Effect of Increasing Root Radius, According to Fastener Material

Material	Method of Threading	Fatigue Life, 10^3 Cycles		Fatigue Life Increase (%)[a]	Tensile Strength Increase (%)[a]
		p/8 Thread	0.224p Thread		
90,000 psi steel, no heat treatment	Cut	30	55	80	4
	Rolled[b]	50	325	650	1
160,000 psi steel, heat-treated	Rolled after ht[d]	40	148	370	6
180,000 psi steel, heat-treated	Cut before ht	20	24	120	3
	Rolled before ht	20	48	240	7
180,000 psi steel, heat-treated	Rolled after ht	50	125	250	5
220,000 psi steel, heat-treated	Rolled after ht	40	1100	2800	6
260,000 psi steel, heat-treated	Ground after ht	22	40	80	8
	Rolled after ht	90	400	440	4
150,000 psi titanium (4Al-4Mn)	Rolled after ht	60	430	720	12
140,000 psi A-286 steel	Rolled after solution treatment before age-hardening	20	44	220	9
					5[c]

[a]From flat root (p/8) to R = 0.268p (55% thread).
[b]Threads rolled into blank, extruded directly to final diameter size.
[c]At 1200°F.
[d]ht = heat treatment.

At the thread runout into the grip, a certain amount of incomplete profile must be permitted. For high strength bolting (260 UTS), a maximum of two pitches and a minimum of one pitch runout is permitted [184], which must include the extrusion angle for rolled threads. At the end of the bolt, the threads adjacent to the chamfer may be incomplete at major and pitch diameters but must be complete at a distance of two pitches from the flat end. General limits on the surface finish are: 32 RMS for the sides of the thread and root area, head to shank fillet, and for the shank and underside of the head; other surfaces may be as rough as 125 RMS.

The head to shank intersection is another highly critical location for fatigue cracks, so that the size and shape of the fillet are very important. The fillet radius should be as large as possible, and tests have indicated a seven times increase in life upon changing the radius from a sharp corner to 0.038 in. Circular fillets are most commonly used and are relatively simple to produce, but there must be a trade-off between the radius and the bearing area. More effective fillet size without loss of bearing area under the head can be obtained with elliptical or compound profiles. Fatigue resistance of the fillet is also much improved by cold working or prestressing in compression; in one test the number of cycles to failure was raised from 7×10^5 to 8×10^6. The cold-rolled elliptical fillet is that type most suitable for high strength, high fatigue bolting.

Lugs or Clevises. The single-pin joint or lug represents a simple form of bolted joint, and its fatigue strength may well govern the fatigue behavior of the entire structure. Despite the apparent geometrical simplicity of lug shapes, correlation of theory and data had been ineffectual for design until Heywood [5] evolved the following procedure based on a "standard" lug with a 1-in. diameter hole and a K_t of 3.

It has been observed from data that the fatigue strength of the lug assembly is extremely low in comparison to that of the materials. For the failure of steel lugs in 10^7 cycles, S_a is only about $\pm 4\%$ of the tensile strength, and for high strength aluminums only $\pm 2.5\%$. This condition points to the theoretical stress concentration factor as the primary influence on the fatigue life. The stress concentration factor is defined as the ratio of the maximum stress to the average stress on the minimum cross-sectional area at the hole:

$$K_t = \frac{\sigma_{max}}{P/(D - d)t} \tag{122}$$

Considering K_t as a function of geometry, we note that as d/D approaches a value of 1, the lug tends to act as a flexible strap around the pin, thus reducing the stress concentration factor toward a value of 1 also. One immediate problem that often arises is that of determining the optimum strength of lugs for a given overall width.

The stress concentration factor for this condition may be redefined as the ratio of maximum stress to the average stress acting on the full section:

$$C_t = \frac{K_t}{1 - d/D} = \frac{\sigma_{max}}{P/Dt} \tag{123}$$

The curves in Fig. 3.149 show a distinct saddle at d/D of about 0.4, indicating a minimum stress concentration factor or a maximum strength at this ratio.

A number of other parameters may affect the stress concentration factor: (1) the effect of "waisting" the lug as shown in Fig. 3.150, (2) pin bending, (3) clearance or interference fits, (4) stress distribution around the hole, and (5) pin material and lubrication.

The effect of bending and shear of the pin is to increase the maximum stress near the faces of the lug. The increase depends primarily on the size of the pin compared to the thickness of the lug, d/t; photoelastic tests have indicated that the increase is fairly small. For a semicircular-ended lug having ratios of overhung length to width H/D of 0.5 and a pin diameter to width d/D of 1/3, the results in Table 3.36 were obtained. The evidence appears to confirm the comparatively small influence of the pin-bending effect on fatigue. For a very thick lug, it is an

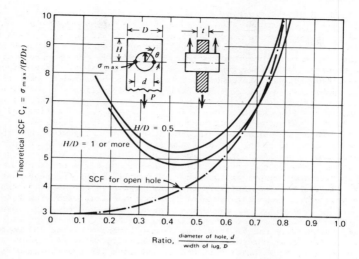

Fig. 3.149   Theoretical stress concentration factors for loaded lugs based on the gross area of the lug. Small clearance between hole and pin. From R. B. Heywood [5], Fig. 9.2.

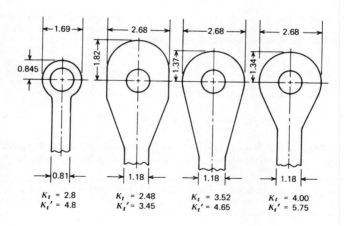

Fig. 3.150   Stress concentration factors for "waisted" lugs. For all cases, the diameter of the hole is 1 in.; for K_t, a neat fit pin is used; for K_t', diametral clearances are 0.019 in. for the first case, and 0.015 in. for other cases. From R. B. Heywood [5], Fig. 9.4.

TABLE 3.36 ΔS_s in Pin

Ratio d/t	0.51	0.72	1.27
Increase in maximum (shear) stress due to pin bending, %	17	10	7.5

advantage to use a large diameter pin, and it has been suggested that the optimum strength is obtained when the ratio of the pin diameter to the width of the lug d/D is about 0.6.

Greater stress concentration factors are obtained with larger clearances between the pin and the hole. The data show that the factor is dependent not only on clearance, but also on the magnitude of the load applied, with higher loads reducing the factor. The points of maximum stress are shifted from their usual position on the transverse diameter for a neat fitting pin toward the loaded side of the pin. The effect of clearance can be attributed to an inward movement of the two sides of the lug when no restraint is offered by the pin. Photoelastic investigations of the maximum shear stress at the hole boundary (on the transverse diameter) have indicated that the interference reduces the rate of increase of stress with load, for small values of load, but gives the normal rate for the higher loads, as would be obtained for the push-fit case. Thus, the stress concentration factor is not constant for a given interference but varies with the pin load. The shear stress distribution around the hole boundary is not constant, and, for the push-fit case has been found to show a maximum at about 30° from the transverse diameter on the loaded side of the pin (angle θ in Fig. 3.149; see also Fig. 3.41). For interferences of 0.003 in. to 0.006 in., the maximum shear stress occurred very close to the transverse diameter ($\theta \sim 0°$). The hoop tension is more symmetrically distributed, with the maximum occurring on the transverse diameter. At the highest pin load this maximum may be substantially reduced by increasing the interference.

Data indicates that the stress concentration factor can be reduced by about 0.25 through the use of a high film-strength lubricant such as colloidal graphite grease. Another reduction of about 0.30 can be obtained by substituting an aluminum pin for a steel one for the case of an aluminum lug in which pin-bending effects are negligible. Even with good lubrication, it is believed that some fretting always occurs and a factor $(1 + d)\beta$ has been used to account for the resultant increase in the stress concentration factor.

A comprehensive analysis of test data on many types of lugs under various conditions resulted in expressions of fatigue behavior in terms of strength reduction factors:

$$K_m = \frac{\sigma_m}{\sigma_{mn}} = K_s + (K_t - K_s)\left[1 - \frac{\sigma_{mn} + \sigma_{an}}{\sigma_{tn}}\right]^2 \tag{124}$$

and

$$K_a = \frac{\sigma_a}{\sigma_{an}} = K_s + [\beta(1 + d)K_t - K_s]\frac{n^4}{b + n^4} \tag{125}$$

where K_m, K_a = the maximum values of the strength reduction factors, for 10^7 cycles or more

σ_m = mean stress, smooth specimen

σ_{mn} = mean stress, notched specimen

K_s = static stress concentration factor, here equal to 1

d = diameter of hole, in.

b = a constant, $\cong 1000$

β = a constant, ~ 2.7 for steel and 1.85 for aluminum

Both expressions apply for notched specimens but Eq. 125 is modified to include the maximum stress concentration factor by the term $\beta(1 + d)K_t$, the fretting factor effect.

A correlation has been made by interpreting the various results for different designs to a standard condition of $K_t = 3$, representing a near-optimum strength design, and a hole diameter of 1 in. The standard σ_{mn} and σ_{an} are derived from the observed σ_{mn} and σ_{an} of any lug from Eqs. 124 and 125, for the condition of equality between the peak σ_{mn} and σ_{an}. It is also assumed that the static stress concentration factor, K_s, equals one, leading to

$$\frac{\sigma'_{mn}}{\sigma'_t} = \frac{\sigma_{mn}}{\sigma_t} \left\{ \frac{1 + (K_t - 1)[1 - (\sigma_{mn} + \sigma_{an})/\sigma_t]^2}{1 + 2[1 - (\sigma'_{mn} + \sigma'_{an})/\sigma_t]^2} \right\} \quad (126)$$

and

$$\frac{\sigma'_{mn}}{\sigma'_t} = \frac{\sigma_{mn}}{\sigma_t} \left[\frac{1000 + K_t(1 + d)\beta n^4}{1000 + 6\beta n^4} \right] \quad (127)$$

where σ'_{mn} = standard mean stress for $K_t = 3$ and $d = 1$ in.

σ'_{an} = standard alternating stress for $K_t = 3$ and $d = 1$ in.

σ'_t = tensile strength

n = log N

Results are plotted in terms of the ratio σ'_{an}/σ'_t for the steels, as in Fig. 3.151; the absolute stress σ'_{an} for the aluminums is given in [5], Fig. 9.9.

In designing a lug for a given fatigue life, the simplest procedure is to start with Fig. 3.151, taking into account the indicated range of scatter. These curves give the magnitude of the standard alternating stress σ'_{an}, which would be obtained for any given number of cycles. This stress may then be interpreted to the alternating stress for the lug under consideration by substitution into Eq. 127. An allowance for mean stress is hardly needed if $\sigma_{mn} > \sigma_{an}$, but if not, Eq. 126 may be applied.

Example

A lug is required to withstand a repeated load of 10,000 lbs. for 100,000 cycles. What dimensions should be used when the material is (1) a steel of 160 ksi tensile strength, or (2) a high tensile aluminum alloy, assuming that safety factors for both load and cycles have already been applied?

Figure 3.149 shows that optimum strength should be obtained when the ratio of hole diameter to the width of lug is about $d/D = 0.43$. This ratio is reduced slightly if one takes account of the size effect implied in Eq. 125, giving an optimum

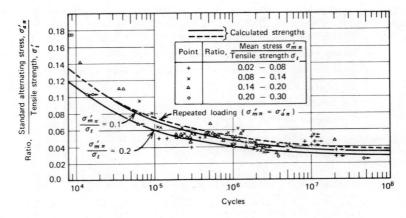

Fig. 3.151   "Standard" fatigue strength of steel lugs as a function
of mean stress. $K_t = 3$; hole diameter = 1 in.   Clearance
pin.   From R. B. Heywood [5], Fig. 9.8.

ratio of, say, d/D = 0.4.   The associated theoretical K_t is 2.9 for a value of H/D
of 1 or more.   For repeated loading the nominal mean and alternating stresses are
of equal value, or

$$\sigma_{mn} = \sigma_{an} = \frac{P}{1(D - d)} = \frac{5}{1.5dt}, \text{ ksi}$$

for d/D = 0.4 and P = 5000 lbs.

For steel, Fig. 3.151 indicates that a representative value of the standard alter-
nating stress $\sigma_{an}^{\prime}$ at 10^5 cycles is $\sigma_{an}^{\prime}/\sigma_t = \pm 0.07$.   Substitution into Eq. 127
gives

$$\sigma_{an} = 0.07 \times 160 \left[\frac{1000 + 6 \times 2.7 \times 5^4}{1000 + 2.7(1 + d)2.9 \times 5^4} \right] = \frac{21.1}{1 + 0.83d}$$

Equating the two expressions for σ_{an}, we find that

$$\frac{5}{1.5dt} = \frac{21.1}{1 + 0.83d}$$

giving

$$t = \frac{0.158}{d} + 0.131$$

This expression shows that there is an infinite series of solutions of values for
t and d that satisfy the conditions.   Typical solutions are listed in Table 3.37.

It is not advisable to use thicknesses of lug appreciably greater than the diameter
of the hole, as strength becomes adversely affected by pin-bending effects.   Thus,
the pin diameter should exceed 0.5 in.   The weight of the lug probably depends on
the cross-sectional area Dt, and for this case, the above figures show that the
smallest diameter pin gives the least weight.   The best design is thus represented

approximately by the dimensions with 1/2 in. pin diameter. All lugs in Table 3.37 should have equal strengths, and it is interesting to note that the nominal stress σ_{an} varies with the size of the pin, since a size effect is involved in Eq. 125.

TABLE 3.37

d	0.25	0.5	0.75	1
t	0.76	0.44	0.34	0.29
D	0.625	1.25	1.87	2.5
Area Dt	0.47	0.56	0.64	0.72
$\sigma_{an}[=5/(D - d)t]$,ksi	17.6	15.0	13.1	11.6
Bearing stress (maximum) = 10/dt,ksi	53	45	39	35

Use of Bushings.   From a fatigue-strength point of view, the pin-bushing combination is roughly equivalent to a pin of the same outside diameter as the bushing when there is clearance at the bushing outside diameter. For lugs with interference-fit bushings (0.003–0.010 in. per in. of diameter) tests have shown consistently longer lives than with interference-fit pins, up to four times as long.

Interference-fit Pins.   Interference-fit pins also show greater lives than clearance-fit pins. Nonfailure at 10^7 cycles can usually be obtained for all aluminum pin-lug combinations at interferences of at least 0.004 in. per in. of diameter, and similarly for steels with about 0.003 in. per in. minimum.

Hole Surface Conditions.   These conditions appear to have approximately the same effect on lug life and strength as they do for bolted joints, but the scatter in both lives and strengths is very much greater. It is generally worthwhile to use a lubricant. The effect of the method of machining the hole is inconclusive: drilled holes were superior to reamed holes for steel lugs, the reverse was true for aluminum. Anodizing the bores in aluminum lugs is quite harmful to life but has little effect on static strength. Cadmium or zinc-plating the pin gives some improvement in life, but much more improvement if both pin and lug are plated, particularly with zinc. Small sample tests indicate chromium plating may be very harmful, depending on the degree of relief of embrittlement. While these statements may appear uselessly qualitative, attempts at correlation of present data do not yield more rigorous conclusions. Designs of optimum geometry and minimum stress concentration factors, together with interference bushing fits, seem to be the best approach to long life. (See also discussion in Sec. 3.2.4.)

Riveted Joints.   The design of both butt splices and lap joints is based necessarily on the tension-shear-bearing ratio, the value of which may vary widely [15, 136]. Riveted joints are infrequently designed to carry any significant part of the sheet load by friction induced by tension in the driven rivet because it is essentially impossible to specify this tensile force. Measurements of the tensile force in hot-driven steel rivets [137] indicate variable consistency from lot to lot, and, as expected, an increasing force with increasing grip. "Proper" clamping force or tension in the driven rivet is usually obtained by trials on scrap stock with correctly dimensioned rivets to determine a proper driving pressure. Rivets are, of course, loaded in tension whenever the condition is unavoidable in the joint design; with the application of conservative judgement in the form of the restrictions below, success is usually attained.

1.   The tension on rivets should be restricted to conditions in which the tension load is incidental to the major shear-carrying purpose of the rivet. When it is difficult to determine if the tension component is incidental or major, a bolt should be used. The tensile capacity of a rivet is:

$$B_{ult} = AS_{ult}$$

where A is the undriven area. S_{ult} is usually taken as 60 ksi for A502 Gr.1, and 80 ksi for Gr.2 rivets.

2.   The ratio of shear to tensile strength for rivets is about 0.75 and is independent of grade (for steels), diameter, grip length and driving procedure.

3.   Rivets loaded in both shear and tension should be checked for combined stresses, using the interaction equation,

$$x^2/(0.75)^2 + y^2 = 1.0$$

where x is the ratio of the shear stress on the shear plane to the tensile strength, and y is the ratio of the tensile stress to the tensile strength.

4.   A sufficient number of rivets shall be used to insure that failure of any one rivet because of improper installation, cracked head, and so forth, shall not result in the failure of the structure.

5.   If there is no load reversal on the assembly, the tension allowables given in [15], Tables A, B, and C, p. D1.28, may be used. If there is load reversal on the assembly, the tension load on the rivet should not exceed 25 percent of the values in the tables. Tension allowables have not been standardized, as for shear and bearing, so it is necessary that the design practice be similar to that for the test joint.

6.   Do not use rivets to fasten aircraft control brackets to a supporting structure.

The following are examples of joints in which rivets are considered to be satisfactory for tension loading:

1.   Skin attachment to ribs and frames.
2.   Attachment of sheet panels to beam flanges and stringers, where inter-rivet buckling or diagonal sheet wrinkling produce tension loads on rivets.
3.   Skin attachment on a pressurized nacelle or body.

For conventional joints, the load is taken primarily by shear in the rivets and bearing in the sheet, the latter condition being that which makes the joint so sensitive to stress concentrations in the sheet. As noted earlier, the values of K_t for slightly out-of-round holes can easily rise to the order of 20. Results of axial fatigue tests of lap joints in aluminum indicate that, at the lower stress, longer life levels, failure is more frequent in the sheet, the mode being tension cracks starting at stress concentrations on hole surfaces. In the higher stress, shorter life tests, the failure may be either shear of the rivets or sheet cracking, while in static tests the usual failure mode is rivet shear. Together, these conditions emphasize that stress concentrations are generally more effective at lower stresses and higher numbers of cycles.

The effect of mean stress has not yet been rigorously determined, the only generalization which appears possible at this time is that riveted joints seem less

sensitive to mean stress than the sheet material. Double-strap butts show somewhat more variation in life with mean stress than do lap joints, but in either case
the general lack of a strong functional relationship is probably a result of the
rise of the local stresses to near yield during the first few cycles. The local
stress controls the life; since change in the mean stress does not vary the local
stress appreciably, the value of the mean stress becomes relatively important.

Rivet spacing has been examined as a factor in joint fatigue strength, and some
generalizations are possible:

1. For single-row joints (3 rivets in a row), fatigue strength
   increases with decreasing pitch.
2. For three-row joints (1 rivet in a row), fatigue strength
   increases with increasing spacing of rows.
3. For multi-row joints (3 rivets in a row), fatigue strength
   increases with increasing numbers of rows.

For more specific results, Table 3.38 gives the following data.

TABLE 3.38   Fatigue Strength of Riveted Joints in 2024-T3
Alclad Sheet

Number of Rows	Spacing of Rows	Rivet Position	Ave. Static Strength, lb/rivet	Fatigue Strength (max load), lb/rivet	
				10^5 cycles	10^7 cycles
1	---	---	692	282	170
2	3/16	Staggered	668	222	156
2	3/8	Staggered	690	266	133
2	11/16	Staggered	705	290	137
2	3/4	In line	725	288	135

SOURCE: H. J. Grover, S. A. Gordon, and L. R. Jackson [9].
NOTES: (1) Rivets, AN426-AD5, (2) pitch = 3/4 in., (3) edge distance = 3/8 in.,
(4) R = 0.25, (5) specimen width = 5 in.

Variations in the surface finish of the hole seem to have little effect on fatigue
life. In [138], for heavy construction, any of the conventional methods of hole
preparation is permitted such as punching, drilling, or reaming; but qualitative
statements regarding fit-up and drifting are added. However, for thin sheet, an
order of decreasing fatigue strength for the methods is given in [9]: coil-
dimpling, spin-dimpling, drilling, and machine-countersinking.

Joints with flush-head rivets have generally shown poorer fatigue behavior than
those with protruding heads; some interesting work by Whaley [135] gives a major
reason. Axial tension tests were made on 7075-T6 Alclad 0.190-in. thick with
plain and with countersunk holes (NAS 100%). A plot of life, from 1000 to 6000
cycles vs. the ratio of edge distance to hole diameter (E/D), showed a life for
plain holes at least twice that for countersunk holes, at any value of E/D between
1.0 to 2.0. A photo-stress study of stress concentrations at various points on
the surfaces of countersunk holes resulted in K_t's at the base of the countersink
(taper) from 13 to 23 percent greater than K_t's for the plain hole, with values
ranging from 2.7 to 3.2. A relationship for the accumulative effect of the several
K_t's was developed, for this geometry, as:

$$K_t = 1 + \frac{(K_{t_1} - 1)(K_{t_2} - 1)(K_{t_3} - 1)}{4}$$

The detailed dimensions of the hole are important to the life of the joint, especially in the top sheet. The taper should never extend through the sheet to produce a feather edge with its extremely high K_t. Also, the production method used may significantly alter the life. Efforts to produce a more controlled fit of the rivet in the hole resulted in the "shock-wave" riveter which electromagnetically generates a stress wave through the rivet. This device is not an electromagnetic hammer, but induces a dynamic wave of controllable form: stress-amplitude vs. time. The net result is to expand the rivet uniformly into the hole, not barrel-shaped, and with an interference of some 0.003 to 0.005" for a 1/2" diameter rivet. The fatigue behavior if the usual double-shear joint so riveted is significantly better than that with conventionally driven rivets, and about the same as that with the taper-lok type fasteners, or over-squeezed rivets. No data seem to be available for riveted joints under combined bending and axial tension, but for the latter condition of loading, Figs. 3.152 and 3.153 are the most extensive, for aluminum alloys. The material, type of joint (double-shear) and its size are

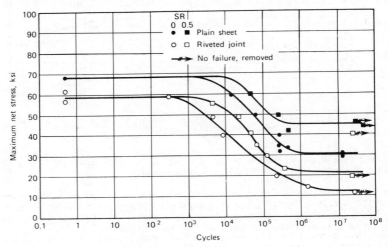

Fig. 3.152 Axial-stress fatigue strengths of 2024-T3 plain sheet
specimens and riveted joints..

illustrative of best modern practice. The data is further useful not only from the wide variety of alloys included, but also for the comparison between the lives of plain specimens and those of actual joints made up of the same material. The joints were designed with a tension/shear ratio of about 1.0/1.0 (bearing not given). The tension/shear/bearing ratio appears to be somewhat less critical for riveted joints of aluminum than for steel, and the fact that the fatigue strength reduction factor for these joints was generally below the K_t reinforces the observation. The fatigue strength reduction factor, K_f, is defined here as

$$K_f = \frac{\text{fatigue strength of plain sheet specimen}}{\text{fatigue strength of riveted joint}}$$

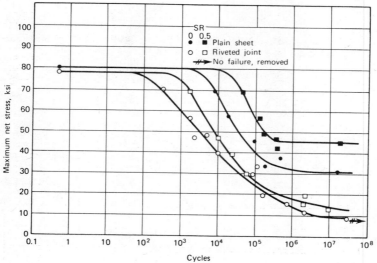

Fig. 3.153 Axial-stress fatigue strengths of 7075-T6 plain sheet
specimens and riveted joints.

and ranged from 1.04 (for 2014-T6 at 10^3 cycles) to 4.16 (for 7178-T6 at 2.5 x 10^7 cycles). The values of K_t for these joints was 3.2.

For heavy aluminum construction as in bridges and buildings, the Structural Division of the American Society of Civil Engineers has written suggested specifications [138]. Section G-4 of [138] on fatigue gives allowable stresses for riveted or bolted joints, as well as some good advice on fatigue design: "Tests indicate that riveted members designed in accordance with the requirements of these specifications and constructed so as to be free of severe re-entrant corners and other unusual stress raisers will withstand at least 100,000 repetions of maximum live load without fatigue failure, regardless of the ratio of minimum to maximum load. Where a greater number of repitions of some particular loading cycle is expected during the life of the structure, the calculated net section tensile stresses for the loading in question shall not exceed the values given in Table 9."

In considering fatigue action on structures, it is well to bear in mind the following points [138]:

1. "The most severe combination of loadings for which a structure is designed (dead load, maximum live load, maximum impact, maximum wind, etc.) rarely occurs in actual service and is of little or no interest from the standpoint of fatigue.

2. The loading of most interest from the fatigue standpoint is the steady dead load with a superimposed and repeatedly applied live load having an intensity consistent with day-to-day normal operating conditions.

3.   The number of cycles of load encountered in structures us usually small compared with those encountered in fatigue problems involving machine parts. For example, 100,000 cycles represents 10 cycles every day for 27 years; 10,000,000 cycles represent 20 cycles every hour for 57 years. Care must be taken not to overestimate grossly the number of cycles for any given load condition.

4.   Careful attention to details in design and fabrication pays big dividends in fatigue life. When a fatigue failure occurs in a structure, it is usually at a point of stress concentration where the state of stress could have been improved at little or no expense."

It may be useful to point out that the "steady dead load" of item 2 above is, for some types of structures, (especially suspension bridges), most probably a pulsating load resulting from seismic and wind forces. The mass, length, and stiffness of many spans are such as to bring their first mode vertical frequency into resonance with that of the natural forces, 0.1 Hz, very approximately; thus, the so-called dead load actually controls the life of the structure. Support for the observation comes from two recent failures, at Tacoma Narrows and at Point Pleasant on the Ohio River, the latter exhibiting a life to fracture of 6.5×10^6 cycles, within 10% of the writer's prediction.

A summary of many tests on both plain carbon and alloy steels for heavy construction emphasizes the crucial point that, despite the wide range of static strengths available from about 60 to 100 ski, the fatigue strength of the joints at 2,000,000 cycles was the same, about 26 ksi. This condition may well raise the question of the economics of specifying the higher strength, and higher cost, nickel or silicon steels, unless some other property such as corrosion resistance is required. Because of the importance of the tension/shear/bearing ratio, Fig. 3.154 is given to show the wide variation in life for different ratios. Note particularly, the lower life for the higher bearing values, another indication of the tremendous effect of the stress concentrations at the hole surface on the allowable bearing strength. This condition is most effective in promoting the use of high strength bolts instead of rivets for such work. In a bolted joint of proper design it is possible to carry the shear load by friction, thus avoiding these particular difficulties.

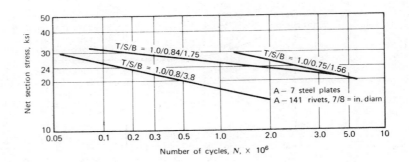

Fig. 3.154   Fatigue life as a function of tension-shear-bearing ratio.
From [136], Figs. 6, 7.

WELDED JOINTS

Arc and Gas Welds

The three factors which predominantly control the fatigue life of weldments are: 1) joint geometry; 2) size of initial defect; and as always 3) the stress range. Within the joint geometry, the flank angle, Fig. 3.155, is probably the most effective factor and, for useful lives, the angle should be minimized by careful control of the bead deposition for a smooth, low profile, or preferably by machining the reinforcement off flush. Typical values of life vs. angle are shown in Fig. 3.155.

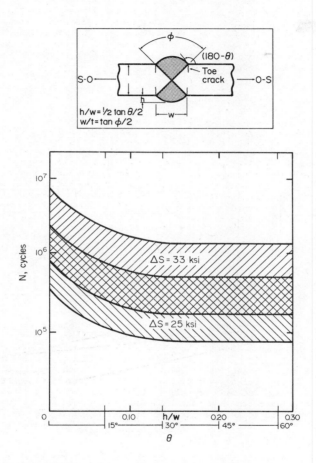

Fig. 3.155 Fatigue life vs. flank angle for butt welds in
A36 and A441 steels (from Ref. 457).

The above presumes no undercut, overfill or other errors of technique. The design of the joint is best taken from the American Welding Society's Structural Welding Code, AWS D1.1-75 or the ASME Pressure Vessel Code. The Codes are quite similar— design for high performance and efficiency is a fundamental—but the latter may be more attentive to some details, especially for heavy sections. Fig. 3.156 shows the ASME categorization of joint types, and samples of designs.

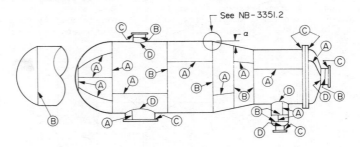

Fig. 3.156a Welded joint locations typical of Categories A, B, C and D.

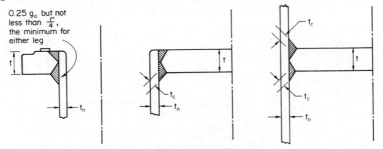

Note: t and t_n are nominal thicknesses

Fig. 3.156b Acceptable full-penetration weld details for Category C joints. Typical type 2 corner welds.

The design of the weldment itself is of importance, with emphasis needed on the attachments and cover plates. Fig. 3.157 gives results of extensive tests of high- way bridge girders. One major detail is the termination of a cover plate, and there is some evidence to show that the weld bead should not be run across the cover plate end because the inevitable shrinkage crack at the toe will be oriented normal to the bending tension. However, the bead termination at the sides then becomes controlling.

A particularly interesting effect of attachment detail has recently surfaced in welded beams supporting crane rails. Separation of the upper flange from the web at the fillet toe, for lengths up to 15 feet, was found to originate at the corners where the vertical web stiffener terminated in a fillet weld to the upper flange, even when generously coped. Local cyclic tension in this generally compressive region was sufficient to fail the beams at times far short of design life. One

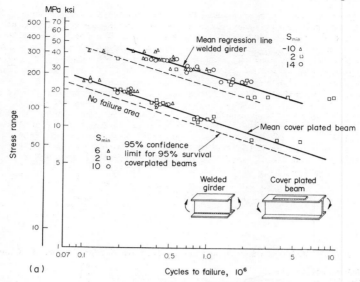

(a)

Effect of minimum stress on stress range-cycle life relationship

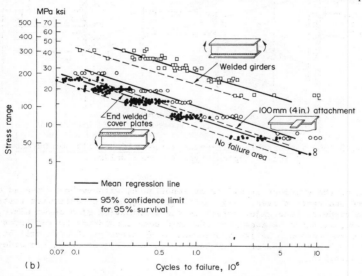

(b)

The effect of attachment length

Fig. 3.157 Fatigue behavior of welded beams (from Ref. 459).

solution would seem to be the elimination of the tensile component by designing without the stiffeners, and requiring sufficient stiffness in the web as a plate alone.

Welding defects are of prime importance in determining fatigue life since they constitute an <u>initial</u> defect or crack prior to service, and spokesmen for such prestigious organizations as the British Welding Institute and the U. S. Air Force have indicated emphatically that it is simply imprudent to neglect their presence. Internal defects—porosity, slag, etc.—are not so effective as those on the surface, except in heavy sections (> ~ 3"). One major exception to this statement is incomplete penetration, Fig. 3.158, for this condition often constitutes a sharp crack normal to a tensile stress. As a matter of good design and fabrication practice, partial-penetration welds should be avoided; both Codes permit them but only under restricted conditions, see AWS Paragraph 9.12, etc., and ASME Paragraph NB 3337.3, 3352.4, etc.

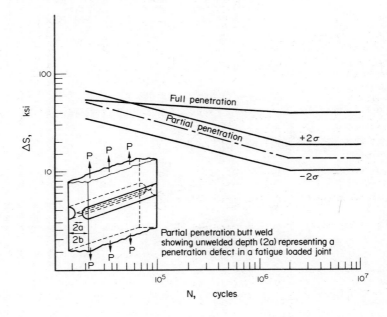

Fig. 3.158 Fatigue life vs. penetration for butt welds in
A36 steel (from Ref. 458).

The general effect on life of surface flaws such as center-bead and toe cracks is shown in Fig. 3.159 and it is immediately noted that the "usual" size of 0.050 - 0.070" is sufficient to reduce the life to uncomfortably low values, except at inefficiently low stresses. Thus, it is incumbent on both the designer and fabricator to minimize the restraint during cooling, a major cause of cracking. Considerable data indicate that the crack growth rate, da/dN, in properly deposited weld metal is at least not lower than that in the parent metal and, with care, the rate for the HAZ can be kept to similar values. It will be further noted that in

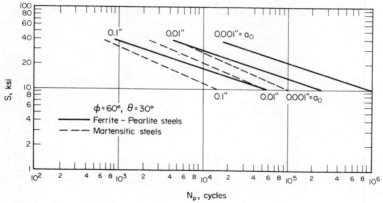

Fig. 3.159   Effect of initial flaw size on fatigue crack propagation
life (from Ref. 457).

Fig. 3.159 the different steels produce essentially the same lives (except at very
low stresses), and that in Table 3.23, the crack growth rate for the three general
classes of steels--austenitic, martensitic and ferritic/pearlitic--are not widely
different. Thus, a general conclusion is that the fatigue life of a steel weldment
is not much affected by its metallurgical structure or static strength. While the
same is generally true of the non-ferrous alloys, some exceptions exist.

The following excerpts from the Summary in Ref. 459 form excellent advice for the
designer of weldments:

"Variations in cover-plate geometry for each end detail had no significant effect on the fatigue strength except for beams with cover plates
wider than the beam flange at the end without transverse fillet weld.
Existing specification provisions that limit the thickness of the cover
plate (or total thickness of multiple cover plates) on a flange to 1-1/2
times the flange thickness can be liberalized. Cover plates may also
be attached singly or in multiples without any difference in strength
resulting.

When cover plates are wider than the flange to which they are attached,
a decrease in fatigue strength will result unless transverse end welds
are used. Hence, transverse end welds should be required on wide cover
plates. The end weld may be returned around the beam flange or stopped
short of the flange toes. Cover plates, narrower than the flange to
which they are attached, may have the end weld omitted if desired.

Plain welded beams do not exhibit the same fatigue strength as base
metal, as currently assumed in the AASHO provisions. A separate provision should be added that reflects the decreased strength of plain
welded beams. Care should also be taken to control the smoothness of
the flange tips. Obvious notches in the flange tips should be removed
by grinding, as the growth of cracks is more severe when initiated
from the tip.

The presentation of allowable fatigue stresses in weldments varies amongst the several agencies, but the information is similar. The AWS gives a form such as Table 3.39, and available tables cover the other constructional steels. The AISC (American Institute for Steel Construction) gives listings, from which Fig. 3.160 is excerpted. The ASME Code provides an S-N type curve for each of the general classes of steels and of the non-ferrous alloys used in Code work, e.g., Sec. III, Div. 1, NA App. XIV. The ASCE (American Society of Civil Engineers) gives information such as that in Table 3.40 for 6061-T6. The degree of conservatism varies somewhat with the presentation (the ASME curves are known to be "quite conservative"), and in general, the designer has the duty to look into the problem further than just the selection of a category and a stress. The Modified Goodman Diagram may also be used to present fatigue stresses; Fig. 3.161 is typical.

Regarding the fatigue behavior of the weldable stainless steels, generalizations are next to impossible. Essentially all grades of wrought stainlesses are weldable to some degree, and the problem is not so much a loss of fatigue strength in or near the weld as it is a loss of corrosion resistance (due to local depletion of chromium), particularly should the rate of heat input be high. Because of the latter condition, the arc is preferred to oxyacetylene for many alloys. A leading alloy for Mach 3 aircraft skin is AM-350CRT, a 17Cr-4Ni with Mo. Results of axial tests on transverse butt welds, made up to MIL-W-8611 show that, while the fatigue strength at any combination of temperature, life, and mean stress (40 ksi in those tests) is down considerably from the static values, the absolute value of the fatigue strength remains sufficiently high to make numerous design approaches possible. S-N curves for a precipitation-hardening steel are given in Fig. 4.54.

Another important group of weldable high-strength, high-toughness steels is the 9Ni-4Co with carbon from 0.20 to 0.45%. In the low-carbon grade, yield strengths of 180-200 ksi can be obtained in welded joints without pre- or post-heat treatment, these rather high values being attributable to the inherent toughness of this alloy system and to the beneficial effect of self-tempering on the weld metal and HAZ. Higher strengths are possible in the higher carbon grades with post-welding heat treatment. Austenitic manganese steel castings are being successfully welded by metal-inert-gas techniques and minimizing the heating time. Electron beam welding is developing very rapidly, (see below), but little specific fatigue data seem to be published yet. Much static tensile and impact data on a wide variety of alloys is given in [153].

A wide selection of aluminum alloys is readily weldable; tests of transverse butt welds in 2014-T6 and 6061-T6 (124) indicate an approximate 2 to 1 improvement for removal of the reinforcement, i.e., for the 6061 from 4400 to 9200 psi at 10^6 cycles. Static strength of the same joint was 23,400 psi, giving a stress level of less than 50% static. At an exposure of 10^6 cycles, this trend seems to hold for many of the aluminums, but for the magnesiums it can drop to as low as 12%. Extensive work [159] in the effort to classify internal weld discontinuities as to their effect on fatigue life continues, as do considerations of surface flaws. Butt joints in alloy 5456 were radiographically inspected and identified as Class I, III or V in accordance with NAS 1514. Testing at loads sufficient to cause failure at 10^4 or 10^6 cycles (R = 0.1) yielded results for all three classes which overlapped to such an extent that no conclusion was possible.

TABLE 3.39 Allowable Fatigue Stress For A7, A373 and A36 Steels and Their Welds

	2,000,000 Cycles	600,000 Cycles	100,000 Cycles	But Not To Exceed
Base Metal In Tension Connected By Fillet Welds But not to exceed →	(1) $\sigma = \dfrac{7500}{1-2/3K}$ psi P_t	(3) $\sigma = \dfrac{10,500}{1-2/3K}$ psi P_t	(5) $\sigma = \dfrac{15,000}{1-2/3K}$ psi P_t	$\dfrac{2P_c}{3K}$ psi
Base Metal Compression Connected By Fillet Welds	(2) $\sigma = \dfrac{7500}{1-2/3K}$ psi	(4) $\sigma = \dfrac{10,500}{1-2/3K}$ psi	(6) $\sigma = \dfrac{15,000}{1-2/3K}$ psi	P_c psi $\dfrac{P_c}{1-\frac{K}{2}}$ psi
Butt Weld In Tension	(7) $\sigma = \dfrac{16,000}{1-\frac{8}{10}K}$ psi	(11) $\sigma = \dfrac{17,000}{1-\frac{7}{10}K}$ psi	(15) $\sigma = \dfrac{18,000}{1-\frac{K}{2}}$ psi	P_t psi
Butt Weld Compression	(8) $\sigma = \dfrac{18,000}{1-K}$ psi	(12) $\sigma = \dfrac{18,000}{1-.8K}$ psi	(16) $\sigma = \dfrac{18,000}{1-\frac{K}{2}}$ psi	P_c psi
Butt Weld In Shear	(9) $\tau = \dfrac{9000}{1-\frac{K}{2}}$ psi	(13) $\tau = \dfrac{10,000}{1-\frac{K}{2}}$ psi	(17) $\tau = \dfrac{13,000}{1-\frac{K}{2}}$ psi	13,000 psi
Fillet Welds ω = Leg Size	(10) $f = \dfrac{5100\omega}{1-\frac{K}{2}}$ lb/in	(14) $f = \dfrac{7100\omega}{1-\frac{K}{2}}$ lb/in	(18) $f = \dfrac{8800\omega}{1-\frac{K}{2}}$ lb/in	8800ω lb/in.

Adapted from AWS Bridge Specifications. K = min/max
  P_c = Allowable unit compressive stress for member.
  P_t = Allowable unit tensile stress for member.

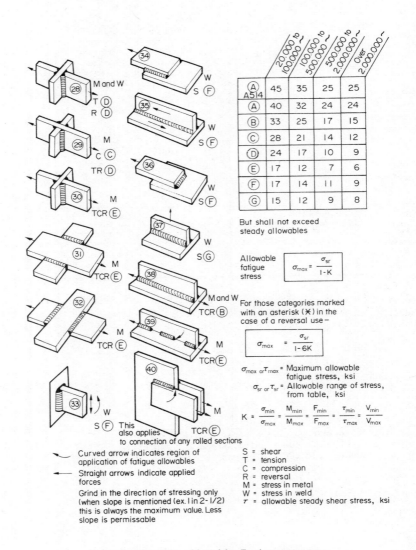

	20,000 to 100,000	100,000 to 500,000	500,000 to 2,000,000	Over 2,000,000
Ⓐ A514	45	35	25	25
Ⓐ	40	32	24	24
Ⓑ	33	25	17	15
Ⓒ	28	21	14	12
Ⓓ	24	17	10	9
Ⓔ	17	12	7	6
Ⓕ	17	14	11	9
Ⓖ	15	12	9	8

But shall not exceed
steady allowables

Allowable
fatigue
stress

$$\sigma_{max} = \frac{\sigma_{sr}}{1-K}$$

For those categories marked
with an asterisk (✱) in the
case of a reversal use –

$$\sigma_{max} = \frac{\sigma_{sr}}{1-6K}$$

σ_{max} or τ_{max} = Maximum allowable
  fatigue stress, ksi

σ_{sr} or τ_{sr} = Allowable range of stress,
  from table, ksi

$$K = \frac{\sigma_{min}}{\sigma_{max}} = \frac{M_{min}}{M_{max}} = \frac{F_{min}}{F_{max}} = \frac{\tau_{min}}{\tau_{max}} = \frac{V_{min}}{V_{max}}$$

This
also applies
to connection of any rolled sections

Curved arrow indicates region of
application of fatigue allowables

Straight arrows indicate applied
forces

Grind in the direction of stressing only
(when slope is mentioned (ex. 1 in 2-1/2)
this is always the maximum value. Less
slope is permissable

S = shear
T = tension
C = compression
R = reversal
M = stress in metal
W = stress in weld
τ = allowable steady shear stress, ksi

Fig. 3.160 AISC allowable Fatigue stresses.

TABLE 3.40   Allowable Tensile Stresses for Repeated Loads in
Welded Aluminum Construction (6061-T6 and 6062-T6)

	Allowable Stress on Net Section, F_{max}[a], ksi	
Number of Repetitions of Load Application	Tension on transverse or longitudinal butt welds and tension at locations adjacent to continuous longitudinal fillet welds	Tension at locations adjacent to transverse fillet welds or ends of longitudinal fillet welds, and shear on fillet welds
20,000	F	F
100,000	$F + 0.67F_{min}$	$0.68F + 0.67F_{min}$
500,000	$0.73F + 0.67F_{min}$	$0.40F + 0.67F_{min}$
1,000,000	$0.67F + 0.67F_{min}$	$0.34F + 0.67F_{min}$
2,000,000	$0.65F + 0.67F_{min}$	$0.29F + 0.67F_{min}$
10,000,000	$0.64F + 0.67F_{min}$	$0.23F + 0.67F_{min}$

SOURCE: Amer. Soc. Civil Engrs., Proc. Vol. 88, No. ST6, Dec. 1962 Papers No. 3341 and 3342 [138], Table 10.

F_{max} = allowable maximum tensile stress on net section for repeated loading, ksi,

F_{min} = minimum stress on net section during loading cycle, ksi (F_{min} is the smallest tensile stress in a tension-tension cycle or the largest compressive stress in a compression-tension cycle. In the latter case, F_{min} is negative).

F = allowable stress for nonrepetitive loading, from Table 6a, 6b, 7a, or 7b [138].

[a]In addition to the limitations given in this table, F_{max} must not exceed F nor the maximum stress permitted by the reversal-of-load rule in Specification G-1, [138].

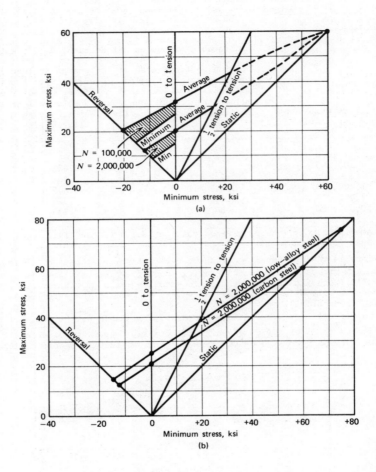

Fig. 3.161 Fatigue strengths of transverse double-V butt-welded
joints (reinforcement on). (a) Structural steel;
(b) comparison of structural steel with high strength
low alloy steel. From [12], Fig. 7.3.

Allowable stresses for aluminum plate construction were given in Table 3.40; Fig. 3.162 indicates specific test results. Note particularly, the near-coincidence of stresses at exposures of 10^5 cycles or higher. Both shot- and hammer-peening improved the fatigue behavior more than did the thermal stress relief. The peening altered the residual surface stress from tension to compression with a delta of about 30 ksi, but it did not affect the average residual stress nor the static strength. Several newer aluminum alloys (X7002, X7106) are being developed for improved weldability, especially for high strengths without post-welding heat treatment, see [144, 145]. The effect of temperature on the fatigue behavior of the aluminums is illustrated in Fig. 3.163 with a notable strength increase in the cryogenic range for both plate and welds. Limited work on the repair welding of premium quality aluminum and magnesium castings has been successful, as a result of tight correlation among design, stress and metallurgical engineering.

The titanium alloy 6Al-4V is being routinely welded by the electrom beam technique; aircraft forgings have shown resulting fracture toughness at the acceptably high level of about 70 ksi $\sqrt{\text{inch}}$. The 8Al-1Mo-1V alloy is readily welded by the TIG process, Fig. 3.164, gives some constant-life diagrams. Static failure occurred in the parent metal, but practically all fatigue test failures were in the bead, thus, this alloy seem to be an exception to the general rule in that it is more sensitive to the metallurgy of the bead than to its geometry.

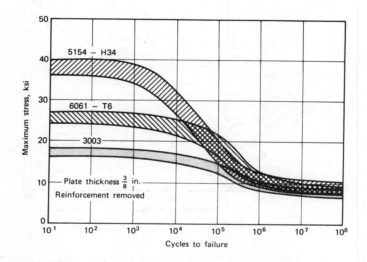

Fig. 3.162   Fatigue of arc-welded joints in aluminum
plate. From [2], Fig. 18, p. 876.

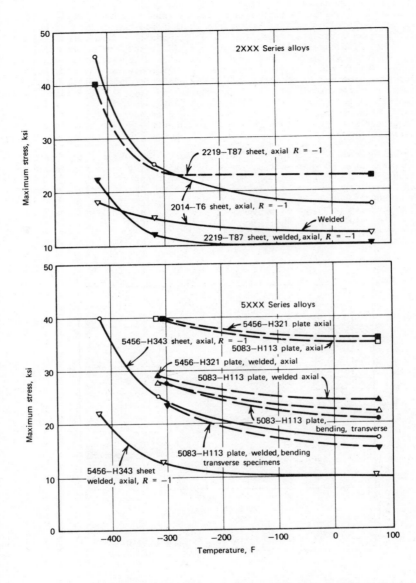

Fig. 3.163 Low temperature fatigue strengths of welded aluminum alloys, at 10^6 cycles. From [146], Fig. 48.

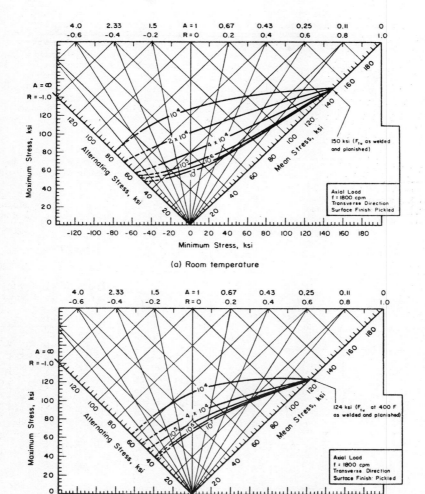

(a) Room temperature

(b) 400° F

Fig. 3.164 Constant life diagrams for welded Ti-8Al-1Mo-1V sheet,
0.050-in. thick, duplex-annealed.

Spot Welds

The fatigue behavior of spot-welded joints is similar to that of riveted joints, the strength being influenced by the type of joint (as lap or butt), the spot pattern, and the weld quality. The mode of failure in fatigue is almost always by sheet cracking, while static failure occurs by the shear of the spot. For joints in steel and aluminum, the fatigue strength averages about 10 to 15% of the static strength, indicating again the strong effect of stress concentrations in sheet bearing. The fatigue strength has been found generally to increase with the increasing size and number of spots in each row and with the number of rows. Double-shear joints are by far the preferred type; in single laps, the bending component decreases both the strength and the reliability. Much improvement in fatigue strength can be obtained through compressive residual stresses created by high pressure or by shot-peening. Although the method has been known for some years, it does not seem to be widely used because of difficulty in controlling the level of residual stress.

Spot welding has been very highly developed as a production assembly method, and design strengths have been determined experimentally for many materials and joint geometries. The static strengths, sheet efficiencies, and so forth, for various joint designs in aluminum are listed in [1], as well as specific S-N curves for both clad and bare alloys. Table 3.41 lists typical fatigue strengths for alclad in pounds per joint, a popular design parameter. From the metallurgical standpoint,

TABLE 3.41   Typical Fatigue Data from Spot Welds in Alclad Sheet

Sheet Material	Spots per Joint	Sheet Gage, in.	Static Strength, lb/joint	Type of Test	Loading condi- tions(R)	Fatigue Strength of Joint (at Given Number of Cycles), lb/joint			
						10^4	10^5	10^6	10^7
24S-T Alclad	4[a]	0.025	335	Axial	+0.25	210	140	110	100
24S-T Alclad	4[a]	0.025	335	Axial	+0.75	---	225	160	140
24S-T Alclad	6[b] (1)	0.025	313	Axial	+0.25	185	150	100	74 (2)
24S-T Alclad	6[b]	0.025	313	Axial	+0.75	---	200	155	145
24S-T Alclad	4[a]	0.032	312	Axial	+0.25	245	170	135	125
24S-T Alclad	6[a]	0.032	325	Axial	+0.25	210	180	120	90
24S-T81 Alclad	1	0.064	1060	Reversed Shear	-1.0	440	350	160	100
24S-T3 Alclad	1	0.064	1220	Reversed Shear	-1.0	450	390	200	110
24S-T86 Alclad	1	0.068	1100	Reversed Shear	-1.0	510	360	160	85
75S-T6 Alclad	1	0.064	1320	Reversed Shear	-1.0	530	380	180	120
14S-T4 Alclad	1	0.064	1210	Reversed Shear	-1.0	480	320	170	100

SOURCE: H. J. Grover, S. A. Gordon, and L. R. Jackson [9], Table 23.

1. Specimens marked [a] had a 1-1/4-in. pitch; those marked [b] had a 3/4-in. pitch.
2. Strength values are in pounds per spot for these specimens.

essentially all the structural aluminum alloys, clad or bare, may be spot-welded in any combination; but corrosion considerations may preclude certain combinations of bare 2014, 2024, and 7075; [15] gives a list of all practical combinations. Although spot-welded aluminum joints are widely used for both static and cyclic loadings, design practice and safety considerations have led to a number of applications that are normally prohibited:

1. Attachment of flanges to shear webs in stiffened cellular construction in wings.
2. Attachment of shear web flanges to wing sheet covering.
3. Attaching of wing ribs to beam shear webs.
4. Attachment of hinges, brackets, and fittings to the supporting structure.
5. At joints in trussed structures.
6. At juncture points of stringers with ribs, unless a _stop_ rivet is used.
7. At ends of stiffeners or stringers, unless a _stop_ rivet is used.
8. On each side of a joggle or wherever there is a possibility of tension load component, unless _stop_ rivets are used.

Axial tension tests at R = +0.25 on ultrasonically welded spots in 0.050-in. aluminum indicate superior strength to standard resistance welded spots at low cycle, high stress, and rough equality at hi-cycle, low stress conditions. The joint was of standard design: seven spots in two staggered rows.

The increasing use of titanium alloys, especially Ti-8Al-1Mo-1V and Ti-6Al-4V, for aircraft and space vehicles has led to the development of much data; Fig. 3.165 illustrates the excellent results of spot welding in mill- and triplex-annealed sheets. A very large number of combinations of the steels may be spot welded; Table 3.42 provides data on several alloy steels for higher performance joints.

Bonded Joints

Both the static and fatigue strengths are functions of the joint factor, defined as: $\sqrt{t}/\ell$ where t is the sheet thickness and l the length of overlap. The fatigue relationship is given in Fig. 3.166 for standard lap joints with the alternating stress in the sheet expressed as:

$$S_a = \frac{1.15/\sqrt{t}}{1 + 1.5\sqrt{t}(1)} \text{ ksi}$$

based on the fatigue strength of the bond material, not of the sheet. Thus, for thin sheets and large overlap, the fatigue load capacity of the bond would be very high, as might be expected; but for extreme values of the joint factor, the equation does not hold, and failure of the sheet could occur at a lower stress than indicated. At the other extreme of joint geometry, thick sheets, and short overlap, failure would occur in the bond at lower stresses, as given by the equation. A typical joint of 1 = 1.0 in. and t = 0.050 in., bonded with an epoxy resin (Araldite), would show a fatigue strength in axial tension of 3.6 to 4.0 ksi, or about 8% of the static strength. For large values of joint factor (small overlaps), the effect of the offset in a single lap becomes important; the adherends bend and reduce the achievable shear stress. For a large overlap (small joint factor), the yield strength of the adherend may be exceeded, leading to a metal (adherend) failure. Yielding of the adherend is equivalent to a reduction in E and increases still further the shear stress concentrations, with more reductions of the achievable mean shear stress. Since adherends are usually ductile alloys, continued increase in

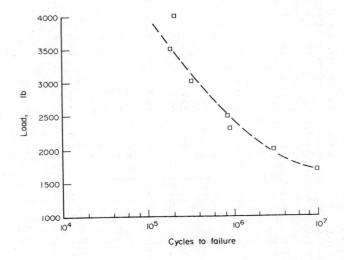

Machine Settings[a]				Tension-shear ratios		
Sheet thickness, in.	Weld phase shift, %	Heat time, cycles	Electrode force, lb	Shear strength, lb	Gross tension, lb	Ratio, %
0.022	35	3	900	920	245	26.6
0.039	40	6	1000	1967	597	30.0
0.062	45	7	1200	3117	1171	34.0
0.20-0.062	45	4	1000	1399	----	---

a RWA Class 5 electrodes, $\frac{5}{8}$-in. diam with a 4-in.-radius dome

Fig. 3.165 Fatigue of spot-welded Ti-8A1-1Mo-1V sheet (From Ref. 148).

TABLE 3.42 Typical Fatigue Data from Spot Welds in Alloy Steel Sheet

Sheet Material	Condition	Spots per Joint	Sheet gage, in.	Static Strength, lb/joint	Type of Test	Loading conditions (R)	Fatigue strength of Joint (at Given Numbers of Cycles), lb./joint			
							10^4	10^5	10^6	10^7
SAEX4130 steel	As welded	4	0.025	1387	Axial	0	850	670	500	---
SAEX4130 steel	Welded, stress relieved 950°F.	4	0.025	2665	Axial	0	960	760	420	---
SAEX4130 steel	Welded, stress relieved 1100°F.	4	0.025	2742	Axial	0	(1100)	880	560	---
SAE 1010 steel	---	---	0.010	2192	Axial	0	900	550	225	---
18-8 stainless steel	---	---	0.020	2783	Axial	0	825	425	---	---
18-8 stainless steel	---	2	0.030	2916	Axial	0	920	500-530	230-290	---
Austenitic Mn-Cr steel	---	2	0.035	2844	Axial	0	920	500-530	230-290	---
Austenitic Mn steel	---	2	0.035	2443	Axial	0	920	500-530 5×10^5	230-290	---
Republic Cor-ten steel	---	1	0.050	2590	Bending	-1.0	---	32.5	27.0	17
Republic Cor-ten steel	---	1	0.050	3020	Bending	-1.0	---	28.0	22.5	16
18-8 stainless steel	---	1	0.050	1185	Bending	-1.0	---	28.5	24.5	20
18-8 stainless steel	---	1	0.050	3225	Bending	-1.0	---	24.0	20.5	17

overlap beyond a characteristic value does not increase the load capacity of the joint--the failure still occurs in the adhesive. Regarding design for elevated temperatures: since the effective shear modulus of the adhesive decreases faster with increasing temperature than does the elastic modulus of the (metal) adherend, a proportionately greater overlap can be advantageous.

For low cycle fatigue (below about 10^6 cycles), failures tend to occur in the bond; but for lower loadings and longer lives, the failure is usually in the adherend. Figure 3.167 shows that generally better performance is obtained by the use of small joint factors (longer overlaps). The stress concentrations at the ends of the (metal) adherends are slightly reduced by increasing the overlap. A typical specification, MIL-A-8331, requires a fatigue strength of 650 psi minimum at 10^7 cycles, irrespective of the joint factor value.

For joints in which the adherends are of different thickness, the thinner of the two should be used to determine the joint factor. This practice is conservative because the thicker sheet takes more of the bending moment, reducing the offset of the thinner sheet.

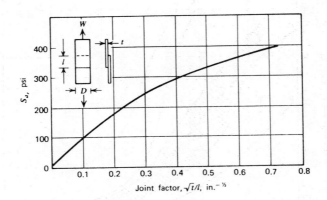

Fig. 3.166   Fatigue of bonded lap joints in axial tension. Aluminum sheets, Araldite or Redux Bond. 10^6 cycles; R = 0.5 to 0.9. From R. B. Heywood [5], Fig. 11.9.

In comparing different designs of joints, it may be said that the ideal scarfed joint has no offset and the tapered adherends have a more uniform strain distribution than does the simple lap, thus improving the shear stress distribution in the adhesive; a tapered lap joint, while retaining some offset, is an improvement over the simple lap; and because of its symmetrical continuous load path, the double-lap joint eliminates bending and provides the highest strength.

In the design of panels and similar structural elements using conventional sheet metal construction, the local buckling stress of a flat sheet is considered as proportional to the square of the thickness-to-width ratio $(t/1)^2$. In bonded construction, it has been found that, over the width of the bond, two sheets act as one of twice the thickness, so that the allowable buckling stress is quadrupled. Compared to riveted or spot-welded assemblies, this condition provides much greater stiffness and reduces considerably the effective width for buckling of adjacent bonded sheets in the same plane. Typical strength requirements are, for epoxy-bonded aluminum, 2500 psi static and 600 psi at 10^6 cycles, with the test conditions as in Mil Spec MMM-A-134, Para. 4.5.5.4.

The general fatigue behavior of a variety of joints, adhesives and adherends is illustrated in Figs. 3.168 and 3.169.

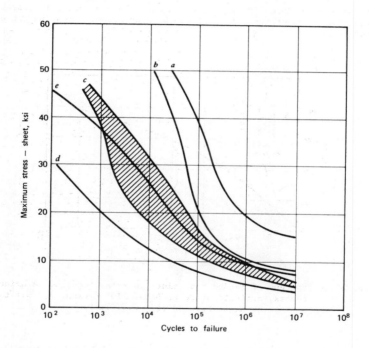

Fig. 3.167 Fatigue curves for bonded lap joints at various
joint factors: a, unnotched sheet; b, lap joints,
$\sqrt{t}/1 = 0.14$; c, lap joints, $\sqrt{t}/1 = 0.28$; d, lap
joints, $\sqrt{t}/1 = 0.36$; e, upper limit for riveted
joints (approx). Alclad 2024-T3 sheet. R = 0.
From [152], Vol. 3, Fig. 49.20.

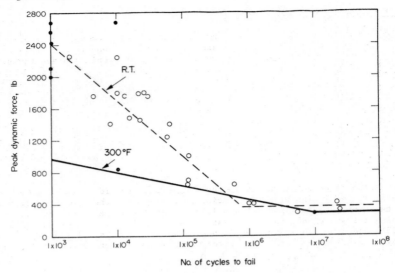

Fig. 3.168 Strength vs. life of bonded panels. 2024 Alum. shim;
EC2186 Epoxy in 1" double lap shear joint. 0 = shim
fails before bond; ● = bond fails before shim.

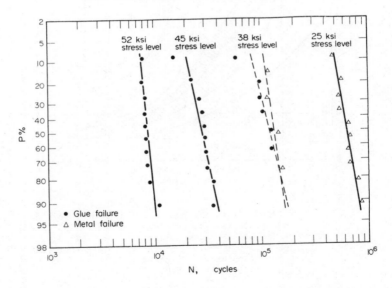

Fig. 3.169 S-N-P Curves for 2024-T3 Alclad Bonded with
FM-123-2 adhesive Double strap joints.
Log-normal probability (From Ref. 397).

3.4.2   TEMPERATURE EFFECTS AND THERMAL FATIGUE

The term "thermal fatigue" refers to that stressing imposed by dimensional changes during thermal cycling as in a turbine or fired pressure vessel, the mean and any working stresses all being additive. "Temperature" cannot, of course, be separated from any consideration of thermal fatigue but the temperature effect is taken to mean simply that change in fatigue strength with change in temperature.

For the conventional structural alloys, an increase in temperature decreases the fatigue strength at any life, a relation that holds over the entire range of practical temperatures, Fig. 3.170. Among the universally negative slopes, mild steel

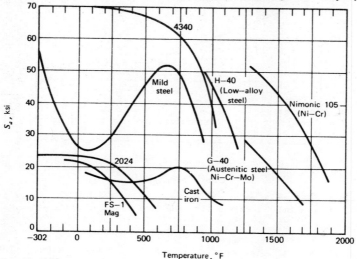

Fig. 3.170   Fatigue strength of various alloys at 10^6 cycles
vs. temperature. From P. G. Forrest [142], Fig. 110.

is the one major exception, showing a strengthening at about 650°F, a condition thought to be due to strain-aging. The ductile to brittle fracture transition with decreasing temperature seems not to be particularly significant in the fatigue of steels, and the strengthening mentioned above is of no practical use. At elevated temperatures, both the alternating and static allowable stresses decrease as the time of exposure increases. The relation between creep rupture strength and time can be shown on the same axes as those for fatigue strength and number of cycles. In Fig. 3.171, the creep rupture strength of N-155 alloy at 1500°F is seen to be lower than the fatigue strength, with the difference increasing with increasing time. The results of superimposing an alternating stress on a static stress are also shown, each curve representing a given value of R, (S_{min}/S_{max}); the value of S_{max} in the cycle is plotted, so that a direct comparison can be made of fatigue and creep rupture strengths. One should note the increase in slope of the fatigue curves with time, an important item implying that high temperature tests must generally be carried on for long times--the curves are not possible of extrapolation.

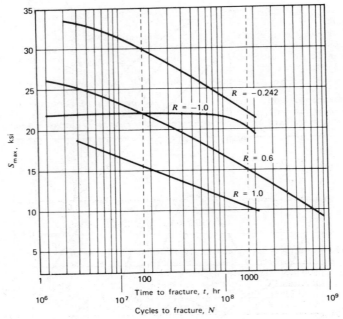

Fig. 3.171   S-N and S-t curves for wrought N-155 iron alloy at 1500°F,
(20% Cr, 20% Ni, 20% Co).   From P. G. Forrest [142], Fig. 112.

To compare various materials at various temperatures, a nondimensional plot similar
to the modified Goodman diagram is convenient, Fig. 3.172, although it should be
noted that other, fixed dimensions (time) are imposed. The important result here
is that the addition of an alternating stress up to about 40% of the mean stress
makes very small change in the rupture strength. The physical cause of this condition is not clear but it has been suggested that metallurgical changes induced
by the alternating stress cause an increase in hardness or promote transcrystalline
slip, which would reduce the effect of stress concentrations at the grain boundaries. Both the Goodman and the Gerber lines generally provide overconservative
estimates of the strength under combined static and cyclic stresses. The appearance of the fractures resulting from this combined loading at high temperatures
depends on the relative magnitude of the static and alternating stresses. As
would be expected, the amount of deformation in the region of the fracture decreases as the ratio of alternating stress to mean stress increases. The basic
problem is, of course, how to apportion the fatigue and mean stresses to prevent
fracture. A very few investigations [340, 341] have evaluated data that, when
plotted as in Fig. 3.173, provide "contours" for both creep and fatigue. Here the
time and temperature are held constant and it may be seen that for practical values
of deformation (up to about 0.2%) the alloy is deformation-critical for all values
of R up to about -0.4. In Fig. 3.174, for the S-816 alloy, only the temperature
was constant; this alloy is deformation-critical for all R values higher than -0.3
and fatigue-critical only at R values less than -0.3.

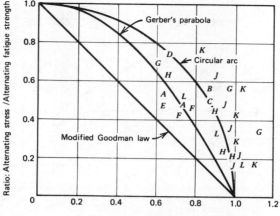

Fig. 3.172 Combined creep and fatigue strengths of high temperature alloys. All results are based on an endurance of 300 hours. A Rex 78, 600°C; B Rex 78, 650°C; C Rex 78, 700°C; D Nimonic 80, 600°C; E Nimonic 80, 650°C; F Nimonic 80, 700°C; G Nimonic 80A, 700°C; H Nimonic 80A, 750°C; J N-155, 649°C; K N-155, 732°C; L N-155, 815°C. From P. G. Forrest [142], Fig. 114.

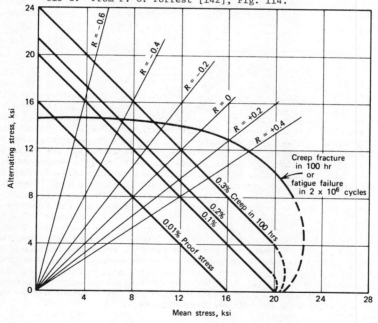

Fig. 3.173 Creep-boundary diagram for aluminum alloy RR59 at 400°F.

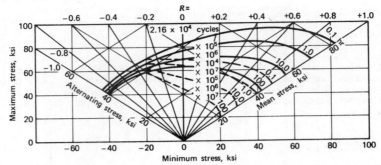

Fig. 3.174 Creep-boundary fatigue diagram for S-816 alloy at 1350°F.
R = min. stress/max. stress. For unnotched specimens
(sinusoidal loading); F = 3600 cycles/min. --- 0.2%
creep (static creep at R = 1.0); ———— fatigue failure
(static rupture at R = 1.0). From [341], Fig. B-6.

The Dorn parameter, as used in fracture mechanics, has been a useful means for relating time and temperature as a function of stress for some specific strain deformation or rupture; its use with data of the type in Fig. 3.174 permits construction of the Master Fatigue Diagram, as in Fig. 3.175. Such a diagram consists of one set of R curves for fatigue rupture and another set for a given value of deformation, covering a temperature range established by the extrapolation permissible from the range of temperatures under which tests are made. Figure 3.175 is

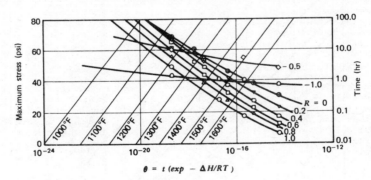

$$\theta = t \, (exp - \Delta H/RT)$$

Fig. 3.175 Master fatigue diagram for S-816 alloy. R = min. stress/
max. stress; F = 3600 cycles/min. Permanent deformation
0.2%; sinusoidal loading; unnotched specimens. From [341],
Fig. B-7.

based on 0.2% deformation and permits one to relate the time to failure, for example, at one stress, temperature, and R ratio to another temperature at the same stress level and R ratio.

The fatigue strength at temperatures below 68°F is increased for both plain and notched specimens. This increase is usually greater for soft materials than for hard ones because of the relatively lower notch sensitivity; it is particularly great for mild steel. For notched specimens, the proportionate increase in fatigue strength is lower than for the unnotched because the notch sensitivity of all alloys generally rises with decreasing temperatures. With regard to the effects of mean stress combined with low temperature, the behavior of plain specimens is quite similar to that at room temperature. with most data falling between the Goodman line and the Gerber parabola; but for notched specimens many points fall below the Goodman line.

Choice of Material

As to materials for optimum fatigue performance at elevated temperatures, the choice is determined essentially by inherent strength, metallurgical stability, and resistance to corrosion. The corrosive environment that inevitably attends hot applications is often the deciding factor in the selection of a specific alloy for a part in a given corrodant. Extensive data on the stainless steels and other heat-resistant alloys subjected to various combinations of temperature, corrodant, and fatigue loadings are provided in [2], pp. 408-536. In general, the light alloys are limited to about 350°F, the titaniums to about 1000°F; while the Ni-base alloys, Ni or Ni-Cr irons and steels, and the refractory metals are selectively useful to around 2200°F.

Changes in the metallurgical structure as a function of temperature are especially important as they may lead to loss of strength, localized strain, and gross distortion. Iron undergoes a phase change from ferrite (b.c.c.) to austenite (f.c.c.) at about 1680°F with a -0.35% linear contraction. Assuming the usual condition that the temperature is not completely uniform throughout the piece, this strain then leads to a nominal stress of about 10,000 psi. Plastic flow may alleviate this value somewhat, but the effect, and its reverse upon cooling, is quite severe in producing distortion. Not only does the volume change at the phase change, but the two phases have different coefficients of expansion; thus, any temperature gradient sufficient to allow two phases to coexist also leads to internal stressing. Additionally, the size and distribution of structural components change with temperature cycling. The carbide particles in the stainless steels will redissolve, but usually incompletely, upon heating into the austenitic range, with reprecipitation upon cooling. Extensive cycling tends to increase the size of the precipitate, especially if there are long holds at high but subaustenitic temperatures, and the particles tend to collect in or near the grain boundaries, a condition that may then lead to intragranular cracking. Certain grades of stainless steel such as 321 and 347 are stabilized against carbide growth by the addition of columbium and titanium. Typical fatigue behavior of a heat-resistant alloy is given in Fig. 3.176.

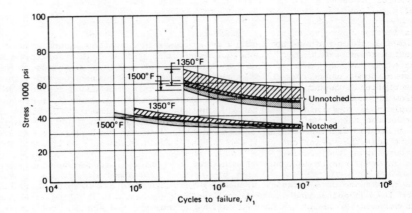

Fig. 3.176 Fatigue of Udimet 700 (nickel-base) alloy

Thermal Fatigue

The term "thermal fatigue" is used to describe the failure of reasonably ductile metals by a repetition of thermally induced stresses. Brittle materials, that is, those with an elongation less than 4 to 5% usually exhibit such poor behavior in thermal fatigue as to preclude their extensive use. The total stress includes any residual stress from fabrication and heat treatment, plus that from temperature gradients; plus that from any constraint of free expansion, and the working or applied stress (which may also be cyclic). The classical examples appear in thermal engines: turbine blades and rotors, combustion chambers, flame tubes; in chemical processing and heat-treating equipment; and in braking devices, particularly railway wheels.

Distinctions between thermal fatigue and that induced by mechanical loading include a wide difference in number and period of the loading cycles, and in the degree of plastic strain. A typical thermal fatigue cycle is performed at the rate of one cycle per day, as in central station turbine operation, and the parameter of hold-time at some, usually maximum, temperature also becomes necessary to complete the description. The number of thermal cycles to failure is also usually much lower, below 10^5 and occasionally below 10^4 for available materials. The extent of macroscopic plastic strain is greater for thermal than for mechanical fatigue; for, during the first cycle, there will usually be some shift in shape and dimensions to relax the elastic strain. This condition, however, is highly complex, involving as it does the variations of yield stress with temperature, the applied and mean temperatures, the gradient and time at temperature, the heating and cooling rate, the external constraint, and the level of applied and mean stresses.

The resistance of a material to thermal fatigue is sufficiently difficult to deduce from its mechanical and physical properties that essentially all information is experimental in nature. The most important material properties are the coefficient of thermal expansion α, the thermal conductivity k, and the resistance to alternating strain ε. The thermally induced strain is, of course, directly proportional to α, but the effect of conductivity depends on the rates of heating and cooling. At low rates where the strains depend only on external constraint, k has no effect. At high rates leading to temperature gradients in the part, a high value of k is usually beneficial because it tends to reduce the gradients, and therefore the strains. However, for heating rates so high that only a thin surface layer reaches the maximum temperature—a condition of very high "thermal impedance"—the conductivity has small effect on the surface strain. It is possible to form the criteria

$$\frac{\varepsilon}{\alpha} \text{ or } \frac{k\varepsilon}{\alpha}$$

as indications of resistance to thermal fatigue, but it is difficult to apply them. Some data [175] have been found to indicate that the resistance to alternating strain, when the temperature and applied strain cycles are in phase, is roughly equal to the resistance when the maximum temperature is maintained throughout the total strain cycling period. This condition apparently obtains because the resistance of many metals to small numbers of high-strain cycles is quite reduced at the higher temperatures.

The total range of cyclic strain that any material can withstand is the sum of the elastic and plastic strains, in an infinitely varying proportion. For a given life, the elastic strain is directly related to the fatigue strength, and the plastic strain is related to the ductility by

$$N^{1/2}\Delta\varepsilon_p = \text{constant}$$

For ductile metals at low endurances, $N < 10^3$, the range of plastic strain is greater than that for elastic strain; but for $N > 10^3$, the latter becomes predominant. In general, therefore, resistance to thermal fatigue for short life depends primarily on ductility, and for longer lives, on strength.

The topic of thermal fatigue is extremely important, not only in design for a given life at temperature and for a life composed of some required number of temperature cycles, but also in the determination of operating procedure details for central station and aircraft turbines, braking cycle time limits, number of various heats allowed for furnace structures, and so forth. A description follows of a highly interesting investigation [177] that yielded some practical limitations on the start-up cycle for steam turbine rotors.

The thermal and geometric characteristics, including particularly the surface stresses, are established for a typical 23-in. diameter, 3600 rpm rotor. Figure 3.177a summarizes the elastic thermal stresses, including the effects of specific stress concentrations. The extreme left-hand scale shown stress per 100°F change in surface temperature, using the value for $E\alpha/(1 - \gamma) = 280$ psi/°F. Figure 3.177b gives corresponding results for a sinusoidal variation of surface temperature, for

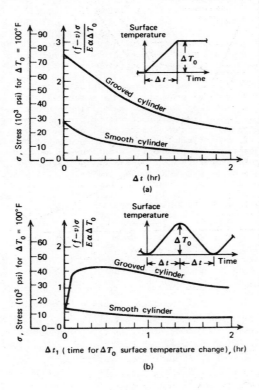

Fig. 3.177   Thermal-stress effects for two temperature vs. time programs.

evaluation of stresses in turbines controlling line frequency where the pattern of load variation may be essentially sinusoidal. The thermal cycle used here is: linear heating rate from T_0 to T_1, equalizing and holding for time t at T_1, linear cooling to T_0, and equalizing and holding at T_0. The stress-strain curve for this cycle, at the first reaction blade groove, is given as the insert in Fig. 3.178. The corresponding sequence is: compressive stressing into the plastic range from A to B during heating, followed by the development of a residual stress of opposite sign as cylinder temperatures equalize at T_1, B to C. Residual stress C remains for time t, until the start of cooling, during which plastic tensile strains appear, C to D, followed by equalization at T_0, D to A. The width of the hysteresis loop so formed gives the plastic strain per cycle, which is a measure of the number of cycles required to initiate a crack. Using the method of [178] a semi-conventional fatigue curve is drawn, relating N to the equivalent elastic thermal stress, S. The curve in Fig. 3.178 is representative of Cr-Mo-V rotor material, and thus of many rotors. Now the data in Fig. 3.177 and 3.178 are combined to form a "cycle-capacity" chart, as in Fig. 3.179. Heating and cooling rates and amounts have

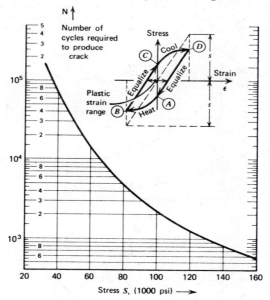

Fig. 3.178 Low cycle fatigue curve for turbine rotor.

been taken equal here for simplicity. For a given rotor-surface temperature change, ΔT_0, which occurs in time Δt, a number of cycles N to produce cracking is determined. Alternately, the time Δt for a given T_0 can be selected, having previously established an acceptable number of cycles as the crack-free lifetime.

As an application, consider the projected operating program given in Table 3.43. Four categories of transients are listed horizontally with their associated changes in rotor surface temperature, and change rates. These values are taken equal to the changes in impulse-chamber steam temperature in the case of load changes, due regard being made of the effect of any concurrent change in inlet steam enthalpy.

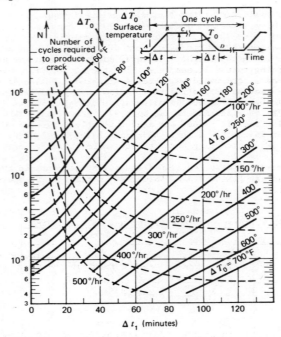

Fig. 3.179 Thermal-stress cyclic capacity chart for typical 3600-rpm rotor.

TABLE 3.43

Rotor Surface Temperature Change (°F) ΔT_0	Temperature Change Rate (°F/hr), dT/dt	Number Cycles Capacity, N	Number Cycles Desired, N'	Damage Increment, $D = \dfrac{N'}{N}$
500	300	1,300	80	0.061
300	300	1,800	1,250	0.695
100	400	10,000	2,000	0.200
50	∞	100,000	3,000	0.030

Total damage index, $\Sigma \dfrac{N'}{N} = 0.986$

SOURCE: W. R. Berry and I. Johnsson [177], Figs. 4, 5, 6 and Table 1.

For start-up, the rotor surface temperature change is taken as the difference between impulse-chamber steam temperature at the final load and the turbine's initial internal metal temperature before admission of steam. Entering the chart with each pair of temperature characteristics, ΔT_0 and dT/dt, yields N, and the damage increment is formed by taking $D = N'/N$. The total damage is then the summation of the increments, according to the linear rule. For the particular numbers in this program, the margin of safety would be rather small, about 1.5%. As an alternate manipulation, the heating or cooling time period may be chosen, instead of the rate. For example, in the second transient, if a time of 40 min were taken instead of dT/dt = 300°F/hr, N would have been about 900 cycles, below the desired value of 1250. Retaining the $\Delta T_0 = 300°$, to meet the desired N' of 1250 would require a time Δt of about 45 min, or a dT/dt of about 380°F/hr. As intuition indicates, long thermal fatigue life requires relatively small values of both T_0 and dT/dt.

Because of the general lack of extensive investigations such as that above, data on thermal fatigue are limited, but Table 3.44 is considered representative of crack susceptibility.

TABLE 3.44 Thermal Fatigue of Inconel 713C and HS-31

Material	Cycles to First Crack	Cycles to Crack 1/8-in. lg.	Temperature Range, °F	Cycle Time, Min
1. Inconel 713C	600–1050	2200–2600	100–1700	1
2. HS-31 grain size				
0.75 in.	300	500–700	100–1700	1
0.50 in.	400–500	600–1000	100–1700	1
ASTM No. 00 to 0000	800–2400	1100–2600	100–1700	1

SOURCE: J. Brit. Iron and Steel Inst., 1951, by permission [208].

A few tentative generalities on thermal fatigue may be attempted:

1.   Thermal-stress fatigue cracks originate at the surface and, with few exceptions, are intercrystalline in origin and propagation. These characteristics thus more closely resemble those of creep behavior than those of mechanical fatigue.

2.   The surface oxidation, which is intergranular in nature (at least in nickel-base alloys) has a significant effect on thermal-stress-fatigue life. The use of an inert atmosphere (argon), rather than air, as the fluidizing gas results in an appreciably longer cyclic life. This, in part, may explain (1) above.

3.   The peak temperature of the cycle is the most important factor governing life. For a given temperature difference between specimen and fluidizing bed or other quenching medium, the cyclic life is greatly reduced by increases in peak temperature.

Design for Thermal Fatigue

Since the basic function of many devices is that of thermal energy exchange or conversion, a change of temperature with a corresponding change of stress level is an inescapable factor in their design. Several general conditions and design features are to be considered for the mitigation of thermal stresses, cyclic or steady. A major distinction is the difference in approach to the reduction of thermal stresses as contrasted to mechanical stresses: the increase in section usually applied in the latter case may be harmful in the form of increasing the constraint and thus raising the thermal stresses. General factors for reduction of thermal stress are: (1) reduction of constraint on deformation, (2) minimize nonuniform expansion, (3) introduction of compressive residual stress, (4) reduction of stress concentration factors, and (5) protection of the surface.

Constraint. The geometry of heat flow by any of the three modes is usually such that the temperature distribution through a section is nonuniform, leading to, as a familiar case, the bowing of a beam heated on one side. If the ends are constrained sufficiently to prevent bowing, the resultant stress distribution through the (uniform) section is linear. Reduction in stress level can be achieved by reduction of the temperature gradient, of the section depth in the direction of heat flow, or removal of the constraint. Functional or mechanical stress requirements often make the first two inapplicable, but floating or semi-restrained constructions can frequently be quite practical and are sometimes the only means of fabrication. Typically, in gas-turbine design, the nozzle diaphragm rings are rigidly attached to the vanes. Cutting or segmenting the outer ring allows some "float" for the vanes and alleviation of the stresses arising from the inevitable temperature gradient. Similarly, in turbine wheels without shrouds or outer rings, slotting the rim permits some individual freedom of motion for the buckets, and with the slot properly keyholed at its bottom, the stress concentration factor is also much reduced. Other forms of floating constructions are the familiar large radius bends in steam piping, and a more complex scheme for heat-exchangers in which the tubes are given a permanent prior bend into a sine wave shape. Then the distance between the head sheets depends only on shell expansion, and any differential expansion between shell and tubes is accommodated by tube movement as bending in the sine wave.

"Contouring" is also a most efficient method of controlling thermal stresses by control of the constraint through the temperature gradient. The term refers to the profile geometry of the load-bearing section, and of course, must be considered as part of the initial design. An interesting example is illustrated in Fig. 3.180, which shows the results of thermal endurance tests on locomotive wheels. Such wheels are not only load-carrying, but also provide the surfaces to which the brake shoes are applied. During the braking operation, the rims rapidly become heated to approximately 1000°F before the temperature at the center of the wheel is appreciably raised. The design problem thus becomes very similar to that of the turbine wheel in which the free expansion of the hot rim is prevented by the cool (and massive) core. Compressive plastic flow occurs in the rim and is followed by residual tensile stress upon cooling. After a sufficient number of cycles of plastic flow, rim cracking occurs, which can spread and cause total wheel fracture. As an experiment in determining the optimum geometric contour, three wheels with progressively reduced webs but the same hub and rim were thermally cycled to fracture. In all cases, including the three conditions of initial heat treatment, the thinnest wheels (1/2) gave the longest life, thus emphasizing the necessity of allowing the heated section to expand. The thermal stress level is directly related to the degree of restraint on this expansion.

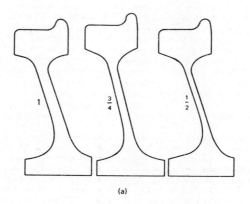

(a)

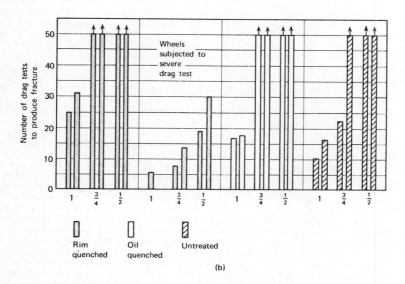

(b)

Fig. 3.180   Effect of web thickness on endurance of locomotive
            wheels. Arrow indicates wheel did not fracture.
            From [211], Fig. 9.7.

Compatibility of Expansion. The matching of lengths and coefficients provides another method of varying the constraint, the phrase meaning a design that results in a reasonable degree of stress uniformity. Thermal stresses arise because of constraint resulting from the expansion of each element according to its temperature and thermal expansivity. Although temperature gradients cannot always be avoided, the compatibility of expansion can still be achieved if the expansion of one member having a specific temperature and specific coefficient of expansion does not interfere with the expansion of another element having a different temperature and different coefficient of expansion. Thus, for example, if the member having the higher temperature can be made of a material having the lower coefficient of expansion, the total expansion can be matched with an element having a lower temperature and higher coefficient of expansion. Such bimetallic construction has been proposed [205] for the wing configuration of a supersonic aircraft. If aluminum is used for the entire wing, the thermal stress turns out to be about 49,500 psi, but is only 25,700 psi for a steel skin with aluminum webs.

In general the best design is that which constitutes the least constraint against the expansion associated with the higher temperatures. Heavy sections with high thermal inertia should be regarded as prime regions for redesign, recalling that the number of cycles applied remains fixed by functional requirement and that the number of cycles allowed may be increased by reduction of the thermally induced stress.

Stress and Strain Concentrators. The effect of these elements in thermal fatigue is similar to that in mechanical fatigue, but often to a more intensive degree. Not only does a change in section from thick to thin introduce a geometrical stress concentration factor, but the strain concentrates in the thin section which also has the lower thermal inertia. The temperature gradient is such as to concentrate the differential strain from expansion in the thin section, thus increasing the applied stress; upon a rise in temperature, the yield strength of the thin section is also reduced more because of the thermal gradient.

In thermal fatigue, the importance of surface finish may often be greater than in mechanical fatigue. The materials usually become embrittled because of the elevated temperatures, thus emphasizing the effect of any geometrical irregularities; quite often the heat-carrying agent is a corrosive fluid, which by chemical reaction tends to further exaggerate the concentration effect. Whenever possible, geometric proportioning should be chosen to avoid the introduction of strain concentration. Cutouts should be avoided and, if absolutely necessary, should be placed in regions of low nominal strain. Fillets should be generous, and changes in cross section gradual.

In the design of weldments, it should be recalled that the beads may contain imperfections that will act as strain-raisers, particularly at the surface. Full penetration welds should, therefore, be used, and they should be located, if possible, away from cross-sectional discontinuities. Figure 3.181 shows possible joint configurations between heavy and light sections. Configurations (b) and (c) are preferable to (a) since they avoid combining the strain concentrations of the weld and that of the cross-sectional change. Particularly important, too, in the welding of heavy sections, whether of equal thickness or not, is the restraint built up by the thermal cycling of the welding operation itself, the sign of the residual stress usually being tension.

Many designs, of course, require changes in cross section; it is often possible to reinforce the region of strain concentration by the addition of metal, by a boss around a cutout. But consideration must be given to the effect of the reinforcement on the thermal conditions; if the additional material increases the temperature gradient locally, the strain will also be increased.

A highly important example of the proper application of reinforcement to improve thermal fatigue life is the building up of the edges of turbine blades and disks. The gradient condition is similar to that noted above for the railway wheels, but here, blunting of the edge reduces the heating rate and tends to equalize the expansion of the cooler core and hotter tip. The results of analysis to determine the effect of varying the leading edge radius of a typical turbine blade show that the blunter edge should have the longer life. This condition follows from the fact that the thermal stress is proportional to the difference between the mean edge temperatures and that this difference is inversely related to the edge radius. Tests on Nimonic 90, Fig. 3.182, show that indeed the thermal fatigue life increases with edge radius but only to a well-defined limit.

Thermal Shock

Problems of adequate test simulation have been quite effective in limiting thermal shock data. The major difficulties have been in the form of the choice of size and shape of the test specimen, of establishing a useful test facility short of the complete engine, and of acceptance of a definition of failure. However, the use of thin-edged cylindrical specimens has become common, for they resemble many practical machine elements, and their geometry favors accelerated failure. Other common shapes are sheet and flat disks. The rapid change in temperature may be accomplished by quenching in water, air, combustion gases, or fluidized beds. Since there is an interchange in the roles of tensile and of compressive plastic flow during heating and cooling shock, it is to be expected that the two different modes of shock may produce different fatigue lives. For heat engines, the most realistic results should be produced by creating gaseous heating shock and air cooling shock. An investigation based on conditions in heating shock [344] gave results for numerous cast and wrought alloys in two specimen shapes. Failure was defined as a visible crack across the entire 1/32-in. edge. The conclusions were generally that the cobalt base alloys gave longer life in thermal shock than the nickel- and iron-base alloys, that the wrought form of the alloy was superior to the cast form (may be in part a result of the smaller grain size of the wrought form), and that, for the wrought stainless steels, thermal shock resistance was greater for the finer grain sizes.

Another investigation [345] used air coolant to provide thermal shock to a thin-edged (0.040 in.) cylinder, the major result being, however, great irregularity in the rating of the specimen materials over a temperature range of 1600-2000°F. Batealloy generally survived for the greatest number of cycles; other alloys included 310 and 347 stainless steels, Inconel B and C, S-816, N-155, and HS-21. Such a wide variation in performance firmly emphasizes the need for extreme care and completeness in design and in the definition of the environment.

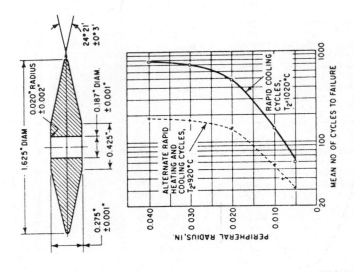

Fig. 3.182 Variation in thermal endurance with edge radius of tapered disks in Nimonic 90. From M. Cox and E. Glenny [180]. Engineering, Sept. 1960.

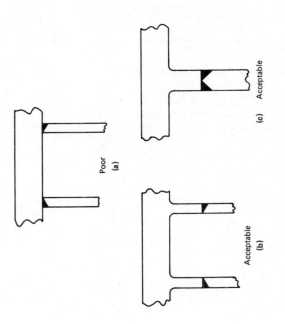

Fig. 3.181 Weld locations and section changes.

3.4.3 Pressure Vessels and Piping

The term "pressure vessel" is here taken to mean a container under either internal or external pressure cycling and, for many applications, under simultaneous thermal cycling: boilers, turbine casings, reactor shells, cryogenic storage tanks, etc. The fatigue life of such vessels and connection piping has only recently been taken into serious consideration as a design parameter. Rarely has the design life needed to exceed 100,000 cycles of changing pressure and/or temperature—more normal range of life for industrial and aerospace vessels is 5000 to 25,000 cycles, Typically, MIL-R-8573 for air reservoirs requires 18,000 cycles from 0 to related charging pressure, plus 2000 cycles from rated to maximum fight pressure. Welded aerospace tanks for cryogenic service have life requirements in the 200 to 500 cycle range, but, during development of some of the complex welded joints, test cycling has resulted in failures in a range of 4 cycles to about 2500, depending on joint details and temperatures.

The discussion is based primarily on the forged and/or welded steels, but many other alloys are finding wide application depending upon the type of service, especially the temperatures. Popular steels are the A2xx and A3xx series, particularly A285, plus T-1 and HT-80, and stainless types 301 and 310. Nonferrous alloys include 2014 and 5083 aluminums; the Inconels; Ti-5Al-2.5Sn and Ti-6Al-4V titaniums, and many more are being investigated for weldability, fracture toughness, and strength at very low temperatures.

The need to verify the design and fatigue life of pressure vessels and piping to a relevant code is essentially universal, but serious differences exist amongst the Codes, (Fig. 3.183). These differences are being resolved but slowly, as the following discussion on the ASME and British Codes illustrates, [453].

ASME Code Section III. The philosophy of the design methods of Section III of the ASME code has been described in an official ASME publication [454].

The fatigue-life calculation process is:

(1) An elastic stress analysis of the vessel is performed.
(2) At each stress-concentration point, the three principal stresses σ_1, σ_2, and σ_3 and the stress differences $S_{13} = \sigma_1 - \sigma_3$, etc. are determined for the two extremes of the loading cycle using elastic theory. If the directions of the principal stresses do not rotate during the loading cycle, the alternating stress intensity S_{ALT} is defined as the largest of $0.5 \Delta S_{12}$, or $0.5 \Delta S_{31}$. If the principal stress directions rotate, a procedure is given for determining S_{ALT}.
(3) The allowable number of cycles for the calculated stress intensity is determined from the appropriate design curve.

The code originally contained two fatigue-design curves, one for austenitic stainless steels and nickel-chrome-iron alloys and a second for carbon and low-alloy steels. In the current version, the former curve is unchanged, but the permitted endurance for carbon and low-alloy steels is dependent upon the specified, minimum, ultimate tensile strength of the material. Two fatigue-design curves are presented, one for less than 80,000 lb/in.^{2} UTS, and one for greater than 115,000 lb/in.^{2} UTS. Linear interpolation between the two curves is specified for steels of intermediate tensile strength. The design curves are derived from strain-controlled fatigue tests, incorporating at all points a safety factor of at least 2 on stress and 20 on life, together with a factor taking into account mean stress effects. Although based on high strain-fatigue tests, the design curves are presented in terms of alternating stress intensity, i.e., the alternating strain multiplied by Young's modulus. (It should be noted that there is the implicit assumption of no crack-like flaws, especially in Article NB-3222.)

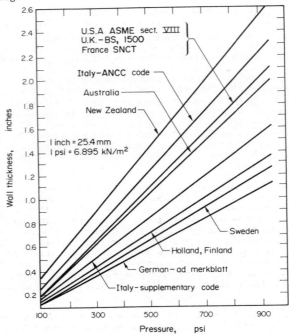

Figure 3.183 Comparison of required wall thicknesses using
various design codes.

BS Code 1515. Part 1 of BS 1515 is restricted to carbon and low-alloy steel pressure vessels, and Part 2 deals with vessels fabricated from austenitic stainless steels. For pressure cycling, the standard presents the formulas:

$$N = \left[\frac{2000(3000 - T)}{2 K f_r - f} \right]^2 \quad \text{(carbon and low-alloy steels)}$$

$$N = \left[\frac{2000(3000 - T)}{1.8 K f_r - f} \right]^2 \quad \text{(austenitic stainless steels)}$$

These formulas represent exemption criteria, and stresses higher than those calculated from the appropriate equation are permitted, provided that they are justified by a fatigue test or are based on previous satisfactory operating experience.

The BS 1515 fatigue-life equations may be derived from the low-cycle fatigue relationship as in Eq. 134, below.

Comparison of the Design Methods of the Two Codes. There is considerable similarity between the fatigue-design methods for pressure cycling of BS 1515 and Section III of the ASME code. Both methods implicitly assume that the maximum strains may be predicted using elastic stress analysis, with failure being determined from low-cycle fatigue considerations. The ASME Section III design-fatigue curves are based on actual fatigue tests with three curves, two for carbon and low-alloy steels, and a third for austenitic stainless and nickel-chromium-iron alloys. It is shown that the BS 1515 pressure cycling fatigue-design equations are based on the low-cycle fatigue relationship $N^{1/2}\varepsilon_p = C$ and that the value of C incorporated is conservative for any steel meeting the standard's elongation requirement.

The ASME Section III design curves incorporate a safety factor, the greater of 2 on stress or 20 on life, on the mean curve from the fatigue tests, together with a mean stress correction at higher lives. It is shown that a safety factor of at least 4 on stress and, therefore, at least 16 on life is incorporated in deriving the BS 1515 equation for carbon and low-alloy steels. For austenitic steels, the factors are 3.6 and 13, respectively. For ductile steels, an additional safety factor results from the conservative value taken for the constant C. It should be noted that the factors quoted in this paragraph relate to parent material properties; in the region of welds, the ductility may be significantly lower.

An additional margin of safety should result from the crack-propagation stage in the fatigue-failure process. In a strain-controlled fatigue test of a plain specimen, a macrocrack initiates toward the end of the fatigue life and propagates rapidly. In a pressure vessel, a fatigue crack initiates in a region of stress concentration and then propagates into a region where the nominal stress is lower. Therefore, the crack-propagation rate in the pressure vessel will be slower than in an equivalent plain fatigue specimen subjected to the strain history occurring at the point of maximum strain concentration in the vessel. Therefore, crack propagation in the vessel will require more cycles than in the equivalent plain specimen, and this should provide an additional margin of safety for both design codes.

It should be noted that the actual factor of safety inherent in pressure vessels designed to the codes will be lower than the factors included in the derivation of the design curves.

In Figs. 3.184 and 3.185, the ASME Section III fatigue-design curves for carbon and low-alloy steels are compared with peak stress values derived from the BS 1515 fatigue-life equation. As the latter is dependent upon the material design stress, two curves are given corresponding approximately to the highest and lowest strength steels in nuclear pressure vessel use. The ASME Section III design curves are shown to permit considerably higher fatigue-design stresses than given by the BS 1515 equation.

Summarizing the comparison of Codes:

Section III of the ASME Boiler and Pressure Vessel Code and BS 1515 should be credited for containing methods of designing against fatigue failure.

The fatigue design curves of ASME Section III for carbon and low-alloy steels have no margin of safety when compared with some pressure-vessel fatigue test results.

An additional safety factor, the more conservative of 3 on life and 1.25 on stress, applied to the ASME Section III carbon and low-alloy steels design curves would introduce an adequate safety margin.

The BS 1515 fatigue design equation guards against fatigue failure caused by pulsating pressure for vessels manufactured from the higher strength steels, but it is unduly conservative for low-strength steels for endurances between 10^4 and 10^6 cycles.

The experimental results show that there is no relationship between the ultimate tensile strength and the fatigue strength of welds in pressure vessels.

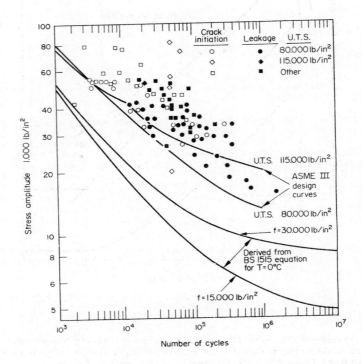

Fig. 3.184 Pressure vessel fatigue tests. Failures
originating in parent metal.

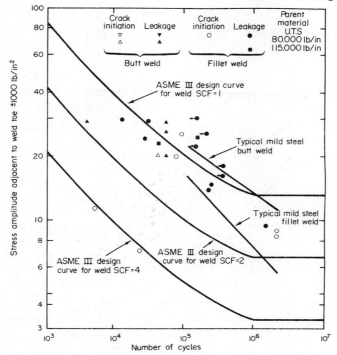

Fig. 3.185 Pressure vessel fatigue tests. Failures
originating at welds.

Both BS 1515 and ASME Section III are in error by permitting higher weld fatigue
stresses in high-strength rather than in low-strength steels. The effect in BS
1515 is lessened because of its inherently higher safety factors.

The stress-concentration factor for the weld profile of an undressed butt or fillet
weld may be taken as 2 for endurances up to 10^6 cycles. The value should be used
in conjunction with the local stress adjacent to the weld toe and the BS 1515
fatigue-life equation or the ASME Section III fatigue-design curve for carbon
steels up to 80,000 lb/in.^{2} UTS.

The ASME Code and Fracture Mechanics. The toughness requirements and application
procedures based on Fracture Mechanics were introduced to the Nuclear Sections of
the ASME Code (III and XI) only recently, 1974, and the latter are contained in
non-mandatory appendicies.

For application to vessels under Section III, the technical approach has the objective of developing toughness criteria to prevent (brittle) failure. It is addressed
to Section III stress limits and materials to 700°F, and it considers normal, upset
and test conditions only. The Procedure involves: 1) a reference fracture toughness (K_{IR}) as a function of temperature, indexed to a nil-ductility-temperature,

NDT(RT_{NDT}) from an adjusted drop-weight test; 2) a postulated surface flaw, semi-elliptical of depth equal to 25% of the wall thickness in sections less than 2.5 inch; and 3) the primary, secondary and fabrication stresses. The leak-before-burst situation is obtained by arranging to have the critical crack size greater than the wall thickness.

The Materials Qualification Test Requirements (NB-2300) include: 1) the Procedure for Determining RT_{NDT} for thick sections; 2) Charpy values (CVN), and the basis for the CV lateral expansion values.

The Toughness Requirements and the determination of K_I are given in Appendix G for operational conditions, and for shop and system hydro-testing.

For Piping Materials (NB-2332) and Bolts (NB-2333), these Articles give the lateral expansion values (CV); and the reference flaw size requirements.

For <u>In-Service Inspection of Nuclear Components</u>, Section XI applies to flaws detected during operation. It gives the rules for transforming flaw indications (ultrasonic) to simple geometries; the standards for those acceptable flaw sizes not requiring analysis, and an analytic method, with acceptance criteria, for verifying the design with those flaws which have grown larger than the basic standard size.

The methodology (Appendix A) is based on linear elastic Fracture Mechanics, and considers:

1. Critical Flaw Sizes:

    a_f – maximum size due to growth by remaining service loadings.

    a_c – minimum initiation size for normal conditions.

    a_i – minimum initiation size for non-arrest under emergency and faulted conditions.

2. K_I Calculation – basic equation and correction factors for flaw type and geometry, free surface effects and type of loading.

3. Material Properties:

    K_{Ic} and K_{Ia} versus temperature

    da/dN versus ΔK, (air and water environment)

    Radiation embrittlement effects

4. Includes all future anticipated loadings and transients.

    Requires transient temperature, stress and K_I calculations.

5. Acceptability Evaluation:

    For normal conditions: $a_f < 0.1a_c$

    For emergency and faulted conditions: $a_f < 0.5a_i$

Thus, the ASME Code presently has requirements and a methodology for design verification by both the older "fatigue damage" approach, and by crack propagation, (Fracture Mechanics). Reference [454] discusses the Pressure Vessel Research Committee recommendations in its section on Fracture Control Plans.

The discussion above is elementary in the extreme, and the dedicated reader is urged to consult the Code directly in all its detail. However, it is not recommended that design verification be attempted with the Code as the sole source of information. The complexity of the problem and the criticality of the component usually make imperative the more extensive consideration of the geometry, of the NDE evaluation, and of the possible need for specific testing and, of course, the depth of the stress analysis should match the requirement.

The full analysis of a vessel design for loads, stresses and concentrations, thermal gradients and transients, is a monumental task and very often does not get done. Recent efforts to simplify the tasks, while retaining sufficient comprehensiveness, have taken advantage of the fact that the fatigue life of most vessels falls into the low cycle range; thus the work of Coffin [88, 175] and others is applicable to design. Their results demonstrate that the fatigue behavior in the high stress, low cycle regime at temperatures below the creep range is properly expressed in data from strain-controlled tests by an equation of the form:

$$C = \epsilon_p \sqrt{N} \qquad (134)$$

where C is a constant and ϵ_p the plastic strain. Langer [178] has performed a service to the industrial community by combining this result with considerations of ductility, mean stress, and stress concentrations to derive an expression for the fatigue curve:

$$S = \frac{E}{4\sqrt{N}[\ln 100/(100 - RA)]} + S_e \qquad (135)$$

(where $S = 1/2E\epsilon_p$, RA = % (reduction of area) in the tensile test and S_e is the endurance limit); and then to establish several design curves. The advantage of this treatment is the evaluation of the decreasing effect of the mean stress, as the applied stress climbs into the plastic range, and the establishment of criteria relative to stress concentration factors upon which a decision may be made as to the need for a full analysis. For the latter, a K_f factor is established as a numerical "quality" factor, describing the shape and quality (workmanship) of the vessel.

Consider first the parameter of mean stress: it is well known that at the high cycle end of the curve, the addition of a steady mean stress reduces the endurance limit, and the amount of reduction can be estimated by means of the Goodman diagram. It has also been demonstrated that at the low cycle end of the curve, where stresses go well up into the plastic range, mean stress has little or no effect on the fatigue life. This condition is due to the yielding that occurs during the first strain cycle, which reduces the effective value of mean stress. When the strain range exceeds twice the yield strain, the mean stress becomes zero. The evaluation of the effect of mean stress is difficult; after having gone through the correct procedure, the designer might well ask whether some residual stress produced by welding, a preservice test, or an accidental overload may not have actually produced a mean stress in the member higher than the one he had calculated. Therefore, it would be much easier, and not unduly conservative to adjust the fatigue curve downward at the high cycle end to allow for the maximum possible effect of mean stress. The required amount of adjustment can be found by means of a modified Goodman diagram, Fig. 3.186, where the mean component of any fluctuating stress is plotted as the abscissa and the alternating component (half total range) as the ordinate. Failure is expected for any stress cycle that falls above the line joining the endurance limit to the ultimate tensile strength, (line ED).

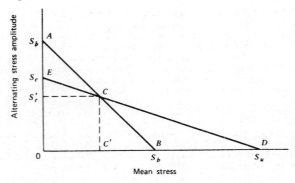

Fig. 3.186 Modified Goodman diagram for low-cycle fatigue.
(Figs. 3.129-3.136 from B. F. Langer [178].)

Attention is called to the 45° line AB that connects the S_b values on the horizontal and vertical axes, S_b is called the "limit of elastic behavior" and may be described as the highest stress amplitude the material can maintain without yielding, even after being cycled several times. It is the yield strength after strain hardening or strain softening, which for practical purposes can be taken as the higher of the 0.2% offset yield strength or the endurance limit. Regardless of the conditions under which any given test is started, the true conditions after the application of a few cycles must fall inside the triangle OAB or on the vertical axis above A, because any test that is started under conditions outside of this region would have a maximum stress greater than S_b and yielding would then reduce the mean stress to a value on AB or on the axis above A.

Now the effect of the mean stress on the amplitude of the alternating stress that is required to produce failure is determined. At zero mean stress, the required amplitude is S_e. As the mean stress increases along OC' in Fig. 3.186, the required amplitude of alternating stress decreases along the line EC. If we try to increase the mean stress beyond C', yielding occurs, and the mean stress reverts to C'. If we apply the mean stress by means of a deadweight load, we can force it to hold a value higher than C', but then failure will be produced more by gross yielding than by fatigue, and the fatigue analysis becomes inapplicable. Therefore, C' represents the highest value of mean stress that has any effect on the fatigue life. The actual value of mean stress in any specific case might well be considerably lower; so the assumption that it has the value represented by point C' is a conservative one. Since S_e' is the alternating stress required to produce failure when the mean stress equals C', S_e' is the value to which the high cycle end of the fatigue curve must be reduced if we are safely to ignore the effect of mean stress. From the geometry of Fig. 3.186, it can be shown that:

$$S_e' = S_e \left[\frac{S_u - S_b}{S_u - S_e} \right], \qquad \text{for } S_e < S_b \qquad (136)$$

For finite numbers of cycles, S_e becomes S_a, as determined from a fatigue curve such as Fig. 3.187, and the reduced value S_a' becomes

$$S_a' = S_a \left[\frac{S_u - S_b}{S_u - S_a} \right], \qquad \text{for } S_a < S_b \qquad (137)$$

When the number of cycles decreases to the point at which $S_a \geq S_b$, then $S_a' = S_a$ and mean stress has no effect on the allowable stress amplitude.

Figure 3.187 shows experimental data for unnotched austenitic stainless steel. The yield strength of this material is usually about 30,000 to 40,000 psi, so that $1/2 E \varepsilon_p$ exceeds the yield strength in the whole range of interest. Therefore, it

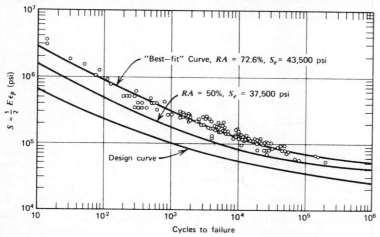

Fig. 3.187 Fatigue data and design curve for austenitic stainless steel.

may be concluded that this material cannot sustain a mean stress at any cyclic stress level that would produce failure, and no correction for the effects of mean stress need be made. Experimental verification of this conclusion for four steels-- Type 304 SS; 9% Ni; A517 Gr F and A201 indicates that the fatigue life is essentially independent of the mean <u>strain</u>, [455]. Figure 3.187 also shows a design fatigue curve for austenitic stainless steel based on the best-fit curve with a safety factor of either 2 on stress or 20 on cycles, whichever is more conservative at each point. It is believed that these safety factors are sufficient to cover the effects of size, environment, surface finish, and scatter of data. Several fatigue tests and simulated service tests on models of components have confirmed this belief. Service experience has also been good but is probably not lengthy enough to be cited as a strong confirmation of the proposed methods.

It may be found that for heat-treated steels, for which the yield strength is not far below the ultimate strength, Eq. (137) gives an unduly conservative result. When this occurs, it is advisable to calculate the mean stress that will actually occur and not assume that it is always at its maximum possible value. Take, for example, a heat-treated bolting steel for which S_u = 150,000 psi, $S_b = S_y$ = 130,000 psi, and S_e = 75,000 psi. Assume also that S_a = 130,000 psi at 2000 cycles and S_a = 95,000 psi at 10,000 cycles. The mean stress must be considered at any number of cycles above 2000, since this is the number of cycles for which $S_a = S_b$. At 10,000 cycles, Eq. 137 gives

$$S_a' = 95,000 \; \frac{150,000 - 130,000}{150,000 - 95,000} = 34,600 \text{ psi} \qquad (138)$$

For this case it would be worthwhile to calculate the actual mean stress, rather than accept such a large penalty in reduced endurance limit.

In the continuation of the attempt to simplify the fatigue analysis of pressure vessels, the stress concentrations are treated as one factor for the assembly, and thus establish a quality level for the assembly and its fabrication. In establishing the fatigue-strength-reduction factor, K_f, the usual expression involving notch sensitivity,

$$q = \frac{K_f - 1}{K_t - 1} \tag{139}$$

is not used because it does not account for the fact that q is not a true material property. Instead, a material property in the form of a dimension delta, δ, is introduced; δ is the distance below the surface of a hole or other discontinuity at which the local stress first reaches the endurance limit, rather than, if δ were zero, the peak stress in the hole surface reaching the endurance limit. This concept can be developed to accommodate the effect of subsurface defects and their depth; it can be shown that a subsurface crack of length 2b is equivalent, in K_f terms, to a surface crack of length b. A relationship of this type, as shown in Fig. 3.188, can be used to estimate the required sensitivity of a method of nondestructive testing such as radiography. When the radiographer uses a penetrameter, having a thickness of, say, 2% of that of the section he is inspecting, he must realize that defects smaller than 2% will be undetected. If sensitivity is expressed in percent of the section thickness, Fig. 3.188 shows the strength reduction factor that may be produced by hidden defects in various thicknesses of the material. The quality of a vessel design can be characterized by a factor K_f, which is the ratio between the highest peak stress at a discontinuity, hole, fillet, and so forth, to the average stress intensity in the shell. If the desired vessel quality corresponds to $K_f = 3$, then it is seen from Fig. 3.188 that 2% sensitivity is adequate up to almost 3-in. thickness, and 1% sensitivity is adequate up to 5-in. thickness because the hidden defects will then not be the controlling factor. If a lower quality vessel is being built, say, $K_f = 5$, then 2% sensitivity is adequate up to 9-in. thickness.

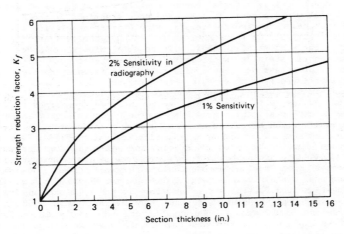

Fig. 3.188 Fatigue strength reduction factors for protection from hidden defects.

In the field of low cycle fatigue, it has been common practice to use lower stress concentration factors for small numbers of cycles than for large numbers of cycles. This is reasonable when the allowable stresses are based on stress fatigue data but is not advisable when strain fatigue data are used. Figure 3.189 shows a typical relationship among S vs. N curves from (A) strain cycling tests on unnotched specimens, (B) stress cycling tests on unnotched specimens, and (C) stress cycling tests on notched specimens. The ratio between the ordinates of curves B and C decreases with decreasing cycles to failure; this is the basis for the commonly accepted practice of using lower values of K_f for lower values of N. In C, however, although nominal stress is the controlled parameter, the material in the root of the notch is really being strain-cycled, because the surrounding material is at a lower stress and behaves elastically. Therefore, it should be expected that the ratio between curves A and C should be independent of N and equal to K_f. The experimental verification of this relationship is not adequate, but some data does exist. The use of a K_f independent of N, together with strain cycling fatigue data appears to be both logical and conservative, even though it is contrary to the most commonly accepted practices.

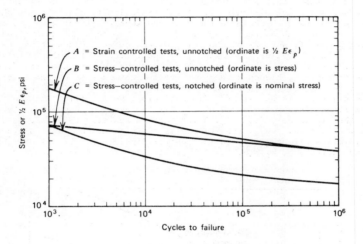

Fig. 3.189 Typical relationships among stress, strain, and cycles to failure.

In applying stress concentration factors to the case of fluctuating stress, that is, a cycle that has both a mean and an alternating component, it has been the common practice to apply K_f to the alternating component only. A careful study of what happens to fluctuating stresses when they enter the plastic range results in a better method of using the stress concentration factor. Take, for example, the case of a material with 40,000 psi yield strength and 30×10^6 psi modulus made into a notched bar that has a $K_f = 3$. The bar is cycled between nominal tensile stress values of 0 and 20,000 psi. Common practice would call the mean stress 10,000 psi and the alternating component $(1/2) \times 3 \times 20,000 = 30,000$ psi. But, from strain fatigue data, one may also calculate the value of $E\varepsilon/2$ as 30,000 psi. The basic value of mean stress is also 30,000 psi, but yielding causes an adjustment down to 10,000 psi. Therefore, the yielding during the first cycle is seen to be the justification for the common practice of ignoring K_f when calculating the mean stress component. It so happens that, for the case chosen, the common practice gives exactly the same result as the proposed method described here. The common practice, however, would have given the same result regardless of the yield strength of the material, whereas the proposed method gives different mean stresses for different yield strengths. For example, if the yield strength had been 50,000 psi, the adjusted value of mean stress would have been 20,000 psi, and the common practice would have given an unconservative result.

Determination of Need for Fatigue Evaluation.* Since most pressure vessels are subjected to limited numbers of pressure and temperature cycles during their lifetime, considerable design effort could be saved by defining the conditions that do or do not require that a fatigue evaluation be made. In order to define these conditions, several factors must be considered. For pressure cycling only, we need to know:

1.   The expected range of pressure, which may be conveniently stated as a fraction of the design pressure; we define this fraction as F.

2.   The expected number of pressure cycles in the required lifetime of the vessel.

3.   The average stress intensity in the vessel wall produced by the design pressure. This will normally be about the same as the allowable stress value for the design temperature tabulated in the ASME Code for the material being used. For the purposes of the present discussion, we call this stress value S_m.

4.   The applicable design fatigue curve for the material, Figs. I-9.1 through 9.4 of Appendix I in Section III, NA.

5.   The ratio of the highest calculated (or estimated) local stress intensity in the vessel to S_m; ratio defined as K.

If thermal cycling as well as pressure cycling is involved, we also need:

6.   The maximum expected temperature difference between two adjacent points. (The term "adjacent points" means that the points must be spaced not more than about two times the thickness of the material under consideration. For example, the difference between the temperature of a nozzle and that of the shell to which it is attached is significant. The temperature range across the thickness of a shell plate or flange is also significant. On the other hand, the temperature difference between the top and the bottom of a long vessel is not significant because there is sufficient flexibility between these two points to accommodate the differential expansion.)

7.   The expected number of thermal cycles during the required lifetime of the vessel. A thermal cycle is the establishment and removal of the temperature difference described in (6) above.

First, consider the case of pressure cycling. The amplitude (half range) of the highest local stress will be

$$S_{alt} = 1/2 K S_M F$$

---

*The ASME Code now provides certain guidelines: Section III Div. 1, NB-3222, and Section VIII Div. 2 Alt. Rules, AD-160 and Appendix 5.

Since S_{alt} must be held within the limits of S_a on the fatigue curve, Fig. 3.187, it is possible to construct families of curves showing the pressure range vs. the allowable cycles for various values of K and S_m. For example, if the design details and inspection methods are chosen with sufficient care to keep K down to a value of 3, Fig. 3.190 shows the relationship between the pressure range and the cycles for various values of S_m. A similar family is shown in Fig. 3.191 for various K values at S_m = 20,000 psi. The latter is of particular value to the designer because it tells him how low K must be kept for the required life. This factor measures inversely the quality of the inspection, and of the design details such as fillet radii, discontinuities, and nozzle attachments. If a high K value is permissible, the cost of the details and inspection can be correspondingly lower. Any combination of pressure range and expected cycles that falls below the limits shown in Fig. 3.190 or 3.191 does not require further fatigue evaluation. A combination that falls above the applicable curve is not necessarily unsatisfactory but requires more detailed study.

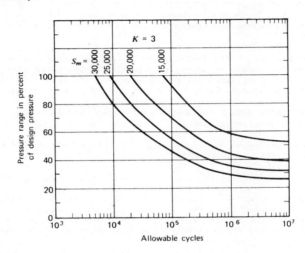

Fig. 3.190   Allowble pressure transients for K = 3 and various values of S_m.

A similar generalization can be derived for the case of fluctuating thermal stress. Thermal stresses are characterized by the expression:

$$S_{th} = HE\alpha T \tag{140}$$

where   E = modulus of elasticity, psi
   α = coefficient of thermal expansion, in./in./degrees F
   T = max temperature difference between adjacent points, degrees F
   ΔT = change in T during a cycle
   H = a factor depending on geometry
   S_{th} = thermal stress intensity

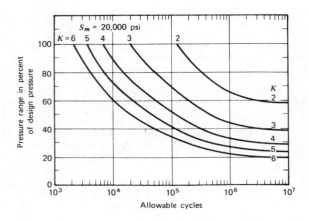

Fig. 3.191 Allowable pressure transients for S_m = 20,000 psi
and various values of K.

The factor H has a magnitude of the order of unity but usually lies between 0.5 and
2; assume, conservatively, that H = 2. The thermal stress intensity amplitude is
then

$$S_{alt} = 1/2(K)2E\alpha\Delta T = KE\alpha\Delta T \qquad (141)$$

or

$$\Delta T = \frac{S_{alt}}{KE\alpha} \qquad (142)$$

Here, again, S_{alt} must be kept within the S_a limits of the fatigue curve, Fig.
3.187. For the case of a carbon steel vessel with good design details, $E\alpha$ = 180
psi/°F and K = 3. Therefore

$$\Delta T = \frac{S_a}{540} \qquad (143)$$

Figure 3.192 shows the allowable thermal transients for carbon steel vessels with
K_f^t = 3 and K_f^t = 5. If another material had the same fatigue curve and the same
modulus but a higher coefficient of thermal expansion, the curves would be lower.
Any case that falls below the applicable curve does not require fatigue analysis.
In using Fig. 3.192, it should be noted that the ordinate is ΔT, the change in T
during the given cycle, which is not always the same as T itself. For the startup-
shutdown cycle, ΔT = T. During operation, however, there may be many cycles that
do not bring the temperature differences to zero, so that the most damaging transient might be one for which ΔT < T but the number of cycles is large.

Even simpler sets of rules can be derived if the type of service allows additional
assumptions to be made. Suppose, for example, that carbon steel vessels are being
designed to a maximum K value of 3 and a maximum nominal stress (S_m) less than
30,000 psi. The service is such that startup and shutdown will not occur more than
5000 times, and significant thermal transients during operation will not occur more
than 100,000 times during the lifetime of the vessel. Since pressure can change
rapidly, no assumption is made regarding the maximum number of pressure cycles.

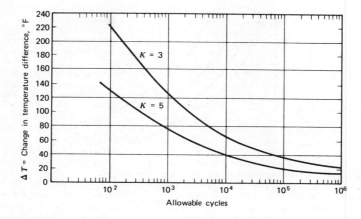

Fig. 3.192 Allowable thermal transients for carbon steel vessels.

From Fig. 3.190, it is seen that an infinite number of pressure cycles up to 25% of design pressure can be imposed without danger of fatigue damage. The 5000 expected startup cycles at 100% of design pressure also give a point below the applicable curve. Referring now to Fig. 3.192, we see that the 5000 startup cycles should not be allowed to involve a temperature difference of more than 80°F and the 100,000 thermal transients should not involve a change in temperature difference of more than 40°F. The rule for these operating conditions can therefore be stated as follows.

A fatigue evaluation is required if, during normal operation,

1.   The pressure fluctuates through a range exceeding 25 percent of the pressure.

2.   The maximum temperature difference between any two adjacent points of the vessel exceeds 80°F.

3.   The temperature difference between any two adjacent points of the vessel changes by more than 40°F.

It may be noted that in the foregoing discussion no consideration has been given to the cumulative effect of various cycles, nor to the additive effect of pressure and thermal cycles. Therefore, the curves and rules given above imply that,

1.   The pressure cycles and the thermal cycles do not produce their highest stresses at the same point in the vessel and therefore need not be added.

2.   All significant pressure cycles can be conservatively represented by a single value of pressure range F applied for a given number of cycles.

3.   All significant thermal cycles can be conservatively represented by a single value of ΔT applied for a given number of cycles.

If these conditions are not all met, it is still possible to avoid the complete fatigue analysis by means of a simplified cumulative damage evaluation. The most general case is that of a vessel that will be subjected to pressure transients F_1, F_2, F_3,..., F_n for cycles numbering n_{p_1}, n_{p_2}, n_{p_3},..., n_{pn}, respectively, and thermal transients ΔT_1, ΔT_2, ΔT_3,..., ΔT_n for cycles numbering n_{t_1}, n_{t_2}, n_{t_3},..., n_{tn}, respectively. The pressure transients will always produce stresses that are additive among themselves, but the thermal stresses may be additive either to the pressure stresses or among themselves. The allowable cycles for each value of F and ΔT can be read directly from the applicable curves of Figs. 3.190, 3.191, or 3.192 to give N_{p_1}, N_{p_2}, N_{p_3},..., N_{pn}, and N_{t_1}, N_{t_2}, N_{t_3},..., N_{tn}. The usage factors n_{p_1}/N_{p_1}, n_{p_2}/N_{p_2}, and n_{t_1}/N_{t_1}, n_{t_2}/N_{t_2} can then all be calculated. Each group of usage factors that produces additive stresses must be summed up and, if the sum of any group exceeds unity, a fatigue study is needed.

The procedure can be illustrated by an example: An autoclave with a flanged closure is to be used in a process requiring 2000 psi and 600°F. The operation is continuous during the week with shutdown over the weekends. Ten batches of material per week are produced. The reloading operation between batches involves the reduction of pressure to 1000 psi and the reduction of temperature to 300°F. Thermocouple readings taken on similar vessels have shown that, during normal steady stage operation, the flange temperature runs lower than the body temperature by as much as 100°F, but during reloading the temperature difference is reduced to 50°F. The material used in the vessel is ferritic steel; the design stress is 20,000 psi. The design details have been chosen to keep the ratio of peak stress to nominal stress down to a value not greater than 3. The required life of the vessel is 20 years at 50 weeks of operation per year.

This operation violates the general requirements stated above in that (1) the pressure fluctuates during normal operation through more than 25% of the design pressure, (2) the maximum temperature difference exceeds 80°F, (3) the change in temperature difference during normal operation exceeds 40°F, and (4) the pressure and thermal stresses are probably additive at the flange to shell junction. Therefore, a more detailed study is needed.

The cycles to be considered are:

1.   100% pressure fluctuation from startup and shutdown. For this cycle, $F_1 = 1$, $n_{p_1} = 20 \times 50 = 1000$, N_{p_1} (from Fig. 3.190) = 20,000, $n_{p_1}/N_{p_1} = 0.050$.

2.   50% pressure fluctuation from reloading. For this cycle $F_2 = 0.5$, $n_{p_2} = 20 \times 50 \times 10 = 10,000$, $N_{p_2} = 450,000$, $n_{p_2}/N_{p_2} = 0.022$.

3.   100°F temperature difference during startup. For this cycle $n_{t_1} = 1000$, N_{t_1} (from Fig. 3.192) = 2500, $n_{t_1}/N_{t_1} = 0.400$.

4.   50°F change in temperature difference during reloading. For this cycle, $n_{t_2} = 10,000$, $N_{t_2} = 35,000$, $n_{t_2}/N_{t_2} = 0.286$.

The sum of all the usage factors is 0.758, which is safely below unity. Therefore, no detailed fatigue evaluation is required. The example given is a borderline case, however, and it would not have taken much increase in the severity of the thermal transients to make a fatigue evaluation necessary. For example, if the startup were too rapid and produced a transient temperature difference of 150°F, N_{t_1} would have been 550 cycles and n_t/N_{t_1} would have been 1.82. In that case, it would have been necessary to calculate the thermal stress at the flange to shell junction to see if there really was any danger of fatigue failure. It may also be observed that poor design details, such as sharp fillets, cannot be tolerated for service of this type. Figure 3.192 shows that, if K = 5, a ΔT of 100°F can be allowed for a total of only 350 cycles without detailed analysis.

Pressure piping has been investigated, especially with regard to welded bends and mitre joints. From the results of studies, and cyclic stress testing in the plane of bending and transverse to it [220], the mitre joints appear at best to behave somewhat like bends of very sharp curvature; but usually the local stresses introduced by the sudden directional change overshadow this effect and lead to even earlier failure.

It has been concluded that, for an identical number of cycles to failure, a long-radius welding elbow can safely be stressed close to 20% higher in bending in the plane, or 70% higher in bending transverse to the plane, as compared with a double-mitre bend of the given proportions. The comparison is even more forceful when expressed in terms of fatigue life under the same bending moment. The welding elbow is found to sustain about 2.4 times as many cycles of full stress reversal in-plane bending or 13.5 times as many cycles transverse bending, as the mitre bend. Figure 3.193 shows the test specimens.

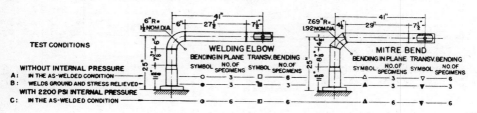

Fig. 3.193 Pipe welding test specimens, conditions, and symbols.
(Figs. 3.193-3.195 from A. R. C. Markl [220].).

A collection of many test results from in-plane and transverse bending is given as the modified Goodman diagram of Fig. 3.194. The basis for the plot is the Moore-Kommers failure criterion [221] in the form:

$$S_e = \frac{S_a + S_m}{3}$$

(144)

where S_m is the constant longitudinal stress due to internal pressure. It is to be noted that all test points for both elbows and mitres fall with ±25% of the median line applicable to straight pipe.

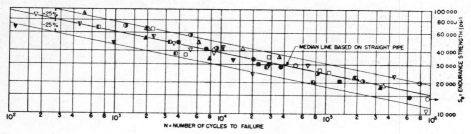

Fig. 3.194 The correlation of test results by Moore-Kommers rule.

A correlation of these results with the Hencky-von Mises combined stress theory is given in Fig. 3.195. An interpretation of the ASA Code for Pressure Piping (ASA B31.1) permits the use of a torsional shear stress $f_t = 4/3S_t$; a longitudinal (bending + pressure) stress $f_1 = 4/3\beta S_b + S_c$, and a circumferential (pressure) stress $f_c = 2S_c$ in the equation

$$S = \sqrt{3f_t^2 + f_1^2 + f_c^2 - f_t f_c} \qquad (145)$$

to yield

$$S = 4/3 \sqrt{3S_t^2 + (\beta S_b)^2 + 1.6875 S_c^2} \qquad (146)$$

and the plot is made on this basis. The correlation is again within ±25% of the median line and accordingly is quite satisfactory, considering that the specimens consisted of commercial pipe and fittings of widely varying physical properties.

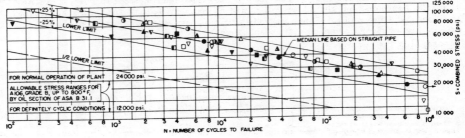

Fig. 3.195   The correlation of test results by the Code for Pressure Piping.

As a last step, these calculated "combined stresses" will be compared with the stress limitations specified in the Code for Pressure Piping (paragraphs 620(g) and (h)). For oil piping within refinery limits, the maximum allowable stress range for carbon steel to ASTM A106, Grade B (which stress would be applicable for operating temperatures up to about 800°F) is 24,000 psi for only mild cyclic conditions and one half that value for definitely cyclic or vibratory conditions. These limiting stress ranges appear in Fig. 3.195 as horizontal lines; the upper one intersects the line defining the lower limit of the test data at an abscissa representing a life equal to about 100,000 cycles, while the lower line may be assumed to be near the endurance limit (normally taken as the endurance strength obtained for at least 2,000,000 and preferably 10,000,000 cycles). Considering the many imponderables in piping design (assumptions of end fixation, effect of intermediate supports, weight and wind stresses, corrosive influences, etc.), and the fact that damage already occurs below the endurance strength, a safety factor of about 2 would appear indicated; even this probably is too low for definitely corrosive conditions. On this basis, the normal allowable stress range specified in the code should apparently not be applied where more than 3000 cycles of major stress change are anticipated.

Case Study PV-1, below, is recommended for its relevant application of Code methodology to the design of welded joints in steel pipe.

CASE STUDY PV-1

Fracture Mechanics Analysis of Steel Pipe Welds (from Ref. [449])

Note: This case study was chosen because it so well illustrates the varieties of application of Fatigue and Fracture Mechanics and the limitations in the evaluation of their several results. Parts A and B satisfy the ASME Boiler Code requirements, but A is strictly an S-N "damage" approach with no propagation of cracks, while B is also a "static" treatment showing only that the type of failure is ductile, $a_{cr} > t$. Part C is a dynamic Fracture Mechanics Analysis, accommodating the presence of cracks prior to service and providing the numerical growth of those cracks during service, all with conservative assumptions (in this case). It is especially to be noted that "even though the NDE indications were small, analysis assumed $a_i = 0.1$, etc.", an illustration of the practice of not taking the NDE numbers too literally, (unless the job is big enough to support specific tests). Note also that the toughness of the alloy, K_{Ic}, did not appear as a controlling parameter because the alloys were highly ductile and tough resulting in $a_{cr} > t$.

Description

> Design Life = 10^4 cycles
> Design Temperature = 70° - 375° F
> Design Stress Range = Zero to max., (values below)
> Materials: A515 Gr65 Carbon Steel
>   Type 410 stainless
>   Electrodes: AWS E7018 & E702-1, E309

There were five types of shop welds used in the cross-around piping carrying steam from the high pressure turbine to the moisture separation reheaters (MSR) and from the MSR to the low pressure turbines. They were: standard girth weld in straight sections; the girth weld attachment to a valve or ring; the true mitre joint weld; the longitudinal seam weld; and the attachment weld between pipe and turning vane assembly elliptical girders. Typical configurations of these pipes and welds are shown in Fig. 3.196. The longitudinal seam welds were made by an automatic process and the other four types by manual shielded arc. The girth welds connecting carbon steel pipes were AWS E7018 and E7021. Stainless steel E309 welds were used for joining the 410 stainless steel turning vane to carbon steel pipes equivalent to ASTM A515 Gr 65. The mechanical properties of these welds are given below:

TABLE 3.45   Mechanical Properties of the Pipe Welds at Room Temperature

	Stainless Steel Welds	Carbon Steel Welds
Yield Strength, Min. (ksi)	30	60
Ultimate Strength, Min. (ksi)	75	72

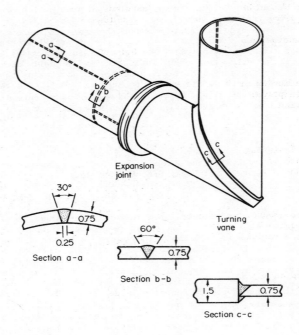

Fig. 3.196 Typical Crossaround Pipe Assembly.

The weld joints were inspected visually; by magnetic particle and by radiography, indicating flaws in both the stainless and carbon steels. The types of flaws were categorized into five groups as shown below:

TABLE 3.46

Type of Defects in Welds Detected by X-Ray		Stainless Steel Welds	Carbon Steel Welds
1.	Spherical inclusion or void, $r =$	0.1 in.	0.1 in.
	$K_f =$	2	2
2.	Circular inclusion or void of elliptical cross section $r =$	0.01 in.	0.03 in.
	$t =$	0.1 in.	0.01 in.
	$K_f =$	3.3	2.7
3.	Long elliptical slag line or lack of fusion line $r =$	0.01 in.	0.03 in.
	$c =$	0.1 in.	0.1 in.
	$K_f =$	4.8 in.	3.9 in.
4.	Long surface groove, weld undercut and lack of penetration $r =$	0.03 in.	0.03 in.
	$K_f =$	2.9	2.7
5.	Weld buildup or protrusion $r =$	0.03 in.	0.03 in.
	$h =$	0.75 in.	0.75 in.
	$K_f =$	1.7	1.6

where r = minimum radius of defects in welds
t = circle of radius of elliptical inclusion
c = 1/2 of major diameter of ellipse
h = weld buildup width

Since no standard ANSI or API standard could be strictly applied to this case, the safety as well as the need for any repair work of these defect-containing welds was evaluated using the ASME Pressure Vessel Code fatigue design curve, and Fracture Mechanics Analysis. The stress analysis included a combination of thermal expansion, dead weight, and local stress effects due to the presence of defects.

Procedure

A. Cyclic Life Evaluation according to ASME Pressure Vessel Code

Assuming no crack propagation, the cyclic lives of various defect-containing welds were obtained from the Code Section VIII, Division 2 Fatigue Design Curve (Fig. 3.197), and Table 3.46.

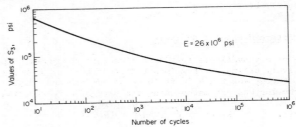

Fig. 3.197 Design Fatigue Curve for Series 3XX, Ni-Cr and Ni-Cu alloys for T < 800°F.

The effective fatigue notch factor for cyclic loading, K_f, is:

$$K_f = q(K_t - 1) + 1 \tag{1}$$

where   q = notch sensitivity factor
          K_t = theoretical elastic stress concentration factor, (from the minimum radii of the notch).

The cyclic lives of the stainless steel welds containing five types of flaws were:

a) Spherical inclusion or void:

Since $\sigma_1 = 12.2$ ksi in the axial direction, and $K_f = 2$, N_1 is found from Fig. 3.197 as 1×10^6 cycles.

$\sigma_\theta = 17.9$ ksi in the tangential direction, and $K_f = 2$, N_θ is also found from Fig. 3.197 as 1×10^5 cycles.

Similarly, the lives for the other cases were calculated as below:

b) Circular inclusion, or void of elliptical cross-section:

$N_1 = 6 \times 10^4$ cycles: $N_\theta = 1 \times 10^4$ cycles

c) Long elliptical slag line, or lack-of-fusion line:

$N_1 = 1.1 \times 10^4$ cycles

d) Long surface groove, weld undercut, and lack of penetration:

$N_1 = 1.1 \times 10^5$ cycles

e)   Weld buildup, or buildup plus undercut:

   $N_1 > 10^6$ cycles

Hence, according to Code, the E309 stainless steel welds are safe for the design life (10^4 cycles) even in the presence of these types of welding defects. Similarly, the cyclic lives for carbon steel welds are much longer than the design life and can safely be used.

B.   Static Fracture Mechanics Analysis

In the absence of any available test data, fracture toughness values of the welds were conservatively estimated from impact test data. The room temperature Charpy V-notch impact energy was CVN = 90 ft-lb obtained from the open literature. Fracture toughness value was estimated using the Barsom-Rolfe relation:

Upper shelf, $K_{IC} = \sigma_y \sqrt{5 \ (CVN/\sigma_y - 0.05)}$ (2)

   $\cong 115$ ksi-in.$^{1/2}$

Transition region, $K_{IC} = \sqrt{2E(CVN)^{3/2}} \cong 206$ ksi-in.$^{1/2}$ (3)

With $E = 25 \times 10^6$ psi at the maximum operating temperature of 375° and $\sigma_y = 30$ ksi, the toughness was found conservatively as, $K_{Ic} = 115$ ksi $\sqrt{in}$. Although this approach is not strictly applicable to such ductile austenitic stainless steel, it is shown later that brittle fracture is not of concern, and therefore, great accuracy in estimating K_{IC} is not essential.

The critical flaw size for brittle fracture was calculated from:

$$(a_{cr})_B = (1/M)(K_{IC}/\sigma_{max})^2$$ (4)

for the assumed surface and internal cracks, where M = component geometry and flaw shape parameter (as given with Case 1 under Stress-Intensity Factors in Section 3.2.4).

   $M = 1.21 \ \pi/Q$, (surface crack)
   $M = \pi/Q$, (internal crack)
   $Q = 1$

Using the ASME Code procedure of multiplying the stress by a factor of 2 to obtain the alternating stress resulted in a conservative value of 35.8 ksi. This was the maximum applied stress employed in the calculation.

According to equation (4), it was found that $(a_{cr})_B = 2.71$ in. for surface crack and $(a_{cr})_B = 3.28$ in. for internal crack. The brittle fracture critical crack sizes in all cases were greater than the weld thickness (0.75 in.). Therefore, brittle fracture was impossible in the temperature range considered. The critical flaw size for ductile rupture, $(a_{cr})_D$, was obtained as follows:

   Assume the crack propagates to $(a_{cr})_D$

   Assume the remaining weld thickness is t,

   Assume the original weld thickness = 0.75 in. = $\left[(a_{cr})_D + t\right]$

Then t is obtained by requiring that the stress on weld does not exceed the tensile strength. Using a minimum tensile strength of 75 ksi gives:

$$t = 0.358 \text{ in.}$$

and critical flaw size for ductile rupture was $(a_{cr})_D = 0.392$ in.

C. Cyclic Fracture Mechanics Analysis

Assume that, in the worst case, all the defects presented were surface and internal cracks. The amount of crack growth in 10^4 cycles was calculated as follows:

Using an empirical fatigue crack growth rate equation

$$da/dN = C_o (\Delta K)^n \tag{5}$$

and a standard relation for stress and stress-intensity:

$$\Delta K = \Delta \sigma (Ma)^{1/2} \tag{6}$$

the number of cycles for an initial crack to grow into a critical size was obtained by integration of Eq. 5.

Since a/2c ratio is approximately the same during crack growth, it is reasonable to to assume a constant M in the integration, thus

$$N = \frac{2}{(n-2)C_o M^{n/2}(\Delta \sigma)^n} \left[a_i^{(2-n)/2} - a_{cr}^{(2-n)/2} \right], \quad n \neq 2$$

and

$$a_f = \left[a_i^{(2-n)/2} - N(n-2)C_o M^{n/2}(\Delta \sigma)^{n/2} \right]^{2/(2-n)}, \quad n \neq 2$$

where N = number of cycles for an initial crack to grow into a critical size.

    n = slope of log da/dN vs. log ΔK curve

    C_o = intercept constant

    $\Delta \sigma$ = applied cyclic stress range

    a_i = initial crack size

    a_f = final crack size

(the expression for N (or N_f) will be recognized as a variation of Eq. 6 in Section 3.2.4.)

Even though the actual indications were small, the analysis was made assuming that the welds contain surface cracks with $a_i = 0.1$ in., $a_i/2c_i = 0.1$; and internal cracks with $2a_i = 0.2$ in., $a_i/2c_i = 0.1$. Conservative upperbound fatigue crack growth rate data for austenitic stainless steel gives $C_o = 3 \times 10^{-10}$ and n = 3.25.

The number of cycles N_θ, N_1) for a given crack to grow into a ductile fracture critical size $(a_{cr})_D$ in tangential and axial directions, and the final crack size: $[(a_f)_\theta, (a_f)_i]$ after 10^4 cycles of loading were calculated as follows:

(a) For a surface crack with a_i = 0.100 in.

    i)   surface crack

        N_θ = 1.3 x 10^5 cycles N_1 = 4.7 x 10^5 cycles

        $(a_f)_\theta$ = 0.107 in. $(a_f)_1$ = 0.102 in.

    ii)   internal crack

        N_θ = 1.07 x 10^5 cycles N_1 = 3.9 x 10^5 cycles

        $(a_f)_\theta$ = 0.105 in. $(a_f)_1$ = 0.101 in.

(b) For a surface crack with a_i = 0.25 in.

        N_θ = 3.1 x 10^4 cycles N_1 = 1.14 x 10^5 cycles

        $(a_f)_\theta$ = 0.285 in. $(a_f)_1$ = 0.259 in.

Thus, if the initial crack is a_i = 0.1 in., it will take more than the design number of cycles to propagate it to critical size; and less than 0.007 in. of maximum crack growth will occur in 10^4 cycles in axial and tangential directions. Even if the initial surface crack is 0.25 in., it will take longer than design life to grow to critical crack size; and less than 0.035 in. of crack growth will occur in both axial and tangential directions under 10^4 cycles of loading. Carbon steel welds similarly showed adequate safety in the calculation.

3.4.4 Effects of Radiation

For most of the constructional alloys the general effect of nuclear radiation is
an increase in strength and a decrease in ductility. The degree of change is usually insignificant, but for certain alloys, temperatures and fluences (flux intensities), it may be sufficient to warrant consideration. The latter is the case
for the aluminums at very high fluences, at least for prior irradiation, as in
Fig. 3.198. For beryllium the data does not correlate completely, but at high
temperatures and fluences, noticeable damage has occurred from the production of
helium and hydrogen bubbles with a resultant expansion, or swelling, of the metal.

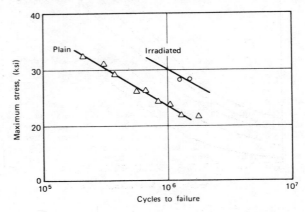

Fig. 3.198   Effect of prior irradiation on unnotched 7075-T6 aluminum
alloy (rotating beam): nvt ~ $2 \cdot 10^{18}$ fast neutrons/cm^2.
From [341], Fig. B-13.

For nuclear components, the effect is significant and must be controlled in both
design and material choice. In the "nuclear" steels, the general effect is a rise
in the brittle-ductile transition temperature, and a fall in the "upper-shelf"
Charpy energy, as well as a rise in yield strength. For a typical, 4-loop reactor
pressure vessel the maximum neutron dose received in the belt-line region (the
most intensively radiated part) is estimated to be $\sim 2 \times 10^{19}$n(>1 MeV)/cm^2 in a 32
year full-power lifetime. This is the dose at the inner surface and it falls off
with increasing depth through the vessel wall. A clear correlation between
irradiation-induced rise of transition temperature and copper content of reactor
steels has been shown to exist, and as a result, the ASME Code Section XI recommends
the use of curves constructed from Charpy V-notch data on SA533B plates, welds and
HAZ's to predict the dependence of RT_{NDT} upon fast neutron dose and Cu content,
Fig. 3.199.

Less than 10% of the data points from tests on SA533B plate and HAZ's exceeded the
shifts predicted from these curves, which therefore appear to be reasonably safe
and accurate guides, especially at the lower copper contents.

A general view of the fatigue behavior of Type 304 SS and of Incoloy 800 is given
in Figs. 3.200 and 3.201. It should be noted that irradiation reduces the allowable fatigue stress at any life, probably because of an increase in the effective
K_t or K at the crack tips.

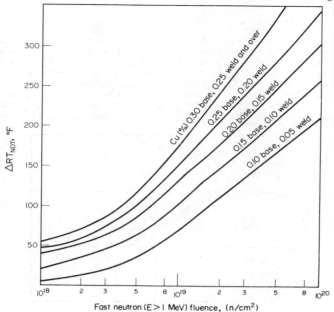

Fig. 3.199   Effect of fast neutron fluence and copper content on shift of RT_{NDT} for reactor vessel irradiation at 550°F.

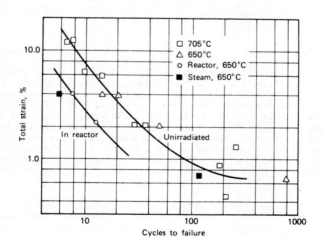

Fig. 3.200   Strain fatigue life of type 304 stainless steel tubing in reactor and out of reactor.   In-reactor: nvt ~ $7 \times 10^{19} n/cm^2$.

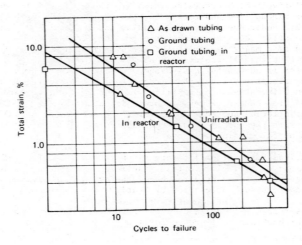

Fig. 3.201 Strain fatigue life of Incoloy 800 tubing at 700°C.
In reactor: nvt ~ 7 x 10^{19}n/cm^2.

3.5   Machine Parts

3.5.1 Bearings

Modern antifriction bearings of any precision class are the results of many extensive efforts in design, metallurgy and manufacturing engineering. The tremendous load, life and speed capacities available provide the designer with a wide latitude in choosing configurations, and it is not often that a design needs to be modified by bearing restrictions. To provide these capacities, the bearings have become assemblies of relatively high precision which are therefore readily susceptible to damage. After mishandling and faulty maintenance, the most probable causes of bearing failure are under-speeding, over-loading and inadequate design of the mounting. The relative velocity between the rolling element and race has become widely recognized as a major factor in the establishment and retention of the lubricant film and thus the allowable speed has joined the allowable load as an essential parameter in successful design and operation. The manufacture of precision bearings, of almost any level above those for roller skates, is sufficiently exacting in talent, investment in facilities and the need for research and development activity that the number of manufacturers is rather small. These companies organized the Antifriction Bearing Manufacturer's Association (AFBMA) to present the bearing user with reasonably consistent and conservative data on performance, and with methods of calculating load and life.

Standard calculations for fatigue life are given in the literature of all major manufacturers; [186] is typical. Certain practices, parameters and levels of conservatism have become accepted as a result of the evaluation of much accumulated data. These parameters for calculation of bearing performance are:

1.   The basic load rating, C, defined as the load that 90% of a group of apparently similar bearings will endure for 1,000,000 revolutions of the inner race.

2.   The median life of a group of bearing, L_{50}, defined as the life resulting in a 50% survival rate.

3.   The rating or catalog life, L_{10}, defined as the life resulting in a 90% survival rate.

The results of standard calculations for life and load involving various catalog factors are characterized by a high degree of conservatism; the validity of these results is supported by a tremendous amount of data and experience. However, applications occasionally arise for which a failure rate lower than the standard 10% must be guaranteed. Several procedures are available for handling this "nonstandard" requirement, depending on the form of the problem statement.

The usual assumption is that bearing failures are fatigue failures. Since the latter are random in nature, they will tend to follow statistical predictions; but, as bearing failures do not occur at a constant rate, the basic Poisson distribution is set aside in favor of the Weibull, expressed as

$$P_S = \exp\left(-\frac{1}{\theta}\right)^b \qquad (147)$$

where P_S = probability of bearing survival without failure for a given time

  t = time

  θ = multiplier for design life, in hours; a constant

  b = the Weibull function exponent. If b is set equal to 1, the Poisson equation results.

The Weibull curve is established on the coordinates of failure percent (ordinate) and the ratio of operating time to the B_{10} design life (abscissa). Two ordinate values are available from catalog data on life, $P_s = 0.90$ and $P_s = 0.50$; two corresponding abscissa values are known for each ordinate. Many manufacturers and the USASI Standard B3.11 give the ratio of median to design life as 5/1, but data from [187] indicate the acceptable but slightly less conservative value of 4.08/1. The two curves are drawn as in Fig. 3.202. Then the probability of survival of the bearing, for which the median life of 5 (or 4.08) times the B_{10} design life, d, has already been determined, is given as

$$\frac{t}{d} = (9.49 \ \log_e P_s)^{0.746} \quad \text{for } \frac{t}{d} = 4.08B_{10} \tag{148}$$

and

$$\frac{t}{d} = (9.49 \ \log_e P_s)^{0.856} \quad \text{for } \frac{t}{d} = 5.00B_{10} \tag{149}$$

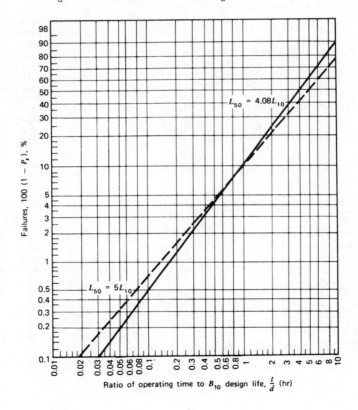

Fig. 3.202 Weibull plot of operating life of bearings. From E. Schube [185], Fig. 1.

The required design life B_{10} in hours is, then,

$$d = \frac{t}{(t/d)}$$

Normally, the reliability P_s and the required time t are known; the curves and Table 3.47 are set up for convenience in determining the design life.

TABLE 3.47   Ratio of Operating Time to B_{10} Design Life for Various Probabilities of Survival

Probability of Survival for Time t P_s	Ratio of Operating Time to Design Life	
	t/d for median life = 4.08d	t/d for median life = 5d
0.995	0.1030	0.0740
0.99	0.1785	0.1395
0.98	0.2910	0.2440
0.97	0.396	0.3460
0.96	0.492	0.4445
0.95	0.584	0.5405
0.94	0.672	0.6341
0.93	0.756	0.7261
0.92	0.840	0.8191
0.91	0.921	0.9101
0.90	1.000	1.000
0.85	1.383	1.450
0.80	1.750	1.900
0.75	2.120	2.364
0.70	2.435	2.835
0.65	2.860	3.331
0.60	3.250	3.850
0.55	3.600	4.340
0.50	4.080	5.000
0.45	4.540	5.650
0.40	5.03	6.345
0.35	5.56	7.150
0.30	6.16	8.040
0.25	6.84	9.040
0.20	7.65	10.06
0.15	8.64	11.80
0.10	10.00	14.00
0.05	12.40	17.55
0.01	16.40	25.15

SOURCE: E. Schube [188], Table 1.

Another common requirement is the relation of the reliability and equivalent radial load at required life to the catalog tabulation of basic load ratings. The Weibull equation is rewritten to the form

$$\frac{R_e}{C} = \left(\frac{\theta}{L_1}\right)^{1/a} \left(\ln \frac{1}{P_s}\right)^{1/ab} \tag{150}$$

where R_e = equivalent radial load, lb.

   C = basic load rating, lb.

   θ = an exponential parameter, as in Eq. 147

   L = life in millions of revolutions

   P_s = reliability, or probability of survival, as in Eq. 147

   a,b = exponential constants

For a ratio of median to rating life of 5, and P_s = 0.50 and 0.10, θ has been evaluated as 6.84 and b as 1.17. For ball bearings, a = 3.00; for roller bearings, a = 3.33. Then Eq. 150 becomes

$$\frac{R_e}{C} = \frac{1.898}{L^{0.333}} \left(\ln \frac{1}{P_s}\right)^{0.285} \tag{151}$$

and

$$\frac{R_e}{C} = \frac{1.780}{L^{0.3}} \left(\ln \frac{1}{P_s}\right)^{0.257} \tag{152}$$

for ball and roller bearings respectively. Equations 151 and 152 are plotted as Fig. 3.203, and the similarity to Fig. 3.202 is immediately noted. In the latter, the reliability is related to ratio of lives, while in the former, required life is related to the ratio of loads for a range of reliabilities. Figures 3.203a and 3.203b are particularly useful because they relate all the normally required parameters and the curves have been adjusted by the data of [190] in the region $P_s > 0.9$, where the Weibull distribution becomes somewhat inaccurate. In general, if reliability is to be increased, life expectancy must be reduced, according to the classical relationship of Fig. 3.204.

Effect of Preload on Bearing Life*. The reasons most often given for preloading bearings are that this action increases the bearing stiffness and decreases lost motion or play. Preloading, either axial or radial, also affects life because the circumferential loading are arc of the bearing is a function of the total load.

The longer the arc, the greater the number of rolling elements it will contain, over which the total load is distributed, thus reducing the unit load on each element. The proper preload for maximum life is that load which with the applied load will extend the arc just to 180°. An additional preload will simply increase the stress in each element and thereby shorten the life.

---

*Much of the material in this section is based on the work of Dr. Tedric Harris, Supervisor of Bearing Technology, SKF Industries, Inc., King of Prussia, PA.

The calculation of preload and life has been extensively considered in [191]. The length of the circumferential arc or "loading zone" is obtained in the form of its projection on the pitch diameter, ε, by simultaneously solving

$$F = ZKJ \left(\delta - \frac{c}{2} \right)^n \tag{153}$$

and

$$\varepsilon = \frac{1}{2} \left(1 - \frac{c}{2\delta} \right) \tag{154}$$

where F = applied load

Z = number of balls or rollers

K = the deflection constant defined below

c = diametral clearance (sometimes called the radial clearance by AFBMA)

J = a radial load function given in Fig. 3.205

n = an exponent = 1.5 for ball, 1.1 for roller bearings

The deflection constant for ball bearings is

$$K = 1.53 \times 10^7 (D)^{0.5} \tag{155}$$

and for roller bearings,

$$K = 5.28 \times 10^6 (L_r)^{0.89} \tag{155a}$$

where D = diameter of the balls

L_r = effective length of rollers

Equations 153 and 154 are solved for ε by trial-and-error:

1.   Assume a value of ε; in good design practice ε is usually between 0.5 and 1.0.

2.   Find J in Fig. 3.205

3.   Solve Eq. 153 for δ.

4.   Substitute δ in Eq. 154 and solve for ε.

5.   Compare ε with assumed value and iterate until the difference is sufficiently small, about 0.01.

The relation between ε and the life adjustment factor, λ , is given in Fig. 3.206; is applied to the L_{10} or rated fatigue life by Eqs. 156 and 157:

$$\text{for ball bearings:} \quad L_{10} = \left(\frac{10^6 \lambda}{60N} \right) \left(\frac{C}{F} \right)^3 \tag{156}$$

$$\text{for roller bearings:} \quad L_{10} = \left(\frac{10^6 \lambda}{60N} \right) \left(\frac{C}{F} \right)^{10/3} \tag{157}$$

Basically, λ accounts for the effect on fatigue life of diametral clearance, either positive or negative. In nonpreloaded bearings the clearances are positive and the corresponding values of λ and 0.7 to 1.0; thus λ is often called a reduction factor. For preloaded bearings, values greater than 1.0 are easily obtained, indicating the

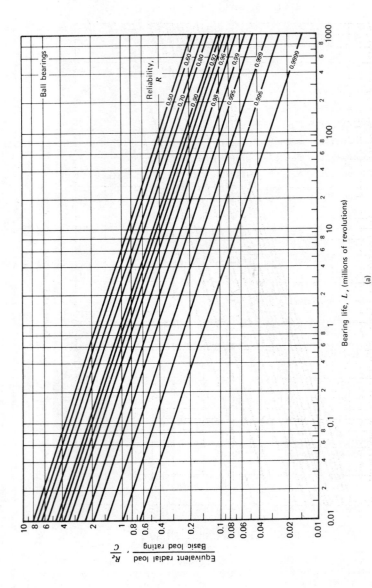

Fig. 3.203   Life curves for anti-friction bearings. a) Ball bearings; (b) Roller bearings.
From C. Mischke [189], Figs. 1, 2. Continued on following page.

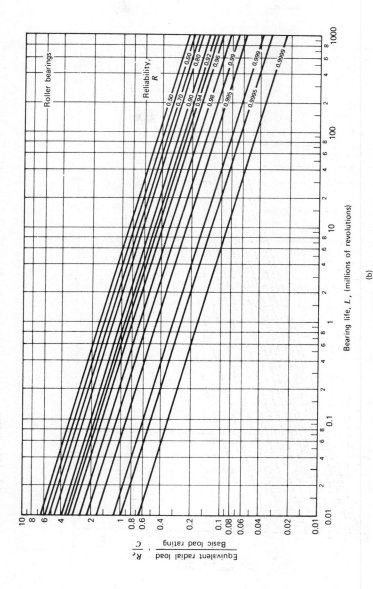

(b)

Fig. 3.203   Life curves for anti-friction bearings.   a) Ball bearings;   (b) Roller bearings.
From C. Mischke [189], Figs. 1, 2.

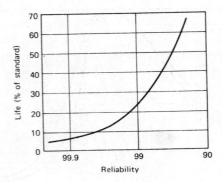

Fig. 3.204   Bearing fatigue life as a function of reliability.
Figs. 3.145–3.151 from T.A. Harris [191 & 192].

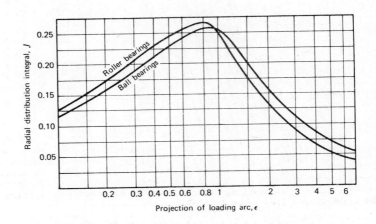

Fig. 3.205   Radial load function, j, vs. zone of loading projection, ε.
For best results, design for $0.5 < \varepsilon < 0.9$.

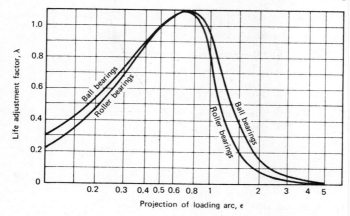

Fig. 3.206 Fatigue life reduction factor, λ vs. ε. Since the most desirable values of λ are greater than 1.0, ε should be between 0.5 and 0.9. Reprinted from *Product Engineering*, July 19, 1965. Copyright 1965 by McGraw-Hill, Inc.

benefits of proper preload but it must be noted that with increasing preload, above $\varepsilon \sim 1.0$, λ starts to decrease rapidly, indicating the sharply lowered life from overpreloading.

Example 1   Radial Bearing, Nonpreloaded Life, [191]

A single-row, deep-groove ball bearing (SKF bearing number 6309 with a loose C3 fit) has a basic dynamic load rating of 9120 lbs. This bearing supports a radial load of 2000 lbs. at a shaft speed of 1000 rpm. According to the catalog, the bearing contains eight balls of 11/16-in. diameter and has a mean diametral clearance of c = 0.001 in. Without any preload, what is the radial deflection and estimated L_{10} fatigue life? From Eq. 155,

$$K = (1.53)(10^7)(0.6875)^{0.5} = 1.269 \times 10^7$$

From Eq. 153,

$$2000 = (8)(1.269)(10^7)\left(\delta - \frac{0.001}{2}\right)^{1.5} J \qquad (158)$$

$$1.97 \times 10^{-5} = (\delta - 0.0005)^{1.5} J$$

From Eq. 154,

$$\varepsilon = 0.5 \left(1 - \frac{0.001}{2\delta}\right)$$

$$= 0.5 \left(1 - \frac{0.0005}{\delta}\right) \qquad (159)$$

We assume a value for ε (a good starting point for nonpreloaded bearings is $\varepsilon = 0.4$). Use the ε value in Fig. 3.205 to determine J, solve for ε in Eq. 158, and then solve for ε in Eq. 159 to see how close it is to the assumed value. This procedure finally yields

$$\varepsilon = 0.402$$

$$\delta = 0.00254 \text{ in.}$$

Now compute the predicted bearing life. At $\varepsilon = 0.402$, from Fig. 3.206, $\lambda = 0.9$ and L_{10} from Eq. 156 becomes

$$L_{10} = \frac{10^6 (0.9)}{60(10^3)} \left(\frac{9120}{2000}\right)^3$$

$$= 1438 \text{ hr.}$$

Example 2   Radial Bearing, Preloaded Life, [191]

Now consider that the bearing of the previous example (Bearing No. 6309) is mounted with a press fit on the shaft and in the housing such that the resultant clearance is 0.0005 in. tight, providing a light radial preload. What radial deflection and L_{10} fatigue life can now be expected? From Eq. 153,

$$2000 = (8)(1.269)(10^7) \left(\delta + \frac{0.0005}{2}\right)^{1.5} J$$

$$0.0000197 = (\delta + 0.00025)^{1.5} J \tag{160}$$

From Eq. 154,

$$\varepsilon = 0.5 \left(r + \frac{0.00025}{\delta}\right) \tag{161}$$

Solving Eq. 160 and 161 with the aid of Fig. 3.202 yields:

$$\varepsilon = 0.577$$

$$\delta = 0.0016 \text{ in.}$$

At $\varepsilon = 0.577$, from Fig. 3.206, $\lambda = 1.055$. Thus, from Eq. 156,

$$L_{10} = \frac{(1,000,000)(1.055)}{(60)(1000)} \left[\frac{(9120)}{(2000)}\right]^3$$

$$= 1660 \text{ hr.}$$

Hence this bearing, when mounted with 0.0005-in. interference, deflects 0.0009-in. less and has a 15% increase in fatigue life.

Axial preloading is also a means of increasing the bearing stiffness or reducing the deflection caused by an applied thrust load. The spring constants for ball bearings with or without preload are nonlinear, but the advantage of using preload is that the rate of change of the slope of the δ/F curves for preloaded bearings is much greater than that without preload. The spring constants for roller bearing are nearly linear, so that there is less to be gained by the use of preload.

The amount of preload is obtained by forcing a pair of duplex bearings on the same shaft to absorb a prescribed amount of axial offset between their inner and outer races. The offset may be accomplished by axial deflection of the face of one race, by adding a shim or, for bearings spaced far apart, by using an inner and outer spacer of slightly different lengths. As an example in computing the amount to be ground off, or the axial deflection, consider the usual pair of angular contact ball bearings mounted back to back, as in Fig. 3.207. Including the applied thrust load, T, equilibrium requires that

$$T = F_1 - F_2 \tag{161a}$$

where F_1 and F_2 are the thrust loads on bearings 1 and 2. For no preload, $F_1 = F_2$. Equations 162 and 163 now relate the axial offset, δ, and the preload, F:

$$F_j = ZD^2G \sin \alpha \left[\frac{\cos\alpha_0}{\cos\alpha} - 1 \right]^{1.5} \tag{162}$$

$$\delta_j = BD \left(\frac{\sin \alpha - \alpha_0}{\cos \alpha} \right) \tag{163}$$

where α_0 = initial contact angle under zero load

α = final contact angle under zero load

G = the axial deflection constant related to B by the curve in Fig. 3.208

B = a constant derived from the inner and outer raceway curvatures, f_i and f_0

$$B = f_i + f_0 - 1 \tag{164}$$

Subscript j relates to the specific bearing No. 1 or 2 in a duplex set.

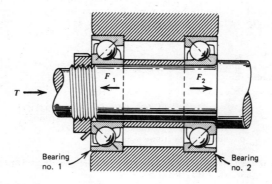

Fig. 3.207 Preloaded set of duplex bearings subjected to an external thrust load, T.

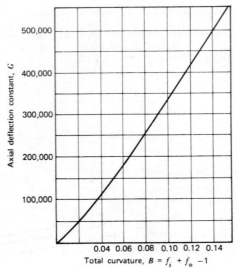

Fig. 3.208   Axial deflection constant, G, as a function of the curvatures of the inner and outer ball grooves, f_i and f_o.

Values for Z, D, and α_0 in the above equations may be obtained from the bearing manufacturers. The axial preload, F, is usually known or assumed from the application requirements. From Fig. 3.208, obtain a value for G based on the computed value for B (from Eq. 164); from the chart of Fig. 3.209, obtain other necessary factors as follows.

1.   We calculate a constant, t, from the known factors in the first part of Eq. 162 by making t equal to

$$t = \frac{F}{ZD^2G} \tag{165}$$

2.   In Fig. 3.209, we locate the point of intersection of the line for t and the radial line for α_0. On the curves, the example is t = 0.01 and α_0 = 40°.

3.   We swing a radius about the right-hand origin through the located point.

4.   At the intersection of this arc and the abscissa line (where α_0 = 0), we locate the value of δ/BD. In the example, δ/BD = 0.089.

5.   We align a straight-edge through the intersection of t and α_0 lines such that the straight-edge is parallel to identically numbered markers of the upper and lower $\alpha - \alpha_0$ scales.

In the example, we locate $\alpha - \alpha_0$ = 3.6°. These parameters substituted approximately in Eqs. 162 and 163 provide values for F and δ. A sample procedure follows.

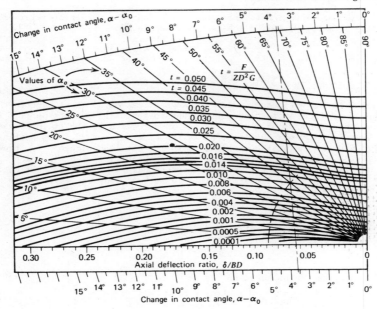

Fig. 3.209   Design chart for computing the axial deflection, δ, under load, and the resultant change in contact angle α. Dashed lines are for Example 3.

Example 3   Axial Preload on Duplex Pair, [191]

It is desired to obtain an axial preload of 500 lbs. from a set of duplex, angular contact ball bearings. The bearings have 52% inner and outer raceway groove curvatures, an initial contact angle of 40°, and a complement of 15 balls of 0.5-in. diameter. How much stock must be ground from the inner ring face of each bearing? From Eq. 164,

$$B = 0.52 + 0.52 - 1 = 0.04$$

From Fig. 3.208 for a value of B = 0.04, G = 110,000. Then, from Eq. 165,

$$t = \frac{500}{[15(0.5)^2 110,000]}$$

$$= 0.0012$$

From Fig. 3.209, δ/BD = 0.022. Hence,

$$\delta_p = (0.022)(0.04)(0.05)$$

$$= 0.00044 \text{ in.}$$

OFD - M

Subscript p was added to denote that the deflection is a result of axial preloading alone. Also from Fig. 3.209, $\alpha - \alpha_0 = 0.9°$. Hence,

$$\alpha_0 = 0.9 + 40 = 40.9°$$

Example 4   Axial Preload Plus External Thrust on a Duplex Pair, [191]

Suppose now that the preloaded duplex bearings of Example 3 are subjected to external thrust load and it is necessary to obtain the axial deflection. It should be recalled that, because of the difference in direction of the preload in the separate bearings only one bearing resists the external thrust. Designating the axial deflection of the bearing set due to the thrust load alone as δ_t, then at bearing 1, Fig. 3.209, the effective axial deflection is $\delta_t + \delta_p$ in which δ_p is the axial deflection due to preloading. Correspondingly, the axial deflection at bearing 2 is $\delta_p - \delta_t$. The latter condition exists as long as bearing 2 is not relieved of all load. Therefore, according to Eq. 163:

$$\delta_p + \delta_t = BD \left[\frac{\sin (\alpha_1 - \alpha_0)}{\cos \alpha_1} \right] \qquad (166)$$

$$\delta_p - \delta_t = BD \left[\frac{\sin (\alpha_2 - \alpha_0)}{\cos \alpha_2} \right] \qquad (167)$$

These two equations can be added to yield:

$$2\delta_p = BD \left[\frac{\sin (\alpha_1 - \alpha_0)}{\cos \alpha_1} + \frac{\sin (\alpha_2 - \alpha_0)}{\cos \alpha_2} \right] \qquad (168)$$

Also, according to Eq. 162 and 164

$$T = ZD^2G \left\{ \sin \alpha_1 \left[\left(\frac{\cos \alpha_0}{\cos \alpha_1} \right) - 1 \right]^{1.5} - \sin \alpha_2 \left[\left(\frac{\cos \alpha_0}{\cos \alpha_2} \right) - 1 \right]^{1.5} \right\} \qquad (169)$$

Equations 167 and 168 must be solved simultaneously for the unknown contact angles α_1 and α_2. This procedure is best done on a digital computer; however, several graphical techniques can be employed in the following manner.

Determine the axial deflection caused by an external 1000-lb. thrust load applied to the preloaded duplex bearing set of Example 3.

Establish for each bearing the load deflection curve caused by preloading alone. This is done by selecting a schedule of loads from Table 3.48A: from 200 to 1600 lbs. Then by using Fig. 3.205 and the method in Example 3, a series of deflections, δ_j, are obtained.

Now plot the values of δ_j vs. F_j (Fig. 3.206). Note that, at the preload of 500 lbs, δ_p on each bearing is 0.00044 in. This value checks with that of Example 3.

Next, with the aid of Fig. 3.210, compute the effect of additional axial deflections caused by external axial thrust. Thus, in Table 3.48B, an additional deflection of 0.0001 results in a total deflection of 0.00054 on bearing 1, and 0.00034 on bearing 2. From Fig. 3.210, the corresponding bearing loads are 670 and 340 lbs., respectively. These loads act against each other; hence, the algebraic sum is T = 670 - 340 = 330 lbs.

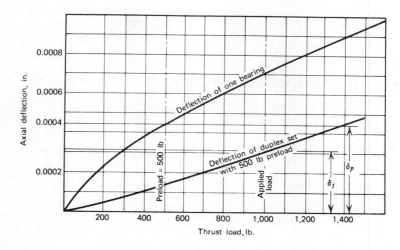

Fig. 3.210 Axial deflection vs. thrust load for sample calculations. The chart is constructed for one set of conditions, then employed to determine the deflection of duplex sets such as the bearings illustrated in Fig. 3.206. Reprinted from _Product Engineering_, Oct. 10, 1966. Copyright 1966 by McGraw-Hill, Inc.

Finally, plot T vs. δ_t in Fig. 3.210. From the latter curve at 1000 lbs applied load, the deflection of the duplex bearing set in the loading direction is 0.0003 in. The thrust load on bearing 1 is 1070 lbs and on bearing 2 is 70 lbs. If a single bearing were used to support the 1000-lb. load with no help from preloading, the axial deflection would be 0.0007 in.

An approximate, but simpler, graphical method is also available. Note that to relieve the preload on bearing 2 in Example 4, the axial deflection δ_t must be equal to the preload deflection δ_p. Therefore, in Fig. 3.210 we find that, at $\delta_1 = 2\delta_p = 0.00088$, the thrust load T = 1380 lbs. Now, T = 1380 lbs. and $\delta = 0.00044$ in. are the coordinates that describe the point on the load deflection curve of the duplex bearing set at which bearing 2 becomes unloaded. Draw a straight line in Fig. 3.210 between the origin and the point just determined. This straight line approximates the deflection curve of the duplex bearing set. At T = 1000 lbs., find $\delta_t = 0.00033$; this value is practically identical to that previously determined.

In the foregoing examples, the bearings in the duplex set were identical, a condition chosen to simplify the calculations. However, it is sometimes desirable to use duplex bearings that are different in ball complement, ball size, groove curvature, and contact angle. In that case, curves of δ_j vs. F_j must be determined for each bearing. Thereafter, the calculation procedure is identical.

TABLE 3.48   Load-Deflection Calculations for the Examples

TABLE A

F_j	$t = F_j/ZD^2G$	δ_j/BD	$\alpha_j - \alpha_0$	δ_j
200	0.0005	0.012	0.5	0.00024
400	0.0010	0.019	0.8	0.00038
600	0.0015	0.025	1.05	0.00050
800	0.0019	0.029	1.2	0.00058
1000	0.0024	0.034	1.4	0.00068
1200	0.0029	0.039	1.6	0.00078
1400	0.0034	0.044	1.8	0.00088
1600	0.0039	0.049	2.0	0.00098

TABLE B

δ_j	$\delta_p + \delta_t$	F_1	$\delta_p - \delta_t$	F_2	T
0.0001	0.00054	670	0.00034	340	330
0.0002	0.00064	870	0.00024	200	670
0.0003	0.00074	1070	0.00014	100	970
0.0004	0.00084	1290	0.00004	20	1270

TABLE C

δ_i	$\delta_{p1} + \delta_t$	F_1	$2F_1$	$\delta_{p2} - \delta_t$	F_2	T
0.0001	0.00038	400	800	0.00034	340	460
0.0002	0.00048	570	1140	0.00024	200	940
0.0003	0.00058	760	1520	0.00014	100	1420

TABLE D

Contact angle, degrees	20	25	30	35	40
Y factor (effect of preload on bearing life)	1.00	0.87	0.76	0.66	0.57

Effects of Speed, Lubrication, and Materials. The effects of these parameters on fatigue life of ball and roller bearings are interwoven in a highly complex manner. Two general levels of attack on the problem seem to exist: (1) the conventional hand computation of the major loads, deflections, and so forth, leading to a life estimate of probably unknown confidence, often followed by the application of a liberal factor of safety, such as 3 on load or 10 on life; and (2) the more comprehensive and sophisticated approach to include the effects of centrifugal force on contact angle, ball gyroscopic moments, and so forth. Harris [192] indicates that the evaluation of a high speed bearing under combined, noncoplanar loading requires solution of more than 85 simultaneous nonlinear equations--quite obviously a computer task. The magnitude of a great many bearing design problems lies in the gray area between these two extremes and attempts to specify a bearing for high speeds and loads, mounted perhaps in a housing of unknown elasticity, contain a considerable potential for total failure. For all but the simplest of cases, the selection of the proper bearing should be left to the practical experts; except for a few semiquantitative points of interest below, this discussion is terminated with the recommendation of any of several excellent works on bearing design and application [186-193].

The effect of increasing speed on fatigue life is chiefly an increase in the operating load and the clearance due to the centrifugal force, with a corresponding reduction of the loading zone. Contact angles also change, that at the inner raceway increasing and that at the outer raceway decreasing with rising speed. Angular contact bearings are particularly sensitive to this effect, and in the extreme case of unsymmetrical loading, it can turn out that the contact angles and ball loads are different at each ball location in the radial plane. Figure 3.211 illustrates the variation in the L_{10} fatigue life for a given angular contact bearing as a function of speed and load.

In the choice of a lubricant for a bearing and its environment, it is often a rather subjective action but it follows from basic concepts that the film should be of sufficient strength to separate the elements and of sufficient thickness to absorb the surface roughness. Were these Utopian conditions always available, bearings would indeed show very extended lives. However, in practical cases, it is often possible to obtain full film lubrication, with resulting lives of 100% or more of standard ratings. The Weibull plot of Fig. 3.212 shows the relative superiority of three common lubricants. The numerical differences obtained by the two test rigs emphasize the difficulty and nonrigorousness of lubricant testing, and the fact that quite different lives may be obtained from the same bearing and lubricant under different loadings. Temperature is probably the most important environmental parameter to affect bearing life by varying the lubricant properties. The reduction in load ratings at the B_{10} and B_{50} lives at room temperature and at 450°F was found to be significant [204], as shown in Table 3.49. The data were taken on 45-mm bore ball bearings and the ratings expressed as a percentage of the basic AFBMA capacity for normally lubricated 52100 steel. The lubricant at room temperature was a mineral-based grease, and that at 450°F a synthetic diester.

Solid lubricants, particularly graphite and molybdenum disulfide (MoS$_2$), have been used with reasonable success at both room and elevated temperatures. The general results indicate that MoS$_2$ provides a longer life than graphite by about one order of magnitude. The form of the lubricant is important: particle size should be in the submicron range and the particles should be suspended in a conventional lubricant. Dry powders often show no improvement in life, and water or alcohol suspensions have shown decreased life. Specific suspensions that extended life are: MoS$_2$ in soap-thickened silicone, MoS$_2$ paste in mineral oil, and graphite up to 10% in mineral oil. There seems to be some preference for MoS$_2$ over graphite, especially for use in vacuum. And although not strictly a solid lubricant, nylon blocks impregnated with octyl adipate have performed successfully in vacuum for both ball

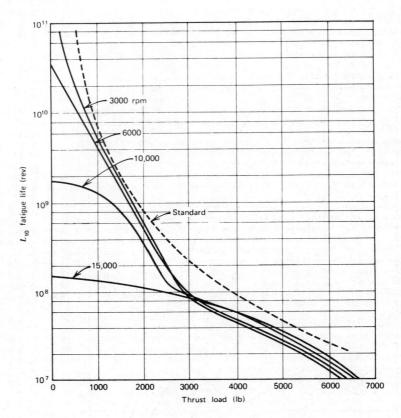

Fig. 3.211   The effect of speed and thrust load on an angular
contact bearing (No. 218).

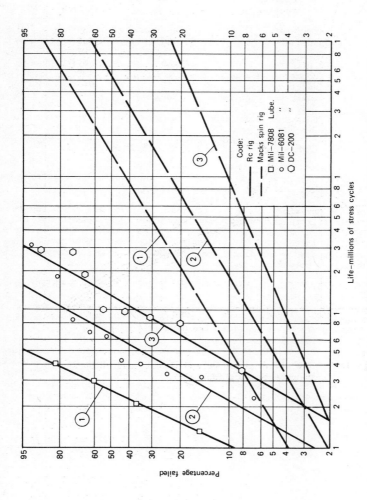

Fig. 3.212 Weibull plot of fatigue data on M1 tool steel. Tested on Macks and on RC spin rigs, with three lubricants. Hertz stress level; 725,000 psi max; room temperature. From R. A. Baughman [195], Fig. 7.

TABLE 3.49 Temperature Effects on Bearing Load Ratings

Bearing Material (steel)	Melting Practice	Temperature °F	Rating as % AFBMA Basic Load	
			10% Life	50% Life
M-1	Basic arc	Room	83	91
M-1	Basic arc	450°F	60	54
M-10	Basic arc	Room	77	70
M-10	Basic arc	450°F	64	64
M-1	Induction vacuum	Room	132	122
M-1	Induction vacuum	450°F	64	77
M-50	Consumable electrode vacuum	Room	144	126
M-50	Consumable electrode vacuum	450°F	93	100

SOURCE: [194], Metal Progress, Nov. 1967, by permission ASM.

bearing and electric motor brush lubrication. The enclosure is sealed to vacuum by a labyrinth, the lubricant vapor pressure provides a flowing mist.

Much theoretical and experimental work continues. In [196] the technique of measuring hydrodynamic film thickness by X-ray techniques is discussed; in [197] the fatigue life of a bearing as an assembly is correlated with life data on the material alone. In the latter it is interesting to note that a function of the hertzian stresses appears to the ninth power in the life equation, strongly emphasizing the deleterious effect of any overloading.

The materials for antifriction bearings are presently in a state of high perfection. Efforts to improve both composition and quality have been very extensive, resulting in the extremely high quality alloys produced by consumable electrode vacuum melting and/or vacuum degassing. The effect of these alloys on bearing design has been to permit an increase in the basic L_{10} life from 500 to 1500 hr, or from 1 to 3,000,000 revolutions, at the original ABFMA load rating. The alloys have remained of the basic compositions: AISI 52100, M-50, stainless 440-C, and 8620 for case-hardened elements; the effect of vacuum melting (and remelting) has been to lower the content of inclusions. Thus, the number of nucleii for pits, spalls, flakes, and cracks is reduced and fatigue life increased. Table 3.50 gives the relation between the inclusion size and number, and the B_{10} life; Fig. 3.213 relates this life to the number of remelts.

TABLE 3.50   Relationship of Inclusion Content to B_{10} Life of
Roller Bearings

| Supplier | Inclusion Size Level[a] | | | | | | | B_{10} Life, hr |
	2	3	4	5	6	7	8	
A	4.0	3.1	1.6	0.7	0.5	0.3	0.2	56
B	7.4	3.2	1.0	0.4	0.2	0.2	0.1	64
C	4.0	1.2	0.7	0.4	0.2	0.1	0.1	153
D	6.5	2.0	0.7	0.2	0.1	0	0	265
E	3.0	1.0	0.4	0.1	0	0	0	247
F	1.0	0.1	0	0	0	0	0	512
G	0.8	0.1	0.1	0	0	0	0	In test
H	0.2	0.1	0	0	0	0	0	500+

SOURCE: J. A. Erickson [194], Table 1.

[a]The 20 bearings representing each supplier were rated ultrasonically (smaller inclusions--levels 2, 3, 4, and 5) and visually (larger inclusions--levels 6, 7, and 8). Numbers represent averages of inclusions (per bearing) of the given size level.

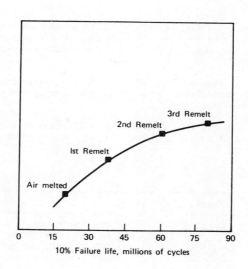

Fig. 3.213   The effect of number of remelts on failure life of bearings in 52100 steel. From R. A. Baughman [195], Fig. 11.

Sleeve Bearings. Bearings of this type using liquid lubrication and operating below 50,000 rpm do not seem to have been the subject of the extensive fatigue considerations accorded the antifriction types. Accepting the inevitable wear due to sliding friction, it appears that a properly designed and lubricated sleeve will run indefinitely under its rated load. The shaft displacement resulting from the wear is usually accepted or taken up by the removal of shims.

The fatigue strength as a bulk property of the sleeve material does not seem to be very important. A fatigue mechanism can become operative if debris gets into the bearing. Penetration of the sleeve will take place as the harder foreign particle (metal chips, dirt as silica, etc.) is forced in by the shaft. Although dirt is often considered as the predominant cause of bearing distress, it is essentially impossible to design for it as a failure mechanism because of lack of knowledge: what is the local stress, the value of K_t, the number of particles, the frequency of stressing, etc.? Particularly, the last item has no relation to shaft speed. Good practice dictates the use of clean assembly, seals and lubricant filter, and normally results in satisfactory, if unpredictable life. The term "distress" was used above in distinction to "failure" because sleeve bearings usually have a relatively long operating period from the start of the malfunction to final seizure or collapse. During this period there may be an increase in temperature and friction, the loss of shaft alignment; the state of failure may be difficult to describe. A major exception to this comment is the failure of waste- or wick-lubricated bearings by the action known as a "waste-grab", which often damages the shaft as well as the sleeve and results in the stalling of the machine, or in complete collapse.

Certain design practices may be used to minimize bearing malfunction or fatigue:

1. Shaft geometry and finish must be adequate to permit film lubrication. Concentricity and taper must be controlled; finish is usually satisfactory and about RMS 125 maximum.

2. Housings must be sufficiently stiff to support the sleeve and shaft loads within the allowable deflection; bore finish must be good enough not to have an effect on the sleeve thickness, especially in thin-wall sleeves or shells; again about RMS 125 maximum.

3. Sleeve wall thickness must be sufficient to carry the compressive load within the allowable deflection. To be effective, the sleeve wall and the housing should be designed together.

4. The length-to-diameter ratio, L/D. Evaluation of much operating data has led to the ranges of values in Table 3.51 as rule of thumb guides.

TABLE 3.51

L/D	Speed, rpm	Load, psi	Lube Method
1.0-3.0	1000	20,000	Grease
0.7-1.3	2500	2,000	Splash or pressure
0.2-1.0	2-4000	2-5,000	Splash or pressure
0.2-0.4	4-10,000	Low	Pressure
0.3 max	10,000	Low	Pressure

3.5.2 Gears

The fatigue failure of toothed gearing may occur in any one or a combination of three basic modes:

1.   Tooth bending, which, if continued for a sufficient number of cycles results in fracture at the root.

2.   Surface distress as a result of mesh stresses exceeding the compressive fatigue limit on some localized area of the tooth face. This condition may be caused by misalignment or excessive deflection of shaft or housing. If caused by an initial defect, the defect may have been on the surface, or subsurface but within the volume of highly stressed material.

3.   Excessive tooth deflection and overloading caused by resonant vibration, particularly in thin—web, thin—rim types.

From a rather approximate evaluation of data on gear failures in land transportation equipment (which does not include the "thin" designs), we may conclude that flexural fatigue accounts for about 25% of the failures; surface destruction 40%; overloading, probably with shock, 15%; of the remaining 20% of miscellaneous causes, half might be attributed to massive "foreign" objects going through the mesh.

The vibrational modes of failure, (3) above, occur almost exclusively in the very high speed, highly loaded gearing, such as that in the drive for a rocket engine turbopump, with the speed becoming critical for spur gears above about 8000 feet per minute (fpm) pitch line velocity, and at 15,000 fpm for helicals. The modes and typical frequency relations are given in Fig. 3.214; it should be noted that the design of the train and of each gear (hub, web, rim, and teeth) must be such that none of these frequencies resonates with a tooth mesh frequency. The hazard is, of course, significantly reduced for the heavier—structured, comparatively lighter—loaded marine and industrial gears; even their lowest frequency, the "first diameter," would (probably) be sufficiently high so as to detune automatically from any but the most accidental harmonic of a tooth mesh.

Power Gears. Gear design has been the object of very extensive work, the efforts of Lewis, Buckingham, Barth, and many others being too well known to review again here. Designs based on their equations have usually performed satisfactorily, but most of the investigators up to about 1940 were primarily concerned with strength and not necessarily with predictability or reliability in the numerical sense. They evolved designs to operate indefinitely at stresses calculated to be less than an endurance limit. These calculated stresses, however, varied according to numerous assumptions, especially those relating to stress concentrations, shape factors, and load sharing among teeth. More recently, investigations have been made on the basis of a unit load that is the equivalent load in pounds per in. per in. of face width on a tooth of 1—pitch, in the normal plane. The parameter is not a stress, dimensions notwithstanding; it may be related to root stress by a gear tooth stress equation. An expression for unit load in spur gears is

$$U_1 = \frac{W_t P_d}{F} \tag{170}$$

and in helicals

$$U_t = \frac{W_t P_d}{F \cos \psi} \tag{171}$$

Gear Vibration Modes

Mode	Normalized Frequency, cps	Schematic	Cross Section
1st Diameter	1		
2nd Diameter	2		
1st Cymbal	2.67		
3rd Diameter	3.66		
4th Diameter	6.2		
5th Diameter	9.04		
2nd Cymbal	9.3		
6th Diameter	12.		
7th Diameter	12.6		

SOURCE: D. W. Dudley [200], Table 2.

Fig. 3.214   Vibration Modes in Gears for Rocket Pumps.

where W_t = tangential driving force, lb = pinion torque/pinion pitch radius

   F = contacting face width, in.

   P_d = diametral pitch

   ψ = helix angle

$P_d/\cos\psi$ = P_{nd} = normal diametral pitch

From the general equation for spur gear tooth strength in AGMA 255.01,

$$S_t = \frac{W_t K_o P_d K_s K_m}{K_v F J} \qquad\qquad (172)$$

the unit load is derived in stress terms as:

$$U_t = S_t \frac{K_v}{K_o} \frac{J}{K_s K_m} \qquad\qquad (173)$$

where S_t = root stress, psi

   K_o = overload factor

   K_t = size factor

   K_m = load distribution factor

   K_v = dynamic factor

   K_f = root stress concentration factor

   J = geometry factor = 1/stress factor

K_f/Y = root stress factor

   Y = factor relating geometry of tooth as a beam with nominal root stress; a "shape" factor

The unit load seems to be a most useful and nearly universal parameter by which to indicate gear capacity and life but it does have the limitation of being associated with a definite load position. The best procedure is to refer all unit loads to a "worst" position usually taken as the highest point of single tooth contact (HPSTC). This operation requires the evaluation of the proportionate effect of difference in load position by means of a root stress equation. Thus, if one takes a simple expression as $S_t = U_1(K_f/Y)$, the root stress varies as K_f/Y. Since both K_f and Y vary with the position of contact, the unit load comparison must be made according to values of the K_f/Y ratio.

The condensed results of very extensive tests [199] are given in Table 3.52.

A plot of the stress factor vs. the load position is given as Fig. 3.215. For the fatigue limit determination, the load positions (the "A" distance) were 0.100 in. and 0.010 in., providing a stress factor ratio of 3.0/4.25. The value 22,000 divided by this ratio then gave the corrected value of 31,200. In an actual set of gears, the highest point of single tooth contact is also affected by the number of teeth and, therefore, the root stress varies somewhat.

TABLE 3.52   Gear Tooth Root Stresses

	Highest Point of Single Tooth Contact Loading	Near-Tip Loading	Tip Corrected to Highest Point of Single Tooth Contact Loading
Fatigue Limit	31,000 psi	22,000 psi	31,200 psi
Unit Load	47,000	30,000	42,500

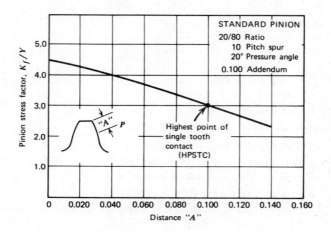

Fig. 3.215 Root stress factor vs. load position for standard test pinion. From AGMA 225.01 (Oct. 1959):

$$\frac{1}{Y} = \frac{\cos\phi_L}{\cos\phi}\left(\frac{1.5}{X} - \frac{\tan\phi_L}{t}\right)$$

$$K_f = H + \left(\frac{t}{r_f}\right)^J \left(\frac{t}{h}\right)^L$$

A general collection of unit load values is given in Table 3.53 for a wide range of "quality," from medium quality industrial gears to those for rocket pump drives, the "Corrected" column refers to test methods and results in [199]. Best estimates of tooth fatigue limit in optimized materials and heat treatments are given in Table 3.54 in terms of the 10 and 50% failure lives, similar to the B_{10} and B_{50} lives for bearings; and Fig. 3.216 provides standard S-N curves. Some interesting data are also available for marine gearing: Broersma [201] determined the optimum load conditions for the teeth of carburized, hardened, and ground pinions of single helix, which were incorporated into successful gears. The optimum condition is derived from the equality in fatigue life or "strength" in flank compression and in root flexural stress. For four typical pinions the unit loads (design) were 19.5, 20.1, 18.4, and 16.0 ksi, values that should be compared with those in Table 3.53, in which 9000 psi is quoted for industrial gears with good reliability, and

TABLE 3.53   Typical Unit Loads for High Performance Gears

Application	Unit Load		Comment
	For design used	Corrected to 20/80 test pinion by 25% reduction	
Aircraft	20,000	15,000	Highest used. Required great excellence in manufacture.
	15,000	11,200	Normal value typical of general use. Does not require extreme accuracy.
Rocket pump drives	40,000	30,000	Highest used. Very critical. Cycles below 10 million.
	25,000	18,000	Normal value. Cycles below 10 million.
Industrial	40,000	30,000	Highest used. Very poor reliability. (Generally a mistake in design.)
	12,000	9,000	Normal value for cases of good industrial quality. Good reliability.

SOURCE: [100], Table 6.

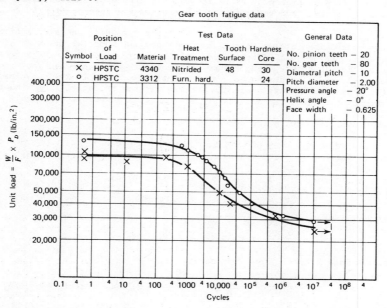

Fig. 3.216   Standard S-N curves showing shape at high loads.

TABLE 3.54   Best Estimate of Fatigue Limit (Tooth Breakage) for
Optimum Material and Treatment

Material and Treatment	Fatigue Limit[a]	
	10% failure	50% failure
Carburized 9310	51	55
Carburized and shot-peened 9310	58	60
Nitrided Nitralloy 135 Mod.	42	
Nitrided Nitralloy N	44	46
Nitrided 5% Ni 2% Al	56	58
Nitrided H 12	50	
Nitrided 4340	34	
Induction-hardened 4340	38	44
Induction-hardened and shot-peened 4340	42	45
Induction-hardened 4140	36	
Induction-hardened and shot-peened 4140	40	
Through-hardened 4340 at 35 Rc	30	
Through-hardened and shot-peened 4340 at 35 Rc	36	
Through-hardened modified 4340 at 42 Rc	41	

[a]In 1000 psi unit load.

where 18,000 psi is quoted as a normal value for rocket drives. Thus, as in the design of all complex elements, there is some difference of opinion as to what constitutes "quality," and the methods of measuring and interpreting it.

As an example of the use of the _unit load_ in design, consider an aircraft gear set with a case-carburized pinion of _real_ aircraft quality material, meaning stock from a vacuum melt. Assume a 10-pitch pinion on a gas-turbine drive to a generator, so that the applied load is smooth; assume further that both tooth accuracy and alignment are excellent. The design life at top rating is to be 10×10^6 cycles and 500×10^6 at 35% top rating; since the stresses at top rating will be almost three times those at the lower rating, the former condition governs the design.

From AGMA 225.01, the following numerical choices are made:

Design stress, S_{at} = 65,000 psi (at 10 million cycles)
Dynamic factor, K_V = 1.0 (excellent tooth accuracy and flexible light-
    weight design)
Overload factor, K_0 = 1.0 (smooth turbogenerator drive)
Size factor, K_s = 1.0 (for 10 pitch)
Load distribution, K_m = 1.0 (excellent alignment)
Geometry factor, $J = \dfrac{1}{\text{stress factor}}$

$$= \frac{1}{3.0} = 0.33 \text{ (from [199], Fig. 4)}$$

Substitution into Eq. 173 gives the result:

$$U_t = 65,000 \times \frac{1.0}{1.0} \times \frac{0.33}{1.0 \times 1.0} = 21,600, \text{ ("design" value)}$$

Data from [199] show that an estimated 90 percent of the best carburized test pinions will carry 50,000 or more unit load for 10 million or more cycles. We can take the 50,000 unit load from this fatigue test and compare it with an industry design value of 21,600 under most favorable conditions.

In the correlation of nonrotating test results with those of running tests, it was determined that, if the dynamic load factor of the run test is about 1.3 (generally considered as a reasonable value for highly precise spur gears run in the Lynn four-square aircraft gear test), the running test would result in a fatigue limit unit load of about 77% as that of the nonrotating test. Also, slight errors in alignment may be present even when the design allows almost no misalignment error and the design is considered as one in which $K_m = 1.0$. The practical designer needs to make an allowance in his allowable stress for a misalignment under worst conditions. This allowance reduces the test value of 50,000 unit load to a potential design value of

$$50,000 \times 0.77 \times 0.75 = 29,000$$

Further consideration must be given to the statistical nature of fatigue, the fact that not all tests are carried to 10^7 or more cycles, and the fact that practical design values will contemplate almost no failures. This leads to a further reduction of the 29,000 unit load to 21,600 as a reasonable design value for the highest quality of case-carburized aircraft parts operating under nearly ideal conditions. The real problem is being assured of quality that will yield the same test data as that used in the calculations.

All present production methods for gears are of the chip-making variety, but roll-forming, as for threads, has been recently introduced. While it is still in the process of transfer from experiment to production, a significant improvement in life is inherent in the presence of the residual compressive stress imparted to the tooth and fillet surfaces.

Shot-peening of the root region can raise the endurance limit load of a tooth in bending by as much as 30%. But, since the faces must be masked, and tooth breakage is only one cause of gear failure, the added expense may well not warrant its use.

<u>Motion Gears</u>. The problem of combating wear and loss of accuracy in instrument gearing has always been severe. The major difference in the design parameters for this lightly loaded, often high speed, gearing from those of power gears is the degree of wear, and fortunately the amount of wear is a reciprocal relation between the gears in the train having the greatest number of contacts (drive pinions) and of least importance and those gears of least number of contacts and greatest importance, for example, final position-resolvers. Most of the investigations leading to methods of life prediction have attempted to evaluate the wear rate of different combinations of materials in meshing gears in terms of S-N or constant life diagrams for the materials in <u>compression</u>, and to account for the effects of speed and load by experimental load life velocity plots, as in Fig. 3.217. By this means it is possible to make a quick check of the meshes in a gear train to isloate potential failures. The only information required is: materials of both gears in the mesh, pitch line velocity in ft/min, and useful load in lb/in. of face width. A conservative gear life estimate can then be made by:

1. Locating the appropriate curve based on the gear and pinion materials, and surface finish.

2. Following the useful load (ordinate) over to the approximate pitch-line velocity (interpolating if necessary), and reading the approximate failure time in hours (abscissa).

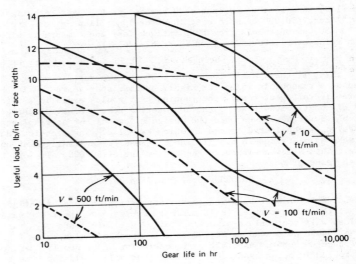

Fig. 3.217 Load/life/velocity curves for aluminum and stainless
steel instrument gears. Mill-finish 2024-T42———;
anodized 2024-T42-----; stainless type 416 passivated.
Lubrication: ANO-6 oil.

It is emphasized that these curves are conservative, permit the elimination of the
gear pairs with acceptable gear life (safely loaded), and are based on the least
favorable geometry encountered in instrument gearing, that is, small pinions, medium size gears, all of 14-1/2° pressure angle. All gears that appear to be too
heavily loaded must be considered in further detail.

3.5.3 Springs

The design of springs is similar in approach to that of gears, in that a number of special, or correction, factors appear in the equations. The conventional expressions for stresses in curved beams apply, provided that at least three such factors are taken into account. It has been shown [218] that, for helical springs of round wire, the shear stress in torsion is

$$S_S = \frac{kTr}{J} = \frac{8kPD}{\pi d^3}$$ (174)

where T = torque = axial force x radius = Pr

P = axial force

r = radius

J = polar moment of inertia

k = Wahl correction factor

Also

$$k = \frac{4C - 1}{4C - 4} + \frac{0.615}{C}$$ (175)

where C = the spring index = D/d

D = outer diameter of coil

d = wire diameter

Wahl's factor k corrects for the stress concentration due to the curvature of the wire, k_c, and for the direct shear stress, k_s; the factor k is then the product of k_c and k_s. The equation for k_s is derived from the expression for shear stress in combined torsion and tension (or compression), initially ignoring the concentration from the curvature:

$$S_S = \frac{Tr}{J} + \frac{F}{A} = \frac{8PD}{\pi d^3} + \frac{4F}{\pi d^2}$$ (176)

where the plus sign applies to the inner fiber, thus the S_S is maximum there. Equation 176 may be rewritten in the form

$$S_S = \frac{8Pd}{\pi d^3}(1 + 1/2C)$$ (177)

thus defining the shear-stress concentration factor as

$$k_S = 1 + 1/2C$$ (178)

Then the stress concentration factor for curvature is

$$K_c = \frac{k}{k_S} \qquad \text{or} \qquad k_c = \frac{C + 2.2}{C + 0.5}$$ (179)

The third stress modification factor arises from any nonsymmetry of loading, particularly from end conditions; and secondarily, from any damage to the wire section or finish by the end-forming tools. Responsibility resides with the designer to see that he does not require forms and conditions that normal shop methods cannot handle or do so with difficulty. Conversely, manufacturing and tooling groups must meet design requirements without inadvertently introducing stress concentrations.

In all the conventional equations for the design of springs, the tacit assumption has been made that none of the springs' resonant frequencies coincide with the frequency of dynamic load application. As an additional task the designer of springs so loaded must determine the first mode resonance and its first few harmonics for checking with the excitation frequency. Coincidence within plus or minus a few percent requires modification in the design.

In the great majority of applications, springs are not loaded in both tension and compression and are thus not subject to reversed stresses. Further, they are usually preloaded, providing a nonzero value of the mean stress. From observation of general constant life diagrams, it is easily noted that, for a given mean stress the lower the value of the stress ratio R, (S_{min}/S_{max}), the greater the life. Were such diagrams available for a spring _configuration_, rather than for only the material, the designer could directly find a life value and estimate the conservatism. In the absence of such individualized plots, a Goodman diagram can be constructed for the spring, based on the worst-case condition as to the effect of alternating stress. Assume that S_{min} is zero, maximizing the range of alternating stress at $S_{max}/2 = S_r/2$. Then, in Fig. 3.218 the point A is located at $S_{max}/2$ on both axes and the applicable portion of the diagram becomes limited to the triangle ABD. (The numerical values are used in the example below.) The variable stress is then written from Eq. 174 as:

$$S_a = \frac{8kD}{\pi d^3} \left[\frac{(P_{max} - P_{min})}{2} \right] \tag{180}$$

and the mean stress:

$$S_m = \frac{8k_s D}{\pi d^3} \left[\frac{(P_{max} + P_{min})}{2} \right] \tag{181}$$

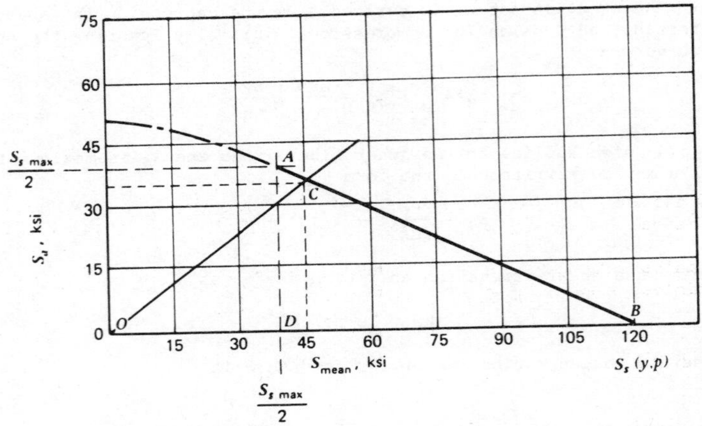

Fig. 3.218   Goodman diagram for a helical spring. From J. E. Shigley [206], Fig. 7.1; _Machine Design_. Copyright 1956 by McGraw-Hill Book Co., used by permission.

Table 3.55 provides experimentally obtained allowable values of S_r for various materials, conditions, and configurations of helical springs; Fig. 3.219 gives a refinement on the curvature factor as a function of geometric or theoretical stress concentration factor, k_t, and notch sensitivity, for steels. Equation 180 may then be rewritten

$$S_a = \frac{8k_f k_s D}{\pi d^3}\left[\frac{(F_{max} - F_{min})}{2}\right]$$ (180a)

where k_f = a reduced value of the correction factor k_c from Fig. 3.215

k_s = 1 + 1/2C, (as in Eq. 178).

TABLE 3.55   Endurance Stress Range R for Helical Springs

Material	Approximate Wire Diameter, in.	Index, C = D/d	Limiting Stress Range R, psi	Notes
Cold-drawn wire	0.13-0.16	10-11	41,000	Tension
Cold-drawn wire	0.16	6	60,000	Compression
Cold-drawn wire	0.135	14	46,000	Compression
Music wire	0.25	8	56,000	Compression
Music wire	0.063	7	76,000	Compression
Music wire	0.148	6-7	70,000	Compression
Music wire	0.148	6-7	115,000	Compression, shot-blasted
Valve-spring wire	0.162	6.5	75,000	Compression
Valve-spring wire	0.162	6.5	115,000	Compression, shot-blasted
Valve-spring wire	0.135	14	68,000	Compression
C steel, cold-wound	0.148	7.4	60,000	Compression
C steel, cold-wound	0.135	14	53,000	Compression
C steel, cold-wound	0.25	8	56,000	Compression
SAE 6150 steel	0.148	7.4	70,000	Compression
SAE 6150 steel	0.148	7.4	115,000	Compression, shot-blasted
18-8 stainless steel	0.148	6-7	45,000	Compression
18-8 stainless steel	0.148	6-7	90,000	Compression, shot-blasted
Phosphor bronze	0.148	6-7	15,000	Compression
Phosphor bronze	0.148	6-7	30,000	Compression, shot-blasted
Open-hearth carbon steel	0.75	5.0	72,000	Compression
Chromium-vanadium steel	0.56	4.8	77,000	Compression
Beryllium bronze	0.56	4.8	33,000	Compression

SOURCE: A. M. Wahl [218], pp. 91-92.
$S_{s,min} \sim 0$

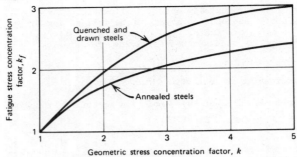

Fig. 3.219 K_f vs. K_t for steels. From J. E. Shigley [206], Fig. 4.24; <u>Machine Design</u>. Copyright 1956 by McGraw-Hill Book Co., used by permission.

Example

A helical compression spring is to be made of music wire and is to have an indefinitely long life. Determine the allowable P_{max} and P_{min} for $P_{max}/P_{min} = 4$ and a factor of safety of 1.8.

$$d = 0.063$$

$$D = 0.187$$

$$C = \frac{D}{d} = 2.98, \text{ or } 3$$

$$S_{syp} = 120 \text{ psi}$$

From Eq. 175,

$$k = \frac{4(3) - 1}{4(3) - 4} + \frac{0.615}{3} = 1.575$$

From Eq. 178,

$$k_s = 1 + \frac{1}{2(3)} = 1.167$$

From Eq. 179,

$$k_s = \frac{1.575}{1.167} = 1.35$$

Entering Fig. 3.219 with this value of k_c, find the corrected concentration factor $k_f = 1.31$. The substitution of $P_{min} = P_{max}/4$, and the above data in Eq. 180a yields

$$S_a = (1.31)(1.167) \frac{8(0.1875)}{\pi(0.063)^3} \frac{3P_{max}}{8} = 1100P_{max} \qquad (181a)$$

Similarly, from Eq. 181,

$$S_m = 1.167 \frac{8(0.1875)}{\pi(0.063)^3} \frac{5P_{max}}{8} = 1400P_{max} \qquad (182)$$

From Table 3.55, the endurance stress range is 76,000 psi, from which $S_{max}/2 =$ 76,000/2 = 38,000 psi. This value is shown as point A on the Goodman diagram, Fig. 3.218. The torsional yield strength S_{syp} = 120 ksi is B and the line BA extended would intersect the S_a axis at 60 ski, the rule of thumb value for fatigue limit, as 1/2 the yield.

The ratio of alternating to mean stress components is

$$\frac{S_a}{S_m} = \frac{1100P_{max}}{1400P_{max}} = 0.785 = R$$

The line OC is constructed with a slope of R = 0.785. Its intersection with AB gives the limiting values of S_a and S_m, scaled at 35 ksi and 45 ksi, respectively. Applying the factor of safety and solving for P_{max} in Eq. 181, we obtain

$$P_{max} = \frac{S_a}{1100(FS)} = \frac{35,000}{1100(1.8)} = 18 \text{ lb.}$$

and

$$P_{min} = \frac{P_{max}}{4} = \frac{18}{4} = 4.5 \text{ lb.}$$

The maximum stress is checked against the yield strength

$$S_{smax} = \frac{8k_sDP_{max}}{\pi d^3}$$

$$= 8(1.167)(0.1875)(18)(\pi(0.063)^3)$$

$$= 40,000 \text{ psi}$$

Compared to the nominal yield in torsion of 120,000 psi this value indicates another safety factor of about 3.

The example illustrates the procedure of simple spring design on an "endurance limit" basis. The conservatism resulting from these particular values may be typical of many designs; the design may be somewhat inefficient if the ratio of the maximum applied stress to the yield is considered (the FS of 3 above). The initial problem statement was "for indefinite life," which is usually taken to mean at least 10^8 cycles. On this basis, then, as a design requirement, the "efficiency" of trying to work the wire at a stress anywhere near its yield becomes irrelevant: the endurance limit becomes the limiting design stress. From Fig. 3.220, it may be seen that, for music wire springs to specification ASTM A228, the design stress at 10^8 cycles is about 80 ksi, which means that the designer would have some freedom in allowing the applied stress to rise somewhat above the present 40 ski. However, for springs with as low a value of index as 3, the stress concentration at the inner radius is fairly high; the small wire diameter (under about 0.100 in.) emphasizes the danger of further concentration of stress from surface defects, so that the indicated design practice is considered quite acceptable.

The type of service is extremely important in considering the design of springs, as is the possible difference in service within some overall assemblies. Automotive valve and transmission springs form two very sharp examples: valve springs would see 10^8 cycles in 10^3 hours of operation, a time easily exceeded by most engines. The number of transmission operations in the same time is less by two to three orders of magnitude. Thus, the spring designer would have two entirely different problems to consider; data in [2] indicate the higher stresses allowed for transmission springs, as well as the lower number of cycles to failure.

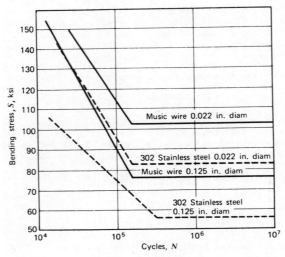

Fig. 3.220 Fatigue curves for spring wire. From Ametek/Hunter
Spring Division [207], Fig. 2.

The choice of allowable stress for dynamically loaded, long life springs requires
considerable attention. The S-N curves such as those in Figs. 3.219 and 3.220 for
specific alloys and spring configurations are a possible starting point, but they
do not seem to be available for the extremely large number of combinations that
might be desirable. Torsional elastic limits or design stresses based on tensile

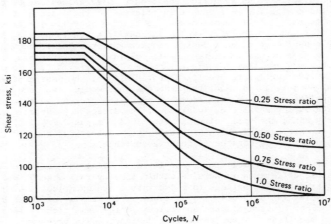

Fig. 3.221 Fatigue curves for music wire at various stress ratios
(diam = 0.022 in.). From Ametek/Hunter Spring Division
[207], Fig. 3.

strengths are provided in [2, 206, 207, 360], and are adjusted for wire size and for values of the spring index. Some Goodman diagrams are available, Figs. 3.222 and 3.223 being typical for helical springs in that they show the allowable range of alternating stress at the semistandardized 10,000,000 cycle life. The horizontal lines indicate the limiting stress at which permanent set occurs.

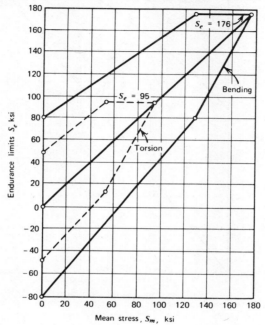

Fig. 3.222   Goodman diagram for SAE 9250 spring steel.
C, 0.50%; Si, 2.0%; Mn, 0.75%. Average
static properties in tension: $S_u = 220,000$
psi; $S_e = 160,000$ psi. From V. L. Maleev
and J. M. Hartman [360], Fig. 14-14.

Helical springs are, of course, widely used and the details of their fatigue be-havior investigated to the extent that, for music wire, a scatter band has been established, as in Fig. 3.224. Considerable scatter in the life of springs must be expected due to the numerous and only partially defined parameters operating on their design. In general, the width of the band is attributed to the nominal dif-ference in both tensile and torsional strength of the wires with changes in diameter, the fatigue limits decreasing as diameter increases. But none of the other parame-ters, including the index, seemed particularly relevant. Of the attempts at cor-relation of these parameters with the results of rotating beam tests and with service tests of complete springs, only very few have been successful in identifying wire from which complete springs would or would not meet service life requirements.

As a result of direct service tests on other than helical forms, Figs. 3.225 and 3.226 show the behavior of torsion bars and leaf springs. Concern had arisen over the wide variation in service life for units made from different heats of steel but

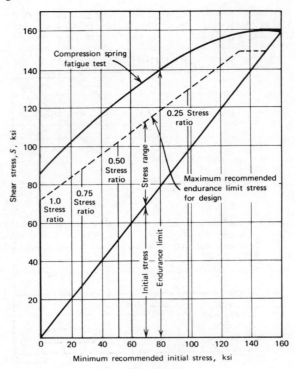

Fig. 3.223 Generalized Goodman diagram for music wire. From [207], Fig. 4.

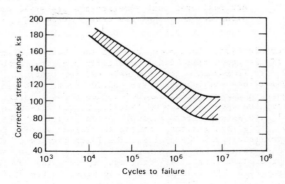

Fig. 3.224 Typical scatter band in fatigue tests of music wire helical springs, stress range zero to maximum. Wire size 0.022-0.048 in.

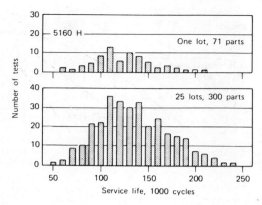

Fig. 3.225   Results of simulated service fatigue tests of front
suspension torsion bar springs of 5160H steel.   The
hexagonal bar section was 1.25 in. across flats.
From [2].

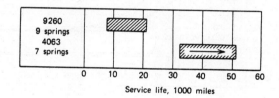

Fig. 3.226   Results of service tests of front leaf springs
on 1-1/2 ton trucks.   (The arrow in the lower
bar indicates that some springs are still operating at the end of the test.)   From [2].

all to the same specification. Bars made from the same heat of steel were regarded
as constituting a single production "lot," possessing a minimum number of variables.
Normally, about 12 bars from such a lot were selected for testing. The life test
results on a typical 12-bar sample varied from 85,000 to 175,000 cycles, with the
average life of the bars approximating 125,000 cycles. For a larger number of tests,
the diagram resembles the upper portion of Fig. 3.225, where results were obtained
on life tests of 71 bars selected from a single production lot. The variation
(55,000 to 215,000 cycles) was greater than for the 12-bar sample, but the average
life was again approximately 125,000 cycles. The lower portion of Fig. 3.225 presents the consolidated results for a number of 12-bar conventional samples taken
from 25 different production lots. This diagram, representing 300 torsion bars,
indicates ranges from 45,000 to 245,000 cycles and peaks in the vicinity of 125,000
cycles. Other data involving springs are summarized in Fig. 3.226. Service failures
of 9260 steel dictated a change in material; the results shown led to its replacement by 4063. Both steels met the original specification, and these data do not
account for the consistent superiority of 4063. Such correlations require extensive

testing of both wire and complete springs; so that, with the determinations of the statistical distribution of lives in the several lots, they become both tedious and costly. Because of the latter conditions, the approach is generally applicable only to large production.

For the special form of spring known as the Neg'ator or constant force type, the design curves of Fig. 3.227 are drawn below all of approximately 200 test points for each material. An upper bound curve, if drawn, would appear at about 1.5 times the ordinate at any life; on the sloping portion, the abscissa displacement or variation in life would be about one order of magnitude. Such results are indeed typical of production manufacture and emphasize the degree to which the unknowns in the melting and processing of alloys affect the performance of the finished part. The limitations of design approach and material specification are still sufficiently severe to preclude an accurate prediction of performance with known confidence. The following discussion of alloys and treatments should be read in this context.

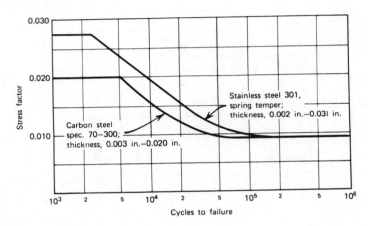

Fig. 3.227 Neg'ator spring stress factors. Stress
factor thickness/winding radius. Maximum
design allowable values. From Ametek/Hunter
Spring Division [207], Figs. 1, 2.

As for any part under cyclic stressing, the fatigue life of springs is highly sensitive to surface defects. The difficulty of describing and measuring the defects leads to a scarcity of numerical data on their effects because of the dual conditions: (1) the impossibility of producing perfectly smooth wire, or strip, and of manufacturing springs with no marks or dents, and (2) the known beneficial effect of fatigue life of inducing compression stresses on the surface, many springs are shot-peened after coiling. The process is often applied to hot-wound, heavy wire units where it acts to relieve the residual tension from forming and to produce surface compression. Data on the probability of failure of springs so treated are given in [2]; Fig. 3.228 distinguishes the benefits of the order of presetting and peening.

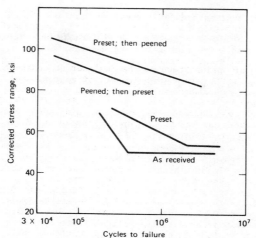

Fig. 3.228   Fatigue curves for helical springs, hot-wound, 0.9% C steel.
Mean stress, 63,000 psi; d = 0.5 in.; D = 2.62 in.   From
R. C. Coates and J. A. Pope [362], by permission of Brit.
Inst. Mech. Engrs. and the authors.

Steels for springs cover a wide range of compositions: carbon steels in the middle
to higher carbon range, AISI 1050–1095 for cold-formed or wound springs, and the
low alloys for hot-wound configurations, typically AISI 4160, 6150H, and 8660H.
The tensile and torsional strengths of the many compositions are not widely different, and the choice of alloy is usually based on freedom from surface defects,
ease of formability, sensitivity to hydrogen embrittlement, and cost.   The popular
music wire, ASTM A228, is the highest quality carbon steel wire for small springs,
Its surface quality is high, comparable to that of valve spring stock (VSQ), and
it is not particularly subject to embrittlement.   Hard-drawn wire ASTM A227 is the
least expensive and of lowest surface quality, usually considered only for low
stresses.   Oil-tempered wire ASTM A229 is heat-treated for the enhancement of its
properties; it has good surface quality, and medium cost.   Valve spring wires are
produced with the highest surface finish obtainable and are relatively costly; the
alloys include A230, plain-carbon; the better A232 (Cr-V) for service to 250°; and
A401 (Cr-Si) to 450°F.

For fair corrosion and heat resistance, cold-drawn A313 (stainless steel type 302)
is a popular choice and has excellent surface quality.   It is available in several
tempers, somewhat more expensive than carbon steel for designs requiring diameters
over about 0.012 in. but less costly than music wire under 0.012 in. because of
the lack of need for plating.   Stainless steel type 316 is a good spring wire
having better corrosion resistance than 302, particularly in salt water, but it is
not so readily available.

Once the required strength level is determined and the environment permits a steel
to be used, the next most important condition is that of surface quality.   In critical applications, the details of surface conditions and inspection may well be a
matter of negotiation between the user and the producer, sometimes for each stock
lot.   The length and the location of seams and laps and the depth of decarburization are particularly important.   Visual, magnetic particle, black-light, and

micrographic are the usual methods of inspection, against the agreed standards. In stock for cold-winding (diameters under about 0.5), occasional seams and decarburization to a depth of 0.002 in. may be tolerated; in hot-winding stock (diameters above about 0.5 in.), more frequent seams and decarburization as high as 0.020 in. are sometimes found. Salt-bath recarburizing and/or shot-peening can often recover or improve the fatigue life of such material. Both decarburization and mechanical defects may be removed by grinding or etching; such finishes represent the ultimate in stock quality and may be specified at extra cost.

For general corrosion resistance and high electrical conductivity, copper-base alloys are often chosen as spring materials. For high stress, long life applications, beryllium-copper (1.7-2.0% Be) is preferred because it can be precipitation-hardened to about Rc40 and thus becomes comparable to many steels in mechanical properties. It also retains its elastic strength to about 600°F, although the rate of loss is high at slightly higher temperatures. Its room temperature fatigue strength at 10^8 cycles ranges from 32 ksi annealed to 42 ksi in the HT (hardened) condition, the tensile strength ranging similarly from about 70 ksi to 200 ksi. Both cartridge brass and the nickel-silvers have a fatigue limit of about 20 ksi, while phosphor-bronze and OFHC copper have somewhat lower limits. However, the conductivity is inversely related to the fatigue limit for these alloys; thus, a compromise must always be made in electrical applications, while corrosion resistance would otherwise control the choice of alloy.

CHAPTER 4

GENERAL FACTORS

4.1   The Mechanism of Fatigue

4.1.1 Summary

Investigations into the physical aspects of fatigue have revealed four salient points which appear common to many, if not all, metals and alloys.

1. Slip on an atomic lattice plane starts at a common value of critical resolved shear stress, approximately 200 psi, and at a common strain of 10^{-4}.

2. The slope of the plastic strain vs. the failure cycles curve, (ε_p vs. N_f) is common at $-1/2$.

3. A value of 1% plastic strain causes failure in about 1000 cycles.

4. The structural metals and alloys divide generally into two groups: those which cyclically strain-harden, and those which cyclically strain-soften; while a few (e.g., copper) will do either depending on their initial state.

4.1.2 Discussion

While progressive cracking as a result of slip on the atomic planes is fairly well-accepted as a description of the fatigue process, discussion continues of the applicability of the several crack propagation "laws".

The terms fatigue and damage (or loss of loading capacity) should be synonymous, and the sequence of words in the title of this Section might imply that the fundamental process or mechanism by which the damage occurred was well known. Nothing could be further from the truth. Despite the enormous expenditure of talent, time and money during the last century or so, the fundamentals have yet to be measured and described mathematically.

Some of this difficulty undoubtedly lies in our insufficient knowledge of the solid state in terms of the detailed properties of the polycrystalline alloys, and some perhaps in our inability to interpret properly the vast amount of data already on hand. But the fatigue investigator and the crystallographer have recently correlated enough of their results to provide us with a fairly adequate picture of the process.

It is now well accepted that the deformation of crystals under a cyclic load occurs by slip or shear displacement of one atom plane past an adjacent plane. The examination of partially fatigued polycrystalline test bars under high magnification shows the result of this motion as slip lines. The location of initial slip on a plane and the choice of a plane are generally determined by the type of crystal and by the presence of a discontinuity in the form of a dislocation (microcrack),

foreign particle, or a void.   Such a break in the regular atomic pattern is a
stress concentration that acts to increase the local stress beyond the shear limit.
A dislocation may be defined as a stable arrangement of atoms such that, in the
region of a few atomic distances, (n + 1) atoms in the slip plane face (n) atoms
across the slip plane.   Cyclic input of strain energy alters the thermal and internal surface energies such that dislocations tend to collect on the planes already
containing one or more of them.   Eventually (n) atoms face (n + 2), (n + 3), or
more atoms, with the result that relative motion of one plane to the other occurs.
Small quanta of strain energy are required to slip one plane one atomic distance
past the other plane; separation normal to the planes does not usually occur.   See
Fig. 4.1.

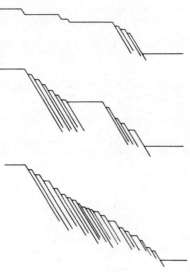

Fig. 4.1   Schematic of Progressive Slip (Slip line spacings vary from
100 Å to 200 Å, approx.)

Continued cycling or input of strain-energy collects more dislocations on these
planes, until the crystal separates into two or more fragments, with interference
and distortion of the adjacent crystals.   Further cycling would eventually cause
the cracks to grow and coalesce into an open area, resulting finally in insufficient
area to support the load, and fracture.   Progressive fracture or cracking is the
commonly applied term, and an apt one it is.   All alloys, glasses, and plastics
contain these discontinuities inherently on the dislocation scale, and, on the
much larger metallurgical scale, voids, foreign particles, crystalline mismatch,
and so forth, as a result of processing.   In copper, iron, and aluminum, electron
micrographs have shown slip displacements of several hundred Angstroms after cycling
to only 0.1% of the engineering fatigue life, plus numerous other discontinuities
at which slip had not yet started.

Because of the high symmetry of metal and alloy crystals, there are many slip systems available for deformation, the "system" comprising one of the crystallographic
planes within the unit lattice plus a single direction in that plane.   For the

body-centered-cubic metals there are 48 theoretically possible systems, but experiment shows that primary slip starts and generally proceeds on only one of them. The acting face is always that one for which the shear stress resolved in the direction of slip is the greatest, by the law of maximum shear stress. Extensive slip starts when the resolved shear stress has reached a certain constant value, S_0, the critical resolved shear stress, as in Fig. 4.2. High tensile or compressive stresses are required to produce the critical resolved shear stress if the slip plane is nearly parallel or perpendicular to the external force direction. The lowest tensile stress, $2S_0$, is required when the slip plane lies at 45° to the force.

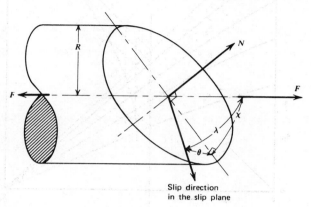

Fig. 4.2   Geometry for critical resolved shear stress. $A = \pi R^2$; $A_{slip} = A/\sin x$; critical resolved shear stress (CRSS) = $F/A \sin x \cos \lambda$.

The critical resolved shear stress should not be confused with any of the critical engineering stresses; it is a material constant depending only on chemical (atomic) composition and temperature. Now, a highly interesting fact appears: for many metals the critical resolved shear stress has nearly the same value, about 200 psi, and the corresponding elastic shear at the yield is about 10^{-4}. The external force and the stress normal to the slip plane may vary very widely, but extensive slip still starts at the same value of critical resolved shear stress, which is obtained independently of whether the external force is tension, compression, or torsion. Thus originate the familiar <u>cup and cone</u> shear faces in the fracture of a ductile tension bar.

Many attempts have been made to derive relationships between the number of cycles to failure, N_f, and stress or strain, some of which utilized the near constancy of the above. In general, those equations based on engineering stress and the S-N curve directly have not been too well accepted; see portions of Section 4.2.4. However, the relationship between the strain and the number of cycles has been more firmly established, except for some residual discussion on accuracy. It was early recognized that both the elastic and plastic components of the total strain should be considered; but that plastic strain was the more important in the low cycle region, and that elastic strain became more important when N was above about 10^4 cycles. Manson published a power law relationship in 1953 as

$$\varepsilon_p = MN_f^z \tag{1}$$

where M and z were material constants.

Coffin [88] showed that M was related to the ductility and z was the slope of the ε-N curve; he rewrote Eq. 1 to the form

$$N^k \Delta \varepsilon_p = C \quad \text{or} \quad \Delta \varepsilon_p = CN^{-1/2} \tag{2}$$

where   N = number of cycles to failure
k = -1/2 = z
$\Delta \varepsilon_p$ = plastic strain range
M = C = a constant related to the fracture ductility

This relationship has been found to hold for a great many metals and alloys, and the constancy of the exponent value at about -1/2 has led to the successful derivation of an equation in stress and N. If we consider that Eq. 2 predicts fracture ductility and that failure occurs in 1/4 cycle, then

$$C = \left(\frac{1}{4}\right)^{1/2} \left(\varepsilon_f\right) = \frac{\varepsilon_f}{2} \tag{2a}$$

and C can be determined from the value of reduction of area in a tensile test:

$$\%RA = \left[\frac{(A_0 - A)}{A_0}\right] \times 100$$

from which

$$\ln\left[\frac{100 - \%RA}{100}\right] = \ln\left(\frac{A}{A_0}\right) = \varepsilon_f$$

where RA = reduction in area
A_0 = initial area
A = final area
ε_f = fracture ductility

Relating the parameter C to the fracture ductility in terms of the reduction in area, we get the expression

$$C = -\frac{1}{2} \ln\left[\frac{100 - \%RA}{100}\right] \tag{3}$$

For application in design the plastic strain range ε_p is converted into the total strain range $\Delta \varepsilon_f$. The difference, $\Delta \varepsilon_t - \Delta \varepsilon_p$, is the elastic portion of the total strain range that can be approximated by the expression $2S_y/E$, where S_y = the yield stress. For interpretation in fatigue data, the yield stress is assumed to be the limiting value of the stress at which the cycles to failure N is infinite. Since it is customary to use the 0.2 percent offset yield strength, and this is usuallly higher than the endurance limit, an unconservative fatigue curve results in the low stress-high cycle range. It has been suggested that a better fit might be obtained if the elastic strain range is taken as $2S_e/E$, where S_e is the endurance limit and E is the modulus of elasticity.

From Eq. 2, Coffin [88] and Langer [279] have derived an expression for the desired fatigue curve in terms of the stress amplitude S:

$$S = \frac{EC}{2N^{1/2}} + S_e \tag{4}$$

where the value of C is found from Eq. 3. Hence S is the stress amplitude, comparable to that obtained from analysis. Equation 4 seems to be an excellent practical form for a generalized fatigue equation. It contains an S_e term to account for the high cycle regime; the first term is derived from ductility and low cycle strain considerations. It represents an attempt to express the nonlinear S–N function over a wide range of N (from 1/4 to ∞). It generally gives conservative stresses, calculated values compared with experimental data showing only slight inaccuracy. The degree of fit between Eq. 4 and experiment for the range of cycling between $N = 1/4$ and $N = 10^8$ depends on three factors.

1.   The reliability of the endurance limit information available for a particular material.

2.   The degree of error in assuming that C in Eq. 2a equals $\varepsilon_f/2$.

3.   The error in calculating the elastic strain range using twice the endurance limit rather than the actual stress range.

The reliability in the endurance limit determines the fit at the high cycle region, and involves no basic obstacle to achieving a fit as good as desired. With respect to the difference between C and $\varepsilon_f/2$, [287] has shown that the agreement was extremely good for a majority of common metals examined, but for some other cases considered, $C > \varepsilon_f/2$. Consequently, the use of $C = \varepsilon_f/2$ in Eq. 4 will be conservative, that is, failure is predicted in fewer cycles than occurs experimentally for a particular stress amplitude S. In strain cycling the stress range is a complicated quantity depending on the material, the strain range, and the number of cycles among other parameters, as discussed in [280]. General use of the actual stress range to determine the elastic strain range is not practical from an engineering point of view. Substitution of the endurance limit introduces simplicity and provides accuracy in the high cycle end of the fatigue curves. However, it also leads to error in the low and middle range of the fatigue curve, since the actual stress range is greater than its assumed quantity, twice the endurance limit.

It has been suggested [281] that most of the (rather small) error could be eliminated by using a more accurate, experimentally determined elastic component of strain for representing high cycle behavior, rather than S_e. Extrapolation of this component, on a plot, could be carried into the low cycle region, but the improvement seems hardly worth the complication. However, the use of S_e presumes the existence of a fatigue or endurance limit, which, in general, is not true, and the S_e in Eq. 4 becomes a point on a curve of nonzero slope at some defined life. The greatest discrepancy between Eq. 4 and data seems to be in the range of 1000 to 2000 cycles, and, coincidentally, it has been noted that, for essentially all the conventional alloys, about 1% strain range causes failure in this range of N. A statistical correlation [283] has shown that, of the several rules of thumb quoted since the establishment of Eq. 4, Peterson's [97] is the best: "±1% strain ($\Delta\varepsilon = 0.02$) causes failure in 1000 cycles."

Fatigue Regimes and Maps. If one considers that the chief cause of fatigue damage is crack propagation, as distinguished from initiation, and that propagation is controlled predominantly by the total strain range and temperature, it is possible to identify the several failure regimes, as:

> Low Cycle Fatigue, controlled by plastic strain;
> High Cycle Fatigue, controlled by stress–intensity;
> Non-propagation, controlled by stress–intensity;
> Oxidation, at high temperature;
> Cavitation, at high temperature.

A preliminary set of fatigue maps for 304 stainless steel in continuous cycling at 0.1 Hz in air is shown in Fig. 4.3, [469] for three crack lengths. This map is cut off at very high temperatures by a metallurgical instability: the dissolution of the chromium carbides. Related maps are being evolved for deformation and fracture, see Section 3.2.4., and it is expected that they will be useful in assessing the operating points for the alloys.

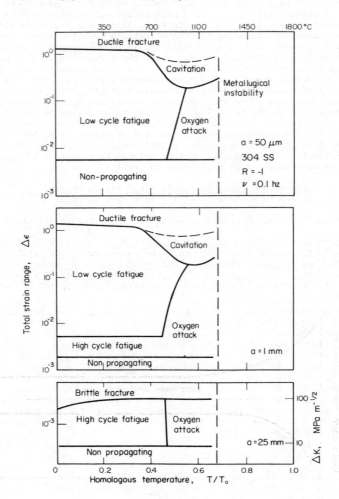

Fig. 4.3   Fatigue mechanism map for 304SS at 0.1 Hz. Crack lengths 50 μm; 1mm and 25 mm (from Ref. 469).

<u>Work-Hardening/Softening</u>. Any attempt to explain the process of fatigue should accommodate the phenomena of work-hardening and work-softening. These processes occur in apparently all polycrystallines, to an extent dependent on the ductility and on the level of stored strain energy at the start of cyclic straining, (that is, whether the fatigue specimen is annealed or cold-worked). When fatigue specimens are cycled between fixed strain limits, the stress range, $\Delta\sigma$ or ΔS generally changes during the test. Experiments by Smith, et al [297] have shown the typical variations of stress range ($\Delta\sigma = \Delta P/A$) with cycles for several test materials at three different values of applied strain range (zero mean strain in all cases). One group of materials including the heat-treatable ferritic alloy steels, hardened AM 350 stainless steel, and titanium (6A1-4V) is characterized by a stress range that decreases from the initial value. Since the stress required to produce a fixed strain decreases in successive cycles, and since hardness tests indicate a coincident softening, these materials are described as cyclic strain-softening. A second group of materials, including stainless steels, Inconel X, 5456-H311 aluminum, and beryllium shows an increase of stress and hardness during strain cycling. These are therefore called cyclic strain-hardening materials. Additionally, for the softening group, the ratio of their ultimate to yield strengths was 1.2 or less; for the hardening group it was 1.4 or more. From this condition it would be expected that an alloy having a ratio value between 1.2 and 1.4 would exhibit relatively little change. Another quantitative measure of cyclic strain hardening and softening has been given as the ratio:

$$\frac{\text{stress to produce 1\% strain after strain cycling}}{\text{stress to produce 1\% strain in virgin material}}$$

Table 4.1 lists the same materials as mentioned above but in different groupings.

One very particular condition has been identified in alloys with a high capacity for work-hardening and/or work-softening--the endurance stress level becomes indistinguishable from the yield strength. This condition is prominent in alloys of high copper content such as OFHC and ETP grades, and in the 304 types of stainlesses Figure 4.4 illustrates the action wherein an alloy "works up to" or "works down to" a mean yield.

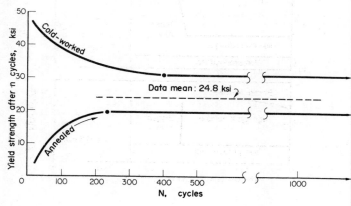

Fig. 4.4 Dislocation saturation strength of OFHC copper, at 70°F.

TABLE 4.1   Quantitative Measure of Cyclic Strain-Hardening
and Strain-Softening

Material	Ratio of Stress to Produce 1% Strain after Strain Cycling to That for the Virgin Material	
AISI 4130 (hard)	0.60	
AISI 4340 (annealed)	0.63	
AISI 4340 (hard)	0.64	
1100 Aluminum	0.68	
AISI 52100	0.72	Cyclic
AISI 4130 (soft)	0.75	strain-softening
Titanium (6A1-4V)	0.78	
AM 350 (hard)	0.89	
2014-T6 Aluminum	0.90	
AISI 304 ELC (hard)	1.07	
Beryllium	1.19	
Inconel X	1.24	
5456-H311 Aluminum	1.39	Cyclic
AISI 310 (annealed)	1.61	strain-hardening
AM 350 (annealed)	2.42	
AISI 304 ELC (annealed)	2.90	

SOURCE: R. W. Smith, M. H. Hirschberg, and S. S. Manson [297], Table 4.

This capacity of many alloys to harden of soften under cyclic loading has been very extensively investigated with the result that a set of "monotonic", "cyclic" and "fatigue" properties are now recognized. Monotonic** tension properties of an alloy can be classed into two groups; engineering stress-strain, and "true" stress-strain properties. "Engineering" properties are associated with the original cross sectional area of the test specimen, and "true" values relate to the actual area while the specimen is under load. The difference between true and engineering properties is insignificant in the low strain region, say up to 1 or 2% strain.

Monotonic stress-strain properties are generally determined by testing a smooth polished specimen under axial loading.

The load, diameter, and/or strain on the uniform test section is measured during the test in order to determine the materials stress-strain response, as illustrated in Figs. 4.5 and 4.6. Properties, most of which are discrete points on the stress-strain curve, can be defined to describe the behavior of a material.

**The term monotonic is used in preference to static, since the test is usually conducted by continuously increasing with time the distance between the cross heads of the test machine (or better still, the strain on the specimen) until fracture occurs.

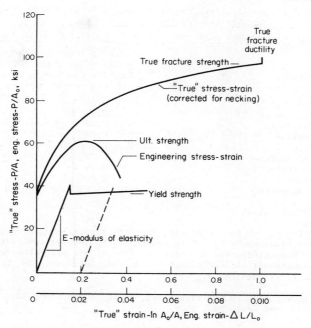

Fig. 4.5   Engineering and "true" stress-strain plot for 1020 HR steel.

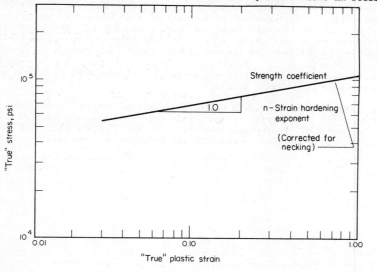

Fig. 4.6   "True" stress-plastic strain plot for 1020 HR steel

Definitions of Monotonic Properties:

Ultimate Strength (S_u) is the engineering stress at maximum load.   In a ductile material, it is governed by necking of the specimen.

where:
$$S_u = P_{max}/A_o \qquad (1)$$

P_{max} = maximum load

A_o = original cross sectional area

True Fracture Strength (σ_f) is the "true" tensile stress required to cause fracture.

$$\sigma_f = P_f/A_f \qquad (2)$$

where:

P_f = load at failure

A_f = minimum cross sectional area after failure.

Tensile Yield Strength (S_{ys}, σ_{ys}) is the stress to cause a specified amount of inelastic strain, usually 0.2%. It is usually determined by constructing a line of slope E through 0.2% strain and zero stress. The stress where the constructed line intercepts the stress-strain curve is taken as the yield strength.   (E = modulus of elasticity.)

Percent Reduction of Area (% RA) is the percentage of reduction in cross-sectional area required to cause fracture.

$$\% \ RA = 100 \left[\frac{A_o - A_f}{A_o} \right] \qquad (3)$$

True Fracture Ductility (ε_f) is the "true" plastic strain required required to cause fracture.

$$\varepsilon_f = \ln(A_o/A_f) = \ln \ (100/(100 - \%RA)) \qquad (4)$$

Monotonic Strain Hardening Exponent (n) is the power to which the "true" plastic strain must be raised to be proportional to "true" stress. It is generally taken as the slope of log σ_p - log ε_p plot.

$$\sigma = K\varepsilon_p^n \qquad (5)$$

Monotonic Strength Coefficient (K) is the "true" stress to cause a "true" plastic strain of unity (see Eq. 5).

Until the test bar begins to locally neck, some simple relationships exist between engineering and "true" stress-strain values.   Eq. 6 gives the relationship between engineering and true strain:

$$\varepsilon = \ln \ (1 + e) \qquad (6)$$

where:

ε = "true" strain

e = engineering strain

Similarly, Eq. 7 relates true stress to engineering stress:

$$\sigma = S(1 + e) \tag{7}$$

where:

    σ = "true" stress

    S = engineering stress

A more detailed discussion and derivation of monotonic and the cyclic stress-strain properties can be found in ASTM STP 465. Figures 4.5 and 4.6 graphically illustrate a majority of these properties.

Cyclic Stress-Strain Properties:

Cyclic stress-strain properties are determined by testing smooth polished specimens under axial cyclic strain control. The cyclic stress-strain curve is defined as the locus of tips of stable "true" stress-strain hysteresis loops obtained from companion test specimens. A typical stable hysteresis loop is illustrated in Fig. 4.7 and a set of stable loops with a cyclic stress-strain curve drawn through the

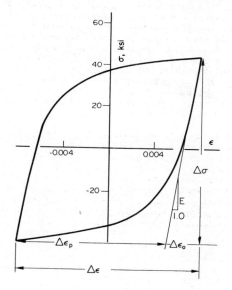

Fig. 4.7 Stable stress-strain hysteresis loop.

loop tips is illustrated in Fig. 4.8. As illustrated, the height of the loop from tip-to-tip is defined as the stress range ($\Delta\sigma$). For completely reversed testing, one-half of the stress range is generally equal to the stress amplitude, while one-half of the width from tip-to-tip is defined as the strain amplitude ($\Delta\varepsilon/2$). Plastic strain amplitude is found by subtracting the elastic strain amplitude ($\Delta\varepsilon_e/2$) from the strain amplitude as indicated in Eqs. 8-10:

$$\Delta\varepsilon_p/2 = \Delta\varepsilon/2 - \Delta\varepsilon_e/2 \tag{8}$$

OFD - N*

According to Hooke's law,

$$\Delta\varepsilon_e/2 = \Delta\sigma/2E \qquad (9)$$

where:

E = modulus of elasticity

Then:

$$\Delta\varepsilon_p/2 = \Delta\varepsilon/2 - \Delta\sigma/2E \qquad (10)$$

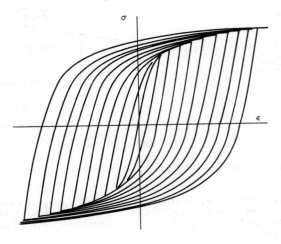

Fig. 4.8 Cyclic stress-strain curve drawn through stable-loop tips.

Definitions of Cyclic Properties:

<u>Cyclic Yield Strength</u> (0.2% a_{ys}) is the stress to cause 0.2% inelastic strain as measured on a cyclic stress-strain curve. It is usually determined by constructing a line parallel to the slope of the cyclic stress-strain curve at zero stress through 0.2% cyclic yield strength.

<u>Cyclic Strain Hardening Exponent</u> (n') is the power to which the plastic strain amplitude must be raised to be proportional to stress amplitude. It is taken as the slope of the log $\Delta\varepsilon_p/2$ - log $\Delta\sigma/2$ plot, where $\Delta\varepsilon_p/2$ and $\Delta\sigma/2$ are measured from cyclically stable hysteresis loops.

$$\Delta\sigma/2 = K'(\Delta\varepsilon_p/2)^{n'} \qquad (11)$$

where: $\Delta\varepsilon_p/2$ = "true" plastic strain amplitude. The line defined by this equation is illustrated in Fig. 4.9.

<u>Cyclic Strength Coefficient</u> (K') is the true stress to cause a plastic strain of unity in Eq. 11.

Stress-strain response of some steels can change significantly when subjected to inelastic strains such as can occur at notch roots due to cyclic loading. When fatigue failure occurs, particularly low cycle fatigue, such inelastic straining is generally present. Hence, the cyclic stress-strain curve may better represent the steel's stress-strain response than the monotonic stress-strain curve.

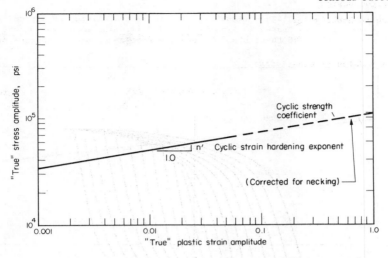

Fig. 4.9   Cyclic stress–plastic strain plot for 1020 HR steel.

In many field test situations, it may be desirable to convert measured strains to stress in order to estimate fatigue life. The cyclic stress–strain curve can be described with an equation using the cyclic properties. Eq. 10 can be rewritten as shown in Eq. 12:

$$\Delta\varepsilon/2 = \Delta\sigma/2E + \Delta\varepsilon_p/2 \qquad (12)$$

Rearranging the terms in Eq. 11 indicates the relationship between plastic strain amplitude and stress amplitude:

$$\Delta\varepsilon_p/2 - (\Delta\sigma/2K')^{1/n'} \qquad (13)$$

Substituting Eq. 13 into Eq. 12 yields an equation relating cyclic strain amplitude to cyclic stress amplitude in terms of the previously defined properties and the modulus of elasticity:

$$\Delta\varepsilon/2 = \Delta\sigma/2E + (\Delta\sigma/2K')^{1/n'} \qquad (14)$$

For a more detailed discussion, see ASTM STP 465.

Fatigue Properties:

Fatigue resistance of metals is generally described in terms of the number of constant amplitude stress or strain reversals* required to cause failure. The properties defined in this section are determined on smooth, polished axial specimens tested under strain control. Stress amplitude, strain amplitude, and plastic strain amplitude can each be plotted against reversals to failure. The plot of log

---

*A reversal is counted each time the stress- or strain-time signal changes direction. In constant amplitude testing, one cycle is equal to two reversals. (See also discussion in Section on the Rain-flow counting Method).

"true" plastic strain amplitude versus log reversals to failure are typically straight lines, as illustrated in Figs. 4.10 and 4.11. The intercept at one reversal and the slope of these straight lines can be described as fatigue properties.

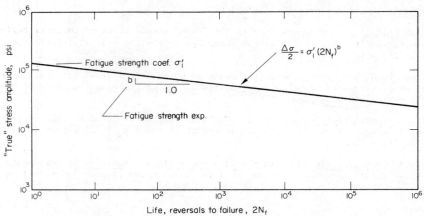

Fig. 4.10 Stress amplitude vs. reversals to failure, 1020 HR steel.

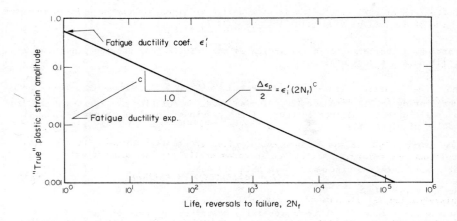

Fig. 4.11 Plastic strain amplitude vs. reversals to failure, 1020 HR steel.

Definitions of the Fatigue Properties:

Fatigue Ductility Exponent (c) is the power to which the life in reversals must be raised to be proportional to the "true" strain amplitude. It is taken as the slope of the log $(\Delta\varepsilon_p/2)$ versus log $(2N_f)$ plot.

Fatigue Ductility Coefficient (ε_f') is the "true" strain required to cause failure in one reversal. It is taken as the intercept of the log $((\Delta\varepsilon_p/2))$ versus log $(2N_f)$ plot at $2N_f = 1$.

Fatigue Strength Exponent (b) is the power to which life in reversals must be raised to be proportional to "true" stress amplitude. It is taken as the slope of the log $(\Delta\sigma/2)$ versus log $(2N_f)$ plot.

Fatigue Strength Coefficient (σ_f') is the "true" stress required to cause failure in one reversal. It is taken as the intercept of the log $(\Delta\sigma/2)$ versus log $(2N_f)$ plot at $2N_f = 1$.

Transition Fatigue Life $(2N_t)$ is the life where elastic and plastic components of the total strain are equal. It is the life at which the plastic and elastic strain-life lines cross.

Discussion:

A metal's resistance to strain cycling can be considered as the summation of the elastic and plastic resistance as indicated by Eq. 15:

$$\Delta\varepsilon/2 = (\Delta\varepsilon_e/2) + (\Delta\varepsilon_p/2) \tag{15}$$

An equation of the "true" plastic strain-life relationship can be written in terms of the above fatigue properties (Fig. 4.11):

$$\Delta\varepsilon_p/2 = \varepsilon_f'(2N_f)^c \tag{16}$$

where: $2N_f$ = reversals to failure.

The "true" elastic strain-life relationship is simply the stress-life relationship divided by the modulus of elasticity (Fig. 4.10):

$$\Delta\varepsilon_e/2 = (\sigma_f'/E)(2N_f)^b \tag{17}$$

Substituting Eqs. 16 and 17 into Eq. 15 gives an equation between "true" strain amplitude and reversals to failure in terms of the fatigue properties:

$$\Delta\varepsilon/2 = (\sigma_f'/E)(2N_f)^b + \varepsilon_f'(2N_f)^c \tag{18}$$

This equation is illustrated in Fig. 4.12.

These properties have been experimentally determined for at least 30 structural alloys, mostly steels and aluminums, in various forms and heat treats. Values for a few selected alloys are given in Table 4.2.

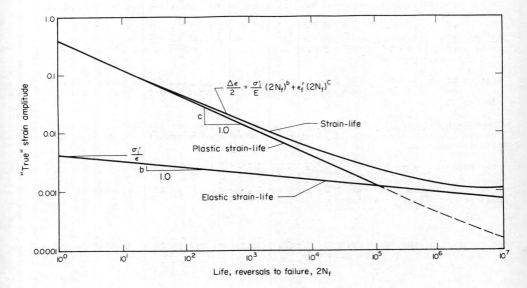

Fig. 4.12 Strain amplitude vs. reversals to failure,
1020 HR steel.

TABLE 4.2a   Monotonic Stress-Strain Properties of Selected Alloys

SAE SPEC	GRAIN DIR	PROCESS DESCRIPTION	ULT STR KSI (MPA)	YIELD STR KSI (MPA)	TRUE FRACT STR KSI (MPA)	% RA	TRUE FRACT DUCTILITY	STRAIN HARD'G EXPONENT	STR COF KSI (MPA)
A-508-A	L	SOL.TR. & AGED	220(1517)	215(1482)	275(1096)	67	1.10	0.030	
1005-1009	LT	H.R. SHEET	52(359)	39(269)	104(717)	73	1.3	0.12	73(503)
1005-1009	LT	C.D. SHEET	68(469)	65(448)	108(745)	66	1.09	0.029	78(538)
30304	L	H.R. & ANNEALED	108(745)	37(255)	228(1572)	74	1.37		
30304	L	C.D.	138(951)	108(745)	246(1696)	69	1.16		
30310	L	H.R. & ANNEALED	93(641)	32(221)	168(1158)	64	1.01		
4340	L	D & T	180(1241)	170(1172)	240(1655)	57	0.84	0.066	229(1579)
1100 AL.	L	AS RECEIVED	16(110)	14(97)		88	2.09		
2014-T6	L	SOL.TR. & ARTIF. AGE	74(510)	67(462)	87(600)	25	0.29		
2024-T351	L	SOL.TR.STRN. HARDEN	68(469)	55(379)	81(558)	25	0.28	0.032	66(455)
2024-T4	L	SOL.TR. & R.T. AGE	69(476)	44(303)	92(634)	35	0.43	0.20	117(807)
5456-H311	L	STRAIN HARDENED	58(400)	34(234)	76(524)	35	0.42		
7075-T6	L	SOL. TR. & ARTIF. AGE	84(579)	68(469)	108(745)	33	0.41	0.113	120(827)

TABLE 4.2b   Cyclic Stress-Strain and Fatigue Properties of Selected Alloys

SAE SPEC	GRAIN DIR	PROCESS DESCRIPTION	MOD OF ELAS KSI (GPA)	CYC YLD. KSI (MPA)	CYC STRAIN HARD'G EXP	CYC STR COF KSI (MPA)	FAT STR COF KSI (MPA)	FAT STR EXP	FAT DUC COF	FAT DUC EXP
A-508-A	L	SOL.TR. & AGED	27000(186)	150(1034)	0.09		240(1655)	-0.065	0.30	-0.62
1005-1009	L	C.D. SHEET	29000(200)	36(248)	0.12	71(490)	84(579)	-0.09	0.15	-0.43
1005-1009	L	H.R. SHEET	29000(200)	33(228)	0.11	83(572)	75(517)	-0.059	0.30	-0.51
30304	L	H.R. & ANNEALED	27000(186)	104(717)	0.36		350(2413)	-0.15	1.02	-0.69
30304	L	C.D.	25000(172)	127(876)	0.17		330(2275)	-0.12	0.89	-0.77
30310	L	H.R. & ANNEALED	28000(193)	50(345)	0.26		240(1655)	-0.15	0.40	-0.57
4340	L	D & T	28000(193)	110(758)	0.14		240(1655)	-0.076	0.73	-0.62
1100 AL.	L	AS RECEIVED	10000(69)	9(62)	0.15		28(193)	-0.106	1.8	-0.69
2014-T6	L	SOL.TR. & ARTIF. AGE	10000(69)	60(414)	0.16		123(848)	-0.106	0.42	-0.65
2024-T351	L	SOL.TR.STRN. HARDEN	10600(73)	62(427)	0.065	95(655)	160(1103)	-0.124	0.22	-0.59
2024-T4	L	SOL.TR. & R.T. AGE	10200(70)	64(441)	0.08		147(1014)	-0.11	0.21	-0.52
5456-H311	L	STRAIN HARDENED	10000(69)	52(359)	0.16		105(724)	-0.11	0.46	-0.67
7075-T6	L	SOL.TR. & ARTIF. AGE	10300(71)	76(524)	0.146		191(1317)	-0.11	0.19	-0.52

4.2   Fatigue Tests and Forms of Data

4.2.1 Objectives, Types, and Evaluations

Machine and structural designers are inevitably concerned with performance testing, which now often includes fatigue runs of models and of deliverable assemblies; while they may not be particularly interested in laboratory tests per se, sufficient background is required to avoid misapplication of the data from either kind of test.

Fatigue tests are made for a wide variety of reasons, among them:

   1.   To determine a material property such as the endurance limit stress or the life at a given stress level; both of which are, of course, affected by the many variables of material condition, mode and sequence of loading, etc.

   2.   To investigate the behavior of a part or an assembly in terms of an allowable fatigue stress at a required life, or the inverse; to check locations of maximum stress and the values of the stress concentration factors; and to see the actual effects of environmental factors as temperature and corrosion.

   3.   To compare designs on the basis of stress or life.

   4.   To determine the residual strength of a damaged article, in which the term "damaged" includes the partial consumption of the conventional fatigue life.

The evolution of fatigue testing has been in the direction of increasing sophistication, urged on by the ever-increasing need to express more properly and completely the material parameters and the service conditions. From the original, simple rotating beams, the samples have progressed to bi- and tri-axially stressed geometries, with and without preloads and gradients; on to very large (i.e., 8 ft x 20 ft) panels of an actual construction with carefully introduced loads, and finally to full-size aircraft and autos of a proposed construction. Of equal concern is the proper expression of the loads, periodic and random, programmed or sequenced, and the test equipment for producing them. In this regard, the leading efforts have been made by the aerospace and automotive industries in cooperation with governmental laboratories, especially ASIP, the Aircraft Structural Integrity Program.

ASIP is generally described in MIL-STD-1530A and 83444 which make reference to the detailed specifications for damage tolerance and durability (fatigue). The new requirements in testing are based on the premise that the most important purpose of a full-scale test is the identification of "hot spots" or regions critical in fatigue and which were missed in the earilier analysis or component testing. This condition makes it imperative to conduct fatigue testing as early as possible in any development program so as to avoid costly downstream modifications and retrofit. The fatigue test is also useful for:

   1.   Demonstrating that the economic operational lifetime does indeed exceed the design service lifetime when subjected to the design loads/environment spectra.

   2.   Providing a basis for establishing when and what special (previously unplanned) inspections or modifications are required for in-service units.

   3.   Providing a basis for verifying initial quality (flaw distribution) estimates used in the original design.

   4.   Verifying the procedures used for analyzing flaw growth.

   5.   Providing partial (and possibly total) compliance to full-scale damage
tolerance testing requirements.

The Industry Advisory Group (IAG) behind this effort is concerned not only with
the myriad items of structural integrity, but also the clear air turbulence problem in terms of gust loads and frequencies, with means of determining residual
strength and of deriving design parameters from the result. Such efforts represent
a tremendous advance in recognition of the scope of the life prediction problem.
To most designers prior to about 1940, "fatigue test" meant running a Wohler sample
to fracture and counting the cycles. While such work will still be needed after
1981, it will represent a decreasing portion of the total effort to calculate the
life of a curved, pressurized panel or a welded truck frame under random loading.

4.2.2 Design of Fatigue Tests

The design of fatigue tests to produce "useful" results is, in general, similar to
that of any set of experiments, but the high inherent scatter in fatigue data makes
the question of statistical significance especially acute. Thus, any set or group-
ing of fatigue tests must be planned as an entity to result in data that can be
treated statistically. One powerful scheme for doing this is to construct factorial experiments and handle the data by the techniques of analysis of variance,
and by regression.

A major effort in this type of activity is the investigation of the fatigue behavior of welded steel bridges by the National Highway Research Board [422]. The
definition for the plan of this work was--"A Statistically-Designed Experimental
Program that includes 374 Specimens Tested under Controlled Conditions so that
Analysis of the Data can reveal the Significance of Several Parameters believed to
be Important in Fatigue Behavior." The specimen type was a welded WF beam, and
the parameters were: 1) type of beam; 2) type of steel; 3) type of welding detail;
4) nominal stress in outer fiber at detail or at point of maximum moment.

After the stress levels and steels for the design variables had been selected, each
beam series was arranged into factorial experiments, Fig. 4.13; the specimen designation is given in Table 4.3.

Each cell of the cover-plated and flange splice beam factorials contained at least
three specimens or replicates. This permitted the variance of each cell to be
estimated. Because the cover-plated beams and the flange splice beams each had
two basic details per beam, information was available for only two locations. It
was for this reason that at least three replicates were provided.

The plain rolled and welded beams only had two specimens assigned to each cell.
Because only one basic geometric configuration existed for each beam, it was considered that more than one fatigue crack was probable between the load points.
This would increase the number of critical locations so that more than one test
value would be available per beam.

None of the experimental factorials was complete. That is, each level of stress
range was not tested at every level of minimum stress. Partial factorials for the
series were developed because of known boundary conditions. The maximum values of
stress had to be limited to stresses near the yield point--otherwise the plastic
strength of the A36 steel beams would be exceeded and the beam would fail under
static loading. Another limitation for some beam specimens was the jack capacity
of the testing facilities. The lower values of stress range were not examined at
all values of minimum stress because the longer anticipated lives would have unduly
extended the testing time. At least 10 million cycles were applied before testing
was discontinued and a fatigue limit was assumed to be reached.

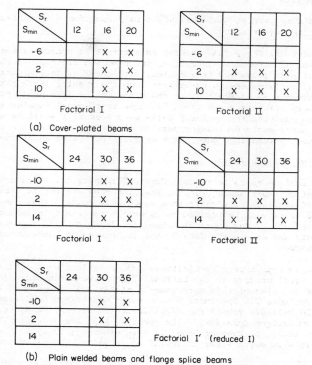

Factorial I   Factorial II

(a) Cover-plated beams

Factorial I   Factorial II

Factorial I′ (reduced I)

(b) Plain welded beams and flange splice beams

Fig. 4.13 Complete factorials for each beam series.

Two complete factorials were usually contained within the basic factorial of each beam series. These are shown in Fig. 4.13. This meant that for the design variables that had been selected, beams were included for each possible combination of these levels. Thus, any level of a particular factor would occur in conjuction with all other possible combinations of levels for the remaining factors.

As is apparent from complete Factorial I for cover-plated beams shown in Fig. 4.13, the dimensions for the CM, CB, CT, and CR-CW beam series for each grade of steel and for each detail were 2 x 3; that is two levels of stress range existed in combination with three levels of minimum stress. Factorial II related three levels of stress range to two levels of minimum stress. The stress levels in the complete factorials were selected to cover the range of life of most interest in design. Comparable complete factorials were designed for the other beam series, typically Table 4.4 for the CR and CW cover-plated beams.

TABLE 4.3   Specimen Designation

Example: CWA312

CW	A	3	1	2
BEAM TYPE	STEEL	S_{min}	S_r	SPECIMEN NO.

Beam type: PR – Plain rolled beam
PW – Plain welded beam
FS – Flange splice beam
CW – Cover plate, welded beam
CR – Cover plate, rolled beam
CT – Cover plate, thickness
CB – Cover plate, width (breadth)
CM – Cover plate, multiple

Steel: A – A36
B – A441
C – A514

S_{min}:   1, 2, and 3 indicate first, second, and third
stress magnitude for S_{min}.

S_r: 1, 2, indicate first, second stress range.

Example: CRA311
Cover plate on rolled beam of A36 steel.
Third magnitude for S_{min} (10 ksi).
First stress range S_r (8 ksi).
Specimen No. 1.

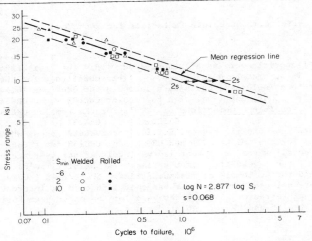

Fig. 4.14   Effect of stress range and minimum stress on the cycle
life for the welded end of cover-plated beams.   A36
steel.

The results of this work are shown (for the cover-plated beams) in Figs. 4.14, 4.15 and 4.16. It should be noted that this statistically significant information is vastly more useful than would have been just a series of S-N curves. Ref. 427 should be consulted for further detail on the statistical design of fatigue experiments.

TABLE 4.4 Experiment Design for CR and CW Cover-Plated Beams

Type Steel	S_{min} ksi	S_r, ksi				
		8	12	16	20	24
A36	−6			CRA131 CWA132 CWA133	CRA141 CWA142 CWA143 CWA144	CRA151 CWA152 CWA153
	2		CRA221 CWA222 CWA223	CRA231 CWA232 CWA233 CRA234	CRA241 CWA242 CWA243	
	10	CRA311 CWA312 CWA313	CRA321 CWA322 CWA323 CRA324	CRA331 CWA332 CWA333	CRA341 CWA342 CWA343	
A441	−6			CRB131 CWB132 CWA133	CRB141 CWB142 CWB143 CRB144	CRA151 ---- ----
	2		CRB221 CWB222 CWB223	CRB231 CWB232 CWB233 CRB234	CRB241 CWB242 CWB243	CWB251 ----
	10	CRB311 CWB312 CWB313	CRB321 CWB322 CWB323 CRB324	CRB331 CWB332 CWB333	CRB341 CWB342 CWB343 CRB344	
A514	−6			CRC131 CWC132 CWC133	CRC141 CWC142 CWC143 CRC144	CRC151 CWC152 ----
	2		CRC221 CWC222 CWC223	CRC231 CWC232 CWC233 CRC234	CRC241 CWC242 CWC243	---- CWC251 ----
	10	CRC311 CWC312 CWC313	CRC321 CWC322 CWC323 CRC324	CRC331 CWC332 CWC333	CRC341 CWC342 CWC343 CRC344	

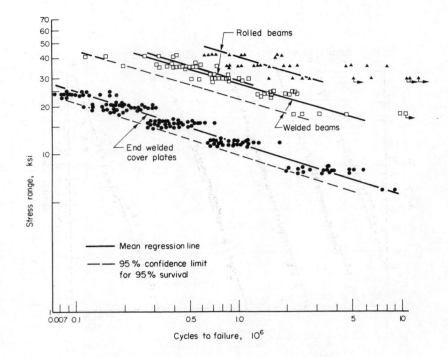

Fig. 4.15 Mean fatigue strength and 95% confidence limits
for rolled, welded and cover-plated beams.

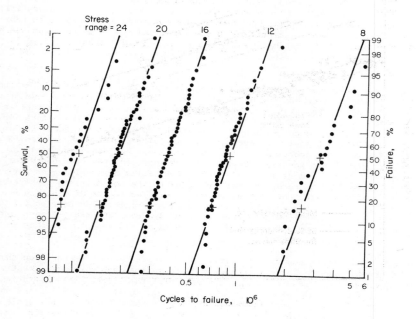

Fig. 4.16 Cumulative frequency diagram--all
beams with end-welded cover plates.

4.2.3   Forms of Data, Equations, and Plots

The experimental relationship between the applied stress level and the numbers of cycles to failure, when plotted on rectilinear coordinates, forms the oldest, most voluminous, and perhaps the least useful of forms of fatigue data. The reasons for this statement are, chiefly, that a separate curve (and set of specimens) must be obtained for every material, condition, shape, size, and so forth, and that such a curve must also be obtained for each parameter of testing, particularly that of stress range. The existence of a curve means that someone, usually unidentified, has already chosen a most likely path among the scattered data points, thereby performing the most important step in the data reduction. This condition has been set aside in part by the establishment of widely accepted "bands," limited by curves through the nearly extreme points, as the S-N figures in [1] and numerous other references.

It is important, however, to note that development of data from which S-N-P curves, Fig. 4.25 may be derived has not been extensive, and that in general these curves are not available. The shift to a log scale for the abscissa was introduced early for convenience in plotting; during the evolution of thought on the nature (mostly on the rate) of damage accumulation as a possible exponential function, the log ordinate also became useful. However, the attempts to plot the data as log S-log N soon showed that the function was not only nonlinear but nonexponential, at least if any reasonable accuracy was to be maintained, and that the hoped-for constant exponent in the damage law equation was not really a material constant after all. This condition is one of several which led to the more serious consideration of the strain-cycle relationship, and which in turn led to the distinction amongst the monotonic, cyclic, and fatigue properties of materials, (see Section 4.1.2). In general, a regression line on a strain-N plot is very acceptably linear (especially in the low-cycle regime where changes in slope are important), thus making possible the derivation of useful relations.

But in S-N terms, there seems to be little or no relief from the condition that a separate curve is needed for every condition of specimen and of testing, but the effect of the level of mean or steady stress and of the stress range on fatigue life has been inescapable almost from the beginning. Quite apparently, any specimen has a total strength capacity, above which no combination of mean and alternating stresses may be allowed to go without inducing failure. Wohler [84] was the first to point out that the number of cycles to failure depended on the stress range, S_r, and the value of S_r to cause failure at any given N decreased as the average or mean stress S_m increased. Figures 4.17 and 4.18 illustrate these effects.

Based on Wohler's data, Gerber [85] proposed a parabolic relationship between S_a and S_m:

$$S_a = S_e \left[1 - \left(\frac{S_m}{S_u} \right)^2 \right] \tag{5}$$

shown in Fig. 4.19.

This curve may be plotted in many ways; for instance, as absolute stress values or as a fraction of the static ultimate strength. The endurance limit curve only is shown in Fig. 4.19a; however, similar parabolas may be constructed representing curves of constant life, as in Fig. 4.19b. The highly useful constant R curves may also be superimposed on the chart. With test data from one particular R curve, it is possible to calculate the equation of a parabolic curve passing through the test data points as follows: from the general equation for a parabola,

$$y = ax^2 + b \qquad \text{or} \qquad S_R = aS_m^2 + b$$

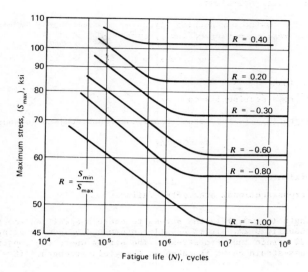

Fig. 4.17 Constant stress ratio curves (for steel). From
H. J. Grover, S. A. Gordon, and L. R. Jackson
[9], Fig. 19.

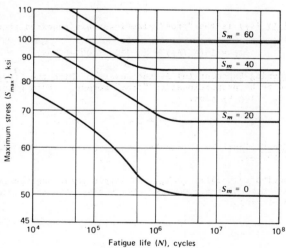

Fig. 4.18 Constant mean stress curves (for steel). From
H. J. Grover, S. A. Gordon, and L. R. Jackson
[9], Fig. 19.

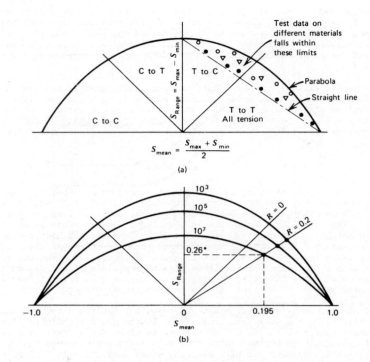

Fig. 4.19 Gerber diagrams. (S expressed as fraction of UTS.)

when

$$S_R = 0, \quad S_m = {}^{+}_{-}1.0$$

and when

$$S_R = 0.26, \quad S_m = 0.195, \quad \text{typically}$$

then

$$0 = a + b \quad \text{or} \quad a = -b$$

and

$$0.26 - a(0.195)^2 + b$$

solving

$$a = -0.27$$
$$b = +0.27$$

or

$$S_R = -0.27S_m^2 + 0.27, \quad \text{for } 10^7 \text{ cycles (for this example)}$$

This process may be repeated for other test data, resulting in a group of related curves. By superimposing lines of constant R, the corresponding coordinates of stress may now be taken from the predicted curves and S-N curves plotted for other ranges of stress. The accuracy of such calculations is of course dependent on the amount and precision of the available test data, and its relevance to the item under design.

From the Gerber diagram one sees that, in completely reversed bending, the mean stress is zero and $S_a = S_e$; or if the mean stress is equal to the ultimate, the $S_a = 0$ and the condition is that of static failure in one cycle. Noting that essentially all data points fell within the parabola and that it, therefore, gave very conservative results, Goodman [83] proposed the straight line relation:

$$S_a = S_e \left[1 - \left(\frac{S_m}{S_u} \right) \right] \tag{6}$$

In its original form, the Goodman relation was based on a number of rather restrictive assumptions, the chief of which was that the fatigue limit was one-third of the ultimate tensile strength; but in its present, modified form, it has been widely accepted. The major difference between Gerber and Goodman seems to be that the latter is slightly, but safely, less conservative. Söderberg has proposed an even more conservative relation based on the yield stress rather than the ultimate strength, and which apparently refers to gross yielding. This concept is in some conflict with the presently established idea that yielding, at least on a local or submacroscopic scale, is neither avoidable nor necessarily harmful, and that it may induce a redistribution of loads with a net beneficial effect.

The Goodman diagram is a useful means of presenting fatigue data, but in the general form of Fig. 4.20a the upper and lower range limits are for infinite life. The necessary modification to indicate finite lives is given for a steel in Fig. 4.20b. The need to include the stress ratio R as well as the mean stress S_m, resulted in the evolution to the Constant Life Diagram, shown schematically in Fig. 4.21 and for two typical alloys in Figs. 4.22 and 4.23. These plots indicate the behavior in terms of constant life lines for all conditions of mean and alternating stress that are possible to apply. The information is contained entirely within the triangles--regions outside the static boundaries represent stresses greater than the ultimate strengths. The diagrams may be regarded as a collection of sections made up of S-N curves, each at the appropriate mean stress, Fig. 4.24 and stress ratio, R. Heywood [5] found all meaningful conditions satisfied upon a mathematical investigation of the necessary and sufficient conditions that the diagrams should exist as closed triangles and that the curves obey both the data and the test of reason.

It is to be noted, Fig. 4.21c, that the subtriangle on the compression side beyond $R = \infty$ represents no practical conditions. Consequently, the trend is to simplify the classical diagram by eliminating this area; in many cases, the S_{min} and S_{max} axes are superimposed. Values of the A ratio, $A = S_a/S_m$, have also been added, and it is becoming common practice to include both the notched and unnotched lifelines on the same diagrams, as in Fig. 4.23.

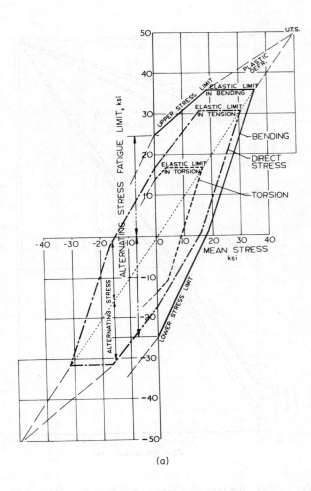

Fig. 4.20a Goodman diagrams. Original form.

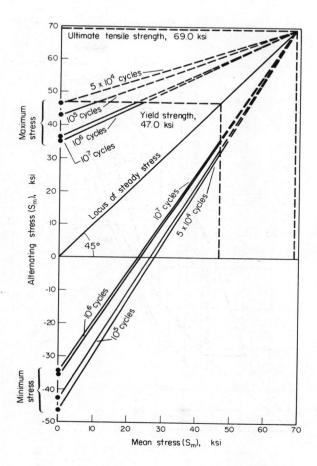

Fig. 4.20b Goodman diagrams. Modified form (for K-20
        steel, as an example).

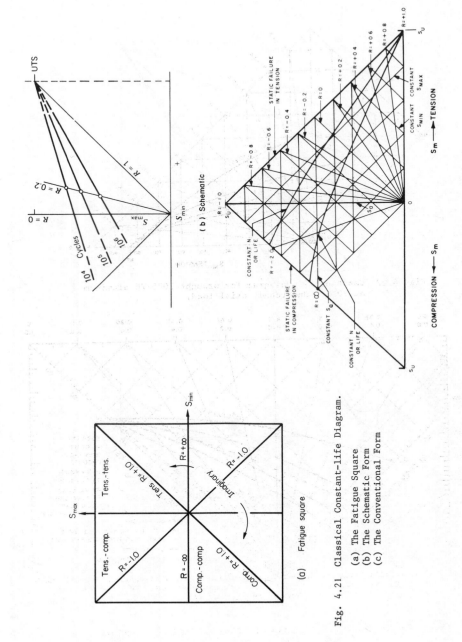

Fig. 4.21 Classical Constant-life Diagram.

(a) The Fatigue Square
(b) The Schematic Form
(c) The Conventional Form

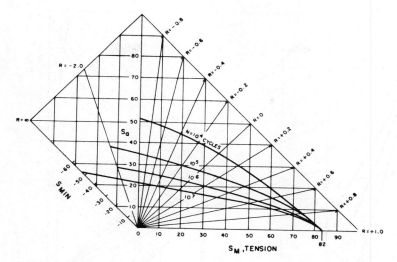

Fig. 4.22   Constant life diagram for wrought 7075-T6 aluminum at 70°F.   Unnotched, axial load.

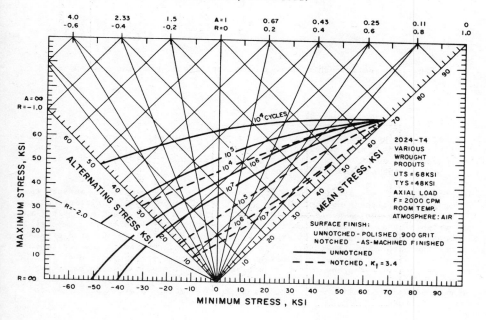

Fig. 4.23   Constant life diagram for 2024-T4 aluminum, complete form.

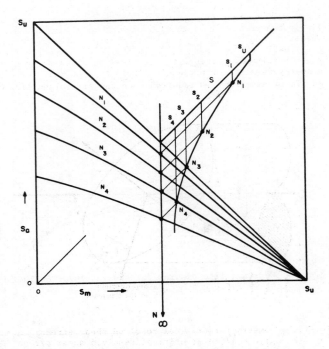

Fig. 4.24 Relationship between constant life diagram and S-N curve.

Scatter in Fatigue Data. The existence of scatter in test results is not peculiar to fatigue tests. Some variation can usually be found in repeated observations of any measurable quantity, but there are particular points that may be noted in connection with the scatter in fatigue data.

First, fatigue tests are destructive; therefore, it is impossible to measure the fatigue limit of the same specimen several times. Accordingly, variation observed in test results includes effects of possible errors in loading and measuring, and of variability of specimen selection and preparation. It is similarly impossible to measure the fatigue limit of a part by testing it to failure and then expect to use the part in a structure. Therefore, the use of results of fatigue tests (or of any destructive tests) in design is based upon these assumptions: (1) the test specimens are representative of a class of samples, and (2) the part to be used in service is of the same class. The validity of such assumptions and the corresponding application of the test results for design is often difficult to defend rigorously.

Other destructive tests, for example, static tensile tests, also show scatter, but the <u>degree</u> observed in <u>fatigue tests is generally large</u>, compared to that in static tensile tests. Figure 4.25 depicts the scatter in an axial loading teston 36 unnotched specimens of 2024-T3 aluminum alloy. The importance of scatter is emphasized by the fact that the dotted line is a mean curve estimated on the basis

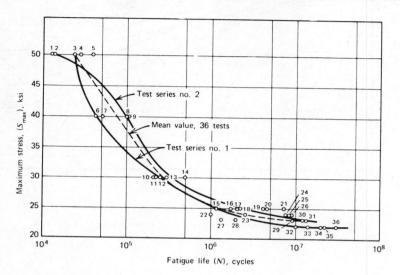

Fig. 4.25 S-N curves for a group of 36 specimens of 2024-T3
aluminum under axial load. R = -1.

of all 36 specimens, while the two solid lines are curves that might have been obtained from two experiments in which only 7 out of the 36 specimens were used in each experiment. Figures 4.26 and 4.27 are typical of the few available curves which are statistically based, while Fig. 4.28 shows life vs. percent failures.

It has been observed that most of the scatter observed in finite fatigue life testing of metals is an inherent characteristic of the material and does not nec-essarily indicate poorly adjusted machines or improper testing techniques. The distribution of the fatigue life of groups of similar specimens is quite varied, but has been found in several instances to approximate closely a logarithmic nor-mal distribution. The amount of scatter, which is usually measured by means of the standard deviation in life, is related to the stress amplitude, being smallest at high stresses and largest at stresses just above the fatigue limit. Therefore, one cannot legitimately speak of the fatigue limit in terms of a single value, but must regard the S-N curve more as a family of curves, each of which indicates a definite probability of failure, P, as shown in Figs. 4.28b and 4.29. In the latter figure it is obvious that, even under carefully controlled laboratory con-ditions using duplicate test specimens from the same bar, it is not unusual to encounter ratios of 10/1 in cycles to fracture at a given stress. Thus, the pre-cision with which a life may be specified is limited. Due to these effects of the inherent scatter, some type of safety factor or other expression of confidence must be used.

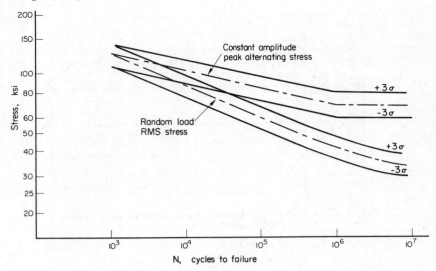

Fig. 4.26 Statistical S-N curves for 4340 steel. R = -1.0.

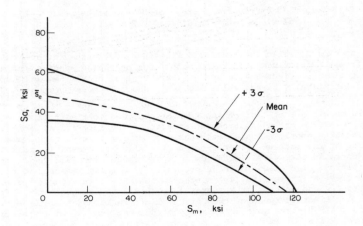

Fig. 4.27 Statistical Goodman Diagram for 4340 steel at 10^6 cycles.

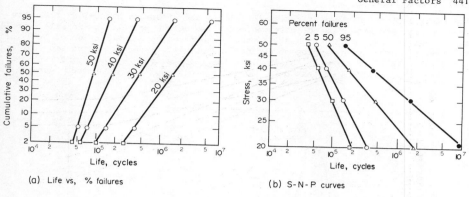

(a) Life vs, % failures

(b) S-N-P curves

Fig. 4.28 Results of a typical fatigue test; data having a log-normal distribution.

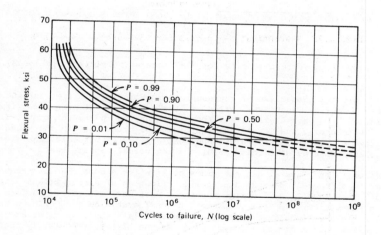

Fig. 4.29 S-N-P curves for 7075 aluminum, $R = -1$, $K_t = 1.0$.

The distributions of times or cycles to failure and of stress at a given life have been extensively investigated, with concentration on the normal, log-normal, Weibull and Gumbel extreme value. Figure 4.30 shows the frequency distribution of life for ST-37 steel in rotating-bending as decidedly skew with respect to numbers of cycles, but when analyzed in terms of the log life the results conform to a symmetric normal distribution. Analysis of this data, according to Weibull, also showed a good fit.

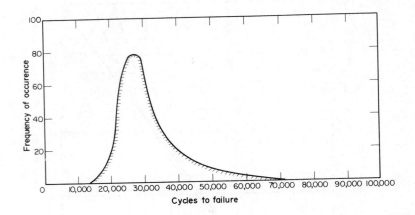

Fig. 4.30 Fatigue test results for ST-37 at ±45.5 ksi (235 specimens).

Figure 4.31 gives a comparison of representations of a large set of fatigue data (one stress level; all specimens failed) in terms of four different distributions, each defined by two parameters, i.e., the normal and two parameter Gumbel distributions of both cycles and log cycles. It is to be noted that in the range of probabilities of survival between 90% and 10%, all four representations give good agreement, and, that outside this range, the log normal appears best.

A distinction in the form of distributions of life for smooth and for notched specimens seems to be needed, because a normal distribution of either life or log life fits smooth data equally well but much notched data does not really conform to either representation. There is, however, a considerable body of opinion to the effect that the distribution of stress at constant life could be regarded as normal while the distribution of log life, at least at the higher stresses, is nearly normal, see Fig. 4.32.

Reference 426 bases a "Maximum Likelihood" technique for estimating the mean and standard deviations of the parent population of a sample of fatigue test results on a log normal distribution of that population, and comparison with the results of related techniques shows the Maximum Likelihood to give very reasonable approximation to the population parameters. (See also Refs. 427-9 for further discussion and detail.) At the present time, expressions of the probability of failure at a known confidence level, as derived by the techniques of Analysis of Variance and Regression Analysis, are widely accepted. Since these procedures for data reduction may become complex and the result may possibly be obscured by the assumptions, the designer should be guided by a statistician. Further discussion of these points is given in Section 3.2.3; see particularly Fig. 3.66.

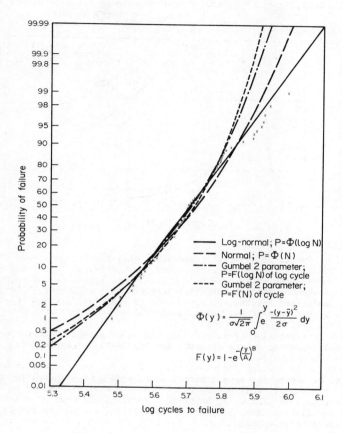

Fig. 4.31 Comparison of 4 two-parameter representations
of the fatigue test results of 100 specimens
of 0.22% C steel at ±21 tsi.

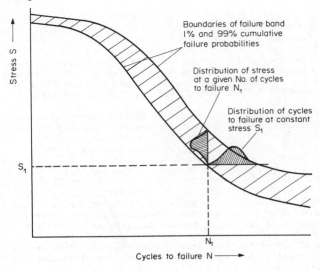

Stress S

Boundaries of failure band
1% and 99% cumulative
failure probabilities

Distribution of stress
at a given No. of cycles
to failure N_1

Distribution of cycles
to failure at constant
stress S_1

S_1

N_1

Cycles to failure N ——▶

Fig. 4.32 Distributions of fatigue failures.

In addition to the stress amplitude, numerous other parameters affect the scatter in life, particularly the stress ratio R and for nonconstant amplitude tests, the sequence with which the amplitudes are applied. One aspect of the results of various load sequences was discussed in reference to the Miner rule, Fig. 3.1; in general, the data from program or spectrum tests show less scatter than that from constant amplitude tests at any value of R.

The problem of presentation of fatigue data is also serious. In the United States, the universally accepted source of metal strength data, MIL-HDBK-5 [1], is undergoing a slow transition in three respects:

1. Use of statistics to refine the guaranteed minimum strength, the A and B values.

2. Evaluation of what, if any, statistical quantities are to be presented along with the A and B values.

3. The accumulation of fatigue data in statistically meaningful quantity.

In the past, fatigue properties have been shown in the form of typical S-N (stress vs. cycles of lifetime) curves for available testing conditions. As more data become available, these are being replotted as modified Goodman, or constant lifetime, diagrams to allow interpolation to the exact loading conditions required. The data could be presented as average curves or as curves of some stated probability of survival. As a practical matter, the constant lifetime diagrams so far have presented average curves, since available data generally are insufficient for meaningful statistical analysis. Likewise, creep data in the past have been shown as typical stress-time curves for various values of strain. Several empirical creep equations are now being studied for the purpose of combining data for various stresses, strains, times, and temperatures, and presenting these in a simple creep nomograph.

As this effort proceeds, probability criteria will undoubtedly be added, subject, however, to the same potential limitations that exist for fatigue data. Fracture toughness indices (K_c, K_{1c}, J_c, etc.) have not yet been included in MIL-HDBK-5. Much of the delay can be attributed to the lack of agreement on how to determine these indices and how to use them in design. At the present time, it appears that reliable and useful K_{1c} values can be provided for selected alloys and products. These will be termed indicative or typical values at first; however, these indices may soon be brought under the classification of design allowables.

The first two items refer generally to the static, room-temperature yield and ulti-mate and, regarding these parameters, the sample statistics of greatest interest for near-normal distributions are sample size (if less than about 300), sample mean, and sample standard deviation. The population (product, property, test direction, etc.) to which each value applies must be indicated clearly. For non-normal dis-tritions, attempts should be made to separate the populations into subpopulations having a normal distribution or transform them to log-normal distributions for which sample statistics can be presented. Otherwise, only frequency-distribution data and minimum values should be presented for non-normal distributions. The manner of presentation of the relevant quantities has not yet been decided, but it appears that they will eventually become available in an auxiliary document.

As an example of the statistical data treatments coming into use to refine the A and B values, tensile test data for 17-7PH stainless steel sheet, strip and plate have been analyzed in four steps:

    1.   Define the populations of strength values with respect to: (a) heat-treated condition; (b) property; (c) test direction; and (d) thickness.

    2.   Compute the mean $\bar{x}$, and standard deviation s_x for each of the samples drawn from the populations defined in Step 1.

    3.   If the sample statistics in Step 2 vary with thickness, express these as a functions of thickness.

    4.   Use the samples to estimate the form of the population distributions.

Ranking tables were established for all required quantities, and F_{tu} was computed from these sample statistics, assuming a normal distribution. Also, F_{ty} was com-puted for both distributions, normal and unknown, and the latter found to give the more conservative results, probably due to the skewing toward low strength. The computed values are given in Table 4.5.

In general support of such work a computer program has been developed for sample sizes. The program will sum the first i items of a binomial expansion of

$$(1 - q) = [p + (1 - p)]^n$$

where $q = 0.95$,

        $p = 0.90$ or 0.99

        n = the sample size

        i = the rank of the determining test point.

The sample size n was determined for successively higher values of i and n within the range ($i \geq 1$, $n \leq 3000$). Unit increments of n were utilized to determine n_B within the range of i ($1 \leq i \leq 188$) and units of 10 were utilized to determine n_A within the range of i ($1 \leq i \leq 21$). The error introduced by incrementing in units of 10 is considered acceptable.

TABLE 4.5   Computation of A and B Values for 17-7PH

Heat Treatment	Property	Assumed Distribution	Computed,[a] Values ksi		Proposed Values ksi	
			A	B	A	B
RH950	UTS(L)	Normal[b]	212.6	217.9	210	217
	UTS(T)	Normal	211.3	217.7	210	217
	TYS(L)	Unknown[c]	<202.5	206.1	190	201
	TYS(T)	Unknown	190.5	201.2	190	201
TH1050	UTS(L)	Normal	178.1	184.1	177	184
	UTS(T)	Normal	176.8	185.1	177	184
	TYS(L)	Unknown	<165.5	172.5	150	169
	TYS(T)	Unknown	151.4	169.2	150	169

[a]A and B values are defined as follows: At least 99% of the population exceeds an A value, with confidence of 95%. At least 90% of the population exceeds a B value, with confidence of 95%.

[b]For the normal distribution, A or B = $\bar{x} - ks_x$, where k is the one-sided tolerance limit factor for the normal distribution.

[c]If the distribution is unknown, the values of ranked test points are used to determine A and B values, as tabulated below:

Rank of Test Point

n	A Value	B Value
87	---	4
340	1	25
585	2	45

Investigation continues of methods for the determination of residual strength in terms of the load at fracture (or at gross yielding) of a cracked, partially fatigued specimen, especially in close relation to the fail-safe approach to design. An example of one of the methods was given in Section 2.4.2 and others are to be found in [58],[61] and [382]. But the bulk of recent work is involved in the fundamentals of crack propagation rates, critical lengths, and the characteristic size of initial defects, as well as in the relationships between stress and stress-intensity. Such efforts are usually considered under the title of Fracture Mechanics that now-well-developed approach to Life Prediction, which has evolved from the older Fatigue (sometimes called "S-N" approach.) Discussion and examples of Fracture Mechanics considerations are given in Section 3.2.4.

Material testing is usually done at constant load or constant displacement of the specimen, for simplicity in design of both the specimen and the test machine. This condition leads to simple stresses: bending with low torsion (the rotating beam), partially or fully reversed flexure (for sheet and strip), uniaxial tension-tension, and simple torsion. Special shapes of specimen and loading devices can be made up for combined stresses, but most of the published S-N type data derives from one of the simple stress modes above. There is a growing tendency toward the testing of

sub- or full-assemblies because of the great difficulty in modeling a geometrically complex part, and of the reverse process also: that of extrapolating test
results from the specimen to the part. Test rigs have been built for cycling full
size aircraft wings, fuselages, and pressure vessels, with some dimensions up to
200 ft., a far cry from the 0.3-in. diameter test beam, and correspondingly more
demanding in the interpretation of results.

The standard, constant amplitude tests are usually applied when specimens are
expensive, the material limited, or for full-size sections; a single sample is
tested at each stress level. Such a procedure gives very little information on
the variability of the material, the specimen, or the procedure--it is much more
satisfactory to test a group of specimens at each of several different stress
levels. To obtain even the most primitive probability curves and to estimate data
variability, each group should contain at least four specimens, and ten or more
specimens are preferable to give an indication of the distribution of life values,
all at three or more stress levels. The drastic variation in mean fatigue limit
stress with number of specimens was illustrated in Fig. 3.66. Groups of supposedly
identical specimens are often run at constant amplitude in tests of the material
for evaluation of the scatter in properties.

In the "Probit" method, one or more groups of specimens are tested for a given
number of cycles at several stress levels distributed about the level of interest,
usually the mean fatigue strength. It is often used to find the fatigue limit for
a material, defined as that stress at which 50% of the specimens will fail prior
to the preassigned cycle life, N, and 50% will survive. The "staircase" or "up-
and-down" method is a variant of the Probit requiring fewer specimens. The first
is tested for the prescribed number of cycles at a stress equal to the estimated
mean fatigue stress. If it does not fail, the second is tested at the next higher
increment of stress. If it had failed, the next specimen would be tested at the
next lower increment of stress. The process is illustrated schematically in
Fig. 4.33. Testing time may be saved by dividing a simple long staircase into
several shorter sections and running the several tests simultaneously.

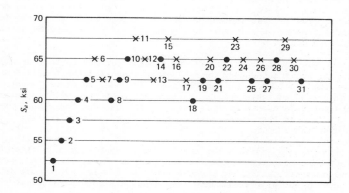

Fig. 4.33 "Staircase" or Probit method: x = failure;
● = no failure; N constant for each test.

The increasing-amplitude methods include the "Step" testing of a single specimen to obtain an estimate of fatigue limit. This sequence is sketched in Fig. 4.34, and it should be noted that the specimen accumulates a very large number of cycles at stresses below its limit, 5×10^7 in the sketch. Thus, the validity of the estimate depends largely on this understressing and its effect on the particular metal through strain-aging.

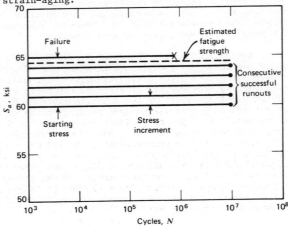

Fig. 4.34   "Step" method of fatigue testing.

The "Prot" and "Locati" variants of the Staircase Test can be used advantageously when only a few test specimens are available and when some information concerning their fatigue characteristics (e.g., the slope p, or m) exists. These methods can also be used to give a first indication of the fatigue limit as a starting point for more precise tests. But in general, they are not as precise as either the Probit or Staircase.

Two of the most difficult questions in any fatigue (or other) test are (1) the proper loads and their sequencing or randomizing, and (2) the required degree of completion of the specimen. For the first, it is usual to choose among the results of the formal stress analysis for the maximum and most frequently occurring loads, taken in an acceptable combination if they are not one and the same. Or the designer may utilize results of previous tests modified as required for the present case. Regarding the specimen, unlike the "one-horse shay," the members of an assembly are not all going to fail at once. One member is more critical than the others, and even if the designer has that supreme capability of picking out the most critical member and redesigning it for improved life, then the next most critical member becomes crucial. This process could be carried out almost indefinitely but is, of course, not practical. How then does this condition affect the choice of a test specimen? Obviously, the most completely simulated structure would yield the most reliable results but is rarely available, for the usual reasons of cost, schedule, and so forth. Thus, the decisions on both loads and specimen often devolve upon the engineer, with concurrence by the program manager and the customer.

Comparative Testing. Tests of this nature are relatively simple and are capable of producing highly useful results, not just data. They may be used to determine directional strength properties, the effect of different heat treatments and of attachment methods. As the name implies, they may be run on similar or different designs to permit a direct comparison of lives or of allowable loads at a given life. Actual values of service loads and completeness of structure need not always be duplicated as the comparative test is used to indicate relative improvements. On a quantitative basis, the comparative test can, however, sometimes indicate service lives with a hopefully known degree of approximation.

This type of test is a most practical way for determining the best attachment methods for joints and has been very useful in evaluating the effects of production tooling on life. As an example, a riveted lap joint, with the holes formed by machine countersinking, was cracking visibly at about 10,000 hours of service, somewhat less than half the required life. Of the other feasible methods of hole forming, the use of hot dimpling dies was thought possibly to be superior. Fatigue tests in axial loading were run on joints of identical design but with holes formed by these two methods, with results as in Fig. 4.35. Now, from the Linear Rule, one might expect that the service life is proportional to N, and write $N_D = N_m(n_D/n_m)$. The stress analysis for this particular joint indicates that the working stresses vary from a maximum of 25,000 psi to a minimum of 10,000 psi. For the purpose of life calculations, it is assumed that average maximum working stress causing the greatest amount of fatigue damage is $1/2(25,000 + 10,000)$ or 17,500 psi. Taking the median values of n from the test curves at this stress level, the life of the joint with dimpled sheet is calculated as:

$$N_D \sim T_D = 10,000 \, \frac{32,000}{14,000} = 22,800 \text{ hr.}$$

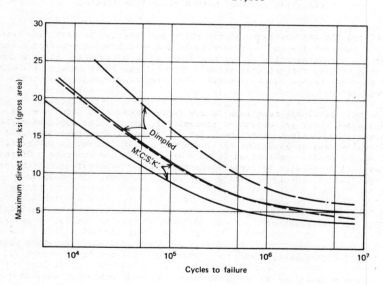

Fig. 4.35 S-N curves from a comparative test of riveted lap joints. R = +0.2.

Note that roughly the same life would result whether one took the minimum, median, average, or maximum values from the curves. Unless all service environments are reproduced in the test conditions, the comparative test method yields only that type of result--a comparison or relative standing--not absolute values.

Tests on Components. The calculations for service life of components to designs similar to existing items of known life may be predicated on past service experience, with little or no knowledge of working stresses; or a comparative test may be run as above. The fatigue analysis of new designs requires a thorough understanding of the behavior of the various materials, the working stresses most frequently applied, and the statistical probability of frequency and magnitude of service loading. Further, the engineer must realize the limitations with which laboratory test results from simulated structures may be equated to the actual structure loaded in service. This work involves the testing of full size and subsize geometrically similar components.

A full size test may imply that either the actual-size member is to be tested, or that the full size, build-up structure containing many members is to be tested. The disadvantage of the built-up structure is the cost. However, if the stress analysis is accurate enough and possibly supplemented with strain-gaged, static-load tests, then the designer may feel that a fatigue test of the single member, based on the loads derived from stress analysis will give reliable answers.

The testing of subsize specimens is often a most practical method when very large parts are under consideration, but a word of caution is necessary here. Experimentally, it has been shown that both the static and the fatigue behavior of large specimens and goemetrically similar subsize specimens are different. The basic size effect is to establish an inverse relation between finished section thickness and allowable static stress. The degree of variation is strongly dependent on the susceptibility of the material to work-hardening or strengthening and, therefore, on the total amount of working, chiefly the reduction in thickness from the original ingot to the final form. The work-hardening characteristics of most commercial alloys are such that the size effect would not show up to an appreciable degree if, say, an 8 in. x 8 in. billet or a 16 in. x 22 in. ingot were reduced to a 2-in. diameter bar, any intermediate heat treatments notwithstanding. But, if 2-in. diameter bars, one each from the billet and from the ingot were further reduced to, say, a 0.1-in. diameter wire, the size effect in terms of the increase in strength would be significant, that is, about 50% for mild steel from 1.5-in. diameter to 0.5-in. diameter. Specifically, for fatigue strengths, it has been shown by test that the region of the greatest increase in fatigue strength is in the small sizes, about 0.6 to 0.2-in. in diameter.

Figure 4.36 shows the test results on the light alloys for different types of loading and for notched and unnotched specimens, permitting the following conclusions:

1. There is a pronounced increase in both notched and unnotched bending or flexural fatigue limit with a decrease in size.

2. Under torsional loading, both notched and unnotched specimens show the increase in fatigue strength.

3. Unique among the different types of loading, unnotched specimens in direct tension do not exhibit this increase in fatigue strength with decrease in size. For notched specimens in direct tension, however, there is an increase which is explained by the stress gradient effect.

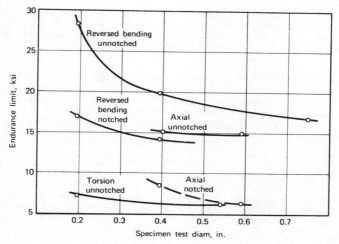

Fig. 4.36 Scale or size effect in light alloys. R = -1.0.

The greater fatigue strengths of the small cross section specimens arise from the lower probability of defects in the smaller volume and from the stabilizing effect of the lower stressed inner fibers on the higher stressed outer fibers, an action generally considered as dependent on the stress gradient. The theoretical treatment of this problem is based on the assumed stress distribution shown in Fig. 4.37. It can be seen that with decrease in size, the fatigue strength could not increase indefinitely but only to a value corresponding to the distribution shown in Fig. 4.37c. From this distribution, upper limits for flexural and bending fatigue strengths are obtained as equal to 1.5 times the direct stress fatigue limits for square bars and 1.7 times for circular bars.

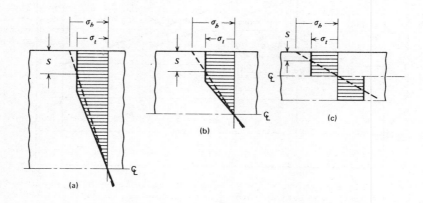

Fig. 4.37 Stress distributions. σ_b = reversed flexure; σ_t = reversed (direct) tension.

A further assumption was that the depth, s, of the small stabilized layer was a constant dependent upon the shape of the cross section and the type of loading. The calculated values for s, and for a few materials and loading modes, proved to be in agreement with the test results, listed in Table 4.6

TABLE 4.6

s-in.	Material and Loading
0.12	Steels--bending and flexure
.06-0.08	Steels--torsion
0.04	Light alloys--bending

The following example indicates a method of predicting the fatigue strength of a large section from known values for a geometrically similar but smaller section. For shouldered shafts in rotating bending, the empirical relation below had been derived for the fatigue strengths at lives between about 10^3 and 10^6 cycles:

$$S_L = S_s \left(1 - \frac{(D_L - D_S)}{15} \right)$$

For the sizes sketched in Fig. 4.38, there resulted

$$S_L = 0.89 S_s$$

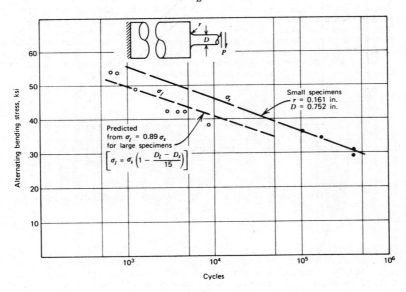

Fig. 4.38   Size effect in 2014 aluminum forgings.   R = -1.0.   o--Actual tests on large parts.   r = 0.50 in.; D = 2.25 in.   From [274], Fig. 55.

and the prediction line was drawn by reducing the ordinates of the test curve for the small specimens by this amount. Then, as a check, the full size forgings were tested (the open points), indicating some lack of conservatism at the longer lives. The geometrical similarity between the large and small specimens was controlled by making R/D ratios constant:

$$\frac{R_L}{D_L} = \frac{R_S}{D_S} = \frac{0.50}{2.25} = \frac{0.161}{0.752} = 0.22$$

Full Scale Testing. Much as been accomplished by the testing of full scale major subassemblies such as wings, tails, wheels, and so forth. Typical work is reported by [159], in which the objective was to examine the applicability of the cumulative damage rule to the design of a nose landing gear; General results from this investigation are that, for R > 0, the Miner criterion gives conservative estimates of life under spectrum loading, while, for R < 0, it resulted in unconservative estimates. Figure 4.39 plots the load vs. the median life.

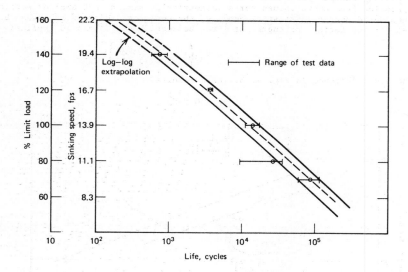

Fig. 4.39 Median life to failure of F-3A landing gear under constant amplitude tests. From M. S. Rosenfeld and R. J. Zoudlik [159], Fig. 3.

Full scale testing is increasing both in the size of the assembly tested and the frequency of use as the need for such testing becomes more widely recognized. The United States Air Force is now requiring full scale testing/(per [143]) of aircraft whose structure is designed to be fail-safe, or damage tolerant. As an example, the C-141 transport was designed to a goal of 30,000 hours of flying, combined with 20,000 fuselage pressurizations, and 12,000 landings. These figures represent lifetime factors from 1.5 to 4.0 depending on the component or region of the structure involved. Typical preliminary results are shown in Fig. 4.40, which, after suitable modification, were tested with complete satisfaction.

In contrast to the minimum-weight type of design discussed above, heavy, large section equipment has also undergone recent evaluation. A small but increasing number of wheel failures on railway freight cars has suggested that the limit of safe performance is being approached in some cases. Knowledge of stresses from specific wheel loading and brake shoe heating that exceed the fatigue limit should enable the railway managements to become aware of potentially dangerous operating conditions. The residual, thermal, and loading stresses have been evaluated with respect to the fatigue strength for a standard B33 steel wheel [162]. Major conclusions were:

   1.   The worst stress condition occurred with a heated rim, a vertical load, and a lateral load on the back rim, as in Fig. 4.41. Even the 100,000-cycle fatigue strength can be exceeded under these conditions; see Fig. 4.42b.

   2.   A high overload is not damaging when a high front face lateral load is present, but it can be damaging in the absence of a normal lateral load or with a back face lateral load.

   3.   In the absence of rim heating, fatigue failure is highly unlikely.

The residual stresses from manufacture by roll-forging and heat treatment were obtained by strain-gaging several production wheels, then sawing out two blocks from each wheel, one for radial and one for tangential strain measurement. Residual stresses for six production wheels averaged about 4000 psi tension at the back face N1-position (BFN1) and 11,000 psi tension at front face N2 (FFN2). These are the critical locations from the standpoint of fatigue because all known wheel plate failures in service have originated in one of these two positions.

Stresses from vertical and lateral loads were obtained from data taken in static tests, the forces being applied by standard rail sections as mandrels in the test machine. The stress patterns produced are best illustrated by considering the maximum loads used. In one wheel revolution, vertical loading stresses would fluctuate from -19,000 psi to + 4500 psi, a stress range of 23,500 psi. Maximum FFN2 stresses from the 20,000-lb. thrust loads, as in the vertical loading, are found in the radial direction at the 0 degree loading position. Loading on the front rim face results in a tensile stress of 19,000 psi at 0° and a compressive stress of 13,500 psi at the 180° position for a range of 32,500 psi. The lateral load stresses are opposite in sign to the vertical loading stresses at any point so that combined vertical and lateral loads at each position would result in lower net stresses. Loading on the back rim results in 24,200 psi compression at 0° and 9700 psi tension at 180°, a stress range of 33,900 psi. It is important to note that these stresses are in the same direction as the vertical loading stresses; therefore, the combined stresses would be additive and can be very substantial. The temperatures developed at wheel rims for operational braking periods were known from associated work. Thus, the thermal rim stresses were obtained by heating production wheel rims; at the critical locations BFN1 and FFN2, the radial stresses were 38,000 and 43,000 psi, respectively. The fatigue strength of the forged wheels was found by axial tension tests on 2-in. wide radial sections, five sections from each of the three wheels. The stress ratio was -1 with peak stresses set at 30 or 37.7 ksi; the data indicated a fatigue limit as about 22.5 ksi minimum, normal for steels of this general strength class (100 ksi UTS), and with a hot-rolled mill surface.

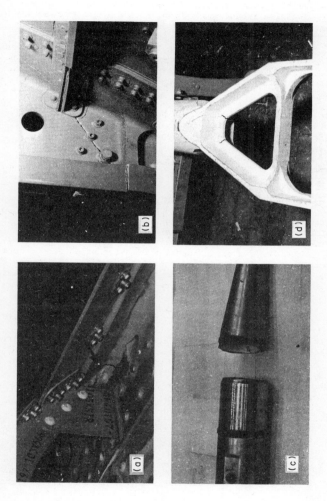

Fig. 4.40   Selected results from preliminary fatigue tests on the C-141 transport fuselage. (a) Crack in dorsal longeron cap; (b) Main frame crack originating at a bolt hole; (c) Trunnion shaft cracked after 18,740 simulated landings; (d) Cracks in aft radii of drag brace.   (Courtesy of the Lockheed-Georgia Co.)

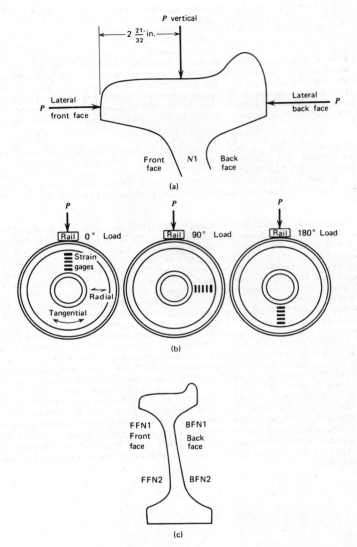

Fig. 4.41 Loading and stress locations on B33 railroad wheels.
(a) Loading positions relative to the wheel;
(b) vertical loading positions; and (c) locations
of high stress.

Combinations of steady and fluctuating stresses that might occur in service loading were analyzed, using modified Goodman diagrams. Data are plotted only for the FFN2 position in the radial direction, since this was found to be in the location on the wheel at which the highest combined tensile stresses occurred, coinciding with possible failure origins. Residual stress was kept constant at 11,000 psi tension for all the cases considered, while vertical loading, lateral loading, and heating stresses were varied to simulate service conditions. The fatigue envelope used for the diagrams was based on a ±22,500 psi fatigue limit at zero mean stress and the wheel plate tensile strength of approximately 110,000 psi. On some diagrams, fatigue strength bands for 100,000 and 1,000,000 cycles are also shown; 240,000 lbs. was used to represent a loaded car or 30,000 lbs. per wheel (910 lb/in. of wheel diameter).

Table 4.7 shows FFN2 steady and fluctuating stresses for several rim heating and loading conditions. A sample calculation illustrates how stresses are combined for the modified Goodman diagrams. It can readily be seen that stresses from vertical and lateral loading on the front rim face oppose each other at 0 and 180° positions. However, lateral loading on the back face results in additive loading stresses.

The data for several simulated loading conditions are plotted as modified Goodman diagrams. Figure 4.42a shows the cases for a fully loaded car (30,000 lbs. vertical) and a 50% overloaded car (45,000 lbs. vertical) combined with a substantial lateral load (20,000 lbs.) on the front rim face. Several trends are apparent here. The steady stress increases with rim heating time, causing the fluctuating stress to become closer to the fatigue limit. As the vertical load increases from 30,000 to 45,000 lbs., the fluctuating stress range contracts further from the fatigue limit, because the vertical and lateral stresses are opposing and the largest stress is due to the lateral loading. Figure 4.42b shows the case in which the lateral load is applied to the back face, which might occur when the wheel impacts against a guardrail. The fluctuating stress increases to exceed both the fatigue limit and 1,000,000-cycle fatigue strength. In fact, the 100,000-cycle fatigue strength is closely approached after 70 min heating for both normally loaded and overloaded cars, clearly a dangerous situation. The data were also studied as to the effect of the absence of rim heating (i.e., no braking) on the range of fluctuating stress at the FFN2 position. This condition provides a mean stress of approximately 12,000 psi tension with the fluctuating stress ranges similar to those of Fig. 4.42a, or about half the fatigue limit, and therefore well within the safe fatigue envelope. The results of this investigation emphasize the severity of the demands placed on the material in this application, which typifies one of the classical trade-offs so often possible (and frequently required) but not always recognized by the designer--why not separate the load-carrying and the braking functions?

TABLE 4.7   Radial Stresses at Location FFN2 in the B33 Railway Wheels

	Steady Stress Source	Stress (psi)
Residual		+11,000
Thermal	30 min heating	+29,200
	70 min heating	+42,500
Mean	45,000 lbs. vertical load	−5,600
	30,000 lbs. vertical load	−3,800
	10,000 lbs. vertical load	−1,350
	20,000 lbs. front face lateral load	+2,750
	10,000 lbs. front face lateral load	+1,100
	20,000 lbs. back face lateral load	−6,250

Fluctuating Stress Source	Fluctuating Stress (psi)		
	0°	180°	Range
45,000 lbs. vertical load	−14,300	+3,000	±8,650
30,000 lbs. vertical load	−9,600	+2,000	±5,800
10,000 lbs. vertical load	−3,600	+900	±2,250
20,000 lbs. front face lateral load	+19,000	−13,500	±16,250
10,000 lbs. front face lateral load	+9,200	−7,000	±8,100
20,000 lbs. back face lateral load	−24,200	+9,700	±17,950

Sample Calculation for 70 min Rim Heating + 30,000-lb.
vertical load and 20,000-lb. front face lateral load

Steady state stress = +11,000 psi (residual)
                      +42,500 psi (70 min heating)
                       −3,800 psi (vertical load mean)
                       +2,750 psi (lateral load mean)
                      ―――――――――
                      +52,450 psi Total

Fluctuating stress at 0° position = −9,600 psi (vertical) + 19,000 psi (lateral)
                                  = +9,400 psi
              at 180° position = +2,000 psi (vertical) − 13,500 psi (lateral)
                              = −11,500 psi
                  Combined = +9,400 psi − 11,500 psi = ±10,450 psi

Sample Calculation for 70 min Rim Heating + 30,000-lb.
vertical load and 20,000-lb. bac face lateral load

Steady state stress = +11,000 psi (residual)
                      +42,500 psi (70 min heating)
                       −3,800 psi (vertical load mean)
                       −6,250 psi (lateral load mean)
                      ―――――――――
                      +43,450 psi Total

Fluctuating stress at 0° position = −9,600 psi (vertical) − 24,200 psi (lateral)
                                  = −33,800 psi
              at 180° position = +2,000 psi (vertical) +9,700 psi (lateral)
                              = +11,700 psi
                  Combined = −33,800 psi + 11,700 psi = ±22,750 psi

SOURCE: J. P. Bruner, G. N. Benjamin, and D. R. Bench [162], Table 8.

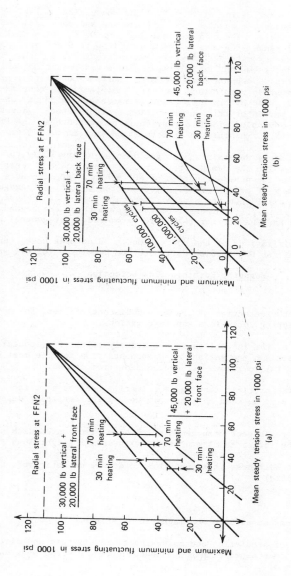

Fig. 4.42 Modified Goodman diagrams for B33 wheels under various loadings. (a) A fully loaded and a 50% overloaded freight car with high lateral loading on the front rim face; (b) a fully loaded freight car with high lateral loading on the back rim face. From J. P. Bruner, G. N. Benjamin, and D. R. Bench [162], Figs. 15, 16.

4.2.4 Detection of Flaws and of Fatigue Damage

Quality control by identification of flaws and the monitoring of fatigue cracks
are operationally inseparable because they utilize the same physical methods and
equipment, excepting x-ray. The conventional NDE (Non-Destructive Evaluation)
methods have a useful range of flaw size of about 0.050 to 0.250 inch, while some
of the advanced laboratory methods can work down to about 0.005 inch. The absolute
minimum detectable size seems not capable of rigorous determination, being a function of many parameters: mechanical and chemical surface conditions; grain size;
details of the equipment and its operation, etc. In Fracture Mechanics terminology
the minimum detectable size of crack is taken as the starting point for propagation
and is a fundamental parameter in life prediction by that method. The tangibility
of crack size as a measurable, linear dimension lends credence to the Fracture
Mechanics approach--further discussion will be found in Section 3.2.4.

The destructive methods of determining fatigue damage, or inversely, of determining
the conditions under which the damage fraction became unity, were described earlier.
This section addresses the non-destructive or NDE schemes, and includes "Flaws" in
its title because flaws or cracks are firmly connected to fatigue damage, even
though they may well exist when the damage fraction is zero. The physics, operation and evaluation of response traces from the many methods are beyond the scope
of this work, Barton and Kusenberger [420] are recommended for general descriptions.

The advantages and limitations of the conventional NDE methods have been admirably
summarized by the British Gas Turbine Collaboration Committee (GTCC), as follows:

Magnetic Particle Flaw Detection

Advantages

1. In addition to the detection of cracks and discontinuities, the method is also
    capable of detecting inclusions and segregations.

2. The efficiency of the inspection is not greatly impaired by contamination of
    the defect. Although the surface of components from service must be cleaned,
    it is not essential to remove all contaminant from cracks.

3. Although magnetic particle testing is primarily used as a surface inspection
    method, it can also be used to a certain degree for detecting defects just
    below the surface.

4. It is possible to inspect components which have been coated with a thin non-
    magnetic coating (e.g., cadmium plated), with only a slight loss in efficiency.

5. Under certain conditions, it is possible to detect magnetic inclusions in non-
    magnetic materials.

Limitations

1. The method cannot be used on non-magnetic materials.

2. The difficulty of predicting and measuring field strength and direction in
    complex shaped components with the consequent risk of leaving areas ineffec-
    tively magnetized.

3. The need to detect flaws in all possible directions necessitates a series of
    different magnetizations in different directions to ensure complete coverage.

4.   Previous magnetizations can adversely influence magnetic particle inspection and initial demagnetization must be carried out.

5.   Due to the abrasive nature of the particles and their tendency to clog in restricted passages, application of the method can only be made when effective post-cleaning can be carried out.

6.   Trained and graded personnel must be employed. Methods of standardization have not been developed.

Liquid Penetrant Methods of Flaw Detection

Advantages

1.   Penetrant methods can be used on any material independent of its physical properties.

2.   They are not seriously affected by the geometry of the part.

3.   Surface defects lying in any plane can be located by a single application of the penetrant process as penetrant flaw detection methods are not directional as are other methods.

Limitations

1.   Can be used to locate only those discontinuities at or reaching to the surface.

2.   Surface openings must be clean and free from contamination. Typical examples of contaminate encountered during manufactur are machining oils, forging and heat treatment scale, welding flux and pickling oils. On components which have been subject to engine running, other contaminants are encountered such as carbonized oils, oxide films, corrosion and paint. In addition to these, some protective coatings have to be removed such as anodic films and cadmium plating, as these will give rise to an unacceptable background against which the defect indications cannot be seen, or phenolic resins and similar coatings where cracks can occur in the base material without cracking the coating.

3.   Porous surfaces such as anodic films and cadmium plating are unsuitable for penetrant inspection, as they cause difficulty of interpretation, and also retain penetrant to give a background colour which reduces the contrast of true defect indications to such an output that fine defect indications are lost in the general background.

4.   Trained and graded personnel must be employed.

Problems Associated with the Application of Penetrants

1.   Penetration

   This is governed by the following factors.

   (a) The physical properties of the penetrant. These are controlled by the manufacturer and it is estimated that the most searching penetrants are capable of entering surface openings in the order of 0.00004 inches wide.

   (b) State and nature of the defect. Contaminants such as carbonized oils, paint films and oxide films will prevent the penetrant entering the defect. Previous operations such as grinding, shot-peening and vapour blasting can close the mouth of the defect and prevent the penetrant entering. These hazards may be reduced by chemical cleaning and/or etching.

(c) Compressive stresses on the surface of the component can hold the mouth of the defect tighly closed and prevent the penetrant entering.

(d) Time required for penetrant to enter a defect, a significant factor which may be governed by the previous factors. However, there are large discrepancies between the minimum contact times quoted by different authorities. These vary from a few minutes, to as long as twelve hours. This factor requires investigation to establish the minimum contact time required to obtain the maximum sensitivity for the various penetrant processes.

2. Removal of Surplus Penetrant from the Surface

This is the most critical operation of the penetrant process. If removal is carried out too vigorously, then penetrant may be partially or completely removed from the surface openings. If too little is removed, the background level will be high and mask indications of fine defects. Both are equally serious and will result in a loss of reliability and sensitivity.

Water washable penetrants can be formulated to give a wide range of rinseability, but to date, experience has shown that rinseability is inversely proportional to sensitivity. Post emulsified penetrants are the most sensitive group, as the emulsification is a separate step in the penetrant process and the rinseability can be controlled within limits for each type of component, thus giving optimum conditions between background and sensitivity.

Thus, the application has to be carefully considered and the most suitable penetrant system selected. To date, there is no one penetrant system suitable for every application.

3. Application of Developer

There appears to be no real problem associated with the application of developers, although care is required to prevent too thick a film being applied which can mask fine defects. Conversely, too thin a film of developer will not draw sufficient penetrant to the surface.

4. Operator Training and Grading

In order to obtain the maximum efficiency from penetrant methods of flaw detection, it is essential that the operators are fully trained and classified in accordance with their skill. In addition, supervision must be highly trained with a sound knowledge of all the factors involved and a wise experience of the various types of penetrant systems available.

Ultrasonic Flaw Detection

Advantages

1. Under favorable conditions, ultrasonic flaw detection is outstanding in its ability to detect small internal defects. This characteristic is maintained over a considerable range of metal thickness, with effective penetration of dense materials and it is generally possible to measure this range with considerable accuracy from any test station. It is, therefore, particularly suited to the inspection of raw materials at an early stage of manufacture

with corresponding economic advantages. The size and orientation of defects can be estimated although not with the precision with which their position can be measured.

2. Ultrasonic tests yield their information instantaneously and materials and components can be scanned rectilinearly, spirally or helically as required. High speed automatic scanning, which eliminates the human element of ultrasonic testing can often be achieved.

3. Ultrasonic methods lend themselves, in certain instances, to "in situ" inspection of simple components without completely dismantling the engine assemblies. Since the probe can be applied to an area away from the point to be inspected, defects can be detected, which might otherwise be hidden to visual examination.

4. A further inspection advantage of ultrasonic testing is the ability of the beam to be focussed, reflected and refracted. Ultrasonic energy can also be passed through many liquids, including water.

Limitations

1. Ultrasonic flaw detection is essentially a method best applied to materials and components of simple form, any changes of section and profile being liable to give an indication which complicates interpretation. In general, the complete examination of finished parts of complicated shape is not practical.

2. The limit on detectability of small flaws is set by the operating frequency of the equipment, the surface finish of the component and by the position of the defect. For most engineering materials operating frequencies of between 2 and 6 megacycles per second are acceptable and these frequencies will give an approximate limit of detectability for defects of greater than 0.015 inches diameter. At the same time, small flaws which are too close to the surface may be lost in the null zone of the probe. This null zone is related to a number of factors including the surface finish of the component being inspected.

3. Since ultrasonics rely on reflection, the best indication of a defect is obtained when the axis of the ultrasonic beam is at right angles to the greatest area of the defect. Thus, the direction of the beam relative to the defect is important and scanning from one direction only can give rise to a risk that only the most favorably orientated defects will be found.

4. Defects which can cause difficulty in detection and size estimating are inclusions which, in addition to being well fused into the surrounding material may be of similar acoustic properties.

5. Interpretation of ultrasonic echo patterns is of prime importance and for inspection to be really effective, the employment of highly trained personnel is essential. As ultrasonic waves are reflected freely at all the boundaries of the workpiece, intricate shapes can considerably complicate interpretation.

Radiographic Methods of Flaw Detection

Advantages

1. The methods can be used on any material independently of its physical properties, provided that absorption is not extremely high.

2. A permanent record is provided by the radiograph.

3. The fact that the image is geometrically related to the object aids interpretation.

4. Variation in the photographic density of the defect image gives information about the nature of the defect.

5. It is often possible to inspect inaccessible parts and to avoid dismantling.

6. Radiography has an established history of operator approval which confers a certain status and has resulted in the provision of training facilities for operators.

Limitations

1. Even under ideal conditions, the methods are insensitive to effective thickness changes of less than about 1 or 2 percent of the total thickness penetrated.

2. The sensitivity is considerably worse if the absorption characteristics of inclusions are comparable with those of the parent material.

3. The detectability of a crack depends on it being of sufficient width; the narrower the crack, the more important it is for its plane to be in the direction of the beam.

4. The methods are such that all features which are superimposed in the direction of radiation form a composite image. As a result, location of a flaw in depth requires views in more than one direction.

5. Skilled personnel are essential for the planning of techniques of radiographic inspection and the interpretation of radiographs.

    NOTE: The need for radiation protection in accordance with statutory requirements although not in itself a limitation of the method, must be understood by anyone undertaking radiography.

Fluoroscopic Methods of Flaw Detection

Advantages

1. Because an instantaneous image on a screen is obtained, all fluoroscopic methods have four major advantages:

    (a) The image is obtained without the use of consumable materials, thereby reducing running costs.

(b) As there is no waiting time for film processing, the speed of inspection is increased.

(c) The image can be viewed, whilst the specimen is moving which in many cases greatly facilitates interpretation.

(d) Using this facility it is possible to orient the specimen into the most favorable position for revealing planar defects.

2. The method can be used on any material independently of its physical properties, provided that absorption is not extremely high.

3. The fact that the image is geometrically related to the object aids interpretation.

4. Variation in brightness of a defect image gives information about the nature of the defect.

5. It is often possible to inspect inaccessible parts and to avoid dismantling.

Limitations

The general limitations of fluoroscopy are:

1. No permanent record is obtained unless additional equipment is employed.

2. The ability to detect defects is generally worse than with film radiography.

Eddy-Current Methods of Flaw Detection

Advantages

1. The test system operates by current induction and no direct contact or couplant is required.

2. The method is capable not only of detecting surface cracks but of giving quantitative information on their depth.

3. The method is capable of measuring changes in permeability and conductivity due to differences in metallurgical structure and, therefore, can be used for indirect hardness and strength determination and for the identification of different alloys.

4. Since the method is dependent only on changes in conductivity or permeability, it is insensitive to any type of non-metallic contaminant in the cracks or on the surface.

5. Because of the inherently high exciting frequency of the system testing speeds can be high, and it is capable of inspecting large quantities of material rapidly. It, therefore, lends itself to automation.

6. By choosing the frequency of the exciting field, it is possible to restrict the inspection to the surface layers or to extend the effect into the underlying material.

7. Consequent on the factors mentioned in 2.6, the system can be used to measure accurately the thickness of thin wall sections and surface coatings.

Limitations

1.   The method cannot be used on non-conducting materials.

2.   The method is geometry sensitive and crack detection is, therefore, limited to simple components and shapes. The detection of defects in complex components might involve complicated scanning systems.

3.   Using encircling coil systems, more complex shapes can be inspected but without the facility for pinpointing defect positions.

4.3   MATERIALS, PROPERTIES AND TREATMENTS

Useful data that may not be readily available in the general handbooks is provided in this Section; the information is gleaned from numerous unclassified sources and the known limitations are noted. However, the user must be aware that much of this data comes from programs still in the developmental stage, from small sample sizes, and from test conditions not necessarily described in full. While the numerical values are, for the most part, generally considered representative, they are far from being verified and guaranteed.

4.3.1 Carbon and Alloy Steels

The first recorded investigations of metal fatigue were made by Wohler [84] over a century ago on steels; and they have been the most extensively studied alloys ever since. Much of the early work might be considered primitive by modern standards, but the results should be taken as representative of the alloys and conditions then obtaining. The sophistication and complexity of modern investigations originates in the growing recognition of both the statistical nature of the mechanism and of the large and still unknown number of variables involved in that mechanism. For example, it has long been thought that the (reversed, axial) fatigue limit of "steel" was rather closely equal to one-half its ultimate tensile strength. No test results have disturbed this rule-of-thumb, but a limit to its applicability has been found; the relation holds very well for tensile strengths up to about 150 ksi (corresponding to a hardness of Rc 35); then the curve starts to break, and at 260 ksi, the average fatigue limit is about 38%, instead of 50%. With reference to yield strength, a linear relation holds well to an S_{yp} of about 200 ksi, the equation of the line being: $S_a = 15 + 0.435 S_{yp}$, in ksi. The fatigue limit in bending is likely to be somewhat greater than that in axial loading, due most probably to the presence of a stress gradient.

Unfortunately, no such simple statements can be made regarding the behavior of notched steels. The fact that the notch sensitivity rises faster than the tensile strength results in only very moderate increases of fatigue limit. Some data tend to indicate that a tensile strength of about 180 ksi should give an optimum fatigue limit; but this cannot be generalized, for refined work on the manufacture and heat treatment of high-strength bolting has shown that the fatigue strength of 260 ksi bolts to be signficantly higher than that of 180 ksi bolts (see Section 4.3.1). The general indications are seen in Fig. 4.43, and specific relationships among K_t, K_f and S_{max} for 4130 steel are given in Fig. 4.44.

The relationship of fatigue strength to composition and microstructure has always been thought to be, generally, that the cleaner steels of smaller grain size and of sufficient alloy content to give the higher tensile strengths would also have the longer life. Increasing amounts of data from refined tests have indicated more and more that this is not necessarily true; and attention is now focusing on the details of related treatments such as melting and deoxidizing practice, mill processing, forming, machining and heat treatment. As a result, the fact is now accepted that, for the service life of the finished part, the effect of these treatments are at least as important as the composition. In other words, the specification of a high tensile strength does not insure long, nor necessarily adequate, service life.

Nonmetallic inclusions have long been suspected of being nuclei for cracks, and the recent developments in vacuum-melted and/or degassed steels have produced much cleaner microstructures. Particularly for 52100 bearing steel, the improvement has been translated into an increase of three times in rated life, or in stress terms, about 50% increase at any life. A general S-N relationship for the 5Cr-Mo-V aircraft steel is given in Fig. 4.45. The behavior of AISI 4340 has been studied in relation to three melting practices: (1) air melt; (2) consumable elctrode-- vacuum arc remelt; and (3) vacuum induction; median S-N curves for these results appear in Fig. 4.46. While the general superiority of vacuum-melted steels is not debated, there is a very considerable scatter in the results. Recent, refined comparisons of air-melted and of vacuum degassed 52100 indicate that, on a Weibull B50 life plot, the confidence bands overlapped noticeably: the center of four of the bands for the six air-melted heats were close (2 x 10³ cycles) to the center of the seven bands for the ten vacuum-melted heats. The other mechanical proper- ties for the two sets of heats were essentially identical, leading only to the conclusion that inclusion content is not the only parameter controlling fatigue life. (See also Fig. 3.209.)

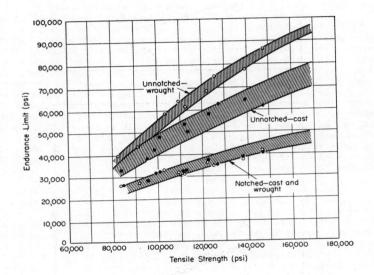

Fig. 4.43 Variation of endurance limit with tensile strength for comparable cast and wrought steels.

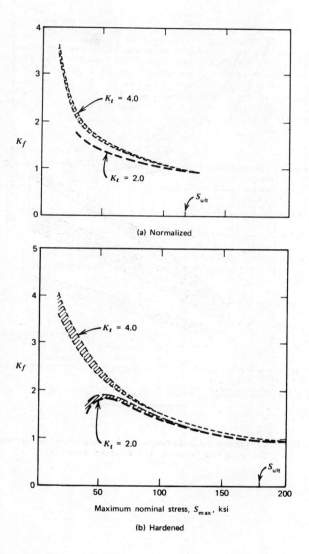

(a) Normalized

(b) Hardened

Fig. 4.44   Relations among K_f, K_t, and S_{max} for notched 4130
steel.   R = -1.   From Walter Illg [226], Fig. 16(c)
and (d).

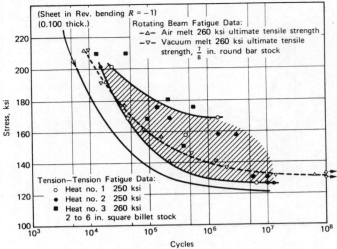

Fig. 4.45 Fatigue curves for 5Cr-Mo-V aircraft steel.

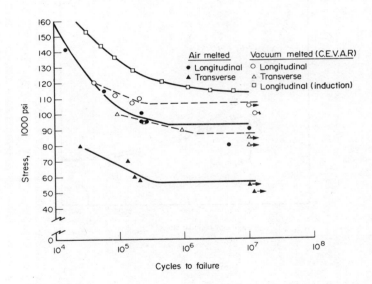

Fig. 4.46 Comparative S-N Curves for AISI 4340 steel
at 190 ksi ultimate strength, melted by
various processes.

Lead is a particularly important form of inclusion because many steels have about 0.2% Pb added for improved machinability, typically C1018, C1045, A4140, and A8620. Test results have indicated no significant differences in fatigue strengths of the leaded or lead-free steels up to about 130 ksi tensile strength. Above this level, the fatigue strengths of the leaded steels fell off until there was a 15% difference at 275 ksi. As a specific example of the fact that the higher tensile strength does not necessarily give longer life, torsional fatigue tests were made on a power take-off shaft, of 4140 plain and of 4140 leaded, since the service life of prior shafts made of 1141 (cold-drawn and stress-relieved) had been poor. The superior number of cycles to failure for either type of 4140 over that for the 1141 (208,000 to 55,000 cycles) is thought to be the result of a lower count of sulfur inclusions, and other, unidentifiable differences in microstructure. The only noted difference between the lead-free and the leaded 4140 was a significant saving in machining time.

The problem of specification of steel for long life is growing in both detail and complexity. It is quite possible that an adequate design may give poor service performance because of inadequate or incomplete specification of the steel to be used. As an example of problems, but not necessarily of solutions, the specification of a vacuum-melted forging steel 9310 has been the subject of discussion. A typical item is the need for better control of the alumina and other oxide inclusions, with the possibility of adding a microcleanliness requirement, perhaps similar to AMS 6265.

Table 4.8 and Figs. 4.47 through 4.53 provide a sample of the data available, particularly on the high-strength steels. Certain effects are to be noted in regard to heat treatment and decarburization, both important, and to irradiation, not so important (see Section 3.4.4).

TABLE 4.8   Effects of Decarburization on the Fatigue Limit of High Strength Steels

| | | | Fatigue Limit, psi | | | |
| | | | Undecarburized | | Decarburized | |
Steel	Core Hardness, Rc	Tensile Strength, psi	Smooth	Notched	Smooth	Notched
AISI 2340 bars[a]	48	250,000	122,000	69,000	35,000	25,000
	28	138,000	83,000	43,000	44,000	25,000
AISI 4140 bars[a]	48	237,000	104,000	66,000	31,000	22,000
	28	140,000	73,000	40,000	32,000	19,000
AISI 5140 bars[a]	48	255,000	125,000	67,000	29,000	19,000
	28	141,000	78,000	42,000	35,000	23,000
H-11 sheet[b]		277,000	108,000		68,000	

[a]Depths of decarburization from 0.080 to 0.100 in.
[b]Depths of decarburization from 0.005 to 0.010 in.

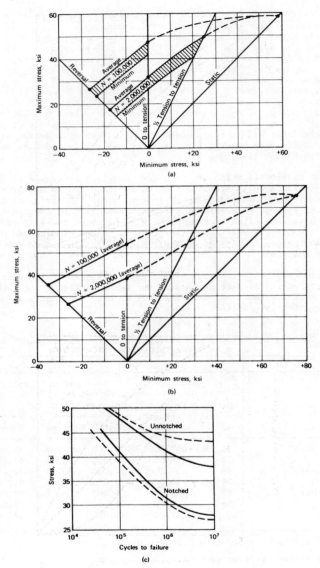

Fig. 4.47 Fatigue strengths of various steels. (a) Low carbon plates. From
[12], Fig. 6.1. (b) High strength low alloy steel ASTM-A242 or
equivalent plates. From [12], Fig. 6.1. (c) AISI 1040 bars,
normalized and tempered. --- Wrought; —— cast. From W. W. Briggs
[222], Fig. 6.

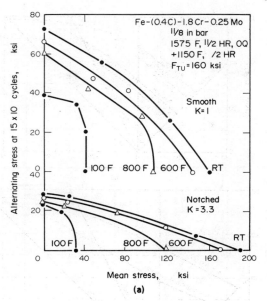

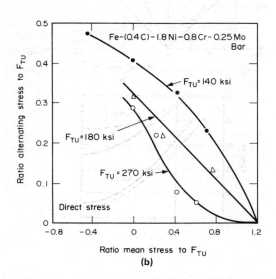

Fig. 4.48   Fatigue behavior of 4340 steel.   (a) Stress range diagrams
at various temperatures.   (b) Stress range diagrams at
various strengths (from Ref. 216).

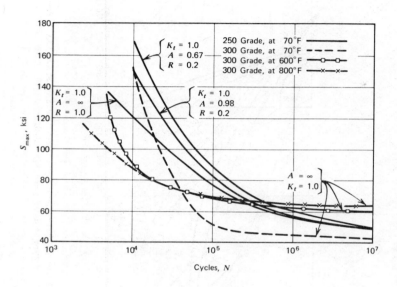

Fig. 4.49   S-N curves for 18% nickel maraging steel (900°F.,
3 hr., air cool).   From A. Graae [213], by permission ASM.

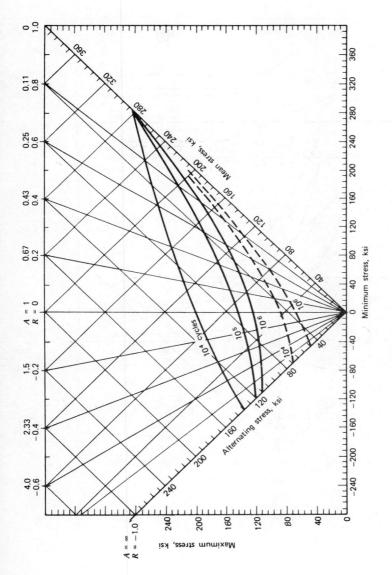

Fig. 4.50 Axial fatigue behavior of 9Ni-4Co-0.54C steels. UTS = 283.0 ksi. Room temperature; air atmosphere. ———— $K_t = 1.0$; ----- $K_t = 3.0$.

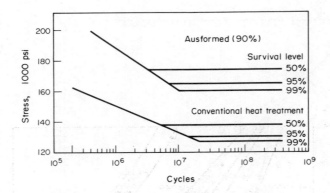

Fig. 4.51 S-N-P curves for ausformed H-11 steel (from Ref. 407).

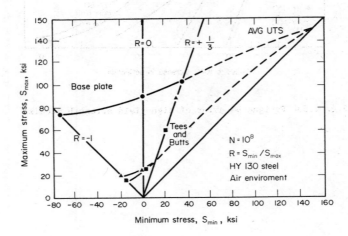

Fig. 4.52 Modified Goodman Diagram for HY-130 steel in flexural loading.

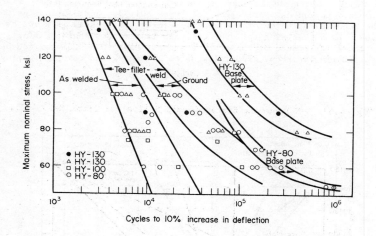

Fig. 4.53 Fatigue behavior of high-yield strength steels.

4.3.2 Stainless Steels

The minimum treatment of the fatigue behavior of the stainless steels automatically becomes encyclopedic, service life being a highly involved function of composition, condition, form, corrosive environment, temperature, etc. For a general coverage of the 300 and 400 series, the reader is referred to [2], pp. 416 ff, and Figs. 4.54 through 4.59 outline the performance of the precipitation-hardenable grades and of the high strength, high temperature alloys. It should be noted that while the values are believed generally representative, they result from a relatively small quantity of data. Prior to any investment in design or shop effort, serious further investigation must be made of the performance of the proposed alloy and form in the given environment. Particularly if extreme temperatures and/or corrosive media are involved, the details of composition, heat treatment, surface finish and joint design must be worked out, and specific tests made as required. Somewhat more detailed data are given for Types 301; 304 and 310 in Figs. 4.60 - 4.62. The behavior of three of the many specialty stainlesses is shown by Figs. . 4.63 - 4.64. As a general note: the designer is reminded that the fine fatigue and temperature resistance of the stainlesses may have to be compromised because of difficulties in the shop, especially as to machining and complex forming.

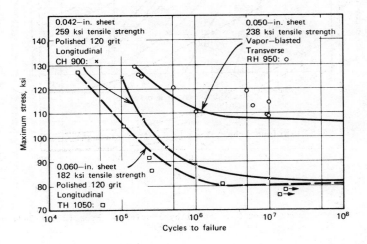

Fig. 4.54   S-N curves for 17-7 PH stainless sheet at room temperature under reversed bending.

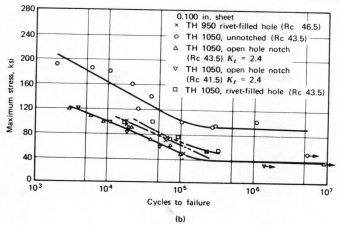

(b)

Fig. 4.55   S-N curves for 17-7 PH stainless sheet at room temperature
under axial tension-tension loading.

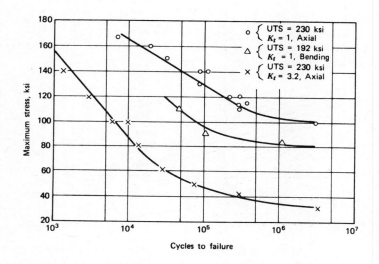

Fig. 4.56   S-N curves for AM 350 (SCT) stainless steel sheet at room
temperature. Sheet 0.070 in. thick.

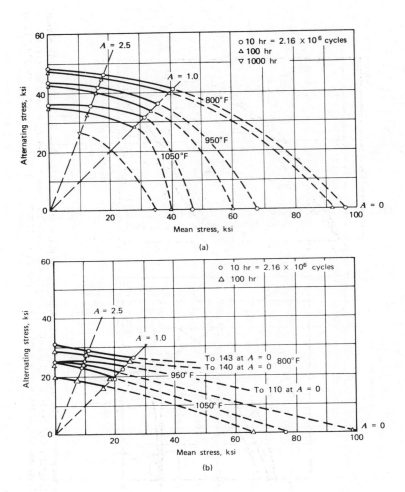

Fig. 4.57   Stress range diagrams for AISI 616 (type 422) stainless
steel.   (a) Unnotched; (b) notched.   From R. G. Matters
and A. A. Blatherwick [219], Figs. 12, 13.

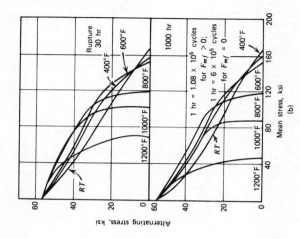

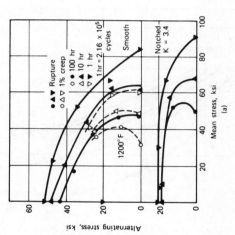

Fig. 4.58   Fatigue behavior of the high strength, high temperature alloys: 16-25-6 and A-286. (a) Stress range diagrams for 16-25-6 alloy, 3/4-in. bar. Hcw (1200°F), 18% + 1200°F, 8 hr. (b) stress range diagrams for A-286 bar and forgings, from 70 to 1200°F, 1650°F, 1 hr., OQ + 1300°F, 16 hr.

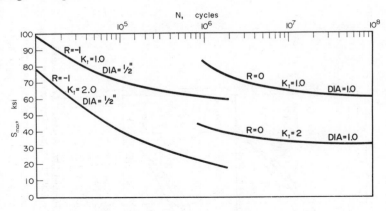

Fig. 4.59a S-N curves for A-286 (A-453) alloy at 90 ksi yield
strength. Small forgings at room temperature.

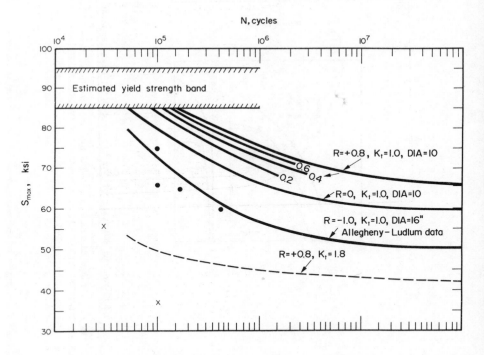

Fig. 4.59b Calculated S-N curves for A-286 (A-453); large forgings;
0.5" dia. test bars.

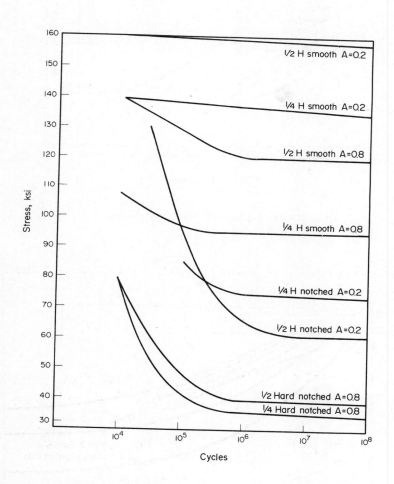

Fig. 4.60 S-N curves for 301 stainless sheet at room temperature.

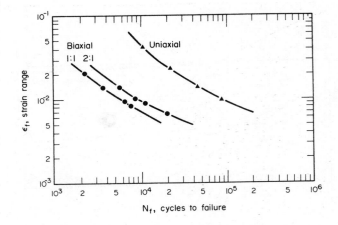

Fig. 4.61 Biaxial fatigue of 301 stainless steel.

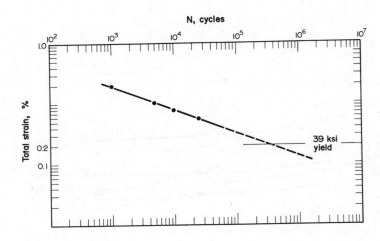

Fig. 4.62 Strain-cycle curve for 304 stainless steel; annealed bars
to 6" dia. (from Ref 409).

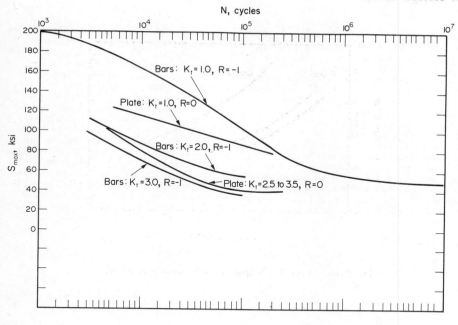

Fig. 4.63 S-N curves for "18-5" type stainless steel;
(18Mn-5Cr-.2N$_2$-V); 175 ksi yield.

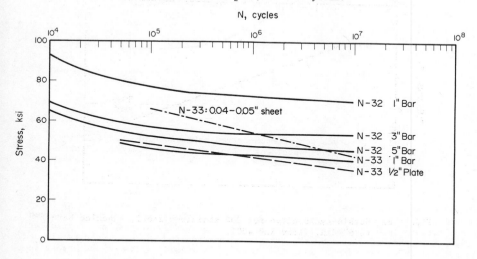

Fig. 4.64 S-N curves for Nitronic 32 and 33; (12Mn-18Cr-.3N$_2$);
R = -1.0; K$_t$ = 1.0; R.T.

4.3.3   Aluminum Alloys

The aluminum base alloys have been extensively investigated as to fatigue behavior
and, in common with the other light metals, their fatigue strengths are generally
low in comparison to their yield and ultimate strengths. Recent developments have
resulted in high ultimates but little improvement in fatigue. In general, the
wrought alloys range in ultimate strengths from 10 to 88 ksi and in endurance
limits from 7 to 24 ksi; the casting alloys exhibit corresponding ranges of 19 to
48 ksi, and of 7 to 21 ksi. Thus, it appears that the endurance limit (S_{max} at
$N = 500 \times 10^6$ cycles) is not a strong function of mechanical working, but chiefly
one of composition and heat treatment. The ratio of endurance limit to ultimate
strength (fatigue ratio) for wrought forms is 0.37 with a range of 0.27 to 0.60;
for castings, it is 0.31 with a range of 0.20 to 0.53. Such average values are
not immediately useful to the designer, but they do point out a way to accommodate
some of the known departures from the above generalities. For example, the ultimate and yield strengths of extrusions are known to be reduced by stretch-forming
and not fully restored by the subsequent heat treatment. Values in the low cycle
range (10^4 cycles), as in Fig. 4.65, may be compared with the usually reported
fatigue strength of 40 to 45 ksi for 2024-T4 smooth bar, indicating a loss of 15
to 25%.

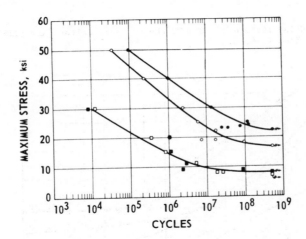

Fig. 4.65   S-N curves for stretch-formed extrusions of 2024-T4 alloy,
longitudinal specimens. Recrystallized: Smooth ○, Notched ▫;
Not Recryst.: Smooth ●, Notched ▪. From J. B. Kaufman et al,
[229], Fig. 7.

Data on the general fatigue behavior of many aluminum alloys are given in [1],
Section 3.3; the information below refers to performance under more specialized
conditions or provides material that has not been widely available. Particularly,
Figs. 4.66 through 4.68 show the effects of notches and cladding, separately and
combined, for sheet forms. Also for sheets, the relationships of the stress reduction factor K_f to K_t and to maximum stress are given in Fig. 4.69.

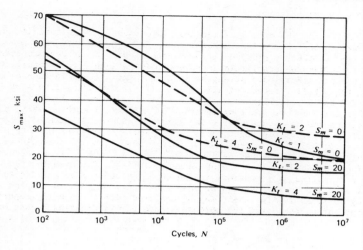

Fig. 4.66 Axial fatigue of 2024-T3 sheet. From Walter Illg [226], Figs. 4, 5, 6.

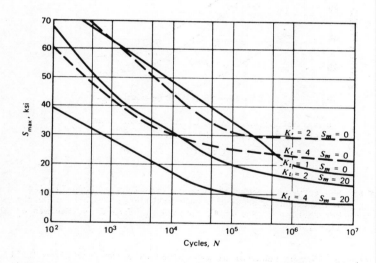

Fig. 4.67 Axial fatigue of 7075-T6 sheet. From Walter Illg [226], Figs. 7, 8, 9.

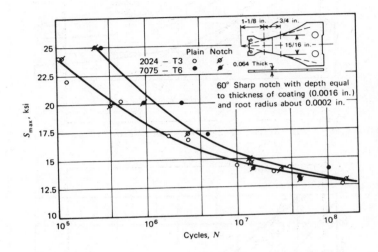

Fig. 4.68 Flexural fatigue of alclad sheet.

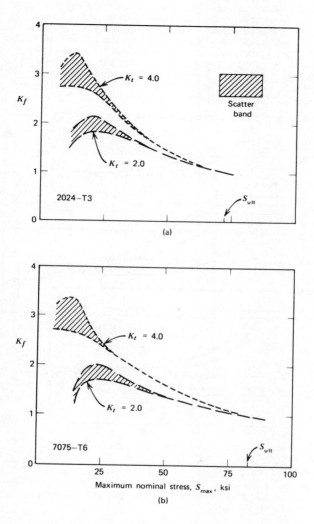

Fig. 4.69   Relations among K_f, K_t, and S_{max} for aluminum sheet. $R = -1$. (a) 2024-T3 alloy; (b) 7075-T6 alloy. From Walter Illg [226], Fig. 16.

The effect of the anodic coating universally used for corrosion protection and of the associated sealants has been extensively investigated by Stickley [230]. A summary of his results for the effect of chromic acid and sulfuric acid electrolytes, with and without sealants, on the rotating beam strength of 7075-T6 rod indicates that the coating thickness seems to be the controlling parameter for reduction of fatigue strength. S-N curves for plain and anodized Alclad Sheets are illustrated in Fig. 4.70; although the anodizing treatment enlarges the root radius of sharp notches, the net effect is to lower the fatigue strength, about 20% up to 10^6 cycles for 7075-T6. Nordmark [231] has compared the relative loss

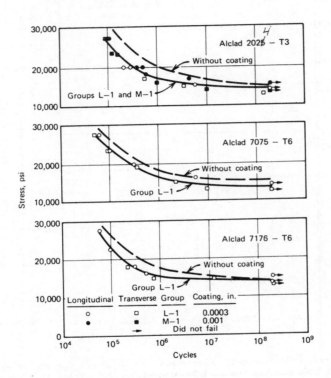

Fig. 4.70   Reversed flexural fatigue data for alclad sheet with sealed anodic coatings made by the sulfuric acid process. From G. W. Stickley [230], Fig. 7.

due to anodizing with that from the inevitable stress concentration factors arising in the fabrication of complex shapes, such as box girders. The results indicate that the differences between bare and alclad fatigue strengths are not likely to show up in fabricated assemblies, at least for those with riveted joints. The stress concentration factors of the joints overshadow the loss due to anodizing. Because the anodic treatment can be applied successfully only to simple shapes such as sheet, rods, and tubes, attempts have been made to provide corrosion protection for complex shapes by spraying them with aluminum. Other work [231] shows that the long life fatigue strength of sprayed plate is about half that of the bare plate, but the sprayed box girders had only slightly shorter lives than the bare ones; there is little difference between the curves for the girders and those for the sprayed axial stress specimens. Generally, the girder failures were initiated at the rivet holes in cover plates under tension. However, two failures were confined to the center section of a cover plate and did not involve a hole. Consequently, it appears that the notches resulting from the surface blasting before spraying were almost as severe as the notches due to the use of rivets.

Closely associated with the loss of fatigue strength from anodizing is that from chemical sizing or thinning of the sheet. The method is very adaptable to the production of complex profiles and is relatively inexpensive, but it does reduce both axial- and flexural-stress fatigue, as shown in Fig. 4.71.

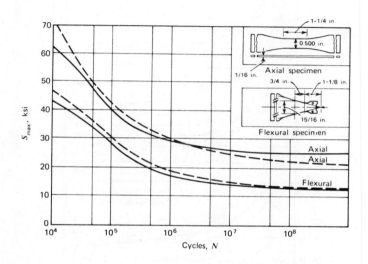

Fig. 4.71 Effect of chemical sizing on aluminum sheet. ——— 2024-T81; ––– 7075-T6.

The fatigue behavior at cryogenic temperatures is important for the weldable alloys; in the 2000 and 5000 series, the fatigue strength at 10^6 cycles is given in Fig. 4.72 as a function of temperature.

The effect of residual stresses in the heat-treatable alloys is usually beneficial because the quench leaves the surface in compression. Thus, stress-relieving may well lower the fatigue performance, see Fig. 4.73, and the comments in the Section on Stress Corrosion Cracking. Details of handling the high-strength die forgings have been investigated; one highly useful result being that if the flash is sanded off longitudinally, instead of transversely, the fatigue life may be increased up to two times. Effects of heat treating details are shown in Fig. 4.74.

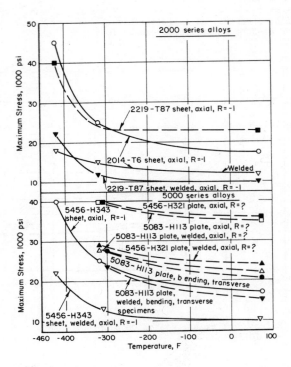

Fig. 4.72 Fatigue strengths at 10^6 cycles for aluminum alloys at low temperature.

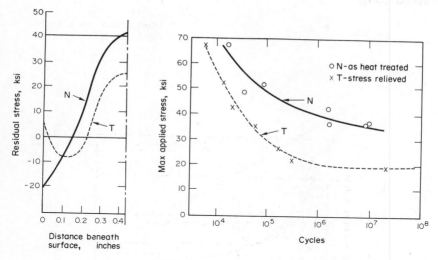

Fig. 4.73   Effect of residual stress on fatigue life of 7075-T6.
R = +.05; dia. = 7/8".

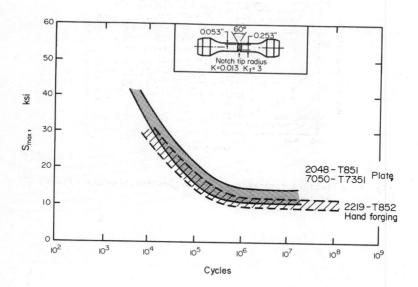

Fig. 4.74   Axial stress fatigue curves for aluminum alloys 2048,
2219 and 7050, at K_t = 3.0 and R = +.013.

The difference in behavior under random and constant amplitude loadings is significant. Figures 4.75 and 4.76 illustrate statistically the lower performance for several higher strength alloys under random loading.

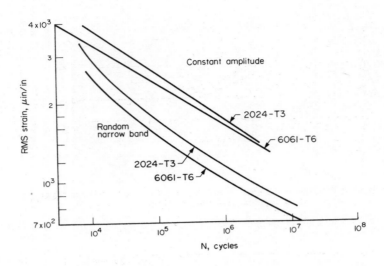

Fig. 4.75   Strain-cycle curve for aluminum in cantilever bending (.063 thick).

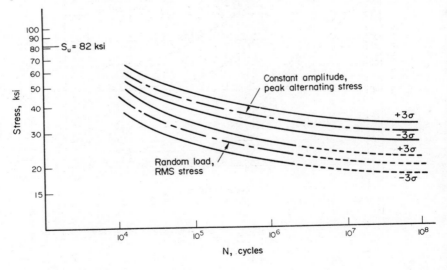

Fig. 4.76   Statistical S-N curves for 7075-T6.   R = -1.0; K_t = 1.0.

Of the casting alloys, those for use in green sand have the lowest endurance limit (7-11 ksi), and die-cast are the highest at 18-21 ksi. As for the magnesium-base alloys, developments have raised the ultimate strength to very high levels (41 ksi for C355-T6), but the fatigue strengths at any life do not show the same improvement. Premium quality castings made of C355 or A356 and cast to MIL-A-21180 will usually show a useful fatigue stress level of about 15 ksi for $N = 10^5$ or 10^6 cycles. The major problem of specifying castings seems to be a proper decision as to the soundness required, for example, the permissible degree of porosity, shrinkage, and gas holes, since no casting is perfect. In this action, negotiation with the foundryman is usually desirable and sometimes imperative, especially in the interpretation of a radiographic standard and the inspection procedure. As to the latter, agreement as to the Grade and Class based on MIL-C-6021 with x-rays read against an agreed level in ASTM E155-60T should insure acceptable castings. Popular sand casting alloys are 43, 220, 355, A355, 356, and 750; permanent mold alloys are 43, 122, A132, 344, 355, and C355; die casting alloys are 13, 43, A214, A360, and A380. S-N curves for 356-T6 sand cast and A356-T6 permanent mold bars are given in Fig. 4.77, while Fig. 4.78 shows the performance of the newer alloy 201-T7, sand-cast.

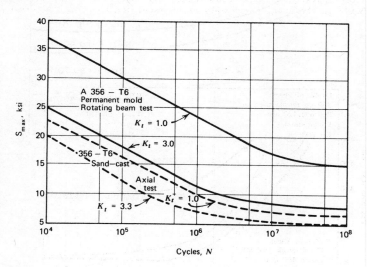

Fig. 4.77   S-N curves for No. 356 alloy castings.   From [233, 234].

The effect of vacuum on the fatigue life of aluminum 1100 has been to prolong it considerably. The life at S_{max} of 9700 psi remained essentially constant at 0.5 x 10^6 cycles from 10^3 to 10^{-1} torr, then rose linearly to about 5.0 x 10^6 cycles at 10^{-5} torr. The reason for this change is not definitely known, but is undoubtedly related to a change in surface energy.

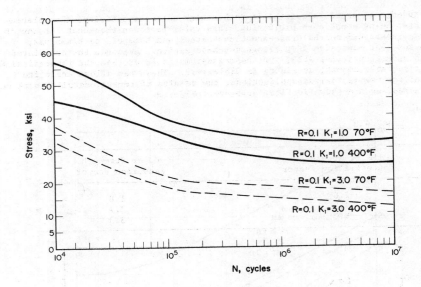

Fig. 4.78   S-N curves for 201-T7 aluminum castings.

4.3.4   Magnesium Alloys

Magnesium alloys are of the class that show no well-defined fatigue limit; although many curves appear to level out at 10^7 or 10^8 cycles, the slope is still finite. The general relation between tensile and fatigue strength shows considerable scatter, especially for the cast alloys. For sheet and extruded forms, the fatigue ratio is about 40%, and for forgings slightly lower, about 35%, but generalization for castings is difficult; see the discussion below.

The extruded and forged forms exhibit rather high notch sensitivity, the alleviation factor being about $\sqrt{\alpha} = 0.015 \sqrt{in}$. This value is lower than that for any of the other constructional alloys, except cast steel and some of the titaniums, which means that magnesium is the more notch-sensitive. Even so, the calculated strength reduction factor (SRF) for the magnesiums is lower than the theoretical K_t by an appreciable amount, as listed in Table 4.9. Thus, the alloys are quite practical for use in fatigue applications; the results of recent, detailed work on ZK60A extrusions are shown in Figs. 4.79 and 4.80.

TABLE 4.9

K_t or Stress Concentration Factor	K_f or Strength Reduction Factor
2.0	1.8
3.2	2.7
5.0	3.3

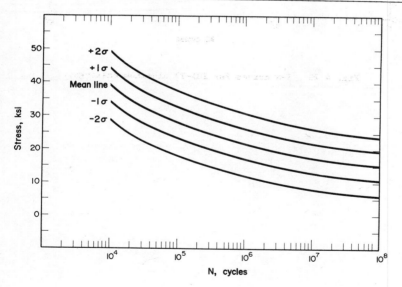

Fig. 4.79   Statistical S-N curves for ZK60A magnesium extrusions. Long direction; tempers F and T5; R = -1.0; K_t = 1.0.

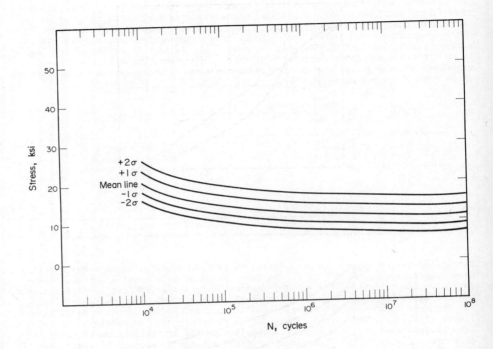

Fig. 4.80   Statistical S-N curves for ZK60A magnesium
extrusions. Long direction, tempers F and
T6; R = -1.0, K_t = 1.3.

The effect of mean stress on the fatigue behavior of magnesium is more marked than for most other alloys; particularly, the modified Goodman relation may not be conservative for all values of mean stress. Figure 4.81 provides a few points indicating the trend for hot-rolled magnesium of the Zn-Zr group. A saving factor may be the limitation of the mean stress to values well below the yield, at such levels as to stay well under the experimental curve.

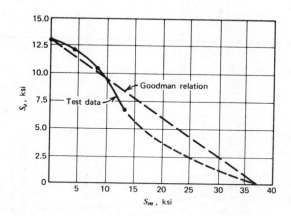

Fig. 4.81   Effect of mean stress on the fatigue strength of
magnesium alloys. From R. B. Heywood [5], Fig. 4.6.

Recent efforts to improve the properties of premium quality, light metal castings have generally been beneficial to the tensile and yield strengths but not necessarily to the fatigue strength. This condition is a specific manifestation of the general situation regarding the light metals in which improvements in metallurgy and processing have produced very high ultimates but little increase in fatigue limit. Table 4.10 represents an attempt to correlate data on castings of AZ91C from various sources. The significant points in the interpretation of these figures are: (1) while the "laboratory quality" tensile is about 25% higher than of the premium quality, the corresponding increase in rotating beam fatigue is only 6%, and flexural fatigue shows no increase, and (2) MIL-HDBK-5 values are already equal to those for premium quality castings. Although the data for (ratio)$_2$ and (ratio)$_4$ are too sparse to permit conclusions, the trend is unmistakable.

Despite the limitations discussed above, the magnesium alloy castings possess a very useful fatigue strength, and the hesitancy on the part of many designers to specify them is unwarranted. Difficulties in the specification and procurement of castings can be effectively eliminated by the proper use of such controls and quality descriptions as are found in MIL-M-46042 and MIL-C-6021. Interpretation and judgment are required: these specifications enumerate grades and classes of stress level in designated areas, the degree of porosity and shrinkage, and so forth. The latter parameters are usually judged against such standards as ASTM E155-60T radiographic standards, and it should be remembered that it is neither necessary nor feasible to specify a "perfect" casting. Forgings and extrusions are also not completely homogeneous nor isotropic.

TABLE 4.10   Properties of Magnesium Casting Alloy AZ91C

Quality	Tensile Strength, ksi	Yield Strength, ksi	Elongation, %	Fatigue Strength at 10^7 Cycles		$(Ratio)_1$	$(Ratio)_2$	$(Ratio)_3$	$(Ratio)_4$
				Rotating Beam	Flexure				
1. Standard, per QQ-M-56 min/ave. for									
-T6 cond.	17/25	12/14	0/0.75						
2. Premius (g)									
-F cond.	23	11		12–15	7–10	0.42	1.23	0.37	0.77
-T4 cond.	32	11		13–17	10–12	0.47	1.37	0.34	0.59
-T6 cond.	32	16		11–15	9–10	0.40	0.81	0.30	0.59
3. MIL-HDBK-5									
-F cond.	23	11		13–17		0.44	1.37	(0.32 ave.)	(?)
-T4 cond.	34	11	7						
-T5 cond.	23	12	2	11–14		0.37	0.78	(0.32 ave.)	(?)
-T6 cond.	34	16	3						
4. Laboratory (g)									
-F cond.	43	26	6	18(est.)	8.3(est.)	(0.42 ave.)	(?)	(0.32 ave.)	(?)

NOTES:
(a) Separately cast test bars.
(b) $K_t = 1$.
(c) $(Ratio)_1$ = rotating beam fatigue strength at 10^7 cycles/tensile strength.
(d) $(Ratio)_2$ = rotating beam fatigue strength at 10^7 cycles/yield strength.
(e) $(Ratio)_3$ = flexure beam fatigue strength at 10^7 cycles/tensile strength.
(f) $(Ratio)_4$ = flexure beam fatigue strength at 10^7 cycles/yield strength.
(g) From M. C. Flemmings and D. Pechner [225].

4.3.5   Titanium Alloys

Much improvement has recently come about in the understanding of the factors which control the properties of the Ti alloys.   In addition to the basic chemistry, these factors are the time, temperature and atmosphere(s) during the entire processing sequence.   A detailed discussion of these factors is beyond the scope of this work, but may be found in Ref. 415, and much of the data below especially that for Ti-6Al-4V results from these extensive investigations.   The 6-4 alloy has emerged as widely applicable and is much favored for forgings, extrusions and plate, (see specific fracture data in Section 3.2.4).

For titanium and its alloys, the "average" fatigue ratio at 0.50 is slightly higher than that for many other alloys, Fig. 4.82.   In addition to the unalloyed titaniums, there are about 18 alloys in three classes, alpha, alpha-beta, and beta, according

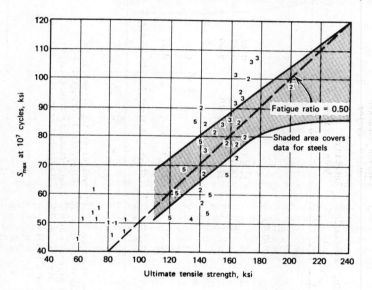

Fig. 4.82   Comparison of fatigue data on titanium alloys and low alloy steels: (1) commercially pure, (2) 6Al-4V, (3) 4Al-4Mn, (4) 3Al-13V-11Cr, (5) 5Al-2.5 Sn.   Rotating beam; K_t = 1.   From [2].

to microstructure resulting from heat treatment and/or working.   The fatigue strength at N = 10^7 cycles of these alloys ranges from 32 ksi for the 2.5Al-16V grade to 124 ksi for aged 4Al-3Mo-1V, Fig. 4.83.   Extensive data are given in [1] on one alpha alloy: 8Al-1Mo-1V; two alpha-betas: 6Al-4V and 4Al-3Mo-1V; and one beta: 3Al-13V-11Cr, while the information below pertains to somewhat specialized alloys or extended conditions.

The choice of a titanium alloy for a given service is an exceedingly complex procedure, chiefly because the benefits and penalties are both rather pronounced.   High values of the strength/density and stiffness/density ratios and retention of strength

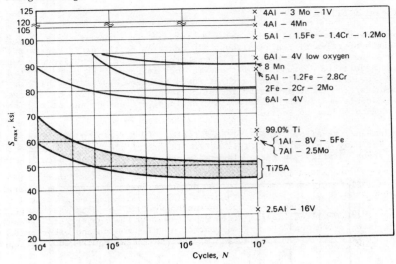

Fig. 4.83   Fatigue data on commercially pure titanium and selected alloys.
Rotating beam; $K_t = 1$.

at elevated temperatures (in the range of about 400 to 900°F titanium alloys will
out-perform many steels), make their use attractive; but rather high sensitivities
to notches and to stress corrosion cracking may be serious disadvantages.

The notch alleviation factor of these alloys is very low, which indicates high
sensitivity to notches. To be more specific, the stress reduction factors have
been determined for two alloys, 4Al-4Mn and 2Fe-2Cr-2Mo, as essentially equal to
the theoretical stress concentration factor K_t; Table 4.11 gives the effect of
notches for the 6Al-4V alloy. A value of K_t at 3.2 reduces the effective fatigue

TABLE   4.11   Effect of Notches and Heat Treatment on the
Fatigue Strength of 6Al-4V Alloy in the Life
Range from 10^6 to 10^7 Cycles; Axial Loading

	Fatigue Strength in ksi					
	Annealed			Heat-Treated		
	A = 0.6	1.7	∞	0.6	1.7	∞
$K_t = 1.0$	115	97	80	135	130	95
$K_t = 3.2$	70	59	45	72	50	42

A = load ratio = (max load − mean load)/(mean load)
Annealed: UTS = 136 ksi; YS = 128 ksi
Heat treated: UTS = 170 ksi; YS = 160 ksi

life to 60% of the unnotched value in the annealed, ductile condition; heat treatment, for a very small increase in tensile strength, intensifies the notch effect to even lower values. The fatigue strength is not a strong function of the load ratio, following the normal reciprocal tendency. In general, it is considered conservative to use the full value of the K_t in design. The fatigue strength reduction effects of various machining processes are given directly in Fig. 4.84 for the 5Al-2.5Sn alloy. Comparison of the 6-4 alloy with a high-strength steel, 300M, shows the lower notch sensitivity (relatively) by the higher strengths in the high-cycle end of the plot.

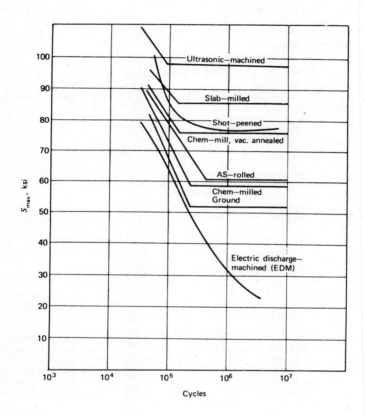

Fig. 4.84 Effect of various machining processes on fatigue of the 5Al-2.5Sn titanium alloy in reversed bending. R = -1.0; K_t = 1.0.

Although fatigue specimens of the titanium alloys show a pronounced notch sensitivity, the performance of assemblies is controlled chiefly by the joint geometry, as it is for other-metal-base alloys. For welded butt joints in 5Al-2.5Sn at R = 0.2 and A = 0.7, S_{max} turns out to be about 50 ksi at 10^7 cycles; for riveted joints with a K_t of 10, S_{max} is about 15 ksi. Under these and similar conditions, then, the reportedly high notch sensitivity of titanium need not be a deterrent to its use. Figure 4.85 illustrates the useful fatigue properties of 8Al-1Mo-IV sheet, fusion-welded, and in bolted joint form. Modified Goodman Diagrams are given for the widely used 6Al-4V alloy in bar and sheet form in Fig. 4.86, while the S-N curves for forgings and plate made by the refined processing mentioned earlier are shown as Figs. 4.87 and 4.88; and for extrusions in Fig. 4.89. Properties of a cobalt-bearing alloy, Ti-6Al-4V-3Co, are given in Fig. 4.90.

In a comparison of performance at elevated temperatures, Fig. 4.91, the general superiority of the 6Al-4V forging alloy is indicated, although several other alloys retain half or more of their room temperature strength to 700°F. Some fatigue data are available for cryogenic applications, alloys 5Al-2.5Sn-2.5V-1.3Cb-1.3Ta and especially 5Al-2.5Sn (ELI) have been investigated down to liquid helium temperature, see Fig. 4.92.

One major problem exists regarding the use of titanium: that of stress corrosion cracking. All titanium alloys are sensitive to the presence of hydrogen, water vapor, and the halogens, particularly chlorine, in that they lose strength and ductility, failing very rapidly in a brittle mode. This action is intensified by elevated temperatures and/or existing stress. The temperature range of about 450 to 600°F seems to be important; below this range, there is little evidence of stress corrosion at any stress level, but above it the effect becomes pronounced at all stress levels. The investigations center on crack growth rates in statically stressed specimens, usually under constant environmental conditions, one variant being the cyclic application of the corrodant and a neutralizer. Despite extensive research efforts, very little fatigue data has been generated, and no rigorously conclusive explanation nor remedy has been forthcoming. Various protective coatings have been tried but all proved to be somewhat deleterious to the fatigue limit. Anodic and cadmium platings were least effective as protectors and degraded the fatigue limit by only some 10% from that of a polished, uncoated rotating beam, while the chemically more efficient nickel and chromium platings dropped the limit by 30 to 50%.

In a comparative sense, 6Al-6V-2Sn seems to be the alloy most susceptible to hot-salt stress corrosion, and 6Al-4V mill-annealed sheet the least. The alloy 8Al-1Mo-1V, a prime candidate for supersonic flight structures, is reasonably resistant but shows serious loss in tensile strength above 550°F. Thus, one may conclude that its fatigue strength is also reduced under these conditions.

Some difficulty in obtaining uniform properties in heavy sections (over 1 in. thickness) by heat treatment has been observed. Since this seems to be an inherent condition, dependent on geometry and composition, efforts to bypass the difficulty have been made in the form of heat-treating thin sections uniformly to the desired strength level, then building up the required thickness by diffusion-bonding. For a 2-in. section of the 6Al-6V-2Sn alloy, the S_{max} at 10^6 cycles was improved from 25 to about 58 ksi.

The casting of titanium alloys is emerging from the laboratory stage, experimental static and fatigue properties of small castings in the 6-4 and 5-2.5 alloys compare favorably with those of equivalent forgings. A comparison of cast structural alloys on the basis of the strength/density ratio vs. temperature shows the Ti alloys generally superior to the maraging steels, Inconel 718, A357 aluminum and AZ91A magnesium

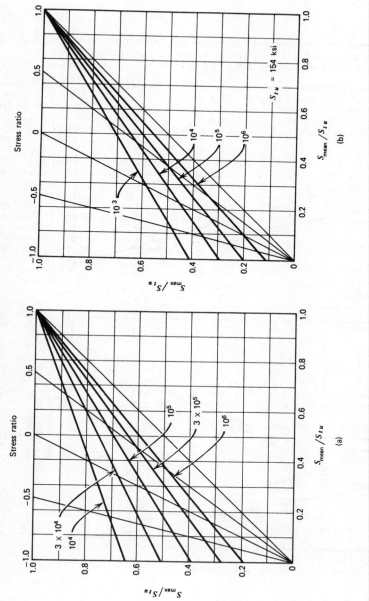

Fig. 4.85   Goodman diagrams for joints in titanium sheet. (a) Transverse butt weld, machined flush. S_{TU} = 145 ksi. (b) Bolted lap joint, hole drilled and reamed, one NAS 1004-2 bolt. S_{TU} = 154 ksi.

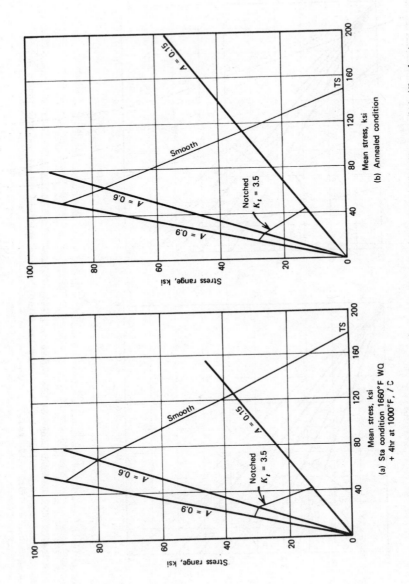

Fig. 4.86 Modified Goodman diagrams for the 10^7 cycle fatigue life of Ti-6Al-4V, sheet.

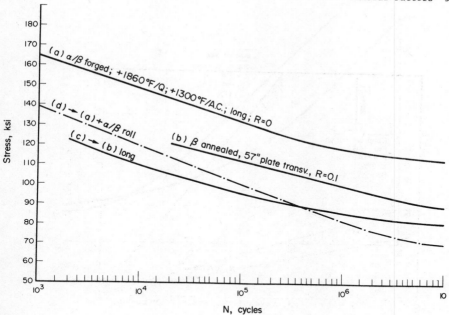

Fig. 4.87   S-N curves for Ti-6A1-4V forgings and plate. K_t = 1.0.

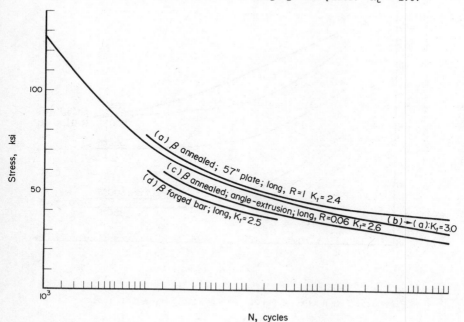

Fig. 4.88   S-N curves for Ti-6A1-4V notched forgings, plate and extrusions.

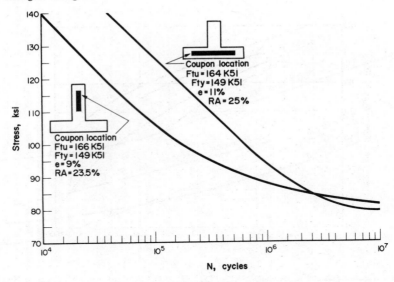

Fig. 4.89   S-N curves for Ti-6Al-4V extrusions.

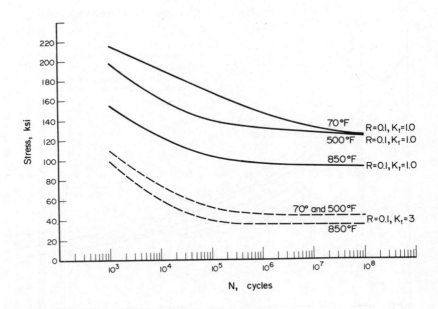

Fig. 4.90   S-N curves for Ti-6Al-4V-3Co bar, axial stress.

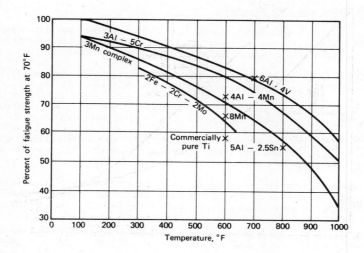

Fig. 4.91   General effect of temperature on fatigue strength of titanium alloys.   Rotating beam; $K_t = 1$.

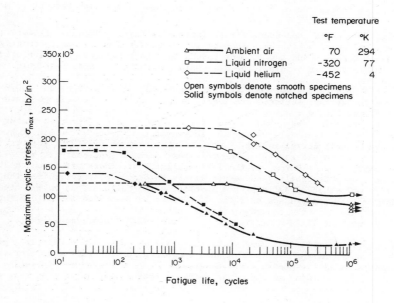

Fig. 4.92   S-N curves for Ti-5Al-2.5Sn at cryogenic temperatures.

4.3.6 Cast Irons and Cast Steels

Cast Irons. All of the engineering cast irons contain free carbon as graphite. The form of the graphite particles and their distribution in the pearlitic or ferritic matrix gives rise to the three common types of cast irons:

1. Gray iron, in which the graphite is distributed as comparatively large flakes in pearlite; used as-cast, possibly stress-relieved.

2. Malleable iron, produced from gray iron by prolonged heating, which results in isolated clusters of graphite in a ferritic matrix.

3. Nodular iron, an advanced "malleable," for which the initial solidification rate or the subsequent annealing is adjusted such that the graphite forms spheres with smooth boundaries, also in a ferritic matrix.

In the progression from gray to nodular iron, the stress-strain curve becomes more nearly linear, the modulus of elasticity rises from about 17 to 25,000,000, the tensile strength increases from a range of 20 to 70 ksi to 50 to 100 ksi, and the ductility improves.

The fatigue ratio for all three irons lies between 0.35 and 0.50. None of the irons is particularly notch sensitive because of the presence of the graphite particles as stress concentrations. The introduction of an external geometric notch seems not to increase the effective concentration over that already present in the microstructure. For gray iron, the strength-reduction factors are only 1.0 to about 1.1, rising with tensile strength; for nodular iron, the stress concentration factor may rise to 2 at a tensile strength of 100 ksi. The Modified Goodman relation gives a conservative indication of the performance of these irons with respect to mean stress.

The information on gray and nodular iron fatigue is fairly extensive [244-247], but malleable iron seems to have been thoroughly investigated only rather recently [243]. The Malleable Research and Development Foundation has generated extensive data on all four ASTM grades in bending; the discussion below is based on this work.

Comparative stress curves for pearlitic malleable ASTM grades 60003 and 80002 show that values for as-cast surfaces are lower than those for unnotched, polished test bars, but higher than for notched, polished bars. Figure 4.93 gives fatigue data for the high strength Grade 80002 under "full cycle" bending (reversed flexure); Figs. 4.94 and 4.95 compare lives of the four grades of iron at two confidence levels. The Weibull curves in Fig. 4.96 indicate the probability of failure for numbers of cycles in excess of the "no failures" number. The best-fit and 2-sigma limit lines are given. Normally, the 2-sigma limits are those within which 95.46% of all events on a probability curve are likely to occur. However, since the requirement is only that of defining the level at which failures may be expected, the upper limit may be ignored and half the difference between 95.46% and 100%, or 2.27%, may be added to the confidence level, making it 97.7%. On the half-cycle curves, however, the limit shown represents a true 95.46% confidence level.

The design engineer will find Figs. 4.93 through 4.95 useful in obtaining a general idea of which of the four ASTM grades of malleable iron will most nearly meet the objectives of life expectancy at the design load. The data in Fig. 4.96, and others in [223], determine more precisely the maximum failure rate that can, with confidence, be expected at that design load and throughout the design life.

As an example, it is desired to specify a grade of malleable iron suitable for a full cycle application (i.e., a wheel hub or a crankshaft); stress analysis indicates 40,000 psi loading; the design life objective is 35,000 cycles and a maximum

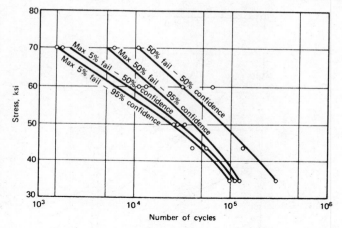

Fig. 4.93   Full-cycle S-N curves for malleable iron, grade 80002;
50% and 90% confidence, 50% and 5% maximum failure
rates. From [243].

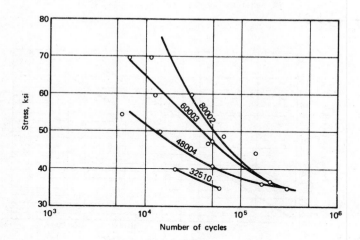

Fig. 4.94   Comparative S-N curves for malleable iron grades
32510, 48004, 60003, and 80002; 50% confidence,
50% maximum failure rated. From [243].

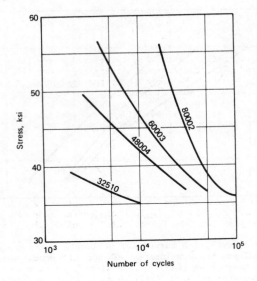

Fig. 4.95   Comparative S-N curves for malleable iron grades
32510, 48004, 60003, and 80002; 95% confidence,
5% maximum failure rates. From [243].

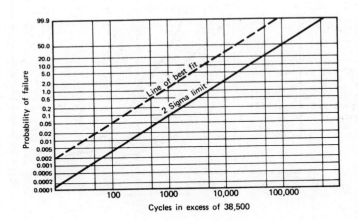

Fig. 4.96   Weibull fatigue life plot for malleable iron grade 80002.
Full cycle tests; median life--145,845 cycles. From [243],
Graph 20.

failure rate of 0.2% is required. Referring to Fig. 4.95, one finds 95% confidence stress curves for all four grades of malleable iron. Checking the performance at 40,000 psi, we note that ASTM 48004 (at 15,000 cycles) and ASTM 60003 (at 25,000 cycles) do not meet the desired life expectancy, while ASTM 80002 exceeds it by better than 10,000 cycles. However, these curves are based on a maximum failure rate of 5%, and will, therefore, indicate a greater life expectancy at a given stress than will actually be the case at the desired 0.2% maximum failure rate. Therefore, the engineer must allow for this by selecting a material that provides considerably more than 35,000 cycles at 40,000 psi (5% failure rate). The obvious choice, then, is ASTM 80002. (Were the application a half-cycle type rather than full cycle, a design life of 1,000,000 cycles or more could be achieved with the same stress level.)

Now one must determine whether ASTM 80002 will provide the needed life expectancy under the given conditions (40,000 psi load and 0.2% maximum failure rate). Reference is made to Fig. 4.96 which plots life expectancy at varying failure rates for ASTM 80002 at stresses of 43,500 to 45,000 psi (slightly in excess of requirements). The solid line is the 50% confidence level, that is, there exists a 50-50 chance that, given a 0.2% maximum failure rate, one can expect a design life of approximately 40,000 cycles--just over the "no-failure" level of 38,500 cycles.

Cast Steel. Data on the fatigue properties of the cast steels are rather sparse, but for the plain carbon and low alloy steels with a carbon content of 0.30 to 0.40%, the fatigue ratio is given in [242] as about 0.45 for unnotched specimens and about 0.29 for those with K_t = 2.2. The cast steels seem not to be particularly notch-sensitive, for a K_t of 2.2 the stress reduction factor is approximately 1.6. This low sensitivity is thought to be related to the comparatively low strength of the unnotched material, rather than that the strength in the notched condition is high. As a result of the great difference in cooling rates at various locations in heavy sections, the size effect is noticeable in reducing both tensile and fatigue strength as samples are taken from the surface toward the center.

Some general data may be found in the Steel Castings Handbook [430], from which Figs. 4.97-4.101 are taken. These curves are of considerable interest because they relate the quality of the casting to its fatigue behavior. Figures 4.97 and 4.98 give values of life or stress according to the Class of Shrinkage as defined by ASTM E71, while the others deal with surface discontinuities. It should be noted that the level or size of discontinuity shown here exceeded all classes of NDE standards: ASTM E71; E97; E125. Thus, these curves represent worst cases when compared with the effect of normally allowable surface defects on fatigue performance.

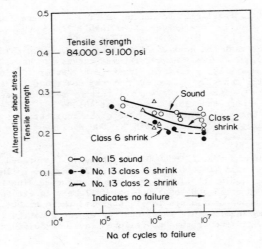

Fig. 4.97   Effect of shrinkage on fatigue life of annealed
8630 cast steel in torsion.

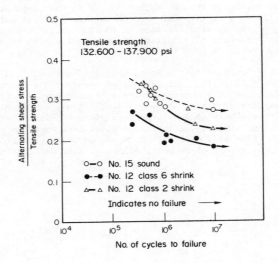

Fig. 4.98   Effect of shrinkage on fatigue life of quenched
and tempered 8630 cast steel in torsion.

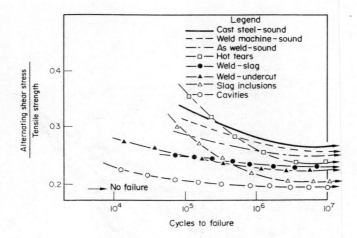

Fig. 4.99   Effect of surface discontinuities on the fatigue life
of normalized and tempered 8630 cast steel in torsion.

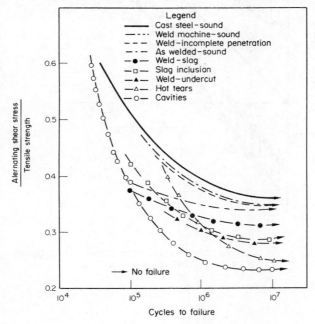

Fig. 4.100   Effect of surface discontinuities on the fatigue life of
normalized and tempered 8630 cast steel in bending.

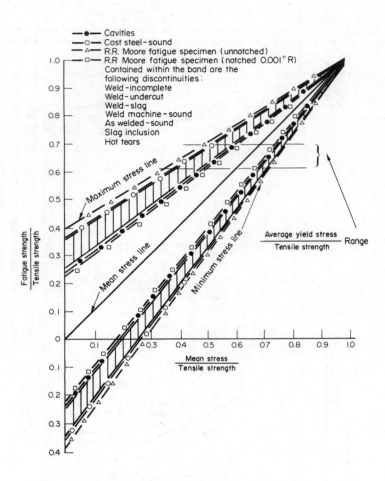

Fig. 4.101   Goodman Diagram for normalized and tempered
8630 cast steel in bending.

4.3.7   General Nonferrous Metals and Alloys

This section collects randomly available data on various metals and alloys. The
designer is reminded that many of the tests were of the "single-run" type, sometimes under special conditions. This data is distinctly not of the guaranteed
value variety (with the possible exception of the fatigue limits of some of the
coppers); generalizations and extrapolations should be undertaken very cautiously,
if at all.

Copper Alloys. These alloys show a fatigue ratio between 0.25 and 0.50 and a
rather wide range of endurance limits, from 11 ksi for OFHC copper to about 40 ksi
for beryllium-copper alloy No. 25. The notch sensitivity of the alloys is hard to
predict. In the soft annealed state, the stress reduction factors are slightly
lower than the stress concentration factors for most alloys; but, as the hardness
and tensile strength rise with temper, the fatigue strength goes up comparatively
little, and the loss of ductility usually means that a greater portion of the
stress concentration factor is applicable. The coppers show anisotropic effects:
the endurance limit for samples prestrained only undirectionally was lower than
for those prestrained by a reversing force. The effect of elevated temperatures
on the copper alloys is a marked loss of tensile strength due to reduction of temper, and a smaller comparative loss of fatigue strength. Although the beryllium-
coppers retain a useful fatigue strength to the highest temperatures of any alloys
in this group, the alpha-beta brasses (60Cu-40Zn) provide good performance, typically 10^4 cycles at 600°F under a mean stress of 6000 psi and an alternating stress
of 2300 psi. The latter alloys are subject to considerable variation in tensile
and fatigue properties, dependent on the dispersion of the alpha phase as a result
of the heat treatment and rolling details.

The commercially pure coppers, OFHC and ETP hsow two distinct modes of failure
in fatigue. In load-controlled tests on ETP, the scatter in life at stresses near
21.5 ksi is much greater than at stresses either higher or lower. Below 21.5 ksi
the crack initiates with a slip-band and its path is dependent on the slip-band
pattern. Above 21.5 ksi that crack started at an L- or Z-shaped nucleus and its
path was random, both inter- and intra-granular, but not slip-dependent. Figure
4.102 illustrates these modes.

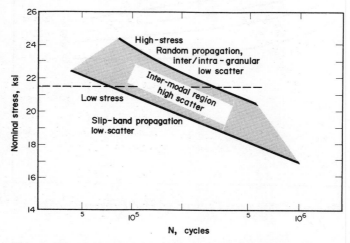

Fig. 4.102 Modes of fatigue failure for annealed ETP copper.
R = -1.0, K_t = 1.0 (from Ref. 412).

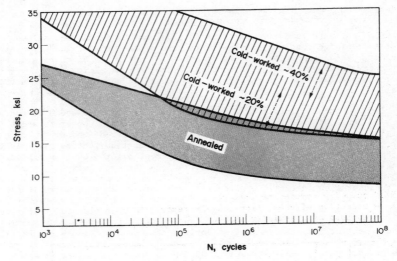

Fig. 4.103 S-N bands for copper alloy CDA-102 ("OFHC").
R = -1.0, K$_t$ = 1.0. Room Temperature.

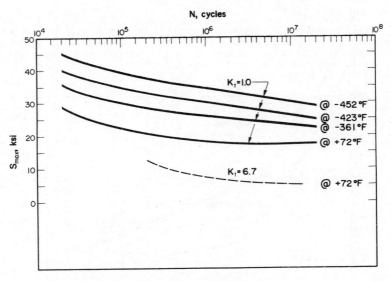

Fig. 4.104 S-N curves for annealed ETP copper at various
temperatures. R = -1.0.

Concern has been growing for the behavior of electrical conductors which are also highly stressed, as in high-speed rotating machinery. These conductors are usually the OFHC or ETP alloys, and considerable test effort has resulted in the curves of Figs. 4.103 and 4.104.

The S-N curves for the popular alloy No. 260, cartridge brass are given in Fig. 4.105. Axial and flexural behavior of beryllium-copper sheet and bar is shown in Figs. 4.106 and 4.107. An S-N curve for 70-30 copper-nickel, Fig. 4.108, extends down to 10^2 cycles, well into the low-cycle, high-stress region. The stress amplitude was derived from the data of constant-strain cycling tests by the "best-fit" curve having the equation:

$$S_a = \frac{E}{4\sqrt{N_f}} \ln \frac{100}{100 - RA} + S_e$$

where S_a = stress amplitude
S_e = endurance limit
N_f = number of cycles to failure
RA = reduction of area, from tensile tests
E = modulus of elasticity

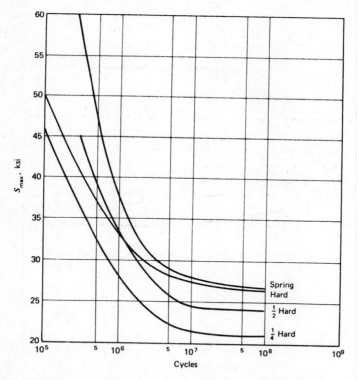

Fig. 4.105 S-N curves for cartridge brass, alloy No. 260. Reversed
flexure. R = -1.0; K_t = 1.0. From C. M. Tyler [250],
Figs. 2, 4.

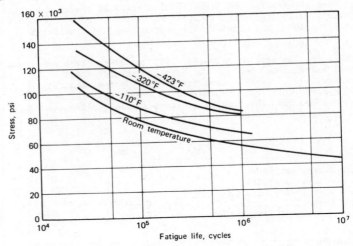

Fig. 4.106 Flexural fatigue of beryllium-copper sheet,
No. 25. 0.078 in. sheet, condition 1/2HT.
Tensile stress = 191,000 psi; R = -1, 1800
cpm at 70°F; -110° and -320°F, 3450 cpm at
-423°F. From [252].

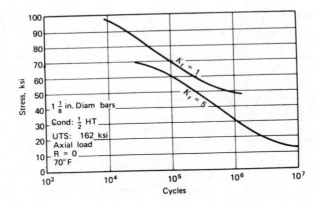

Fig. 4.107 Axial fatigue of beryllium-copper No. 25 bar.
1-1/8-in. diameter bars; condition 1/2HT; UTS:
162 ksi; axial load. R = 0; 70°F. From [252].

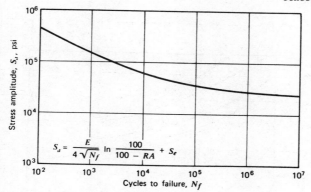

Fig. 4.108 S-N curve for 70-30 copper-nickel alloy. From
W. G. Gibbons [253], Fig. 2.

The endurance limit is about 23 ksi, and the curve and related values are considered suitable for design use.

The room-temperature performance of various coppers in various conditions is given in Figs. 4.109 and 4.110.

S-N curves for the cast copper alloy No. 932 at room temperature are given in Fig. 4.111 for three casting processes.

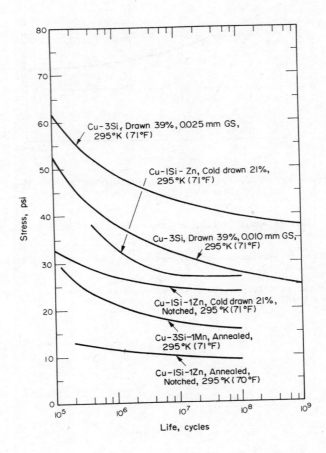

Fig. 4.109 Fatigue behavior of silicon bronze (Cu-Sn and Cu-Sn-Ni).

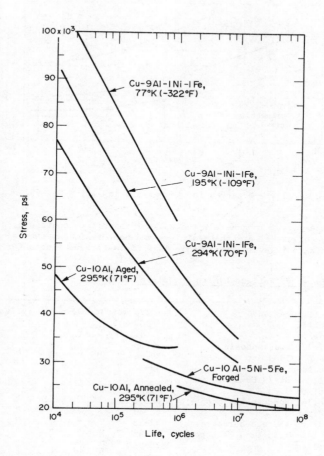

Fig. 4.110   Fatigue behavior of aluminum bronze   (Cu-Al-Fe and Cu-Al-Ni-Fe).

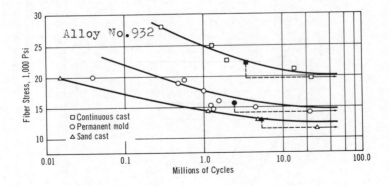

Fig. 4.111 S-N curves for Copper casting alloy No. 932.

Nickel and its Alloys. In general, the bulk of the alloys containing a significant percentage of nickel are the stainless and heat-resistant steels, for which
very extensive data is available. Table 4.12 lists fatigue limits for nickel,

TABLE 4.12 Fatigue Limits for Nickel Alloys

"A" Nickel, annealed	24 ksi at 100 x 10^6 cycles	
Hot-rolled	30 " " " " "	
Cold-drawn	42 " " " " "	
Monel, 67-30, rod, annealed	31-36 " " " " "	
Rod, hot-rolled	39-44 " " " " "	
Rod, cold-drawn	39-47 " " " " "	
"K" Monel, 66 Ni-29 Cu-3 Al, rod,		
Hot-rolled	38-45 " " " " "	
Hot-rolled and age-hardened	45-53 " " " " "	
Cold-drawn	40-43 " " " " "	
Cold-drawn and age-hardened	42-53 " " " " "	
Inconel, 718 Ni-15 Cr-7 Fe annealed	31-45 " " " " "	
Hot-rolled	38-51 " " " " "	
Cold-drawn	40-51 " " " " "	
Inconel 71C, as cast, at 1200°F	45 " " 1 x 10^6 "	
at 1200°F	28 " " 10 x 10^6 "	
at 1200°F	22 " " 100 x 10^6 "	
at 1500°F	47 " " 1 x 10^6 "	
at 1500°F	35 " " 10 x 10^6 "	
at 1500°F	30 " " 100 x 10^6 "	

Monel and Inconel, while Figs. 4.112 and 4.113 provide S-N curves for Nickel Alloy
No. 440, a 2% Be - 0.5% Ti composition for high strength and good temperature
resistance (250 ksi UTS to 600°F), and for Nilvar, a 65% Fe - 35% Ni alloy of the
Invar family. Constant life diagrams for Inconel 718 sheet at room and elevated
temperatures are found in Ref. [1], Figs. 6.3 and 8.28, while strain-N curves are
given in Fig. 4.114 for two temperatures and heat-treats. Cryogenic performance

for the 718 alloy is shown in Fig. 4.115. S-N curves for Inconel 625 bar and weld metal at various temperatures is given in Fig. 4.116. Welded Inconel 82 has been tested in several orientations of specimen axis to weld axis, to 650°F, with no particular trends appearing. Generally, the curves fall within the band for alloy 718.

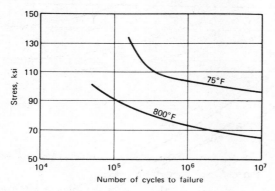

Fig. 4.112 S-N curves for nickel-beryllium strip, alloy No. 440, AT temper. From [254], Fig. 3.

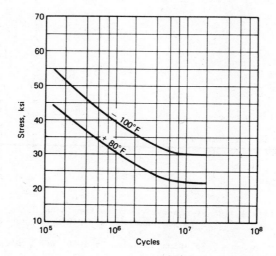

Fig. 4.113 Flexural fatigue of Nilvar sheet, 35% Ni-65% Fe. 0.040-in. thick; hard temper.

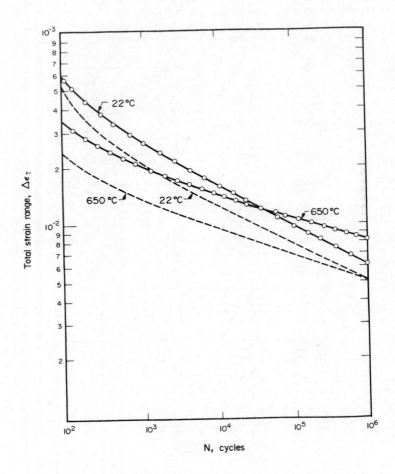

Fig. 4.114 Strain-cycle curves for Inconel 718
Solution annealed at 940°C and duplex
aged 0-0
Solution annealed at 1040° and duplex
aged ---
Strain rate is 4 x 10⁻³; R = -1.0,
K_t = 1.0.

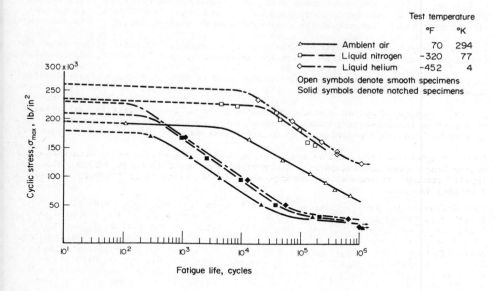

Fig. 4.115   S-N curves for Inconel 718 at cryogenic temperatures.

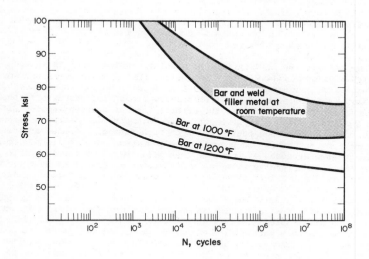

Fig. 4.116   S-N curves for Inconel 625 hot-rolled and solution annealed
bar.   R = -1.0, K_t = 1.0.

4.3.8 Plastics, Laminates, and Composites

Reinforced Plastics. One of the more useful nonmetallic materials is fiberglass
sheet or bar, a composition of glass fibers in a polyester, phenolic, epoxy, or
silicone matrix. Bulk glass generally will carry a given cyclic stress for roughly
the same time as it will carry a steady load equal to the maximum cyclic load. The
phenomenon of _delayed fracture_ originates by chemical (atmospheric) attack on the
glass surface, which together with the inherently high notch sensitivity results
in the usual brittle fracture. Reinforced plastics exhibit the standard fatigue
effect, failing under cyclic loading in a shorter time than under static loading,
in addition to delayed fracture. This condition is thought to arise from failure
of the matrix-glass bond, allowing atmospheric corrosion of the fibers. In fatigue testing at normal rates, 1800 cpm or more, these materials become rather
hot because of their high damping. Thus, caution should be exercised both in
applications requiring high rates of stress cycling and in the interpretation of
test results.

There are numerous variables in the make-up of fiberglass compositions: the type
of glass and size of fiber, the type and percentage of resin in the matrix, the
number and orientation of plies, external finish, and so forth. A general summary
of the static and fatigue strengths of the polyester laminates is given in the
modified Goodman diagrams of Fig. 4.117. Particularly important is the high sensitivity of these materials to tensile mean stress: the test results for all lives
and levels of mean stress lie far below the linear relation. This condition is
related to low values of the stress rupture strength, and is shown in the plot by
terminating the life curves on the abcissa at a _static_ mean stress value obtained
in a time corresponding to that for the fatigue test. Upon sufficient rise in
temperature, the laminates will usually exhibit a decreasing compressive strength.
With the polyester resin, the effect is not noticeable to 300°F; but, at 500°F,
Fig. 4.117c shows that somewhat higher alternating stress levels can be carried
at low mean stress than at zero mean stress; that is, the life curves undergo a
reversal in slope.

The type of glass and of resin in the matrix, and of the resin content, affect
fatigue behavior in that an optimum composition exists in the form of the 181-type
glass fabric in a 26 to 28% pehnolic matrix.

The angle between the loading direction and ply orientation is a strong factor in
determining fatigue life. Data show that a 5° bias angle for a non-woven laminate
provides better performance than the 0° angle by reducing the amount of fiber
splitting. This problem arises from the low transverse strength of the fibers;
another fabrication to minimize the effect, that of cross-orienting plies at 0°
and at 90°, is also some improvement. The "85% parallel" structure consists of a
21 ply, 0° core, plus two facing plies on each side at 90° plus another 0° final
facing ply on each side. In general, the nonwoven, unidirectional laminates will
sustain greater stresses at any life than will the woven fabric reinforcements,
because the latter involve crimping and stress concentrations at the crossing
points within the fabric. About 30 to 40% increase in loading capacity may be
realized with the nonwoven constructions, especially in the high stress, low cycle
region; see Fig. 4.118a. S-N curves for a standard polyester-181 glass fabric,
including the effect of ply orientation and elevated temperature are shown in Fig.
4.118b. The highly promising, large diameter (0.003 in. to 0.005 in.) S-glass
filaments in an epoxy matrix gave a fatigue performance superior to 2024-T3 in the
low stress, high cycle range, but inferior to the 8-1-1 titanium. Further improvements in performance provided by two of the newer glasses are shown in Fig. 4.119.

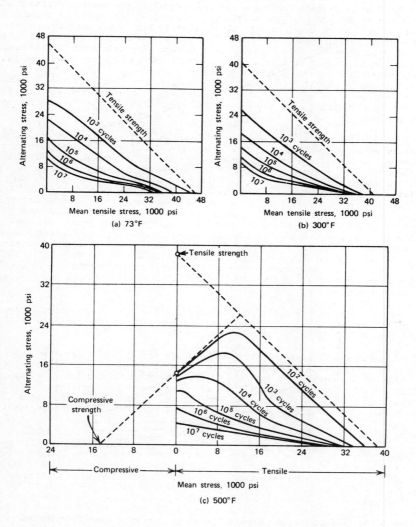

Fig. 4.117   Modified Goodman diagrams for polyester-glass
laminate (181 glass). From [259], Figs. 9,10,11.

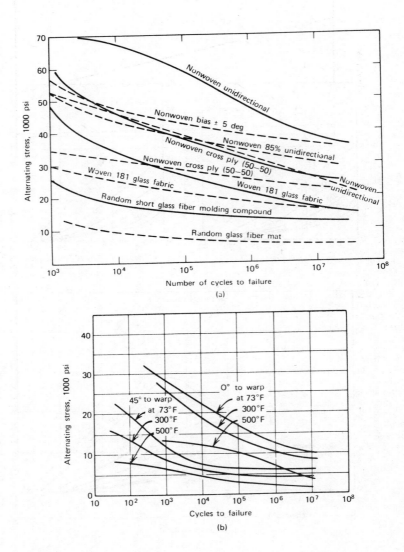

Fig. 4.118 Effect of ply orientation on fatigue of glass-fiber laminates.
(a) Woven, nonwoven, and random weaves. —— Flexural fatigue
(73°F); ---- axial fatigue (73°). From J. W. Davis et al.
[260], Fig. 5. (b) Temperature and direction of loading for
polyester-heat resistant. Reinforcement: 181 glass fabric;
mean stress: 0. From [259], Fig. 5.

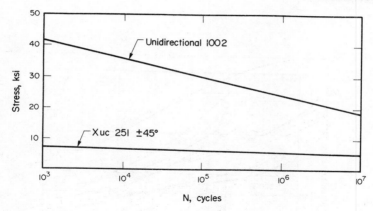

Fig. 4.119   S-N curves for S-glasses, types 1002 and XP251 (from Ref. 403).

Fatigue strength is strongly affected by temperature, as seen from Fig. 4.120, the type of resin being controlling factor. Any specific data are, as noted above, dependent on the rate of loading; the values in these curves were obtained at 15 Hz. The moisture content of the laminate has a highly variable effect on allowable fatigue stress level, as a function both of resin type and of life. For a change in relative humidity of 50 to 100%, the polyester laminates show a reduction in stress by as much as a half at the higher stress levels in the low cycle regime, tapering off to a negligible difference at 10^7 cycles, while the curves for the epoxy laminates deviate noticeably only at the low cycle end.

The internal macrostructure of the fiber reinforced plastics may be regarded as highly redundant. The fibers carry essentially all the tensile and compressive loads, with a somewhat more equal division of the shear between them and the matrix. As an individual fiber fractures, its load is readily transferred to adjacent ones, resulting in a material not highly sensitive to notches. In general, the strength reduction factor is about one-half the theoretical concentration factor.

Much of the original work on these materials in laminate form was done at the Forest Products Laboratory of the U.S. Dept. of Agriculture and reported by Boller [262] and by Werren [263].

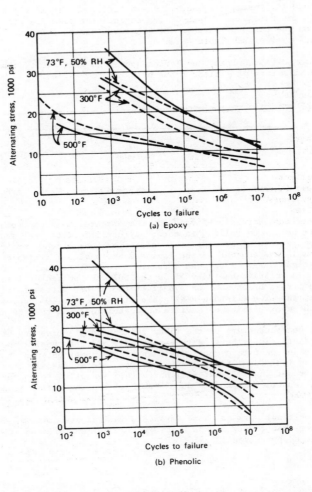

Fig. 4.120 Effect of notching and temperature on fatigue of glass
fiber laminates. Reinforcement: 181 glass fabric;
mean stress: 0; direction: 0° to warp. --- Notched;
—— Unnotched. (a) Epoxy-heat resistant. From [259],
Fig. 5. (b) Phenolic-heat resistant. From [259], Fig. 6.

Nonreinforced Plastics. The fatigue behavior of the homogeneous, or at least not intentionally reinforced, plastics depends on many variables, only a few of which have yet been identified: molecular weight, tensile strength, elastic modulus, and total capacity for strain energy. Essentially all available information is empirical in nature; general data on the engineering plastics are arbitrarily summarized on the basis of whether the S-N curve passes above or below the point 4 ksi and 10^4 cycles, see Table 4.13.

S-N curves for Zytel, Delrin, and Teflon appear in Figs. 4.121 through 4.124 and for the specific form of endless Mylar belts in Fig. 4.125.

TABLE 4.13   Fatigue Strengths for Plastics

Above		Below	
	at 10^4 cycles		at 10^4 cycles
Polyimides	5.5 ksi	Polystyrenes	2.0 ksi
Phenolics, molded	4.8	Vinyls, rigid	3.0
Polysulfones	4.0	Acetals	3.5
Acrylics	7.0	Polyethers	3.0
Polycarbonates	4.5	Polypropylene	3.0
Nylons	7.0	Polyethylene	1.6

SOURCE: [264].

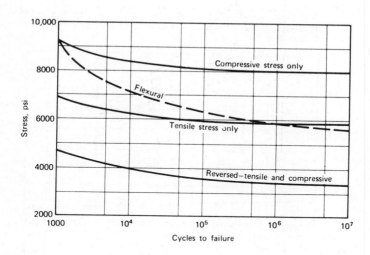

Fig. 4.121   Fatigue behavior of Zytel 101 (nylon) at 73°F and 2.5% moisture. 1200 cpm; no cooling.

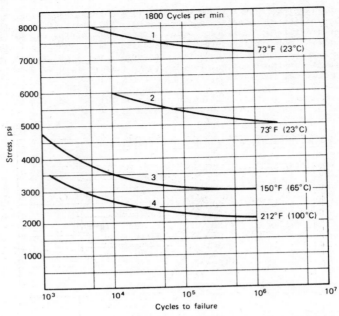

Fig. 4.122 Fatigue behavior of Delrin (acetal). (1) Tensile stress
only, (2), (3) and (4) completely reversed tensile and
compressive stress.

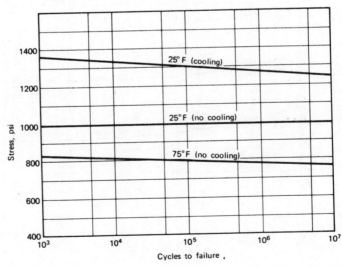

Fig. 4.123 Fatigue behavior of TEFLON 1 (fluorocarbon), reversed stresses.

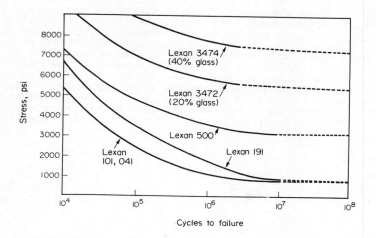

Fig. 4.124   S-N curves for LEXAN resins.

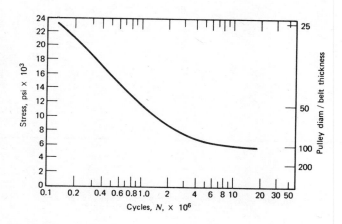

Fig. 4.125   S-N curve for endless belt of MYLAR, (polyester).
Initial tension: 1600 psi.   From J. H. Licht and
A. White [266].

<u>Metal-Matrix Composites</u>   The great interest in these composites arises from the possibility of tailoring a material such that it will have a combination of properties not possible in any single composition. Many objectives exist: very high strength at low density, and superior fatigue performance are two leading items. A discussion of the technology of these composites is beyond the scope of this work; suffice it to say that four forms of fibers are being investigated as the reinforcing agent in both high- and low density metal matrices. The materials and forms are: for wires, Be, Mo, and stainless steel; for filaments, Boron and SiC; for whiskers, Al_2O_3, B_4C_3, and SiC; and for phase particles, Al_3Ni, $CuAl_2$, NiBe, and many more. Figures 4.126 through 4.128 provide a limited view of fatigue behavior of a few composites; again it is to be noted that the values are preliminary, highly dependent on processing variables and distinctly not guaranteed for design.

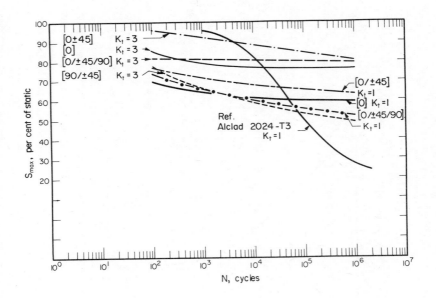

Fig. 4.126   S-N curves for graphite/epoxy type AS3002,
R = 0.05, 70°F (from Ref. 405).

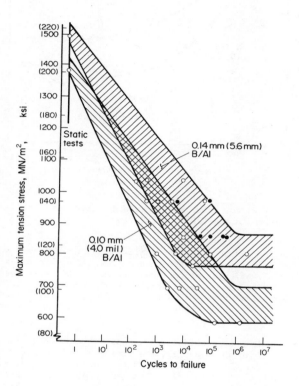

Fig. 4.127 S-N curves for B/Al composite, 50 v/o. Unidirectionally
reinforced. R = +0.1.

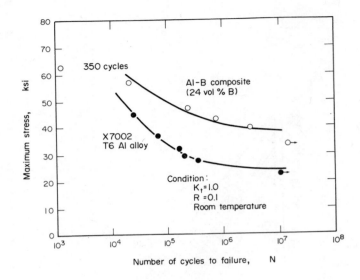

Fig. 4.128 S-N curves for X7002 aluminum, and a composite of 24 v/o boron
filaments in the X7002.   32 Hz.

Bonded Metal Laminates.   Another form of tailor-made material exists as adhesively
or metallically bonded metal sheets and thin plates.   The objectives are varied:
to obtain high damping, or superior fatigue performance by resistance to crack
propagation, or to reduce the anisotropy of the rolled product, etc.   Figure 4.129
shows S-N curves for a laminate of 10 sheets of 0.0035 in. thick 1100 aluminum and
epoxy resin [268].   It is thought that the deviation at the higher stresses was
caused by breakdown of the resin.

Clad metals are found in many combinations.   Copper-steel-copper is available in
several thicknesses: 5-90-5% with a fatigue limit, S_e = 52 ksi; 10-80-10%, S_e =
42 ksi; 2.5-28-39-28-2.5, S_e = 54 ksi.   Stainless-copper-stainless steel at 30-40-
30 has an S_e of 59 ksi.   Diffusion joints between copper-plated sheets of Ti-6Al-
4V provide performance equal to that of the parent titanium, see Fig. 4.130.   More
sophisticated forms are being continually developed--an interesting one involves
4-ply boron-epoxy facings for aluminum core honeycomb.   Steel foil is being produced in limited quantities, which together with acrylic polymer adhesives promises
still another combination in the laminate/sandwich field.

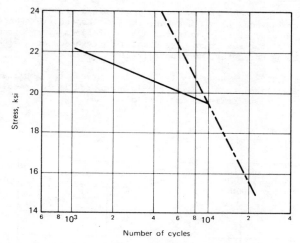

Fig. 4.129   Fatigue life comparison of solid sheet aluminum and a bonded laminate.
--- Solid sheet: ——— laminate; ——— — — solid sheet and laminate.

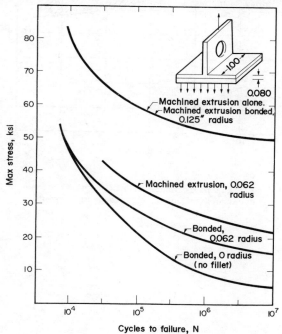

Fig. 4.130   Fatigue strength of copper-plated diffusion joints in
Ti-6Al-4v (from Ref. 401).

<u>Honeycomb Constructions</u>. The literature on the fatigue of sandwich and honeycomb constructions is voluminous [354]; however, the conclusions from these investigations are varied because of the tremendous number of types of specimens and load spectra. Beyond the statement that core density is regarded as a highly important and often controlling parameter, no generalizations are attempted.

Figures 4.131 and 4.132 show possibly typical S-N curves, the implication being that any even semicritical panel design should be tested in full scale, particularly if the major excitation is acoustic noise.

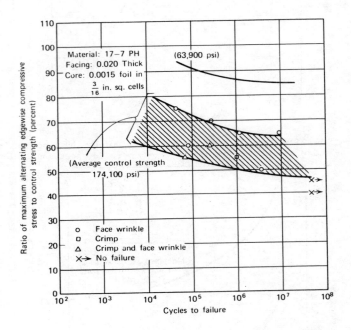

Fig. 4.131 Fatigue performance of stainless steel sandwich construction under edgewise compression.

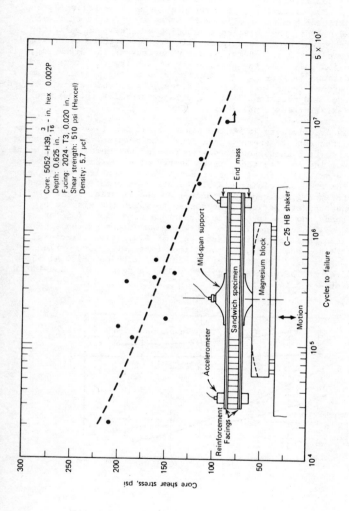

Fig. 4.132 Fatigue properties in shear of adhesive-bonded 5052-H39 aluminum alloy type E sandwich specimens, R = -1.

4.3.9   Glass and Ceramics

The static and fatigue behavior of these brittle materials is controlled by two characteristics that distinguish them from the more ductile metals: (1) the glasses and ceramics are completely brittle; their stress-strain plots show essentially no deviation from linearity up to fracture, as well as extremely low strains, and (2) a significant number of flaws and checks are present on their surfaces as manufactured. Thus, the response to loading will be much different from that of materials showing any measurable ductility. Since there is no yielding or plastic flow, there can be no redistribution of stress at the crack tip; all fracture must be by tension only, and the sensitivity of the fracture stress to details of surface conditions is very great. But, because of the presence of many surface flaws, the difference in ultimate tensile strength of smooth and of notched specimens is seldom large, and occasionally the UTS is greater for the notched specimen due to the triaxial stress condition.

In common with extremely brittle materials, the behavior of the glasses and ceramics depends strongly on the surface energy conditions, the mean stress or steady load, and the strain rate, to a much greater degree than do the more ductile metals. The effect of water vapor in lowering the surface energy of glass has been likened to that of hydrogen in steel. For the latter, the result is usually termed creep or stress-rupture but for the nonyielding glass, the creep deformation is absent. Thus, the delayed fracture of glass under a steady load and of steel under cyclic loading have some common characteristics. The degree to which the glasses are insensitive to cyclic loads and the importance of surface conditions (air or vacuum) are shown in Figs. 4.133 and 4.134. The overwhelming effect of the steady load and its duration is emphasized by the fact that, in Fig. 4.134, fracture occurred at the same stresses and times, although the number of cycles varied by about three orders of magnitude. The behavior of sintered alumina, Fig. 4.135, shows a discernible difference between static and cyclic fatigue, and a considerable difference in fracture stress between grades.

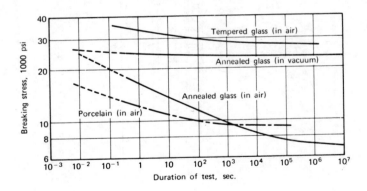

Fig. 4.133   Static fatigue characteristics of glass and porcelain.
From E. B. Shand [357], Fig. 7.

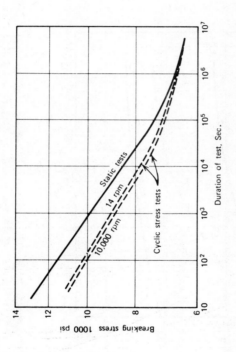

Fig. 4.134   Static and cyclic fatigue characteristics of glass.   From
E. B. Shand [357], Fig. 8.

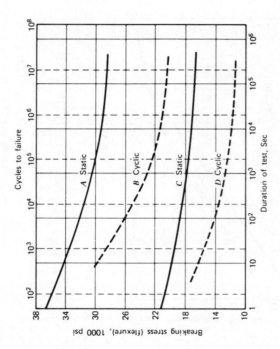

Fig. 4.135   Static and cyclic fatigue characteristics of sintered alumina. Curves A and B: lab-made, high purity, E = 43 x 10^6; curves C and D: commercial grade, E = 38 x 10^6. From L. S. Williams [358], Fig. 9.

4.4   RESIDUAL STRESSES AND SURFACE TREATMENTS

4.4.1   Residual Stresses

There are already in print any number of adequate treatments of "stresses"--simple, combined, mean, residual, alternating, etc.; the objective here is limited to a discussion of the relationships among process variables, the state of residual stress, and fatigue behavior.   The applied load and stress are assumed to be known by proper analysis and/or test, and it is important to recognize that the total (material) strength of the piece is altered by whatever level of residual stress exists as a result of mill processing, heat treatment, fabrication, and surface treatment.   The sign of the residual stress, tension or compression, is characteristic of the integrated effect of all the processing, the part geometry, and the temperature gradient, if any.   Since fatigue failures usually start at a surface location under a tensile stress, any treatment that leaves the surface under compression would be beneficial by reducing an applied tension.   Residual stresses arise from any non-uniform, permanent change in the size or shape of a part.   They are sometimes classified as "macro" or "micro" stresses on the basis of the length or volume in which their magnitudes are observed.

The origins of residual stress in engineering materials are the normal (and any special) thermal, chemical and mechanical treatments and their combinations and sequences, as follows:

1.   Plastic deformation from non-uniform thermal expansion or contraction produces residual stresses, in the absence of phase transformation.   An example is the drastic quenching of steel from a temperature below the critical range to produce compressive surface stresses.

2.   Volume changes from chemical reactions, precipitation, or phase transformations produce residual stresses in various ways.   Nitriding, for example, produces compressive stress in the diffusion region.

3.   Mechanical working that produces non-uniform plastic deformation may be used to impose favorable residual stress patterns.   Examples are surface-rolling or shot-peening of fillets and seats for press fits, and auto-frettage of gun tubes or pressure vessels.   Unwanted tensile stresses from cold-straightening represent the opposite situation.

All mill and shop processes: casting, rolling, welding, heat treatments, machining, and grinding establish their characteristic pattern of residual stress.   Typical stress distributions for carburizing and nitriding are given in Fig. 4.136, and similar gradients result from shot-peening, Fig. 4.138.   When bending or tensile stresses are applied to such gradients, the resultant stresses in the surface layers are considerably reduced.   That is, in fatigue loading, these surface layers experience a much lower peak stress than if the compression layer induced by the peening or heat treatment were absent; Fig. 4.137 sketches the new pattern for a carburized bar after a rotating-bending fatigue test.   For bars with the same residual stress pattern and level, A, two loading levels are shown, B_1 and B_2, with the resultant stress patterns, C_1 and C_2.   The effect of stress concentrations is also shown by the K_t scale, effectively a divisor for the allowable stress scales, for both case and core.   It is thought that an additional change in pattern relaxation of the residual stress may be occasioned in cyclic loading as a result of micro-yielding, and the action outlined in Fig. 4.137 takes place to an extent dependent on the ductility.   If the net stress exceeds the yield strength, local flow takes place until the stress peak subsides into the elastic range.   However, if the core of a large ductile part is under high triaxial tension, plastic flow may be hindered, and the sum of applied and residual stresses can reach the tensile limit, causing

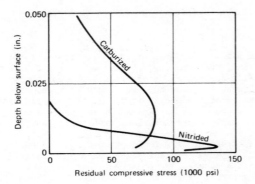

Fig. 4.136 Residual stresses developed in nitrided and carburized steel specimens by expansion of the case during hardening transformations. From C. Lipson [347], Fig. 3.

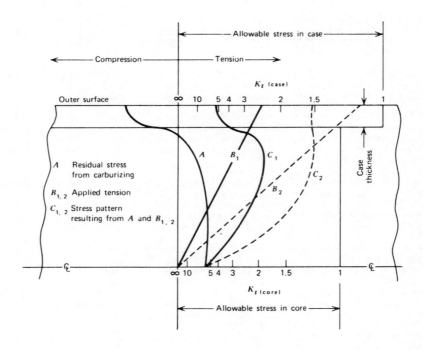

Fig. 4.137 Stress patterns in the tension side of a case-carburized bar in bending.

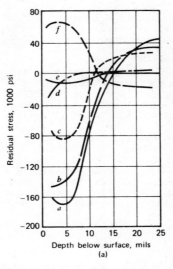

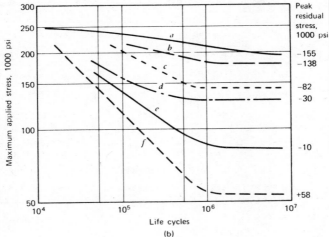

Fig. 4.138 Residual stress and fatigue data for strain-peened, SAE 5160 steel, RC 48. (a) Longitudinal residual stress distribution; (b) one-directional bending S-N characteristics. Curve a, -0.60%; curve b, -0.30%; curve c, 0 (conventionally peened); curve d, preset only; curve e, +0.30%; and curve f, +0.60%. From W. E. Littman [346], Fig. 1.

a crack in the core. Such failures may occur, especially with impact loading. It should also be noted in Fig. 4.136, that the peak tensile stress does not occur exactly on the surface but a few thousandths under it. In nonheat-treated weldments, the stress level can be high and the gradient sharp, depending on the cooling rate and the restraint. Typically, in steel tubing, the circumferential stress varied in the first 1/4-in. from the centerline of the butt weld from 15,000 psi tension to 8000 psi compression, and then required another 3/4-in. to fall to zero, a condition illustrating the high desirability of stress relief.

The major processing factors that affect residual stresses are summarized in Table 4.14; and some qualitative control measures in Table 4.15. The effects of residual stresses on the performance of a part are often difficult to dissociate from strength modifications inherent in their origin, but an investigation by the SAE Iron and Steel Committee (Division 4, Residual Stresses and Fatigue) concluded that:

1. Residual stresses have a similar effect on the fatigue behavior of materials as do mechanically imposed static stresses of the same magnitude.

2. Thus, the significant residual stresses are beneficial if compressive, and detrimental if tensile, particularly in "hard" materials. A "hard" material is one that has been hardened by working or heat treatment (high stored energy). A "soft" material is one that has been annealed or softened (low stored energy).

3. With applied stresses near the endurance limit, the residual stress level remains practically unchanged by fatigue loading.

TABLE 4.14 Processing Factors that Affect Residual Stresses

Residual Stress Introduced or Altered by	Process Temperature Change	Phase Transform.	Cold Plastic Flow	Warpage	Altered Chemistry
Heat-treatment (quenching and tempering)	Yes	Yes	No	Possibly	Possibly
Case hardening (carburizing and nitriding)	Yes	Yes	No	No	Yes
Induction hardening	Yes	Yes	No	No	No
Machining	Possibly	Unknown	Yes	Yes	No
Grinding	Yes	Unknown	Yes	Yes	No
Cold-working (shot-peening, surface-rolling, tumbling, lapping, blast cleaning)	No	No	Yes	Yes	No
Straightening	No	No	Yes	No	No

TABLE 4.15   Residual-Stress Control Measures

Material	Failure Mechanism	Control Measures
Aluminum-alloy steels heat-treated to high hardness	Stress corrosion cracking	Stress relief anneal slightly below tempering temperature. Mechanical surface treatment.
Aircraft and other steels heat-treated to high hardness	Stress corrosion and hydrogen embrittlement	Stress-relief anneal slightly below tempering temperature. Mechanical surface treatment. Bake at 300 to 400°F (hydrogen removal)
Brass and copper alloys subjected to drawing and spinning Carbon and low alloy steels Stainless steels Aluminum alloys	Season cracking and related failures	Stress relief anneal (does not adversely affect hardness). Apply mechanical finishing operations (stretching, compressing, bending, reeling, polishing). Proper combination of depth of draw and corner radius
Alloy steels	Thermal fatigue	Eliminate low melting impurities (if not intentionally added constituents of the alloy). Use proper quenching procedures. Anneal at slow heating rate reduces residual stresses and minimizes thermal stresses)
Age-hardening alloys	Stress precipitation cracking	Mechanical surface treatment

4. At stresses above the endurance limit, residual stresses may relax as an accompaniment of the fatigue process, this effect being greater in "soft" materials and at stresses well above the endurance limit.

5. As a result of (4), the fatigue life at high applied stresses depends very little on the initial residual stresses.

6. The significant residual stress in bending is the peak value near the surface, whether it is tension or compression.

Examples of data that support these conclusions are given in Figs. 4.138 through 4.148. Specimens of a conventional low alloy steel (the SAE 5160) were strain-peened to produce the various residual stress profiles of Fig. 4.138a, then S-N curves were obtained from the same samples, Fig. 4.138b. This fatigue data was combined with standard S-N data for a higher-strength steel, 4340, to form the Goodman diagram of Fig. 4.139. The parallelism of the plots demonstrates that the

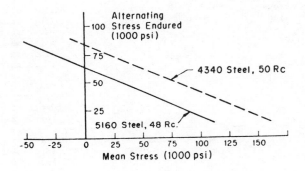

Fig. 4.139   Goodman diagram for 4340 and 5160 steels.   From
W. E. Littman [346], Fig. 2.

effect of the peak residual stress near the surface is like that of a static stress superimposed on the externally applied dynamic stresses. The offset is due to the inherent difference in strength of the two steels; had the same steel been used for both sets of experiments a zero offset would be expected. Similar results for steels of widely different strengths, treated to similar hardness levels, reinforce the conclusion that residual stresses have the same effect as an externally applied stress of equal magnitude.

The interaction between the applied cyclic stress and the prior residual stress is the crux of the problem of predicting the effect of the residual stress on fatigue behavior. As noted above, the general effect of the residual stress is very similar to that of a mean stress; the real question is the determination of the residual stress level. Destructive tests for this value are well established for geometrically simple test coupons, but such procedures are often impractical for machine parts. Based on the now-proven hypothesis that relief of residual stress will occur whenever the combined residual and applied stress exceeds the yield, Rosenthal [131] shows that for metals this condition corresponds to a critical value of the

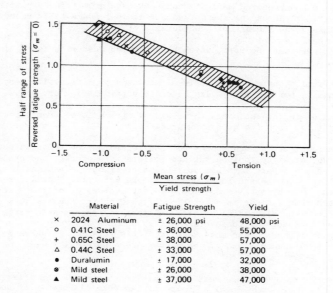

	Material	Fatigue Strength	Yield
×	2024 Aluminum	± 26,000 psi	48,000 psi
o	0.41C Steel	± 36,000	55,000
+	0.65C Steel	± 38,000	57,000
△	0.44C Steel	± 33,000	57,000
●	Duralumin	± 17,000	32,000
⊗	Mild steel	± 26,000	38,000
▲	Mild steel	± 37,000	47,000

Fig. 4.140   The effect of normal stress on permissible alternating
shear stress.   From H. O. Fuchs and E. R. Hutchinson
[352], Fig. 13.

maximum shear stress criterion. Under cyclic tension and compression, the residual stress will be reduced to maintain constant the sum of the local and residual maximum shear stress, the constant being one-half the yield strength. Equations are established relating the maximum and minimum components of the cyclic, mean, and residual stresses that are aligned with the principal directions of the local stress, and the ratio of the tensile yield to the fatigue strength at zero mean stress. Simultaneous solutions are obtained satisfying the above criterion, with typical results (for 6061-T6) being an increase of about 25% in allowable stress for compressive residual stress and a decrease of 30% for tension, that is, the S-N curve for compression residual lies about 25% above the 0 residual curve. The lack of data on the required material constants and geometric complications has tended to restrict the use of the method but it is fundamentally sound. In the absence of this data, the direct procedures are:

1.   Fatigue test the part, which has been produced with controlled process variables or,

2.   Determine the mean and residual stresses by destructive test of the part (trepanning, boring, etc.) with adequate strain-gaging or,

3.   Calculate (or estimate) the residual stress level from the warpage (strain) resulting from the stressing process.

A somewhat more direct statement may be made regarding the relation between residual and fatigue stresses. Since fatigue failures are cracks, and cracks never open unless adjacent particles are pulled apart, it may be assumed that cracks cannot start in a compressed layer nor propagate into it. Detailed research shows to what extent this hypothesis is true. Recent evaluation of the best available data indicates that fatigue is a function of alternating shear stress and average normal (compressive or tensile) stress, and that the permissible range of alternating stress increases as the average stress becomes more compressive and decreases as the average stress becomes more tensile. This relation is shown in Fig. 4.140 by the continuous upward slope of the curves from the tensile side toward the compressive side (abscissae).

Peening or other cold-working overcomes brittleness because the plastic deformation necessary to produce small local adjustments can take place only if the shear stress has a sufficiently high value. Cracking will occur when tensile stress reaches a limit value. Treatment must then permit a high shear stress with a low tensile stress, which is done by providing compressive prestress. The same stress acting inside a part is less dangerous than when it acts at the surface, because the surface is subjected to damaging influences from minute imperfections, traces of corrosion, cracks, and lack of cohesion. Residual stresses remain only as long as the total stress (load plus residual stress) has not exceeded one-half of the yield strength. The yield strength for repeated loading is lower than the statically measured yield strength by 25 to 40%, as shown by the settling of springs.

These considerations explain why shot-peening and similar treatments based on residual stresses become less effective for the higher ranges of alternating stress and must lose their effectiveness when the stress range reaches twice the dynamic yield strength. Under such conditions, residual stress would disappear and only the effect of strain hardening would remain. Testing will, therefore, fail to show benefits from shot-peening if test stresses are appreciably higher than service stresses. Practical solution to such limitations is, of course, found by using material of higher yield strength, such as very hard steels.

Presumably, the above considerations will apply to induction-hardened or flame-hardened parts. But because the nature of the processing is different--heat treatment with no change in chemical composition--a normalized or annealed zone exists between the case and the core; and its hardness and strength may be lower than those of the core. Consequently, in conservative design, the hardness of the transformation zone, rather than that of the core, should be used in instances where the core hardness would otherwise be critical, that is, for smooth or mildly notched specimens. In the presence of sharp notches, the criterion of significant strength will be the hardness of the case, as for carburized parts.

In the case of <u>nitriding</u>, tests have indicated that the endurance limit is unaffected by surface finish for smooth or moderately notched specimens. Only the very sharp discontinuities that cause a large stress concentration will decrease the endurance limit. Consequently, the above considerations for determining allowable stresses are not particularly applicable to nitriding, and recourse should be taken to endurance limits experimentally determined.

There are several methods for determining the level of residual stress:

1. Sach's Boring and Turning
2. Layer Removal
3. Beam Dissection
4. X-Ray Diffraction
5. Hole Drilling

no method, however, being capable of producing the "true" value. From the viewpoint of reproducibility and low sensitivity to local lab technique, (3) and (4) are very good and (1) seems to give equally reproducible results but is less able to cope with sharp stress gradients. The simplicity of (2) is attractive and, with non-mechanical removal, should be as accurate as any of the others.

A selection of values and patterns of residual stress generally indicative of what may be obtained in a variety of alloys and treatments is given below:

a. Welded aluminum plate (5456-H321); as-welded; thermally stress relieved; peened; Fig. 4.141.
b. Stresses in drawn steel rods, roller-straightened; Fig. 4.142.
c. Surface stress in steel cylinders, water-quenched; Fig. 4.143.
d. Stress distribution obtained by boring a large steel rod, Fig. 4.144.
e. Residual stress analysis of a heavy aluminum forging; Fig. 4.145.
f. Residual stress and corrosion resistance of drawn brass tubing; Fig. 4.146.

Distortion, or warpage is the result of changing an existing stress pattern by thermal or mechanical treatment, Fig. 4.147 shows typical effects.

The relief of residual stresses by thermal treatment has been worked out in detail for the ferrous alloys and their forming processes, as well as for many non-ferrous items such as aluminum forgings and magnesium weldments. The objective of these time-temperature schedules is usually the complete relief of the stresses, a condition often difficult to achieve without undesirable side effects such as grain growth or loss of corrosion resistance. Frequently, the degree of completeness may not be known. Table 4.16 lists typical treatments and Fig. 4.148 summarizes the schedule for a specific titanium alloy.

TABLE 4.16 Typical Stress-Relief Treatments

Metal	Temperature, F	Time at Temperature, hr.
Gray cast iron	800 to 1100	5 to 1/2
Carbon Steel		
Less than 0.35% C, less than 3/4-in.	Stress relief usually not required	
Less than 0.35% C, 3/4-in. or greater	(1100 to 1250)	1
More than 0.35% C, less than 1/2-in.	Stress relief usually not required	
More than 0.35% C, 1/2-in. or greater	(1100 to 1250)	1
Specially killed for service at low temperature	(1100 to 1250)	1
Carbon-Molybdenum Steel (All Thicknesses)		
Less than 0.20% C	(1100 to 1250)	2
0.20 to 0.35%	(1250 to 1400)	3 to 2
Chromium-Molybdenum Steel (All Thicknesses)		
2% Cr, 0.5% Mo	1325 to 1375	2
2.25% Cr, 1% Mo and 5% Cr, 0.5% Mo	1350 to 1400	3
9% Cr, 1% Mo	1375 to 1425	3
Chromium Stainless Steel (All Thicknesses)		
Types 410 and 430	(1425 to 1475)	2
Type 405	Stress relief usually not required for thicknesses less than 3/4 in.	
Chromium-Nickel Stainless Steel		
Types 304, 321 and 347	Stress relief usually not required for thicknesses less than 3/4 in.	
Types 316, more than 3/4 in.	1500	2
Types 309 and 310, more than 3/4 in.	1600	2
Welding Dissimilar Materials		
Cr-Mo steel to carbon steel or to C-Mo steel	1350 to 1400	3
Types 410 and 430 to any other steel	1350 to 1400	3
Cr-Ni stainless steel to any other steel	As required for the steel to which Cr-Ni stainless steel is joined	
Copper Alloys		
Copper	300	1/2
90Cu-10Zn	400	1
80Cu-20Zn, 70Cu-30Zn	500	1
63Cu-37Zn	475	1
60Cu-40Zn	375	1/2
70Cu-29Zn-1Sn	575	1
85Cu-15Ni, 70Cu-30Ni	475	1
64Cu-18Zn-18Ni	475	1
95Cu-5Sn, 90Cu-10Sn	375	1
Magnesium Alloys		
M-1, hard-rolled sheet, 1.5Mn	400	1
M-1 extrusions, 1.5Mn	500	1/4
AZ31X hard-rolled sheet, 3Al, 1Zn, 0.3Mn	300	1
AZ31X extrusions, 3Al, 1Zn, 0.3Mn	500	1/4
AZ51X hard-rolled sheet, 5Al, 1Zn, 0.25 Mn	375	1
AZ61X extrusions, 6Al, 1Zn, 0.25Mn	500	1/4
AZ80X extrusions, 8.5Al, 0.5Zn, 0.15Mn	400	1
AZ80X HTA extrusions, 8.5Al, 0.5Zn, 0.15Mn	600	1/4

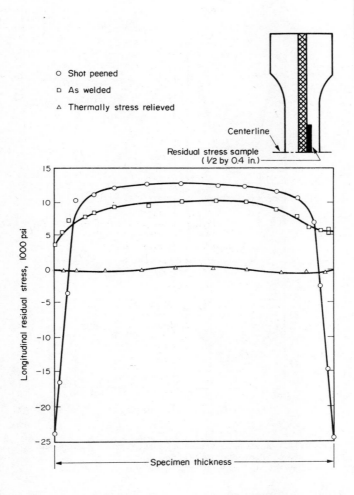

Fig. 4.141   Stress distributions in welded 5456–H321 plate,
by various treatments (from Ref. 417).

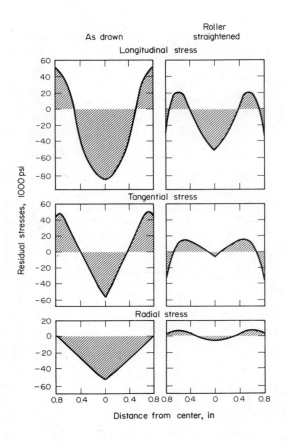

Fig. 4.142   Residual stresses in steel rods before and
after straightening.

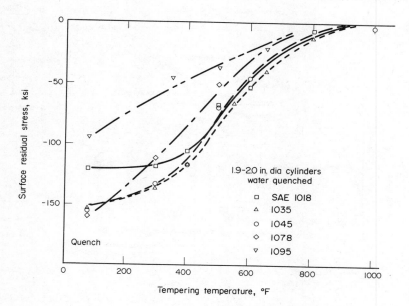

Fig. 4.143   Surface residual stresses as a function of tempering temperature for various steels (from Ref. 418).

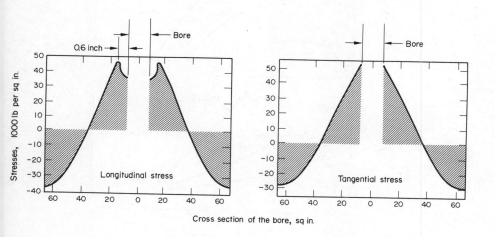

Fig. 4.144   Residual stress patterns in a 10" dia. piston rod as determined by the Sach's Boring-out Method.

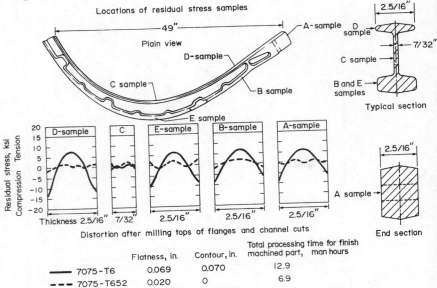

Fig. 4.145 Residual stress analysis for frame forging (from Ref. 416).

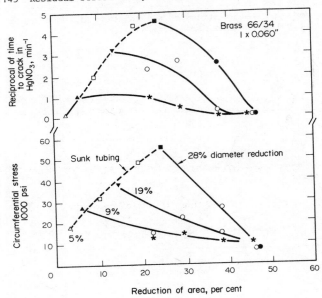

Fig. 4.146 Relationship between residual stress and corrosion
cracking in a 67/33 brass tube.

Butt welds

Distortion 'X' in unrestrained joints

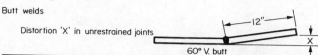

60° V. butt

No. of runs	'X' ins.	No. of runs	'X' ins.	No. of runs	'X' ins.
1	5/16	11	15/16	21	1 9/16
2	3/8	12	1	22	1 5/8
3	7/16	13	1 1/16	23	1 11/16
4	1/2	14	1 1/8	24	1 3/4
5	9/16	15	1 3/16	25	1 13/16
6	5/8	16	1 1/4	26	1 7/8
7	11/16	17	1 5/16	27	1 15/16
8	3/4	18	1 3/8	28	2
9	13/16	19	1 7/16	29	2 1/16
10	7/8	20	1 1/2	30	2 1/8

Fillet welds

Distortion 'X' in unrestrained joints

'W' ins.	'T' inches							
	3/8	1/2	3/4	1	1 1/4	1 1/2	2	2 1/2
	'X' distortion in 64ths. of an inch							
6	3	2	1	1	1	–	–	–
9	6	5	3	2	2	2	1	1
12	9	7	5	3	2	2	2	1
15	12	9	6	5	4	3	2	2
18	15	11	7	5	4	4	3	2
21	18	14	9	7	6	5	3	3
24	21	16	10	8	7	6	4	3

The above figures are based upon the use of $\frac{1}{4}''$ fillets with normal penetration electrodes. For other fillet sizes multiply the above figures by:

$\frac{3}{8}''$ Fillet leg . . . . Factor 1.15

$\frac{1}{2}''$,,   ,, ... ,,   1.50

$\frac{3}{4}''$,,   ,, ... ,,   1.80

If clamped down during welding the distortion will be approximately half the above figures.

Fig. 4.147 Welding distortion in steel plate.

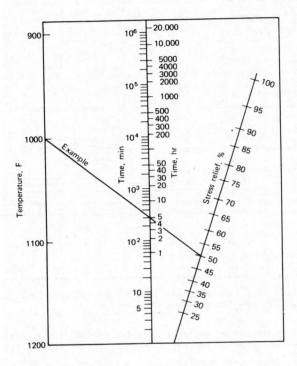

Fig. 4.148 Relationship of time, temperature, and percent stress relief for Ti-6Al-4V alloy.

4.4.2   Surface Treatments

The term "surface" as used here has the connotation of a layer of finite depth,
but usually small compared with the thickness or radius of the part. Each of the
numerous surface treatments (machining, cladding, plating, quenching, etc.) establishes a characteristic state of residual stress in the subsurface layer. From a
somewhat simplified viewpoint, the capacity of this stressed layer to affect the
fatigue life of the part depends on the sign of the residual stress and the fact
that the applied tensile stress will be maximum on the surface, decreasing toward
the center, except in the case of pure uniaxial loading. The residual stress is
always to be algebraically subtracted from the tensile strength of the material in
arriving at an allowable fatigue strength (applying design factors as appropriate),
thus, the benefit of residual compression, with its minus sign.

Of the several processes for inducing surface compression, cold-working in any
of its variants is eminently practical. Shot-peening and cold-rolling are both
widely used, the latter for railway axles, marine shafting, and the like, and the
former on shapes of any complexity, particularly on fillets in forgings, on springs,
axles, and many other machine parts. In the peening process, the spherical metal
or glass shot (flour or rice hulls for fine finishes) are propelled against the
part with sufficient energy to cause plastic indentation. These indentations
result from local plastic yielding; as the deformed volumes try to expand, they
are restrained by the subsurface layer that was not deformed. The tendency of
the deformed surface layer to expand is balanced by the tension in the interior of
the piece. Since the compression layer is thin, the stresses are high; since the
tension layer is thicker, the stresses are lower, resulting in a net balance of
forces. If, in Fig. 4.136, the curves were continued to greater depths, they
would cross the zero stress axis, indicating mild tension. Also, when viewed on
a very fine scale, it should be noted that the curves return to zero at the (theoretical) surface. In general, this description of the states of stress in a
surface-treated piece is valid irrespective of the sign of the stress and the
method by which it originated.

Three parameters determine the "quality" of peening:

    1.   Level and sign of the residual stress, which depend on the yield strength
and on the sign and level of any prestrain during peening.

    2.   Number of impacts per unit area, called the degree of coverage.

    3.   Depth of the peened or compressively stressed layer, which depends on
the hardness and ductility of the part, its state of strain and the type of shot
and of the stream.

The residual stress can reach a level of about half the yield stress when the
specimen is not strained during peening and the coverage is adequate. These integrated effects are best measured by an Almen test strip [106], which is a thin
(0.051 in.) steel strip, peened on one side only. The residual compression causes
convex curvature on the peened side, and the arc height is taken as a measure of
the degree of peening or stress level. The use of this instrument is a good process control for the peening operation but the basic translation from arc height
of the strip to fatigue life of the part must be done by test of the part. This
relationship is shown for a spring steel in Fig. 4.149, which also includes the
effect of exposure time and the size of the shot. It becomes evident that exposure
times must reach a definite point before maximum fatigue durability is realized;
longer exposure is of no particular benefit. The effect of variable air pressure
is also noted to be rather ineffectual above a certain minimum. The considerable
benefit of peening cold-straightened axles, with their extremely poor pattern of

residual stresses after straightening (i.e., high tension) is given for a typical
case in Fig. 4.150. Improvements obtainable by peening after plating are shown
by the S-N curves in Fig. 4.151, and similarly proportionate increases are possible
by peening after grinding.

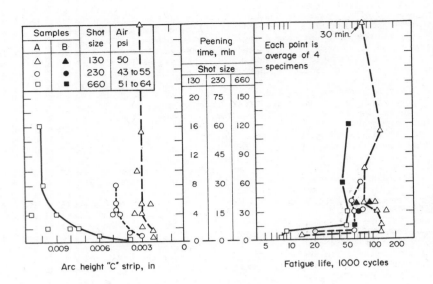

Fig. 4.149 Variation in fatigue life with peening time,
shot size and air pressure (From Ref. 351).

The objective of __strain-peening__ is to combine the advantages of strain-strengthening
and shot-peening. The benefits of this process can be great in certain applications,
but like all combinations, they must be carefully considered to avoid defeating the
objective. Specifically, strain-peening is advantageous in undirectional repeated
loading, as in automotive leaf springs; but it should not be used in reversed, repeated loading. This condition occurs because the material will always deform in
such a manner as to absorb the least energy, that is, it will take on a stress or
strain level of lesser magnitude than initially, indicating a drift toward a residual stress or strain of opposite sign. Thus, it follows that, to induce a residual
compressive stress by peening, the stress applied during peening must be tensile.
This operation is known as presetting; the optimum prestrain seems to be about 50
to 65% of the yield strain. Stress distributions and one-directional S-N curves
for bars were shown in Fig. 4.138, and for coil springs in Fig. 3.224. The increase in fatigue properties results from the very high residual compression on
the surface of the spring, where the service loads produce tension. By the Wohler
method, the latter may be subtracted from the former, so that the actual stress
on the working surface is much reduced.

For __carburized parts__ the major problem is the determination of an allowable stress,
in view of the great difference in hardness and strength of the case and core. The
first step in considering whether case- or core-hardness should be used is to

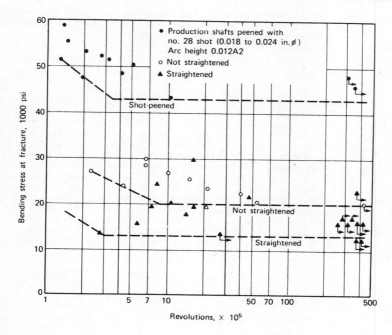

Fig. 4.150   Effect of peening after straightening on fatigue life
of axles. (From Ref. 352.)

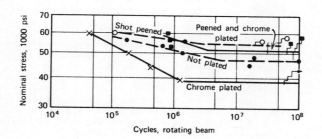

Fig. 4.151   Fatigue strength of chrome-plated steel, peened and
unpeened.   From H. O. Fuchs and E. R. Hutchinson
[352], Fig. 10.

superimpose applied stresses on the allowable stresses. This condition was illustrated in Fig. 4.137, where it may be noted that failure will occur in the case or in the core, depending on the level of loading, the value of K_t, and the relative levels of the allowable stresses. In the case of B_2, for example, it is seen that the maximum applied tension must be the lower of the two values: (case allowable)/1.51, or (core allowable)/1.25.

Plating is applied to surfaces to prevent wear or as a protection against corrosion; but, like most other surface treatments, it affects the fatigue strength of the member. It has been found that softer electro-plating processes such as zinc, lead, and copper have no detrimental effect on the fatigue strength of steel when tested in air. However, when the tests were conducted in fresh water, the plated specimens showed a slight increase in strength over the bare steel, indicating that soft platings offer advantages for members subjected to both corrosion and fatigue. This statement does not apply to hard platings, such as chrome or nickel, when they are deposited in a highly stressed condition, or when the conditions of plating produce hydrogen embrittlement. The detrimental effect is more pronounced with the harder steels. For both nickel and hard-chrome plated steels, the approximate loss in fatigue strength at any lifetime is 30 to 40%; see Figs. 4.151 and 3.52. In the low cycle region, it is noted that the curves converge appreciably toward the yield strength, an observation in line with results of static tests which indicate that electroplating has little effect on the static strength of steels.

As-forged: S_a rises linearly with UTS to 25 ksi

Hot-rolled: S_a rises linearly with UTS to 65 ksi

Ground: S_a rises linearly with UTS to 85 ksi

These values then represent the integrated results of the surface finish and the residual stress pattern and level characteristic of the entire process.

The fatigue results from electrochemical machining (ECM), electropolishing, vapor honing, and combination processes are both sparse and diverse. A typical result is given in Fig. 4.152. There is general agreement, based upon the results of various studies on the effects of these surface-finishing methods on fatigue strength, that:

1.   Various finishing methods impart different textures and stresses to the surfaces of metals. Since fatigue is a surface phenomenon, the various finishing methods will influence fatigue strength differently.

2.   The mechanical polishing methods used in the preparation of fatigue test specimens usually impart compressive stresses to the metal surface, increasing the fatigue strength. Therefore, electropolishing or electrochemical machining, when it removes such stresses by dissolution of the surface layer of the metal, will seem to have lowered the fatigue strength or endurance limit values by about 10 to 25%. Thus, the fatigue strength, as determined for an electropolished or electrochemically machined part with its stress-free surface, represents the true value for the metal, rather than that of the metal plus a particular surface-finishing treatment.

3.   Mild surface treatments, such as hand-honing, vapor-honing, or shot-peening, which impart compressive stresses, will restore completely the "apparent loss" of fatigue strength to the electropolished or electrochemically machined part.

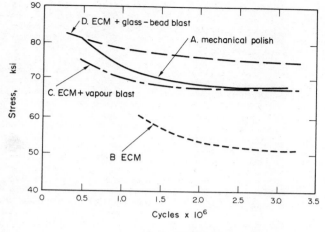

Surface treatment

A. Mill bar on all sides and lightly polish on the long axis to 16 microinch finish.

B. Electrochemically machine (ECM) bar on all sides, surface finish 30-40 microinches.

C. Electrochemically machine to size, followed by vapour blast at 85 psi with 240 mesh grit, final surface finish 15-20 microinches.

D. Electrochemically machine to size, followed by glass bead blast at 84 psi with No. 131-size bead; final surface finish 15-20 microinches

Fig. 4.152   Effect of electrochemical machining and subsequent surface treatments on fatigue life of 403 stainless steel.

4.   Electropolished and electrochemically machined specimens have shown very little scatter in the S-N fatigue data, as compared with mechanically finished specimens which is an indication of their uniformity and reproducibility. Parts so finished are in optimum condition for any further processing such as shot-peening or vapor-honing to obtain the ultimate in fatigue life.

REFERENCES

1. MIL-HDBK-5A, *Metallic Materials and Elements for Flight Vehicle Structures*, Department of Defense, December 1968.

2. *Metals Handbook*, Vol. 1, *Properties and Selection of Metals*, 8th ed., American Society for Metals, Metals Park, Ohio, 1961.

3. *Metals Handbook*, Vol. 2, *Heat Treating, Cleaning and Finishing*, 8th ed., American Society for Metals, Metals Park, Ohio, 1964.

4. A. M. Freudenthal, ed., *Fatigue in Aircraft Structures*, Academic Press, New York, 1956.

5. R. B. Heywood, *Designing Against Fatigue of Metals*, Reinhold, New York, 1962.

6. A. J. Kennedy, *Processes of Creep and Fatigue in Metals*, Wiley, New York, 1963.

7. J. M. Lessells, *Strength and Resistance of Metals*, Wiley, New York, 1954.

8. *A Guide for Fatigue Testing and the Statistical Analysis of Fatigue Data*, STP 91-A, 2nd ed., ASTM, Philadelphia, 1963.

9. H. J. Grover, S. A. Gordon, and L. R. Jackson, *Fatigue of Metals and Structures*, NAVAER 00-25-534, United States Navy, 1954. (See also [348].)

10. Proceedings of Conference on Acoustical Fatigue, WADC Tech. Rept. 59-676, Wright Air Development Division and University of Minnesota, March 1961.

11. Proceedings of International Conference on Fatigue of Metals, Brit. Inst. Mech. Engrs. and ASME, London, 1956.

12. W. H. Munse and L. Grover, ed., *Fatigue of Welded Steel Structures*, Welding Research Council, New York, 1964.

13. S. H. Crandall and W. D. Mark, *Random Vibration in Mechanical Systems*, Academic Press, New York, 1963.

14. F. R. Shanley, *Weight-Strength Analysis of Aircraft Structures*, Dover, New York, 1960.

15. E. F. Bruhn, *Analysis and Design of Flight Vehicle Structures*, Tri-State Offset Company, Cincinnati, Ohio, 1965.

16. W. C. Hurty and M. F. Rubenstein, *Dynamics of Structures*, Prentice-Hall, Englewood Cliffs, N.J., 1964.

17. G. Gerard, *Introduction to Structural Stability Theory*, McGraw-Hill, New York, 1962.

18. N. J. Hoff, *The Analysis of Structures*, Wiley, New York, 1956.

19. S. Timoshenko, *Theory of Plates and Shells*, McGraw-Hill, New York, 1940.

20. R. J. Roark, *Formulas for Stress and Strain*, McGraw-Hill, New York, 1954.

21. D. J. Peery, *Aircraft Structures*, McGraw-Hill, New York, 1950.

22. B.K.O. Lundberg, Discussion of [80] in *Fatigue in Aircraft Structures*, ed. by A. M. Freudenthal, Academic Press, New York, 1956, pp. 341-344.

23. A.J.S. Pippard and J.L. Pritchard, *Aeroplane Structures*, Longmans, Green, New York, 1919.

24. B.K.O. Lundberg and S. Eggwertz, *A Statistical Method for Fail-Safe Design with Respect to Aircraft Fatigue*, The Aeronautical Research Institute of Sweden Rept. No. 99, Stockholm, 1964.

25. E. H. Spaulding, *Observations on the Design of Fatigue-Resistant and Fail-Safe Aircraft Structures*, Proceedings of International Conference on Fatigue of Metals, Brit. Inst. Mech. Engrs. and ASME, London, 1956, pp. 628 ff.

26. E. A. Rossman and W. T. Shuler, *A Critical Review of Fail Safe (Damage Tolerance) and Fatigue Design Methods*, unpublished, Lockheed-Georgia Company, February 1965.

27. H. Giddings, "The Fatigue Problem in Aircraft Design," in A. M. Freudenthal, ed., *Fatigue in Aircraft Structures*, Academic Press, New York, 1956, pp. 347 ff.

28. A. Palmgren, "The Life Expectancy of Ball Bearings," ZVDI, Vol. 68, No. 14, 339; 1924. ("Die Lebensdauer von Kugellagern.")

29. B. F. Langer, "Fatigue Failure from Stress Cycles of Varying Amplitude," *Trans. ASME*, Vol. 59, A160; 1937.

30. M. A.Miner, "Cumulative Damage in Fatigue," *Trans. ASME*, Vol. 67, A159; 1945.

31. E. Gassner, *Effect of Variable Load and Cumulative Damage on Fatigue in Vehicle and Airplane Structures*, Proceedings of International Conference on Fatigue of Metals, Brit. Inst. Mech. Engrs. and ASME, London, 1956, pp. 304 ff.

32. A. M. Freudenthal and R. A. Heller, "On Stress Interaction in Fatigue and a Cumulative Damage Rule," *J. Aero/Space Sciences*, 431 ff., July 1959.

33. S. M. Marco and W. L. Starkey, "A Concept of Fatigue Damage," *Trans. ASME*, 627 ff., May 1954.

34. A. K. Head and F. H. Hooke, *Random Noise Fatigue Testing*, Proceedings of International Conference on Fatigue of Metals, Brit. Inst. Mech. Engrs. and ASME, London, 1965, pp. 301 ff.

35. A. M. Freudenthal, *Cumulative Damage under Random Loading*, Proceedings of International Conference on Fatigue of Metals, Brit. Inst. Mech. Engrs. and ASME, London, 1956, pp. 257 ff.

36. H. T. Corten and T. J. Dolan, *Cumulative Fatigue Damage*, Proceedings of International Conference on Fatigue of Metals, Brit. Inst. Mech. Engrs. and ASME, London, 1956, pp. 235 ff.

37. A. M. Freudenthal, *Life Estimate of Fatigue Sensitive Structures*, United States Air Force Materials Laboratory, TDR 64-3400, 1964.

38. H. Switzky and J. W. Cary, "The Minimum Weight Design of Cylindrical Structures," *AIAA Journal*, Vol. 1, No. 10, October 1963.

39. C. C. Osgood, *Spacecraft Structures*, Prentice-Hall, Englewood Cliffs, N.J., 1966.

40. L. A. Schmidt, Jr., et al., "Structure Synthesis Capability for Integrally Stiffened Waffle Plates," *AIAA Journal*, Vol. 1, No. 12, 1963.

41. P. R. Walker, *J. Roy Aeronaut. Soc.*, Vol. 53, 763, 1949.

42. W. S. Pellini, "Principles of Fracture-Safe Design - Part 1," *Welding Research Suppl.*, March 1971, pp. 91 - 1095.

43. Staff of Battelle Memorial Institute, *Prevention of the Fatigue of Metals under Repeated Stress*, Wiley, New York, 1941, pp. 92-96, 188-206.

44. Discussion of Langer's paper [29] by A. Thum and W. Bautz, *J. Appl. Mech. Trans. ASME*, Vol. 60, A180, 1938.

45. A. M. Freudenthal, *Fatigue of Materials and Structures under Random Loading* in Proceedings of Conference on Acoustical Fatigue, WADC Tech. Rept.-59-676, Wright Air Development Division and University of Minnesota, March 1961.

46. A. A. Griffith, "The Phenomena of Rupture and Flow in Solids," *Trans. Roy Soc.*, Sec. A., Vol. 221, 1920.

47. G. R. Irwin, "Fracture Dynamics," in *Fracturing of Metals*, American Society for Metals, 1948.

48. H. F. Hardrath, "Crack Propagation and Final Failure," Part 2 of *Fatigue of Metals in Materials Research and Standards*, ASTM, February 1963.

49. S. S. Manson, "Fatigue: A Complex Subject--Some Simple Approximations," *Journal of the Soc. for Exp. Stress Anal. (Experimental Mechanics)*, Vol. 5, No. 7, July 1965.

50. C. O.Albrecht, "Statistical Evaluation of a Limited Number of Fatigue Test Specimens, Including a Factor of Safety Approach," in *Symposium on Fatigue of Aircraft Structures*, ASTM STP No. 338, 1962.

51. Federal Aviation Agency, *Civil Aeronautics Manual No. 6*, Appendix A, Supt. of Documents, Washington 25, D.C., October 1959.

52. S. Tanaka, "On Cumulative Damage in Impulse Fatigue Tests," *Trans. ASME, J. Basic Eng.*, 535-38, December 1963.

53. W. Weibull, *The Effect of Size and Stress History on Fatigue Crack Initiation and Propagation*, Proc. Crack Propagation Symposium, 271-86, Cranfield, 1961.

54. A. J. Troughton and J. McStay, "Theory and Practice in Fail-Safe Wing Design," in *Current Aeronautical Fatigue Problems* (ICAF-AGARD Symposium at Rome, 1963), ed. by Schijve et al., Pergamon Press, New York, 1965.

55. *Current Aeronautical Fatigue Problems* (ICAF-AGARD Symposium at Rome, 1963), ed. by Schijve et al., Pergamon Press, New York, 1965.

56. *Full-Scale Fatigue Testing of Aircraft Structures* (ICAF-AGARD Symposium at Amsterdam, 1959), ed. by Plantema and Schijve, Pergamon Press, New York, 1961.

57. B.K.O. Lundberg, "The Quantitative Statistical Approach to the Aircraft Fatigue Problem," in *Full-Scale Fatigue Testing of Aircraft Structures* (ICAF-AGARD Symposium at Amsterdam, 1959), ed. by Plantema and Schijve, Pergamon Press, New York, 1961.

58. R. T. Hunt, "Crack Propagation and Residual Static Strength of Stiffened and Unstiffened Sheet," in *Current Aeronautical Fatigue Problems* (ICAF-AGARD Symposium at Rome, 1963), ed. by Schijve et al., Pergamon Press, New York, 1965.

59. W. J. Crichlow, "The Ultimate Strength of Damage Structure--Analysis Methods with Correlating Test Data," in *Full-Scale Fatigue Testing of Aircraft Structures*, (ICAF-AGARD Symposium at Amsterdam, 1959), ed. by Plantema and Schijve, Pergamon Press, New York, 1961.

60. *Fatigue of Aircraft Structures* (ICAF-AGARD Symposium at Paris, 1961), ed. by Barrois and Ripley, Pergamon Press, New York, 1963.

61. N. F. Harpur, "Crack Propagation and Residual Strength Characteristics of Some Aircraft Structural Materials," in *Fatigue of Aircraft Structures* (an ICAF-AGARD Symposium at Paris, 1961), ed. by Barrois and Ripley, Pergamon Press, New York, 1963.

62. H. Switzky, "Designing for Structural Reliability," *Journal of Aircraft*, Vol. 2, No. 6, AIAA, New York, November-December 1965.

63. C. R. Smith, *Fatigue-Service Life Prediction Based on Tests at Constant Stress Levels*, Proc. Soc. Exp. Stress Anal., Vol. XVI, No. 1, 1957.

64. E. B. Haugen, *Statistical Methods for Structural Reliability Analysis*, Proc. Tenth Natl. Symposium on Reliability and Quality Control, Washington, D.C., 1964. Ed. by IEEE, New York.

65. J. Schijve, "Endurance under Program Testing," in *Full-Scale Fatigue Testing of Aircraft Structures* (ICAF-AGARD Symposium at Amsterdam, 1959), ed. by Plantema and Schijve, Pergamon Press, New York, 1961, pp. 41-59.

66. J. Taylor, "Measurement of Gust Loads in Aircraft," *J. Roy. Aeron. Soc.*, Vol. 57, 78-88, February 1953.

67. W. A. Weibull, "A Statistical Representation of Fatigue Failures in Solids," *Acta Polytechnica*, Mech. Eng. Series, Vol. 1, No. 9, 1949.

68. J. Kowalewski, "On the Relation between Fatigue Lives under Random Loading and under Corresponding Program Loading," in *Full-Scale Fatigue Testing of Aircraft Structures* (ICAF-AGARD Symposium at Amsterdam, 1959), ed. by Plantema and Schijve, Pergamon Press, New York, 1961.

69. F. R. Shanley, *A Theory of Fatigue based on Unbonding during Reversed Slip*, Rand Corporation, Rept. P-350-1, November 11, 1952.

70. H. J. Grover, *Cumulative Damage Theories*, Wright Air Development Center (WADC) Tech. Rept. 59-507, August 1959.

71. C. R. Smith, *Prediction of Fatigue Failures in Aluminum Alloy Structures*, Proc. Sec. Exp. Stress Anal., Vol. 12, No. 2, 1955.

72. B.K.O. Lundberg, "Fatigue Life of Airplane Structures," *J. Aeronaut. Sci.*, Vol. 22, No. 6, June 1955.

73. D. L. Henry, "A Theory of Fatigue Damage Accumulation in Steel," *ASME Trans.*, Vol. 77, No. 6, August 1955.

74. E. Gassner, "Eine Bemessungsgrundlage fur Konstruktionsteile mit statistisch wechselnden Betriebsbeanspruchungen," *Konstruktion*, 6, 97-104, 1954.

75. T. Haas, "Simulated Service Life Testing," *The Engineer*, November 14, November 21, 1958.

76. J. B. Kommers, *The Effect of Overstress in Fatigue on the Endurance Life of Steel*, ASTM Proc., Vol. 45, 1945.

77. F. E. Richart, Jr., and N. M. Newmark, *A Hypothesis for Determination of Cumulative Damage in Fatigue*, ASTM Proc., Vol. 48, 1948.

78. *An Engineering Evaluation of Methods for the Prediction of Fatigue Life in Airframe Structures*, Tech. Rep. No. ASD-TR-61-434 (AD 276249), Flight Dynamics Lab, ASD, Wright-Patterson Air Force Base, Ohio, March 1962.

79. W. A. Stauffer, *A Review of Dynamic Loads Criteria and Design Practice*, AIAA Paper No. 64-532, 1964.

80. F. Turner, "Aspects of Fatigue Design of Aircraft Structures," in A. M. Freudenthal, ed., *Fatigue in Aircraft Structures*, Academic Press, New York, 1956.

81. J. J. Frank, "Structural Fatigue," *Machine Design*, Penton Publishing Co., Cleveland, Ohio, June 18, 1964.

82. R. L. Schleicher, "Practical Aspects of Fatigue in Aircraft," in A. M. Freudenthal, ed., *Fatigue in Aircraft Structures*, Academic Press, New York, 1956, p. 381.

83. J. Goodman, *Mechanics Applied to Engineering*, Vol. 1, 9th ed., Longmans Green, London, 1930.

84. A. Wohler, "Uber die festigkeitversuche mit eisen und stahl," *Z. fur Bauweisen*, Vol. 8, 1958; also Vols. 10, 13, 16 and 20, 1870.

85. W. Gerber "Bestimmung der zulossigen spannungen in eisen constructionen," *Z. Bayer Arch. Ing. Ver.*, Vol. 6, 1874.

86. R. W. Fralich, *Experimental Investigation of Effects of Random Loading on the Fatigue Life of Notched Cantilever Beam Specimens of 7075-T6 Aluminum Alloy*, NASA Memo 4-12-59L, NASA, Washington, D.C., June 1959.

87. J. W. Miles, *An Approach to the Buffeting of Aircraft Structures by Jets*, DAC Rep No. SM-14795, Douglas Aircraft Co., Santa Monica, California, June 1953.

88. L. F. Coffin, Jr., *Low Cycle Fatigue: A Review*, General Electric Research Lab Reprint No. 4375, Schenectady, N.Y., October 1962.

89. A. M. Freudenthal and R. A. Heller, "Accumulation of Fatigue Damage," in A. M. Freudenthal, ed., *Fatigue in Aircraft Structures*, Academic Press, New York, 1956, pp. 146-177.

90. K. D. Raithby, "A Comparison of Predicted and Achieved Fatigue Lives in Aircraft Structures," in *Fatigue of Aircraft Structures*, (ICAF-AGARD Symposium at Paris, 1961), ed. by Barrois and Ripley, Pergamon Press, New York, 1963.

91. *Tables of the Normal Probability Functions*, United StatesNational Bureau of Standards, Applied Mathematics Series 23, Washington, D.C.

92. R. B. McCalley, Jr., "Nomogram for Selection of Safety Factors," *Design News*, September 1, 1957.

93. R. E. Peterson, *Stress Concentration Design Factors*, Wiley, New York, 1953.

94. F. R. Shanley, *Strength of Materials*, McGraw-Hill, New York, 1957.

95. H. Neuber, *Theory of Notch Stresses*, J. W. Edwards, Publisher, Ann Arbor, Michigan, 1946. (Kerbspannungslehre.)

96. G. Gerard, *Some Structural Aspects of Orbital Flight*, Am. Rocket Soc. Paper 729-58, November 17, 1958. (Now AIAA, New York).

97. R. E. Peterson, *Fatigue of Metals in Engineering and Design*, (Marburg Lecture, 1962), ASTM, 1962.

98. R. A. Toupin, "Saint-Venant and a Matter of Principle," *Trans. N.Y. Acad. of Sciences*, Series 11, Vol. 28, No. 2, December 1965.

99. J. W. Carter et al., "Fatigue in Riveted and Bolted Single-Lap Joints," *Trans. ASCE*, Vol. 120, 1353, August 1955.

100. *Hi-Shear Standards Manual*, Drawing No. B352, Rev. 18, January 7, 1963, Hi-Shear Corporation, Torrance, Calif.

101. P. N. Bright, "Structural Design Problems in Gas Turbine Engines," *General Motors Eng. J.*, 15, September 1955.

102. T. J.Dolan and J. H. McClow, *The Influences of Bolt Tension and Eccentric Loads on the Behavior of a Bolted Joint*, Proc. SESA, Vol. 8, No. 1, 29, 1950.

103. E. Chesson, Jr., and W. H. Munse, *Studies of the Behavior of High-Strength Bolts and Bolted Joints*, University of Illinois Exp. Sta. Bulletin No. 469, 1961.

104. W.A.P. Fisher, R. H. Cross, and G. M. Norris, "Pre-tensioning for Preventing Fatigue Failure in Bolts," *Aircraft Eng.* 160, June 1952.

105. J. O. Almen, "On the Strength of Highly Stressed, Dynamically Loaded Bolts and Studs," *SAE J. (Trans.)*, Vol. 52, No. 4, 151, April 1944.

106. J. O. Almen and P. H. Black, *Residual Stresses and Fatigue in Metals*, McGraw-Hill, New York, 1963.

107. W. H. Munse et al, "Laboratory Tests of Bolted Joints," *Trans. ASCE*, Vol. 120, 1299, August 1955.

108. C. W. Lewitt et al., "Riveted and Bolted Joints: Fatigue of Bolted Structural Connections," *J. of Structural Div.*, Proc. ASCE, Vol. 89, No. ST-1, 49, February 1963.

109. W. C. Stewart, "Bolted Joints," Sec. 6.16 in ASME Handbook *Metals Engineering--Design*, McGraw-Hill, New York, 1953.

110. G. N. Mangurian and N. M. Johnston, eds., *Aircraft Structural Analysis*, Prentice Hall, Englewood Cliffs, N.J., 1947.

111. Standard Pressed Steel Co. Tech. Rept., *Basic Design and Manufacture of Aircraft Fasteners for Use up to 1600°F*, Jenkintown, Pa., December 17, 1958.

112. Standard Pressed Steel Laboratories Rept. No. 1325, *Evaluation of a 250,000 psi High Fatigue Fastener System*, Jenkintown, Pa., February 28, 1966.

113. *Design Manual No. 5930*, Elastic Stop Nut Corporation of America, Union, N.J., 1959.

114. J. A. Sauer, *Evaluation of Double-Fatigue-Life Thread Form in Self-Locking Nuts*, ESNA Rep. No. ER-167-1873.2, Union, N.J., October 1959.

115. H. L. Cox, *Stress Concentration in Relation to Fatigue*, Proceedings of International Conference on Fatigue of Metals, Brit. Inst. Mech. Engrs. and ASME, London, 1956, p. 212.

116. M. Hetenyi, *A Photoelastic Study of Bolt and Nut Fastenings*, paper presented a ASME annual meeting, New York, December 1942.

117. Bland Burner Co., Precision Threaded Products Division, Brochure: *Tru-load Fasteners*, Hartford, Conn., 1963.

118. Advertisements, Standard Pressed Steel Co., Jenkintown, Pa., 1962-1963.

119. Advertisements in *Machine Design*, Vol. 35 No. 6, March 14, 1963, by Cooper and Turner, Ltd., Sheffield, United Kingdom.

120. *Report No. 4-1502*, Hi-Shear Corporation, Torrance, Calif. September 20, 1957.

121. *Brochure No. 132*, Huck Manufacturing Co., Detroit, Mich., 1962.

122. *Specification for Structural Joints using ASTM A325 or A490 Bolts*, Research Council on Riveted and Bolted Structural Joints, Engineering Foundation, March 1962.

123. "Fasteners Reference Issue," *Machine Design*, Penton Publishing Co., Cleveland, Ohio, Vol. 39, June 15, 1967.

124. B. King, *The Mechanical Properties of Beryllium*, Brochure, The Brush Beryllium Co., Cleveland, Ohio, March 1963.

125. *Fatigue Investigation of AN-4 Bolts*, Repts. No. 4-1302-1, -3, -5, -9, -11, and -13 Hi-Shear Rivet Tool Co., Torrance, Calif., July 1, 1957.

126. W. A. Hyler and H. J. Grover, *Tension-Tension Fatigue Behavior of Alloy-Steel Huckbolts*, Battelle Memorial Institute, Columbus, Ohio, December 18, 1953.

127. *Effect of Heli-Coil Wire Inserts in 75S-T6 upon the Fatigue Properties of Aircraft Bolts*, Tech. Rept. Nos. 618/c41 and 668/c44, Lessells and Associates, Inc., Boston, August 5, 1958 and January 27, 1960. (For Helicoil Corp., Danbury, Conn.)

128. *Test Rept. No. 1504* (an S-N curve for a Rosan Slimsert No. SR 258L), Briles Manufacturing Co., El Segundo, Calif., November 4, 1960.

129. E. C. Hartman, M. Holt, and I. D. Eaton, *Static and Fatigue Strengths of Aluminum-Alloy Bolted Joints*, NACA Tech. Note 2276, February 1951.

130. "Cryogenic Fastener Materials," *Metals Progress*, July 1966.

131. G. Sines and J. L. Waisman, eds., *Metal Fatigue*, McGraw-Hill, New York, 1959.

132. H. G. Popp, W. A. Hyler, and H. J. Grover, *The Effect of Interference Fit on the Fatigue Behavior of Aluminum Alloy Joints*, Battelle Memorial Institute, Summary Report to Huck Manufacturing Co., April 19, 1957.

133. *Summary of Self-Broaching and Self-Sizing, Huckbolt Fastener Fatigue Study, Parts 1, 2, and 3*, Huck Manufacturing Co., Detroit, Michigan 48207, September 15, 1960.

134. G. E. Nordmark and I. D. Eaton, *Axial-Stress Fatigue Strengths of Plain Sheet Specimens and Riveted Butt Joints in High Strength Aluminum Alloys*, Alcoa Res. Lab. Rept. No. 12-63-24, May 6, 1963.

135. R. E. Whaley, "Stress Concentration Factors for Countersunk Holes," *Exp. Mech.*, Proc. SESA, Vol. 22, No. 2, October 1965.

136. C. W. Lewitt, E. Chesson, Jr., and W. H. Munse, *The Effect of Rivet Bearing on the Fatigue Strength of Riveted Joints*, A Progress Report by University of Illinois, Eng. Exp. Station for Illinois Div. of Highways and Dept. of Commerce-Bureau of Public Roads, Urbana, Illinois, January 1959.

137. W. M. Wilson and F. P. Thomas, *Fatigue Tests of Riveted Joints*, Bulletin No. 302, University of Illinois, Eng. Exp. Station, Urbana, Illinois, updated.

138. *Suggested Specifications for Structures of Aluminum Alloys 6061-T6 and 6062-T6*, Am. Soc. Civil Engrs., Proc. Papers No. 3341 and 3342, Vol. 88, No. ST6, December 1962 (J. Str. Div.).

139. ASME Handbook, *Metals Engineering--Design*, ed. by O. J. Horger, Sec. 6.18, "Welded Structural Joints" by W. M. Wilson, McGraw-Hill, New York, 1953.

140. G. E. Nordmark, *Effect of Peening on the Fatigue Strength of Longitudinal Butt Welds in Alloy 5456-H321 Plate*, Alcoa Res. Lab., Rept. No. 12-62-6, New Kensington, Pa., February 8, 1962.

141. R. P. Newman and T. R. Gurney, "Fatigue Tests on Mild Steel Butt Welds," *Brit. Weld. J.*, Vol. 6, 569, 1959.

142. P. G. Forrest, *Fatigue of Metals*, Addison-Wesley, Reading, Mass., 1962.

143. *United States Air Force Structural Integrity Program Requirements*, ASD-TR-66-57, January 1968.

144. A. I. Kemppinen, W. B. Jenkins, and G. E. Stein, "X7002: A New High Strength Weldable Aluminum Alloy," *Metal Progress*, 100-103, July 1964.

145. H. H. Nuernberger, "Alcoa Aluminum Alloy X7106," *Alcoa Green Letter*, Pittsburgh, Pa., October 1, 1963.

146. R. P. Meister and D. C. Martin, *Welding of Aluminum and Aluminum Alloys*, DMIC Rept. 236, Battelle Memorial Institute, Columbus, Ohio, April 1, 1967.

147. D. R. Apodaca and J. G. Louvier, *Static and Fatigue Properties of Aluminum and Magnesium Premium Quality Castings*, ASM Rept. No. W6-6.3, Amer. Soc. Metals, Metals Park, Ohio, 1963.

148. R. A.Wood, *The Ti-8A1-1Mo-IV Alloy*, DMIC Rept. No. S-10, Battelle Memorial Institute, Columbus, Ohio, April 1, 1965.

149. *Effect of Discontinuities on Fusion-Welded Butt Joints,* Value Engineering Rept. No. VE-510, Lockheed-Georgia Co., Marietta, Ga., December 15, 1964.

150. J. J. Peterson, *Fatigue Behavior of AM-350 Stainless Steel and Titanium-8A1-1Mo-1V Sheet at Room Temperature, 550°F and 800°F,* NASA CR-23, May 1964.

151. T. P. Groenweld, A.R. Elsea, and A. M. Hall, "High Strength Steels Get Tougher," *Materials in Design Engineering,* Reinhold, Stamford, Conn., December 1966.

152. *Welding Handbook,* Vols. 3, 4, and 5, American Welding Soc., New York, 1961.

153. R. Bakish and S. S. White, *Handbook of Electron Beam Welding,* Wiley, New York, 1964.

154. G. Epstein, *Adhesive Bonding of Metals,* Reinhold, New York, 1954.

155. *Stainless Steel Handbook,* Allegheny-Ludlum Steel Corp., Pittsburgh, Pa., 1956.

156. R. H. Sonneborn, *Fiberglass Reinforced Plastics,* Reinhold, New York, 1954.

157. A. C. Grimaldi, *Dynamics Testing of Bonded Joints for Use under Severe Vibrating Stress,* ASME Paper No. 67-VIBR-38, ASME, New York, 1967.

158. C. W. Hamilton, "A Bayes Evaluation of Building Materials Reliability," in *Durability of Adhesive Joints,* ASTM STP 401, ASTM, Philadelphia, Pa., 1965.

159. M. S. Rosenfeld and R. J. Zoudlik, *Determination of Fatigue Characteristics of a Typical Nose Landing Gear,* Naval Air Eng. Center, NAEC-ASL-1079, Philadelphia, Pa., December 21, 1964.

160. J. J. Peterson, *Fatigue Behavior of Ti-8A1-1Mo-1V Sheet in a Simulated Wing Structure under the Environment of a Supersonic Transport,* NASA CR-333, Washington, D.C., November 1965.

161. D. N. Gideon, C. W. Marschall, F. C. Holden, and W. S. Hyler, *Exploratory Studies of Mechanical Cycling Fatigue Behavior of Materials for the Supersonic Transport,* NASA CR-28, Washington, D.C., April, 1964.

162. J. P. Bruner, G. N. Benjamin, and D. R. Bench, "Analysis of Residual, Thermal and Loading Stresses in a B33 Wheel and their Relationship to Fatigue Damage," *Trnas. ASME, J. Eng. Ind.,* May 1967.

163. J. J. Glacken and E. F. Gowen, Jr., *Evaluation of Fasteners and Fastener Materials for Space Vehicles,* NASA CR-357, Washington, D.C., January 1966.

164. Guy Leneman, "The Noise of Rockets," *Space and Aeronautics,* 76-83, October 1965.

165. W. J. Trapp and D. M. Forney, eds., "A Review of Acoustical Fatigue," in *Fatigue--An Interdisciplinary Problem,* Sagamore Army Materials Research Conference Proceedings, Syracuse Univ. Press, Syracuse, N.Y., 1964.

166. M. J. Cote, *Comparison of Approaches for Sonic Fatigue Prevention;* ASD-TDR-63-704, September 1963.

167. *Structural Design for Acoustic Fatigue,* ASD-TDR-63-820, October 1963.

168. H. H. Hubbard, P. M.Edge, and C. T. Modlin, *Design Considerations for Minimizing Acoustic Fatigue in Aircraft Structures,* in Proceedings of Conference on Acoustical Fatigue, WADC Tech. Rept. 59-676, Wright Air Development Division and University of Minnesota, March 1961.

169. J. W.Miles, "On Structural Fatigue under Random Loading," *J. Aeronaut. Sci.,* Vol. 21, No. 11, 753-62, November 1954.

170. A. Powell, "On the Fatigue Failures of Structures Due to Vibrations Excited by Random Pressure Fields," *J. Acoust. Soc. Am.,* Vol. 30, No. 12, 1130-34, December 1958.

171. H. C. Schjelderup, "Prediction of Acoustic Fatigue Life," in *Symposium on Acoustical Fatigue,* ASTM STP No. 284, Philadelphia, Pa., 1964.

172. B. L. Clarkson, "The Design of Structures to Resist Jet Noise Fatigue," *J. Roy Aeron. Soc.,* Vol. 66, No. 622, October 1962.

173. G. Maidanik, "Response of Ribbed Panels to Reverberant Acoustic Fields," *J. Acoust. Soc. Am.,* Vol. 34, No. 6, June 1962.

174. T. J. Dolan, B. J. Lazan, and O. Horger, *Fatigue,* American Society for Metals, Cleveland, Ohio, 1953.

175. L. F. Coffin, Jr., "A Study of the Effects of Cyclic Stresses on a Ductile Metal," *Trans. ASME,* 931-50, August 1954.

176. A. E. Carden, "Thermal Fatigue of a Nickel-Base Alloy," *H, Basic Eng., Trans. ASME,* 237-44, March 1965.

177. W. R. Berry and I. Johnson, "Prevention of Cyclic Thermal-Stress Cracking in Steam Turbine Rotors," *J. Eng. Pwr., Trans. ASME,* 361-68, July 1964.

178. B. F. Langer, "Design of Pressure Vessels for Low-Cycle Fatigue," *J. Basic Eng., Trans. ASME,* Series D, Vol. 84, 389-402, 1962.

179. H. S. Avery, "The Mechanism of Thermal Fatigue," *Metal Progress,* 67-70, August 1959.

180. M. Cox and E. Glenny, "Thermal Fatigue Investigations," *Engineering,* London.

181. *Fundamentals of High-Strength Fasteners,* Standard Pressed Steel Laboratories Rep., Jenkintown, Pa., 1958.

182. E. I. Radzimovsky, "Bolt Design for Repeated Loading," *Machine Design,* 135-46, November, 1952.

183. R. L. Sproat and R. A. Walker, "Radiused-Root Threads--Are They Really Better?," *Assembly Engineering,* Hitchcock Publishing Co., Wheaton, Ill., April 1965.

184. Standard Pressed Steel Co., Specifications SPS-B-280, *Bolt, Tension, 260,000 psi, Increased Fatigue,* and SPS-N-281, *Nut, Self-locking, Nine-spline, External Wrenching, 260,000 psi,* Jenkintown, Pa., November 1, 1965.

185. Standard Pressed Steel Co., *The Root of the Thread,* Form 2434, Jenkintown, Pa., undated.

186. The Barden Corporation, *Engineering Catalog G-3,* Danbury, Conn., 1962.

187. J. Lieblein and M. Zelen, "Statistical Investigation of the Fatigue Life of Deep Groove Ball Bearings," *J. Res. Natl. Bur. Std.* (U.S.), Vol. 57, No. 5, Research Paper 2719, November 1956.

188. E. Schube, "Ball Bearing Survival," *Machine Design*, 158–61, July 19, 1962.

189. C. Mischke, "Bearing Reliability and Capacity," *Machine Design*, 139–40, September 30, 1965.

190. T. A. Harris, "Predicting Bearing Reliability," *Machine Design*, 129–32, January 3, 1963.

191. T. A. Harris, "Preloaded Bearings," *Product Eng.*, 84–93, July 19, 1965.

192. T. A. Harris, "Predicting Bearing Performance," *Machine Design*, 158–62, August 17, 1967.

193. T. A. Harris, *Rolling Bearing Analysis*, Wiley, New York, 1966.

194. J. A. Erickson, "Degassing and Consumable-Electrode Remelting Improve Bearings," *Metal Progress*, ASM, Metals Park, Ohio, 69–73, November 1957.

195. R. A. Baughman, *Rolling Contact Bearing Fatigue Studies*, General Electric Co., Material Development Lab., Flight Propulsion Division, Evendale, Ohio, 1961.

196. *Influence of Lubrication on Endurance of Rolling Contacts*, AD No. 274137, BuWeps United States Navy, Washington, D.C. (SKF Rept. No. A262T004), 9/22/61 to 2/22/62.

197. J. D. Morrow, "Correlation of the Pitting Fatigue Life of Bearings with Rolling Contact Rig Data," *J. Basic Eng. ASME Trans.*, 583–88, September 1966.

198. N. S. Grassam and J. W. Powell, eds., *Gas Lubricated Bearings*, Butterworths, London, 1964.

199. J. B. Seabrook and D. W. Dudley, "Results of a Fifteen-Year Program of Flexural Fatigue Testing of Gear Teeth," *Trans. ASME, J. Eng. for Ind.*, 221–33, August 1964.

200. D. W. Dudley, "Space Gearing: Successes and Failures," *Mech. Eng.*, ASME, 34–37, April 1965.

201. G. Broersma, *Marine Gears*, The Technical Publishing Co., H. Stam, Haarlem, The Netherlands, 1961.

202. M. C. Shaw and Fred Macks, *Analysis and Lubrication of Bearings*, McGraw-Hill, New York, 1949.

203. E. E. Bisson and W. J. Anderson, *Advanced Bearing Technology*, NASA SP-38, 1964.

204. T. W. Morrison et al., "Materials in Rolling Element Bearings for Normal and Elevated (450°F) Temperatures," *Trans. ASME*, Vol. 2, No. 1, 129–46, April 1959.

205. E. W. Parkes, *A Design Philosophy for Repeated Thermal Loading*, AGARD (NATO) Rept. 213, Neuilly-sur-Seine, France, October 1958.

206. J. E. Shigley, *Machine Design*, McGraw-Hill, New York, 1956.

207. *Design Curves for Neg'ator Springs*, Hunter Spring Division, Ametek Corp., Hatfield, Pa., September 14, 1967.

208. P. H. Frith, "Fatigue Tests at Elevated Temperatures," *J. Brit. Iron and Steel Inst.*, 175-81, 1951.

209. "Two-Year Study Yields Data for Evaluating Coated Cable," *Product Eng.*, McGraw-Hill, New York, 78-80, October 10, 1966.

210. Catalog, Form 218, Thomson Industries, Inc., Manhasset, N.Y., 25-26, February 1967.

211. H. R. Wetenkamp et al., "The Effect of Brake Shoe Action on Thermal Cracking and on Failure of Wrought Steel Railway Car Wheels," University of Illinois, *Eng. Exp. Sta. Bull.*, No. 387, Vol. 47, No. 77, Urbana, Ill., June 1950.

212. W. M. Justusson et al., "Half-Million PSI Steels by Ausforming," *Materials in Design Eng.*, Reinhold, New York, May 1964.

213. A. Graae, "How to Nitride Maraging Steels," *Metal Progress*, July 1967.

214. G. Martin, "Maraging Steels Provide Ultra-High Strength," *Materials in Design.*, December 1964.

215. P. E. Ruff, "Hot-Work Toolsteel for Aircraft," *Metal Progress*, March 1959.

216. *Aerospace Structural Metals Handbook*, Vol. 1, ASD-TR-63-741, Air Force Materials Lab., October 1963.

217. D. W. Dudley, *Practical Gear Design*, McGraw-Hill, New York, 1954.

218. A. M. Wahl, *Mechanical Springs*, Penton Publishing Co., Cleveland, Ohio, 1944.

219. R. G. Matters and A. A. Blatherwick, "High-Temperature Rupture, Fatigue and Damping Properties of AISI 616 (Type 422 Stainless Steel," *Trans. ASME, J. Basic Eng.*, June 1965.

220. A.R.C. Markl, "Fatigue Tests of Welding Elbows and Comparable Double-Mitre Bends," *Trans. ASME*, Vol. 69, No. 8, 1947.

221. H. F. Moore, *Textbook of the Materials of Engineering*, McGraw-Hill, New York, 1941.

222. C. W. Briggs, "Carbon and Low Alloy Steels," *Machine Design*, Metals Ref. Issue, Vol. 39, No. 29, December 14, 1967.

223. R. H. Kaltenhauser and A. J. Lena, "New Concepts in Economy," *Metal Progress*, August 1967.

224. *Tables of Fatigue Strength of Sand-Cast Magnesium Alloys*, Metal Products Dept., Code 216, Dow Chemical Co., Midland, Mich., November 1965.

225. M. C. Flemings and D. Peckner, "Premium Quality Castings," *Materials in Design Eng.*, August 1963.

226. Walter Illg, *Fatigue Tests on Notched and Unnotched Sheet Specimens of 2024-T3 and 7075-T6 Aluminum Alloys and of SAE 4130 Steel with Special Consideration for the Life Range from 2 to 10,000 cycles,* NACA TN 3866, September 5, 1956.

227. *Evaluation of Aluminum Alloy 7001-T75,* ATC TR No. 101, Aerospace Ind. Assoc., Washington, D.C., December 1966.

228. H. A. Leybold, H. F. Hardrath, and R. L. Moore, *An Investigation of the Effects of Atmospheric Corrosion on the Fatigue Life of Aluminum Alloys,* NACA TN 4331, Washington, D.C., September 1958.

229. J. B. Kaufman et al., *Fatigue Strengths of Recrystallized and Undercrystallized Heat Treated Aluminum Alloys,* ASTM Proc., 1962.

230. G. W. Strickley, *Additional Studies of Effects of Anodic Coating on the Fatigue Strength of Aluminum Alloys,* ASTM Proc., Vol. 60, 577–88, 1960.

231. G. E. Nordmark, "Fatigue of Aluminum with Alclad or Sprayed Coatings, *AIAA J. Spacecraft,* Eng. Note, 125–27, January 1964.

232. J. L. Miller and B. W. Lifka, *Effect of Chemical Sizing on Some 2024-T81 and 7075-T6 Sheet,* Alcoa Res. Lab. Rept. No. 9-63-6/5-XA-104, New Kensington, Pa., February 12, 1963.

233. *Unpublished S-N Curves for 356-T6 Aluminum Alloy Sand Cast Test Bars,* Battelle Memorial Institute, Columbus, Ohio, October 22, 1962.

234. *Unpublished data on A356=T6 Aluminum Alloy Permanent Mold Test Bars,* Alcoa, Newark, N.J., October 22, 1962.

235. J. Rushing, *Fatigue Testing of Sand Cast KIA Magnesium Alloy,* Metallurgical Rept. MR-65-5, Ryan Aeronautical Co., San Diego, Calif., December 13, 1965.

236. R. J. Jackson and P. D. Frost, *Properties and Current Applications of Magnesium--Lithium Alloys,* NASA SP-5068, Washington, D.C., 1967.

237. R. G. Dermott, "Extending the Possible in Metalworking," *Metal Progress,* 60–66, April 1967.

238. G. E. Martin, "Design and Fabrication of Welded Titanium Wing Leading Edge, *SAMPE J.,* 28–27, January 1966.

239. J. G. Weinberg and I. E. Hanna, *An Evaluation of the Fatigue Properties of Titanium and Titanium Alloys,* TML Rept. 77, July 17, 1957.

240. Properties of Ti-6Al-4V, *Titanium Eng. Bull. No. 1,* Titanium Metals Corp., New York, December 1966.

241. *Fatigue Characteristics of the Ti-5A1-2.5Sn and Ti-6A1-4V Titanium Sheet Alloys,* Tech. Serv. Dept. Titanium Metals Corp., New York.

242. C. W. Briggs et al., *fatigue Properties of Comparable Cast and Wrought Steels,* Proc. ASTM, Vol. 56, 979–1011, 1956.

243. *The Fatigue Life of As-Cast Surfaces of Malleable Iron,* Bull. No. 27, Malleable Research and Dev. Foundation, Dayton, Ohio, February 1966.

244. T. E. Eagan, "Fatigue Strength of Nodular Iron," *Iron Age*, Vol. 168, December 13, 1961.

245. K. B. Palmer and G.N.J. Gilbert, "The Fatigue Properties of Nodular Cast Iron," *BCIRA J. Res. Dev.*, August 1953.

246. F. R. Brotzen, "Fatigue Properties of Grey Cast Iron," *Machine Design*, Vol. 29, December 1957.

247. W. I. Collins and J. O. Smith, *Fatigue and Static Load Tests of a High-Strength Cast Iron at Elevated Temperatures*, Proc. ASTM, Vol. 41, 1941.

248. H. L. Logan, *The Stress-Corrosion of Metals*, Wiley, New York, 1966.

249. H. P. Godard, W. B. Jepson, M. R. Bothwell, and R. L. Kane, *The Corrosion of Light Metals*, Wiley, New York, 1967.

250. C. M. Tyler, "Influence of Temper on the Fatigue Strength of Cartridge Brass (Alloy 260) Sheet," *Materials, Res. and Std.*, ASTM, 59-64, February 1967.

251. D. Y. Wang and S. M. Marco, *Fatigue Behavior of Tantalum and the Effect of Strain Aging*, Proc. ASTM, 595-611, 1964.

252. *Tech. Bulletin No. 1040-A*, Beryllium Corp., Reading, Pa. undated.

253. W. G. Gibbons, "Strain-Cycle Fatigue of 70-30 Copper-Nickel," *Trans. ASME*, 552-54, June 1966.

254. *Descriptive Bulletin No. 1105-B*, Beryllium Corp., Reading, Pa. undated.

255. E. L. Terry and B. King, *Static and Repeated Loading Characteristics of Joints in Beryllium Structures*, Brush Beryllium Co., Cleveland, Ohio, March 1965.

256. R. W. Fenn et al., *Properties and Behavior of Beryllium-aluminum Alloys*, LMSC No. 895380, Lockheed Missiles and Space Co., Palo Alto, Calif., October 1964.

257. *Metal Progress*, 130, 145, April 1954.

258. *Standards Handbook (for) Copper and Copper Alloys*, 5th Ed., Copper Development Assoc., New York, August 1964.

259. "Fatigue Properties of Reinforced Plastics," *Materials in Design Engineering*, Reinhold, New York, September 1966.

260. J. W. Davis et al., "The Fatigue Resistance of Reinforced Plastics," *Materials in Design Eng.*, Reinhold, New York, December 1964.

261. E. L. Strauss, "How to Design Mechanical Joints," Part 1 of "Fastening Reinforced Plastics," *Materials in Design Eng.* M/DE Manual No. 203, Reinhold, New York, February 1963.

262. K. H. Boller, *Fatigue Tests of Glass-Fabric-Base-Laminates Subjected to Axial Loading*, United States Forest Products Lab, Rept. 1823, August 1958. (Information reaffirmed, 1965.)

263. Fred Werren, Supplement to [262], United States Forest Products Lab, No. 1823-B, August 1956. (Information reaffirmed 1962.)

264. "The Long-Term Mechanical Properties of Plastics," *Materials in Design Eng.*, Reinhold, New York, November 1965.

265. "Propylene Plastics," *Materials in Design Eng.*, Reinhold, New York, August 1966.

266. J. H. Licht and A. White, *Polyester Film Belts*, NASA Tech. Note D-668, (AD256504), May 1961.

267. W. J. Crichlow and V. S. Sorenson, "Advancements in Monofilament Structural Composite Technology," *AIAA J. Aircraft*, Vol. 3, No. 5, September-October 1966.

268. R. W. Baird, F. W. Forbes, and H. A. Lipsitt, *Tensile and Fatigue Properties of Laminate Sheet Structures*, Aeron. Res. Lab., WADC, Wright-Patterson Air Force Base, Ohio, June 1959.

269. Correspondence of W. P. and R. H. Wallace in *Metals Prgress*, April 1954.

270. D. R. Harting, "The S-N Fatigue Life Gage: A Direct Means of Measuring Cumulative Fatigue Damage," *Exp. Mech.* Vol. 6, No. 2, 19A-24A, Soc. Exp. Stress Anal., Westport, Conn., February 1966.

271. M. R. Gross, *Low-Cycle Fatigue of Materials for Submarine Construction*, R & D Rept. 91-197D, (ASTIA No. 013146), United States Naval Eng. Exp. Station, Annapolis, Md., February 1963.

272. R. S. Horne, *A Feasibility Study for the Development of a Fatigue Damage Indicator*, AFFDL-TR-66-113, (ASTIA No. AD813344), January 1967.

273. *Applications Manual: The S-N Fatigue Life Gage*, Micro-Measurements, Inc., Romulus, Mich., August 1966.

274. *The Fatigue of Metals and Structures*, a series of lecture notes, Douglas Aircraft Co., Santa Monica, Calif., 1960.

275. Paul Kuhn, *Residual Strength in the Presence of Fatigue Cracks* (presentation to the AGARD Structures and Material Panel, 1967), NASA, Washington, D.C., 1967.

276. S. J. Klima, D. J. Lesco, and J. C. Freche, *Ultrasonic Technique for Detection and Measurement of Fatigue Cracks*, Rept. TN D-3007, (DMIC No. 61812), NASA-- Lewis Research Center, Cleveland, Ohio, September, 1965.

277. M. R. Achter et al., *A Flexural Fatigue Machine for High-Temperature Operation at Resonance in Vacuum*, NRL Rept. 6275, (DMIC No. 62406), United States Naval Res. Lab., Washington, D.C., June 11, 1965.

278. S. R. Swanson, *Random-Load Fatigue Testing: A State-of-the-Art Survey*, Paper presented at ASTM 70th Annual Meeting, June 1967.

279. B. F. Langer, Correspondence with ASME Boiler and Pressure Vessel Committee, the Special Committee to Review Code Stress Basis, and the Task Group on Fatigue, July 17, 1959.

280. L. F. Coffin and J. Tavernelli, "The Cyclic Straining and Fatigue of Metals," *Trans. AIME*, Vol. 215, 1959.

281. S. S. Manson, Discussion of [282].

282. J. Tavernelli and L. F. Coffin, "Experimental Support for Generalized Equation Predicting Low-Cycle Fatigue," *J. Basic Eng., Trans. ASME,* December 1962.

283. J. D. Morrow and T. A. Johnson, "Correlation between Cyclic Strain Range and Low-Cycle Fatigue Life of Metals," Tech. Note in *Materials Res. and Std.,* January 1965.

284. J. J. Burke et al., eds., *Fatigue--An Interdisciplinary Approach,* Proc. 10th Sagamore-Army Materials, Res. Conference, Syracuse University Press, Syracuse N.Y., 1964.

285. T. J. Dolan, *Models of the Fatigue Process,* paper in Proc. 10th Sagamore-Army Materials Research Conference, Syracuse University Press, Syracuse, N.Y., 1964.

286. Discussion of [285] by G. H. Rowe.

287. J. F. Tavernelli and L. F. Coffin, "A Compilation and Interpretation of Cyclic-Strain Fatigue Tests," *Trans. ASM,* Vol. 51, 1959.

288. C. F. Tiffany and J. N. Masters, *Fracture-Toughness Testing and Its Applications,* ASTM, STP 381, Am. Soc. Test. Materials, Philadelphia, Pa. 1965.

289. C. F. Tiffany and P. M. Lorenz, *An Investigation of Low-Cycle Fatigue Using Applied Fracture Mechanics,* (AF ML-TDR-64-53), May 1968.

290. "Progress in Measuring Fracture Toughness and in Using Fracture Mechanics," Sp. ASTM Committee, *Materials and Res. Std.,* Vol. 4, No. 3, ASTM, Philadelphia, Pa., 1964.

291. E. P. Dahlberg, "Fatigue Crack Propagation in High Strength 4340 Steel in Humid Air," *Trans. ASM,* Vol. 58, 1965.

292. G. R. Irwin, "Analysis of Stresses and Strains Near the End of a Crack Traversing a Plate," *J. Appl. Mech., Trans. ASME,* Vol. 24, No. 3, September 1957.

293. Paul C. Paris, *The Fracture Mechanics Approach to Fatigue,* paper in Proc. 10th Sagamore-Army Materials Research Conference, Syracuse University Press, Syracuse, N.Y., 1964.

294. S. H. Smith et al., *Fatigue-Crack-Propagation and Fracture Toughness Characteristics of 7079 Aluminum Alloy Sheets and Plates in Three Aged Conditions,* NASA CR-996, February 1968.

295. J. E. Srawley and W. F. Brown, "Fracture Toughness Testing and Its Applications," STP 381: *Fracture Toughness Testing,* ASTM, Philadelphia, Pa., 1965.

296. K. E. Hofer, "Fracture Mechanics," *Machine Design,* February 1, 1968.

297. R. W. Smith, M. H. Hirschberg, and S. S. Manson, *Fatigue Behavior of Materials under Strain Cycling in Low and Intermediate Life Cycling,* NASA-TN-D1574, April 1963.

298. J. C. Grosskreutz and G. G. Shaw, *Mechanisms of Fatigue in 1100-0 and 2024-T4 Aluminum,* Rept. AFML-TR-65-127, Midwest Research Institute, Kansas City, Mo., DMIC No. 62410, July 1965.

299. W. Pfeiffer, "Bolted Flange Assemblies," *Machine Design*, June 20, 1963.

300. Private correspondence, George W. Bishop, Bishop Engineering Co., Princeton, N.J.

301. D. Kececioglu and D. Cormier, *Designing a Specified Reliability Directly into a Component*, paper in Proc. Sixth Reliability and Maintenance Conference AIAA, New York, 1967.

302. H. C. Shah and T-Y Chow, *Use of Maximum Entropy in Estimating the Damage Distribution of a Single Degree of Freedom System Subjected to Ransom Loading*, paper in Proc. Fifth Reliability and Maintenance Conference, New York, 1966.

303. P. Kluger, *Evaluation of the Effects of Manufacturing Processes on Structured Design Reliability*, paper in Proc. Tenth Natl. Symposium Reliability and Quality Control, IEEE, Washington, D.C., 1964.

304. Rome Air Development Center, *RADC Unanalyzed Non-Electronic Part Failure Rate Data*, Interim Rept. NEDCO-1, RADC TR-66-828, Rome, N.Y., December, 1966.

305. J. Gutman, *Statistical Strength Properties of Common Alloys*, SID Report 63-160, North American Aviation Corp., Downey, Cal., February 1963.

306. Structures Reliability Report, Vol. 1, *Methods and Design Data*, Aerospace Div., Martin-Marietta Corp., Denver, Colo., December 1961.

307. W. P. Geophert, "Variations of Mechanical Properties in Aluminum Products," Aluminum Co. of America, Pittsburgh, Pa., 1962.

308. *A Statistical Summary of Mechanical Property Data for Titanium Alloys*, OTS PB 161237, Defense Metals Information Center, Battelle Memorial Institute, Columbus, Ohio, February 1961.

309. J. D. Murray and R. J. Truman, *The High Temperature Properties of Cr-Ni-Nb and Cr-Ni-Mo Austenitic Steels*, Joint International Conference on Creep, Book 4, Paper No. 61, BIME., London, 1963.

310. S. S. Manson and G. Succop, *Stress-Rupture Properties of Inconel 700 and Correlation on the Basis of Several Time-Temperature Parameters*, ASTM STP No. 174, Symposium on Metallic Materials for Service at Temperatures above 1600°F, 1956.

311. R. M. Goldhoff, "Comparison of Parameter Methods for Extrapolating High Temperature Data," *Trans. ASME*, Vol. 81, 1959.

312. W. J. Crichlow, *The Materials-Structures Interface--A Systems Approach to Airframe Structural Design*, Proc. 10th ASME/AIAA Structures, Structural Dynamics and Materials Conference, New Orleans, La., April 14, 1969, ASME, New York, 1969.

313. T. J. Dolan et al., "The Influence of Shape of Cross-Section on the Flexural Fatigue Strength of Steel," *Trans. ASME*, Vol. 72, 1950.

314. M. R. Raghavan, "Effect of Cross-Sectional Shape on the Fatigue Strength of Steel," *Materials, Res. and Std.*, ASTM, 290-95, June 1964.

315. A. W. Cochardt, "A Method for Determining the Internal Damping of Machine Members," *ASME Appl. Mech.*, Paper No. 53-A-44, 1953.

316. B. J. Lazan, *Properties of Materials and Joints under Alternating Force*, Status Rept. 53-1, WADC, February 1953.

317. W. W. Sanders et al., "Effect of External Geometry on Fatigue Behavior of Welded Joints," *J. AWS*, Weld. Res. Suppl, 49-S to 55-S, February 1965.

318. M. H. Raymond and L. F. Coffin, Jr., *Geometrical Effects in Strain Cycled Aluminum*, ASME Metals Eng. Div., Paper No. 62-WA-231, November 1962.

319. K. Nishioka and N. Hisamitsu, "On the Stress Concentration in Multiple Notches," *J. Appl. Mech.* 575-77, September 1962.

320. N. P. Inglis and G. F. Lake, "Corrosion Fatigue Tests of Steels in River Tees Water," *Trans. Faraday Soc.*, Vol. 27, 803, 1931; Vol. 28, 715, 1932.

321. S. Hara, *Corrosion Fatigue of Marine Propellor Shafts*, Proceedings of International Conference on Fatigue of Metals, Brit. Inst. Mech. Engrs. and ASME, London, 1956, p. 348.

322. H. E. Haven, "Corrosion Fatigue of Streamline Wire for Aircraft," *Trans. ASME*, AER 109, Vol. 54, 1932.

323. B. B. Westcott, "Fatigue and Corrosion Fatigue of Steels," *Mech. Eng.*, Vol. 60, 813, 1938.

324. T. S. Fuller, "Endurance Properties of Steel in Steam," *Trans. AIME*, Vol. 90, 1930.

325. D. G. Sopwith and H. J. Gough, "Effects of Protective Coatings on the Corrosion Fatigue Resistance of Steel," *J. Iron and Steel Inst.*, Vol. 135, 1937.

326. R. Cazaud, *Fatigue of Metals*, 4th Ed., Dunod, Paris, 1949. (See also [386].)

327. T. J. Dolan and H. H. Benninger, *Effect of Protective Coatings on the Corrosion Fatigue Strength of Steel*, Proc. ASTM, Vol. 40, 1940.

328. W. E. Harvey, "Protective Coatings Against Corrosion Fatigue of Steels," *Metals and Alloys*, Vol. 1, 1930; Vol. 2, 1932.

329. B. B. Westcott, Proc. ASTM, Vol. 40, 667, 1940.

330. D. J. McAdam, *Corrosion Fatigue of Non-Ferrous Metals*, Proc. ASTM, Vol. 2711, 1927.

331. H. J. Gough and D. G. Sopwith, "Some Comparative Corrosion Fatigue Tests Employing Two Types of Stressing Action," *J. Iron and Steel Inst.*, Vol. 127, 1933.

332. H. J. Gough and D. G. Sopwith, "Resistance of some Special Bronzes to Fatigue and Corrosion Fatigue," *J. Inst. Metals*, Vol. 60, 1937.

333. D. G. Sopwith, *The Resistance of Aluminum and Beryllium Bronzes to Fatigue and Corrosion Fatigue*, Aeron. Res. Council R & M, 2486, 1950.

334. C. A. Stubbington and P.J.E. Forsyth, *Some Corrosion Fatigue Observations on Al-Zn-Mg Alloys*, RAE Tech. Note, Met. 289, 1958.

335. D. J. Mack, *Corrosion Fatigue Properties of Some Hard Lead Alloys in Sulphuric Acid*, Proc. ASTM, Vol. 45, 1945.

336. A. Beck, *Technology of Magnesium and Its Alloys*, F. A. Hughes and Co., Ltd., London, 3rd ed., 1943.

337. H. J. Gough and D. G. Sopwith, "Atmospheric Action as a Factor in Fatigue of Metals," *J. Inst. Metals*, Vol. 49, 1932.

338. N. J. Wadsworth and J. Hutchings, "The Effect of Atmospheric Corrosion on Metal Fatigue," *Phil. Mag.*, Vol. 3, 1958.

339. A. Thiruvengadam, "High-Frequency Fatigue of Metals and their Cavitation-- Damage Resistance," *Trans. ASME*, 332-40, August 1966.

340. H. J. Tapsell, P. G. Forrest, and G. R. Tremain, "Creep Due to Fluctuating Stresses at Elevated Temperatures," *Engineering (Brit.)*, Vol. 170, 1950.

341. R. H. Christensen and R. J. Bellinfante, *Some Considerations in the Fatigue Design of Launch and Spacecraft Structures*, NASA CR-242, June 1965.

342. S. R. Valluri et al., *Theory of Elevated Temperature Fatigue*, Douglas Aircraft Paper No. 1823, Santa Monica, Calif., 1964.

343. S. S. Manson, *Thermal Stresses and Low-Cycle Fatigue*, McGraw-Hill, New York, 1966.

344. F. L. Muscatell et al., *Thermal Shock Resistance of High-Temperature Alloys*, Proc. ASTM, Vol. 57, 1957.

345. T. A. Hunter, *Thermal Shock Testing of High Temperature Metallic Materials*, ASTM STP 174, 1955.

346. W. E. Littman, "Residual Stresses in Steel," *Machine Design*, February 27, 1964.

347. C. Lipson, "Wear Considerations in Design," *Machine Design*, November 21, 1963.

348. H. J. Grover, *Fatigue of Aircraft Structures*, NAVAIR 01-1A-13, Naval Air Systems Command, United States Navy, Washington, D.C., 1966.

349. *Airlifters*, publication of the Lockheed-Georgia Co., Division of Lockheed Aircraft Corp., April, July, and October 1967, Marietta, Ga.

350. F. S. Ople and C. L. Hulsbos, *Probable Life of Pre-Stressed Beams as Limited by Concrete Fatigue*, Lehigh University Lab. Rept. 223.26A, 1962.

351. R. L.Mattson and W. S. Coleman, "Effect of Shot Peening Variables and Residual Stresses on Fatigue Life," *Metal Progress*, 108-12, May 1954.

352. H. O. Fuchs and E. R. Hutchinson, "Shot Peening," *Machine Design*, 116-25, February 6, 1958.

353. C. Lipson, G. C. Noll, and L. S. Clock, *Stress and Strength of Manufactured Parts*, McGraw-Hill, New York, 1950.

354. N. L. Roust, *Vibration and Fatigue, a Honeycomb Sandwich Bibliography*, Hexcel Corp., Dublin, Calif. January 20, 1968.

355. J. W. Murdock, *A Critical Review of Research on Fatigue of Plain Concrete*, University of Illinois Exp. Sta. Bull. 475, Urbana, Ill., 1965.

356. W. J. Trapp and D. M. Forney, eds., *Acoustical Fatigue in Aerospace Struc-tures*, Proc. 2nd International Conference, 1964, Syracuse University Press, Syracuse, N.Y., 1965.

357. E. B. Shand, *Structural Applicators of Glass and Ceramics*, Corning Glass Works, Corning, N.Y., undated paper.

358. L. S. Williams, "Stress-Endurance of Sintered Alumina," *Trnas. Brit. Ceramics Soc.*, Vol. 55, 287-312, 1956.

359. C. Gurney and S. Pearson, "Fatigue of Mineral Glass under Static and Cyclic Loading," *Proc. Roy. Soc. (London)*, A192, 537-44, 1948.

360. V. L. Maleev and J. M. Hartman, *Machine Design*, International Textbook Co., Scranton, Pa., 1957.

361. O. G. Meyers, "Working Stresses for Helical Springs," *Machine Design*, Penton Publishing Co., Cleveland, Ohio, 135, November 1951.

362. R. C. Coates and J. A. Pope, *Fatigue Testing of Compression-Type Coil Springs*, In Proceedings of International Conference on Fatigue of Metals, Brit. Inst. Mech. Engrs. and ASME, London, 1956, pp. 604 ff.

363. H. G. Rylander et al., "Stress Concentration Factors in Shouldered Shafts Subjected to Combinations of Flexure and Torsion," *J. Eng. Ind., ASME Trans*, 301-307, May 1968.

364. I. M. Allison, "Elastic Stress Concentration Factors in Shouldered Shafts." *The Aeron. Quart.*, Part I, Torsion, Vol. 12, No. 2, May 1961; Part II, Bending, Vol. 12, No. 3, August 1961; Part III, Axial Load, Vol. 13, No. 2, May 1962.

365. H. J. Grover et al., *Axial-Load Fatigue Tests of Notched Sheet Specimens of 24S-T3 and 75S-T6 Aluminum Alloys and of SAE 4130 Steel with Stress Concen-tration Factors of 2, 4, and 5*, NACA Tech. Note 2389 and 2390, June 1951.

366. J. T. Broch, "Peak Distribution Effects in Random Load Fatigue," *Bruel and Kjoer Tech. Rev.*, No. 1, B & K Instruments, Inc., Cleveland, Ohio, 1968.

367. E. J. Richards and D. J. Mead, eds. *Noise and Acoustic Fatigue in Aeronautics*, Wiley, New York, 1968.

368. A. F. Madayag, ed., *Metal Fatigue: Theory and Design*, Wiley, New York, 1969.

369. S. A. Clevenson and R. Steiner, *Fatigue Life under Various Random Loading Spectra*, Proc. 35th Symposium on Shock and Vibration, New Orleans, La., October 1965, USNRL, Washington, D.C. (DOD-DDR & E).

370. J. C. Houbolt, *Interpretation and Design Application of Power Spectral Gust Response Analysis Results*, Proc. 6th Structures Conference, AIAA/ASME, AIAA Publications, New York, 1965.

371. J. S. Bendat, *Probability Functions for Random Responses: Prediction of Peaks, Fatigue Damage and Catastrophic Failures*, NASA-CR-33, Washington, D.C., April 1964.

372. *Aircraft Fatigue Handbook*, Vol. II, *Design and Analysis*, ARTC/W-76, Aircraft Structural Fatigue Panel, Aircraft Industries Assoc. (now Aerospace Industries Assoc., AIA), Washington, D.C., January 1957.

373. L. Kaechele, *Review and Analysis of Cumulative Fatigue Damage Theories*, Memo RM-3650-PR, Rand Corporation, Santa Monica, Calif., August 1963.

374. W. T. Shuler, *Damage Tolerance and Logistic Transport Design*, Paper No. 68-23, Proc. 6th Congress of International Council of Aeronautical Sciences, Munich, September 1968.

375. D. Y. Wang, *An Investigation of Fatigue Crack Propagation and Failsafe Design of Stiffened Large Aluminum Alloy Panels with Various Crack Stoppers*, Proc. 10th ASME/AIAA Structures Structural Dynamics and Materials Conference, New Orleans, La., April 14, 1969; ASME, New York, 1969.

376. E. H. Scheutte, "A Critical Look at Fatigue Equations," *Prod. Eng.*, McGraw-Hill, New York, 150-51, July 1952.

377. F. R. Shanley, *Fatigue Analysis of Aircraft Structures*, ASTIA Document No. AT1210794, July 31, 1953.

378. J. B. Bidwell et al., eds., *Fatigue Durability of Carburized Steel*, American Society for Metals, Cleveland, Ohio, 1957.

379. M. F. Garwood et al., *Interpretation of Tests and Correlation with Service*, American Society for Metals, Cleveland, Ohio, 1951.

380. J. C. Conover et al., *Simulation of Field Loading in Fatigue Testing*. SAE Paper No. 660102. Soc. Auto. Engrs., New York, 1966.

381. W. E. Schilke, *The Reliability of Transmission Gears*, SAE Paper No. 670725, Soc. Auto. Engr., New York, 1967.

382. W. J. Crichlow, *Stable Crack Propagation--Fail-Safe Design Criteria--Analytical Methods and Test Procedures*, AIAA Paper No. 69-215, Am. Inst. Astro and Aero., New York, 1969.

383. G. H. Jacoby, *Fatigue Life Estimation Processes under Conditions of Irregularly Varying Loads*, AFML-TR-67-215, August 1967.

384. E. Gassner, *Verwendung Eines Einheits--Kolletivs bei Betriebsfestigkeits--Versuchen* (Application of a Standard Pattern to Investigations of Service Strength), Tech. Rept. No. 15/65, Laboratorium fur Betriebsfestigkeit, Darmstadt, December 12, 1966.

385. C. R. Smith, "The Effect of Tapered Bolts on Structural Integrity," *Assembly Engineering*, Hitchcock Publishing Co., Wheaton, Ill., July 1967.

386. R. Cazaud, *Fatigue of Metals*, transl. by A. J. Fenner, Chapman and Hall, London, 1953.

387. W. J. Harris, *Metallic Fatigue*, Pergamon, London, 1961.

388. F. J. Kovac and K. B. O'Neil, "Predicting Fatigue Performance of Tires," *Materials Res. and Std.*, Vol. 8, No. 6, ASTM, Philadelphia, Pa., June, 1968.

389. F. Frank and W. Hofferberth, "Mechanics of the Pneumatic Tire," *Rubber and Chemical Tech.*, Vol. 40, No. 1, Div. of Rubber Chem., American Chem. Soc., February 1967.

390. G. J. Lake and P. B. Lindley, "Mechanical Fatigue Limit for Rubber," *Rubber Chem. and Tech.*, Vol. 39, No. 2, Div. of Rubber Chem. American Chem. Soc., March 1966.

391. J. Lippman and W. P. Nanny, *Quantitative Analyses of Enveloping Forces on Passenger Tires*, SAE Paper 670174, 1967.

392. K. L. Floyd, "The Behavior of Rayon Tire Cords under Compressive Flexing," *J. Textile Inst.*, Vol. 53, No. 10, T449-T463, October 1962.

393. R. G. Patterson and R. K. Anderson, "Fatigue Failure in Nylon Reinforced Tires," *Rubber Chem. and Tech.*, Vol. 38, 832-39, 1965.

394. T.A.S. Duff et al., "Elastic and Fatigue Properties of Vacuum Remelted vs. Air-melted Ni-SPAN-C Alloy 902 for Bourdon Tubes," *J. Basic Eng.*, *Trans. ASME*, Series D, Vol. 89, No. 3, 561-569, September 1967.

395. *Machine Design*, Vol. 41, September 11, 1969, pg. 20.

396. J. W. Fisher & J.H.A. Struik, *Guide to Design Criteria for Bolted and Riveted Joints*, Wiley, 1974.

397. D. R. Hamel, et al., "Fatigue Strength Optimization of Bonded Joints," *J. Basic Eng.*, December 1971.

398. M. Stasuiski & T. Endo, "The Fatigue Life of Materials Subjected to Random Strains," *Soc. Mech. Eng.*, *Japan*, March, 1968.

399. N. E. Dowling, "Fatigue Failure Predictions for Complicated Stress-Strain Histories," *J. Mat'ls.*, Vol. 7, No. 1, 1972.

400. J. H. Rondeel et al., "Comparative Fatigue Tests with 24 SOT Alclad Riveted and Bonded Stiffened Panels," *Natl. Aero. Res. Inst. (Dutch)*, Rept. S.416.

401. K. C. Wu, "Resistance of Non-Ti-Bond Joining of Ti Shapes," *AWS Suppl.*, September 1971.

402. E. Taylor, "Multiphase Alloy Environmental Resistance," *Standard Pressed Steel*, Report No. 5718, June 2, 1977.

403. M. B. Cutler et al., "Static and Fatigue Test Properties for Woven and Non-woven S-glass Fibers," *AD 688971*, April 1969.

404. M. S. Hersh, "Fatigue of Boron/Aluminum Composites," *NASA Rept. No. ZZL-72-006* (MP-643-2-3), June 1972.

405. "Advanced Composites Data for Aircraft Structural Design," Vol. IV, Allowables: Graphite - Epoxy, Fig. 227.

406. E. A. Simkovich, et al., "Effect of Decarburization and Grinding Conditions on Fatigue Strength of 5% Cr-Mo-V Sheet Steel," *Trans. ASM*, Vol. 53, 1959.

407. R. Clark, et al., "Ausformed Steels in Automotive Applications," *Metal Progress*, March 1966.

408. E. P. Esztergar, "Creep-Fatigue Interaction and Cumulative Damage Evaluations for Type 304 Stainless Steel," *ORNL 4757*, June 1972.

409. J. B. Conway et al., "Fatigue, Tensile and Relaxation Behavior of Stainless Steel," TID-26135, *Tech. Info. Center, USAEC,* 1975.

410. J. B. Conway et al., "High-Temperature, Low-Cycle Fatigue of Copper Base Alloys in Argon," *NASA CR-121259.*

411. A. Saxena et al., "Low-Cycle Fatigue Crack Propagation and Substructures in Cu-Al Alloys," *Met. Trans. A,* Vol. 6a, September 1975.

412. J. Porter et al., "The Fatigue Curves," *J. Inst. Met.,* Vol. 89, 1960-61.

413. J. J. Esposito et al., "Thrust Chamber Life Prediction," Vol. 1, *NASA CR-134806,* March 1975.

414. Alcoa 467 Process X7475 Alloy; Alcoa Green Letter, May 1970.

415. G. W. Kuhlman & F. R. Billman, "Selecting Processing Options for High-Fracture Toughness Titanium Airframe Forgings," *Metal Progress,* March 1977, pg. 39-49.

416. Alcoa Aerospace Technical Information Bulletin, Series 68, No. 2; Figs. 3 & 4.

417. G. E. Nordmark, "Peening Increases Fatigue Strength of Welded Aluminum," *Metal Progress,* November 1963.

418. D. V. Nelson et al., "The Role of Residual Stresses in Increasing Long Life Fatigue Strength of Notched Machine Members," *ASTM STP 407,* 1970, Fig. 2.

419. R. S. Barker and G. K. Turnbull, "Control of Residual Stresses in Hollow Aluminum Forgings," *Metal Progress,* November 1966. Figs. 3, 4, 5, Table III.

420. J. R. Barton and F. N. Kusenberger, "Fatigue Damage Detection," in Metal Fatigue Damage-Mechanism, Detection, Avoidance and Repair, *ASTM STP 495,* 1971.

421. L. E. Tucker et al., "Proposed Technical Report on Fatigue Properties for the SAE Handbook," *SAE* Paper No. 740279, 1974.

422. J. W. Fisher et al., "Effect of Weldments on the Fatigue Strength of Steel Beams," *Natl. Cooperative Highway Research Program,* Report No. 102, 1970.

423. H. R. Jaeckel, "Simulation, Duplication and Synthesis of Fatigue Load Histories," *SAE* Paper No. 700032, 1970.

424. A. A. Pollock, "Acoustic Emission," *Machine Design,* Vol. 8, No. 8, April 8, 1976.

425. A. M. Stagg, "An Investigation of the Scatter in Constant Amplitude Fatigue Test Results of Aluminum Alloys 2024 and 7075," *Aero. Res. Council,* C.P. No. 1093, 1970.

426. A. M. Stagg, "Parameter Estimation for the Log-Normal Parent Population of Fatigue Failures from a Sample Containing both Failed and Non-failed Members," *Aero. Res. Council,* C.P. No. 1144, 1971.

427. R. E. Little and E. H. Jebe, *Statistical Design of Fatigue Experiments,"* Applied Science Publishers, Ltd., 1975.

428. L. G. Johnson, *The Statistical Treatment of Fatigue Experiments,* Elsevier, 1964.

429. *Handbook of Fatigue Testing,* R. Swanson, ed., ASTM STP 566, 1974.

430. *Steel Castings Handbook,* Steel Founders Society of Amer., 4th Ed., 1970, Cleveland.

431. Jo Dean Morrow et al., "Laboratory Simulation of Structural Fatigue Behavior," in *ASTM* STP 462, 1970.

432. H. Neuber, "Theory of Stress Concentration for Shear Strained Prismatical Bodies with Arbitrary Non-Linear Stress Strain Law," *J. App. Mech., ASME,* December 1961.

433. J. F. Martin et al., "Computer-Based Simulation of Cyclic Stress-Strain Behavior," *TAM Rept. No. 326,* U. Ill., July 1969.

434. P. C. Rosenberger, "Fatigue Strength of Smooth and Notched Specimens of Man-Ten Steel," MS Theses, *TAM* Dept. U. Ill., Urbana, 1968.

435. *Fatigue under Complex Loading,* R. M. Wetzel, Ed., AE-6, SAE, 1977.

436. H. O. Fuchs et al., "Shortcuts in Cumulative Damage Analysis," SAE Paper No. 730565, 1973.

437. K. H. Klippstein & C. C. Schilling, "Stress Spectrums for Short-Span Steel Bridges," Paper at *ASTM Symp.,* Montreal, 1975.

438. George Sih, "Handbook of Stress-Intensity Factors," Lehigh Univ., *Inst. of Fracture & Solid Mech.,* Bethlehem, Pa., 1973.

439. Hiroshi Tada, Paul Paris, Geo. Irwin, *The Stress Analysis of Cracks Handbook,* Del Research Corp., 226 Woodbourne Dr., St. Louis, Mo. 63105, 1973.

440. D. P. Rooke & D. J. Cartwright, *Compendium of Stress-Intensity Factors,* H. M. Stationery Office, PC6 (Room E22), London, EC1P 1BN.

441. *Fatigue and Fracture of Aircraft Structures and Materials,* Proc. Conf. December, 1969, Ed. H. A. Wood et al., AFFDL TR 70-144, AD 719756.

442. S. T. Rolfe & J. M. Barsom, *Fracture & Fatigue Control in Structures,* Prentice Hall, Englewood Cliffs, N.J. 07632, 1977.

443. R. W. Hertzberg, *Deformation & Fracture Mechanics of Engineering Materials,* Wiley, New York, 1976.

444. Bela I. Sandor, *Cyclic Stress and Strain,* Univ. Wisconsin Press, Madison, Wisc., 1972.

445. J. D. Landes & J. A. Begley, *Test Results from J-integral Studies--An Attempt to Establish a J_{Ic} Testing Procedure,* ASTM STP 560, Philadelphia, Pa., 1974, pp. 170-186.

446. John F. Harvey, *Theory & Design of Modern Pressure Vessels,"* 2nd Ed., Van Nostrand Reinhold Co., New York, 1974.

447. *Fracture Mechanics Design Handbook,* U.S. Army Missile Command, Redstone Arsenal, Ala., Tech. Rept. RL-77-5, December 1976, ADA038457.

448. *Damage-Tolerant Design Handbook,* (AFFDL/AFML), MCIC HB-01, Parts 1 & 2, January, 1975.

449. *Case Studies in Fracture Mechanics,* Ed. by T. P. Rich & D. J. Cartwright, AMMRC MS77-5, AD-A045877, June 1977.

450. B.G.W. Yee, P. F. Packman et al., "Assessment of NDE Reliability Data," *NASA* CR-134991, October 1976.

451. "Rapid Inexpensive Tests for Determining Fracture Toughness," NMAB-328, *Natl. Acad. Sci.,* Washington, D.C. 20418, 1976.

452. W. G. Barrois, *Mannual of Fatigue of Structures,* AGARD MAN-8-70, June 1973.

453. K. Jerram, "An Assessment of the Fatigue of Welded Pressure Vessels," *Intl. Conf. Press. Vessel Tech.,* Part II, Delft, 1969, ASME and KIVI.

454. "Criteria of Section III of the ASME Boiler and Pressure Vessel Code for Nuclear Vessels," *ASME,* 1964.

455. J. DuBuc, et al., "Evaluation of Pressure Vessel Design Criteria for Effect of Mean Stress in Low-Cycle Fatigue," *Int'l. Conf. on Press. Vessel Tech.,* Part II, Delft, 1969, ASME and KIVI.

456. *Fracture Mechanics Evaluation of B-1 Materials,* Vols. I & II, AFML - TR-76-137.

457. F. V. Lawerence, "Estimation of Fatigue Crack Propagation Life in Butt Welds," *Weld. Res. Suppl.,* May 1973, pp. 212-218.

458. F. V. Lawerence, "Fatigue Crack Propagation in Butt Welds Containing Joint Penetration Defects," *Weld. Res. Suppl.,* May 1973, pp. 221-225, 232.

459. J. W. Fisher et al., "Effect of Weldments on Fatigue Strength of Steel Beams," *Nat'l. Coop. Highway Res. Program,* Rept. No. 102, NAS/NAE, 1970.

460. "Welding Nuclear Components" (welding req'ts. extracted from ASME Code Sec. III, Div. 1), Arcos Corp., Philadelphia, Pa., 1974.

461. M. M. Leven, "The Interaction of Creep and Fatigue for a Rotor Steel," Murray Lecture, 1972, *Exp't'l. Mech.,* September 1973.

462. D. Broek, "Cracks at Structural Holes," MCIC-75-25, *Metals and Ceramics Information Center,* Battelle, DoDIAC, Columbus, Ohio.

463. L. R. Hall and W. L. Engstrom, *Fracture and Fatigue Crack Growth Behaviour of Surface Flaws and Flaws Originating at Fastener Holes,* Boeing, 1973. (To be published as AFFDL Tech. Rept.)

464. J. C. Newman, Jr., "Predicting Failure of Specimens with either Surface Cracks or Corner Cracks at Holes," *NASA,* TN-D-8744, 1976.

465. R. E. Frishmuth, "Use of Fracture Mechanics Methods for Establishing Inspection Level for Turbine Wheels," *ASME J. Eng. Mats. & Tech.,* Vol. 101, January 1979.

466. H. A. Wood et al., "The Analysis of Crack Propagation under Variable Amplitude Loading in Support of the F-11 Recovery Program," *AFFDL* TM #71-3-FBR, December, 1971.

467. R. M. Engle & J. L. Rudd, "Spectrum Crack Growth Analysis using the Willenborg Model," *J. Aircraft*, Vol. 13, No. 7, July 1976.

468. A. F. Rubio et al., *Fracture Design Practice for Rotating Equipment*, Vol. 5 of "Fracture, An Advanced Treatise," ed. H.Liebowitz, Academic Press.

469. D.M.R. Taplin et al., "The Cyclic Stress Response of Copper Alloys at 100–500°C," *Third Annual Rept. INCRA Proj.* #228B, December 1976, Univ. Waterloo, Waterloo, Canada.

470. B. F. Brown, "Stress Corrosion Cracking Control Measures," *Nat'l. Bur. Stds. Monograph 156,* June 1977.

471. H. T. Corten, *Influence of Fracture Toughness and Flaws on the Interlaminar Shear Strength of Fibrous Composites*, in Fundamental Aspects of Fiber Reinforced Plastic Composites, Ed. by H. T. Schwartz and H. S. Schwartz, Wiley-Interscience, NYC, 1968.

472. D. T. Read & R. P. Reed, *Toughness, Fatigue Crack Growth, and Tensile Properties of Three Nitrogen-Strengthened Stainless Steels at Cryogenic Temperatures*, in The Metal Science of Stainless Steels, Ed. by E. W. Collings * H?W. King., AIME, 1979.

473. *Effects of Radiation on Structural Materials*, Ed. by J. A. Sprague and D. Kramer, ASTM STP No. 683, 1978.

474. J. Y. Mann, *Fatigue of Materials*, Melbourne Univ. Press, 1967.

475. P. C. Paris & F. Erdogan, "A Critical Analysis of Crack Propagation Laws," *J. Basic Eng., Trans. ASME* 85D, 1973.

SELECTED BIBLIOGRAPHY

The following references are supplementary to those listed above, and are included for more detailed information or for further study.

Bibliographies

J. Y. Mann, *Fatigue of Materials*, Melbourne Univ. Press, 1967. (Especially good on the historical development of the Fatigue Problem.)

References on Fatigue 1964, ASTM STP 9-0, 1964.

Annual Reports of the Materials Properties Council, 1977 et seq, MPC 345 East 47th St., NYC 10017

"Weldment Fatigue - A Bibliography of Cited USA References," Int'l. Institute of Welding, Commission XIII, Doc. No. XIII-819-77, 1977, by Julius Heuschkel & Assoc. Ltd., No. Huntingdon, Pa.

Texts

Metal Fatigue, N. E. Frost, K. J. Marsh & L. P. Pook, Clarendon Press - Oxford, 1974.

Fatigue of Welded Structures, T. R. Gurney, Univ. Printing House, Cambridge, 1968.

Fatigue of Materials, J. Y. Mann, Melbourne Univ. Press, 1967.

Fracture of Structural Materials, A. S. Tetelman & A. J. McEvily, Wiley-Interscience, 1967.

SAE Fatigue Design Handbook, Ed. by J. A. Graham, et al., 1968. (Revised edition expected in late 1981.)

Fracture Mechanics Design Handbook, U.S. Army Mat'ls. Command, Redstone Arsenal, Ala., TR-RL-77-5, 1976, AD038457.

Case Studies in Fracture Mechanics, Ed. by T. P. Rich & D. J. Cartwright, AMMRC MS-77-5, ADA045877, June, 1977.

"The Modern View of Fatigue and its Relation to Engineering Problems," Union College and G.E. Co., Schenectady, NY, 1975 et seq. (Extensive notes and reprints from a 5-day course.)

Variations in the approach to Fatigue Analysis are provided by:

Fatigue Resistance, by P. Ye Kravchenko
English Translation by N. L. Day, Pergamon/Macmillan, 1969.

Manual of Fatigue of Structures, W. G. Barrois, AGARD MAN-8-70, June 1973 (Ref. 452).

Fatigue Design of Machine Components, Lazlo Sors; English Translation by S. E. Mitchell; Pergamon Oxford, 1971.

Fundamentals of Machine Design, P. Orlov; English Translation by Yu Travnichev; MIR Publishers, Moscow, 1976.

Special Technical Publications of the A.S.T.M.

These continuing publications result from Symposia, Conferences and other activities of the Committees, E-9 on Fatigue and E-24 on Fracture. Collectively, they provide a coverage both wide and deep on many topics in the fields of Fatigue and Fracture.

"A Guide for Fatigue Testing and the Statistical Analysis of Fatigue Data," STP #91-A, 1963.

"Fatigue Crack Propagation," STP #415, 1967.

"Irradiation Effects in Structural Alloys," STP #484, 1970.

"Effects of Notches on Low-cycle Fatigue," STP #511, 1971.

"Influence of State of Stress on Low-cycle Fatigue of Structural Materials," STP #549, 1974.

"Handbook of Fatigue Testing," STP #566, 1974.

"Effects of Radiation on Structural Materials," STP #570, 1974.

"Cracks and Fracture," STP #579, 1976.

"Composite Reliability," STP #580, 1974.

"Statistical Planning and Analysis for Fatigue Experiments," STP #588.

"Mechanics of Crack Growth," STP #590, 1976.

"Resistance to Plane-stress Fracture of A572 Steel," STP #591, 1975.

"Fatigue Crack Growth under Spectrum Loads," STP #595, 1975.

"Properties Related to Fracture Toughness," STP #605, 1976.

"Thermal Fatigue of Metals and Composites," STP #612, 1975.

"Flaw Growth and Fracture," STP #631, 1977.

"Developments in Fracture Mechanics Test Methods," STP #632, 1977.

"Fatigue Crack Growth," STP #637, 1977.

"Effects of Radiation on Structural Materials," STP #683, 1979.

"Presentation of Data and Control Chart Analysis," STP #15-D, 1976.

Journals

Wohler's original papers, in German, see Ref. 84.

Special Issue of the "Journal of Strain Analysis for Engineering Design" on Fracture and Fracture Testing, Vol. 10, No. 4, October 1975, Mech. Eng. Publications, Ltd. London (for the IME).

Special Issue of the "Journal of Engineering Materials and Technology" on Structural Integrity, Trans. ASME, Vol. 102, No. 1, January 1980.

"Fatigue of Engineering Materials and Structures," Ed. K. J. Miller, Pergamon, Oxford. All special issues.

"Engineering Fracture Mechanics," Ed. H. Liebowitz, Pergamon, Oxford.

"Int'l. Journal of Fracture," Sijthoff & Noordhoff Int'l. Publ., Winchester, Mass. USA, 01890. (Formerly Int'l. Jnl. of Fracture Mechanics.)

Failures and Fractures

"The Face of Metallic Fracture," Vol. I Text; Vol. II Plates; Ed. by E. J. Pohl, Pub. by the Munich Reinsurance Co., Munich, 1964. (English translation of "Das Gesicht des Bruches metalischer Werkstoffe".)

Source Book on Failure Analysis, ASM, 1974.

Failure Analysis and Prevention, ASM Metals Handbook, Vol. 10, Eighth Ed., 1975.

Proceedings of Conferences and Symposia

"Fracture 1977," Proc. of Fourth Int'l. Conf. on Fracture, Vols. I, II and III, Ed. D.M.R. Taplin, Waterloo, Ontario, 1977.

"Fracture 1973," Proc. of Third Int'l. Conf. on Fracture, Vols. I-XI, DVI, Munich, 1973. (In German.)

"Advanced Approaches to Fatigue Evaluation." Sixth ICAF Symp., 1971, NASA, 1972. (Emphasis on airframe design and testing.) (Earlier ICAF - AGARD Symp. 1959-65, see Refs. 55, 56, 60.)

Proc. of Conf. on Acoustical Fatigue, WADC TR 59-676, 1961.

"Fatigue - An Interdisciplinary Approach," Sagamore Army Materials. Res. Conf., 1963, Syracuse Univ. Press, 1964.

Proc. of Int'l. Conf. on Fatigue of Metals, IME London & ASME NYC, 1956, Publ. by IME, London, SW 1. (Very extensive coverage.)

Miscellaneous Items of Special Interest

"Residual Strength in the Presence of Fatigue Cracks," by Paul Kuhn, AGARD/NASA, 1967.

"Role of Applied Fracture Mechanics in the U.S. Air Force ASIP (Airframe Structural Integrity Program): by H. A. Wood, TM 70-5-FDTR, 1970.

"Fatigue Data Bank and Data Analysis Investigation," by Radziminski et al., Univ. of Illinois Eng. Exp. Station, UILU-Eng-73-2025, 1973.

SELECTED CONVERSION FACTORS TO STANDARD INTERNATIONAL UNITS

To Convert From	To	Multiply By
angstrom	meter (m)	1×10^{-10}
atmosphere (normal=760 torr)	pascal (Pa)	1.013×10^5
bar	pascal (Pa)	1×10^5
calorie	joule (J)	4.184
degree Celsius	kelvin (K)	$t_K = t_C + 273.15$
degree Fahrenheit	degree Celsius	$t_C = (t_F - 32)/1.8$
degree Rankine	kelvin (K)	$t_K = t_R/1.8$
dyne	newton (N)	1×10^{-5}
dyne-centimeter	newton-meter (N-m)	1×10^{-7}
dyne/centimeter^2	pascal (Pa)	1×10^{-1}
electron volt	joule (J)	1.602×10^{-19}
erg	joule (J)	1×10^{-7}
egs/centimeter^2	joule/meter^2 (J/m^2)	10^{-3}
foot	meter (m)	3.048×10^{-1}
foot-pound	joule (J)	1.356
gram-force/centimeter^2	pascal (Pa)	9.807×10^1
inch	meter (m)	2.54×10^{-2}
inch^2	meter^2 (m^2)	6.452×10^{-4}
inch of mercury (32F)	pascal (Pa)	3.386×10^3
kilocalorie	joule (J)	4.184×10^3
kilogram-force	newton (N)	9.807
kilogram-force-meter	newton-meter (N-m)	9.807
kilogram-force/centimeter^2	pascal (Pa)	9.807×10^4
kilogram-force/meter^2	pascal (Pa)	9.807
kilogram-force/millimeter^2	pascal (Pa)	9.807×10^6
kip (1000 pounds)	newton (N)	4.448×10^3
kip/inch^2 (ksi)	pascal (Pa)	6.895×10^6
kip/inch$^2 \cdot \sqrt{\text{inch}}$ (ksi$\sqrt{\text{in}}$.)	pascal$\cdot \sqrt{m}$ (Pa$\sqrt{m}$)	1.099×10^6
mil	meter (m)	2.54×10^{-5}
millimeter of mercury (mm Hg)	pascal (Pa)	1.333×10^2
poise	pascal-second (Pa-s)	1×10^{-1}
pound-force	newton (N)	4.448
pound-force/inch^2 (psi)	pascal (Pa)	6.895×10^3
pound-force/inch$^2 \cdot \sqrt{\text{inch}}$ (psi$\sqrt{\text{in}}$.)	pascal$\cdot \sqrt{m}$ (Pa$\sqrt{m}$)	1.099×10^3
torr [mm Hg, (0C)]	pascal (Pa)	1.333×10^2

INDEX